国家出版基金项目
NATIONAL PUBLICATION FOUNDATION

现代兽医基础研究经典著作
世界兽医经典著作译丛

鱼类疫苗

Fish Vaccination

［挪］冉·古廷（Roar Gudding）
［挪］爱特·林汉格（Atle Lillehaug）　　编著
［挪］奥斯坦·伊温森（Øystein Evensen）
战文斌　高迎莉　邢　婧　迟　恒　阎斌伦　主译

中国农业出版社
北　京

图书在版编目（CIP）数据

鱼类疫苗/（挪）冉·古廷，（挪）爱特·林汉格，（挪）奥斯坦·伊温森编著；战文斌等主译．—北京：中国农业出版社，2021.6

（现代兽医基础研究经典著作）

国家出版基金

ISBN 978-7-109-28205-6

Ⅰ．①鱼…　Ⅱ．①冉…　②爱…　③奥…　④战…　Ⅲ．①鱼病—疫苗—免疫学　Ⅳ．①S942

中国版本图书馆 CIP 数据核字（2021）第 082825 号

Fish Vaccination

By Roar Gudding，Atle Lillehaug and Øystein Evensen

ISBN 978-1-118-47206-4

合同登记号：图字 01-2018-8289 号

鱼类疫苗

YULEI YIMIAO

中国农业出版社出版

地址：北京市朝阳区麦子店街 18 号楼

邮编：100125

责任编辑：王金环

版式设计：王　晨　　责任校对：刘丽香

印刷：北京通州皇家印刷厂

版次：2021 年 6 月第 1 版

印次：2021 年 6 月北京第 1 次印刷

发行：新华书店北京发行所

开本：880mm×1230mm　1/16

印张：23.75

字数：595 千字

定价：138.00 元

编著者及单位

Ofer Ashoulin
Dagon, Maagan Michael fish farm, Kibbutz Ma'agan Michael D. N. Menashe, Israel 37805
E-mail: labmad@mmm. org. il

Stéphane Biacchesi
National Institute for Agricultural Research (INRA), 78352 Jouy en Josas, France
E-mail: stephane. biacchesi@jouy. inra. fr

Eirik Biering
Norwegian Veterinary Institute, Tungasletta
2, 7485, Trondheim, Norway
E-mail: eirik. biering@vetinst. no

Michel Brémont
National Institute for Agricultural Research
(INRA), 78352 Jouy en Josas, France
E-mail: michel. bremont@jouy. inra. fr

Andrew Bridle
NCMCRS University of Tasmania, Locked
Bag 1370, Launceston 7250, Tasmania,
Australia
E-mail: andrew. bridle@utas. edu. au

Jarl Bøgwald
Norwegian College of Fishery Science, UiT
The Arctic University of Norway, 9037
Tromsø, Norway
E-mail: jarl. bogwald@uit. no

Duncan J. Colquhoun
Norwegian Veterinary Institute, PO Box 750,
0106 Oslo, Norway
E-mail: duncan. colquhoun@vetinst. no

Roy A. Dalmo
Norwegian College of Fishery Science, UiT
The Arctic University of Norway, 9037
Tromsø, Norway
E-mail: roy. dalmo@uit. no

Arnon Dishon
KoVax Ltd, Bynet Build. Har Hotzvim Ind.
Park, PO Box 45212 Jerusalem, Israel 90836
E-mail: arnon. dishon@kovax. co. il

Diane G. Elliott
U. S. Geological Survey, Western Fisheries
Research Center, 6505 Northeast 65th Street,
Seattle, Washington 98115, USA
E-mail: dgelliott@usgs. gov

Øystein Evensen
Norwegian School of Veterinary Science,
PO Box 8146 Dep, 0033 Oslo, Norway
E-mail: oystein. evensen@nmbu. no

Knut Falk
Norwegian Veterinary Institute, PO Box 750,
0106 Oslo, Norway
E-mail: knut. falk@vetinst. no

Arne Marius Fiskum
Pharmaq, 7863 Overhalla, Norway
E-mail: arne. fiskum@pharmaq. no

Thomas Goodrich
AquaTactics Fish Health, 12015 115th
Avenue NE, Suite 120, Kirkland,
Washington 98034, USA
E-mail: tomg@aquatactics. com

Roar Gudding
Norwegian Veterinary Institute,
PO Box 750, 0106 Oslo, Norway
E-mail: roar. gudding@vetinst. no

K. Larry Hammell
Atlantic Veterinary College, University of Prince Edward Island, 550 University Avenue, Charlottetown, C1A4P3, Canada
E-mail: lhammell@upei. ca

Anja Holm
Danish Health and Medicines Authority, Axel Heides Gade 1, 2300 Copenhagen S, Denmark
E-mail: anh@dkma. dk

Eva Högfors-Rönnholm
Laboratory of Aquatic Pathobiology, Environmental and Marine Biology, Department of Biosciences, Tykistökatu 6, Biocity, Åbo Akademi University, 20520 Turku, Finland
E-mail: ehogfors@abo. f

Jorunn B. Jørgensen
Norwegian College of Fishery Science, UiT The Arctic University of Norway, 9037 Tromsø, Norway
E-mail: jorunn. jorgensen@uit. no

Indrani Karunasagar
Department of Fisheries Microbiology, Karnataka Veterinary, Animal and Fisheries Sciences University, College of Fisheries, Mangalore - 575002, India
E-mail: karuna8sagar@yahoo. com

Viswanath. Kiron
Aquatic Animal Health Unit, Faculty of Biosciences and Aquaculture, University of Nordland, PO Box 1490, 8049 Bodø, Norway
E-mail: kiron. viswanath@uin. no

Phillip H. Klesius
USDA-ARS, Aquatic Animal Health

Research Laboratory, 990 Wire Road,
Auburn, Alabama 36832, USA
E-mail: phillip. klesius@ars. usda. gov

Dag Knappskog
MSD Animal Health, Thormøhlensgate 55,
5001 Bergen, Norway
E-mail: dag. knappskog@merck. com

Erling Olaf Koppang
Norwegian School of Veterinary Science,
PO Box 8146 Dep, 0033 Oslo, Norway
E-mail: erling. o. koppang@nmbu. no

Moshe Kotler
Department of Pathology, The Hebrew
University-Hadassah Medical School,
Jerusalem, 91120, Israel
E-mail: moshek@ekmd. huji. ac. il

Joseph Koumans
MSD Animal Health, Wim de Körverstraat
35, Post code 5381, AN Boxmeer, The
Netherlands
E-mail: sjo. koumans@merck. com

Inger Kvitvang
Pharmaq, 7863 Overhalla, Norway
E-mail: inger. kvitvang@pharmaq. no

Atle Lillehaug
Norwegian Veterinary Institute, PO Box 750,
0106 Oslo, Norway
E-mail: atle. lillehaug@vetinst. no

Biswajit Maiti
Department of Fisheries Microbiology,
Karnataka Veterinary, Animal and Fisheries
Sciences University, College of Fisheries,
Mangalore - 575002, India
E-mail: maitibiswajit@yahoo. com

Sergio H. Marshall
Institute of Biology, Faculty of Sciences,
Pontifical Catholic University of Valparaiso,
PO Box 4059, 2340025 Valparaiso, Chile
E-mail: smarshal@ucv. cl

Emilie Mérour
National Institute for Agricultural Research
(INRA), 78352 Jouy en Josas, France
E-mail: emilie. merour@jouy. inra. fr

Paul J. Midtlyng
Norwegian School of Veterinary Science,
PO Box 8146 Dep, 0033 Oslo, Norway
E-mail: paul. midtlyng. aquamedic. no

Hetron Mweemba Munang'andu
Norwegian School of Veterinary Science,
PO Box 8146 Dep, 0033 Oslo, Norway
E-mail: hetronmweemba. munang'andu@nmbu. no

Stephen Mutoloki
Norwegian School of Veterinary Science,
PO Box 8146 Dep, 0033 Oslo, Norway
E-mail: stephen. mutoloki@nmbu. no

Singaiah NaveenKumar
Department of Fisheries Microbiology,
Karnataka Veterinary, Animal and Fisheries
Sciences University, College of Fisheries,
Mangalore - 575002, India
E-mail: aquas510@gmail. com

Audun H. Nerland
Department of Clinical Science, University
of Bergen, 5021 Bergen, Norway
E-mail: audun. nerland@k2. uib. no

Ken Noda
Food Safety and Consumer Affairs Bureau,
Ministry of Agriculture, Forestry and
Fisheries, 1-2-1 Kasumigaseki Chiyoda-ku,
Tokyo 100-8950, Japan

E-mail: ken_noda@nval. maff. go. jp

Barbara Nowak
NCMCRS University of Tasmania, Locked Bag 1370, Launceston 7250, Tasmania, Australia
E-mail: B. Nowak@utas. edu. au

Sonal Patel
Institute of Marine Research, Nordnesgaten 50, 5005 Bergen, Norway
E-mail: sonal. patel@imr. no

Trygve T. Poppe
Norwegian School of Veterinary Science, PO Box 8146 Dep, 0033 Oslo, Norway
E-mail: trygve. poppe@nmbu. no

Julia W. Pridgeon
USDA-ARS, Aquatic Animal Health Research Laboratory, 990 Wire Road, Auburn, Alabama 36832, USA
E-mail: julia. pridgeon@ars. usda. gov

Praveen Rai
Department of Fisheries Microbiology, Karnataka Veterinary, Animal and Fisheries Sciences University, College of Fisheries, Mangalore - 575002, India
E-mail: raiprav@gmail. com

Linda D. Rhodes
Northwest Fisheries Science Center, National Marine Fisheries Service, 2725 Montlake Boulevard East, Seattle, Washington 98112, USA
E-mail: linda. rhodes@noaa. gov

Espen Rimstad
Norwegian School of Veterinary Science, PO Box 8146 Dep, 0033 Oslo, Norway
E-mail: espen. rimstad@nmbu. no

Jesus L. Romalde
Department of Microbiology and Parasitology, CIBUS-Faculty of Biology, University of Santiago de Compostela, 15782 Santiago de Compostela, Spain
E-mail: jesus. romalde@usc. es

Jan H. W. M. Rombout
Aquatic Animal Health Unit, Faculty of Biosciences and Aquaculture, University of Nordland, PO Box 1490, 8049 Bodø, Norway
E-mail: jhwmrombout@gmail. com

Byron E. Rippke
Center for Veterinary Biologics, USDA, 1920 Dayton Avenue, PO Box 844, Ames, Iowa 50010, USA
E-mail: byron. e. rippke@aphis. usda. gov

Kira Salonius
448 Boulter Loop, RR2, Prince Edward Island, Canada COA 1JO
E-mail: ksalonius@bellaliant. net

Craig Shoemaker
USDA-ARS, Aquatic Animal Health Research Laboratory, 990 Wire Road, Auburn, Alabama 36832, USA
E-mail: craig. shoemaker@ars. usda. gov

Krister Sundell
Laboratory of Aquatic Pathobiology, Environmental and Marine Biology, Department of Biosciences, Tykistökatu 6, Biocity, Åbo Akademi University, 20520 Turku, Finland
E-mail: ksundell@abo. f

Carolina Tafalla
Animal Health Research Center (CISA-INIA), Carretera de Algete a El Casar km 8. 1. Valdeolmos 28130, Madrid, Spain

E-mail: tafalla@inia. es

Jaime A. Tobar
Centrovet Ltda, Avenida Salomon Sack 255,
9201310 Cerrillos, Santiago, Chile
E-mail: jaime. tobar@centrovet. com

E. Scott Weber III
University of California, 2108 Tupper Hall,
Davis, California 95616, USA
E-mail: sharkdoc01@gmail. com

Gregory D. Wiens
USDA-ARS, National Center for Cool and
Cold Water Aquaculture, 11861 Leetown Rd,
Kearneysville, West Virginia 25430, USA
E-mail: greg. wiens@ars. usda. gov

Tom Wiklund
Laboratory of Aquatic Pathobiology,
Environmental and Marine Biology,
Department of Biosciences, Tykistökatu 6,
Biocity, Åbo Akademi University, 20520
Turku, Finland
E-mail: twiklund@abo. f

Rune Wiulsrød
Harbitzalléen 2A, Postbox 267 Skøyen, 0213
Oslo, Norway
E-mail: rune. wiulsrod@pharmaq. no

译 者 序

众所周知，应用疫苗实施鱼类疾病的免疫防控不仅可有效控制传染性疾病发生，而且避免了药物长期使用导致的病原微生物耐药性问题，并有利于提高鱼类养殖的经济效益，保护生态环境，减少养殖风险，保障食物安全以及产业的可持续健康发展。

Fish Vaccination 一书由全球鱼类免疫学、鱼病学、微生物学、病理学、分子生物学、疫苗研发、生产管理等方面的50余位知名专家参与编写，汇聚了众多原创性研究成果，经系统整理汇总，由挪威兽医研究所 Roar Gudding、Atle Lillehaug 和挪威兽医科学学院 Øystein Evensen 编著而成，由 John Wiley & Sons，Ltd 出版发行。

该书的内容几乎涉及与鱼类疫苗相关的各个方面，包括鱼类的免疫系统、疫苗的类型与研发、疫苗的接种策略、疫苗效果与副作用、疫苗的法规与监管等。特别介绍了主要养殖鱼类的重要细菌性和病毒性疾病的疫苗免疫防控，分别对疾病的危害性、发生流行、病原、重要的抗原、疫苗类型、疫苗接种程序、疫苗接种的效果以及可能的副作用等进行了系统论述。基于该书的内容，其中文版书名译为《鱼类疫苗》。

该书的翻译引进，期望为我国鱼类疫苗相关领域的科研、教学人员以及本科生、研究生，同时为鱼类疫苗生产和管理人员，乃至鱼类养殖与疾病防控技术人员提供参考。

由衷感谢江育林先生为译稿提出的宝贵意见，诚挚感谢挪威 Roy A. Dalmo 和 Jarl Bøgwald 教授为该书给予的翻译推荐。特别感谢中国农业出版社王金环编辑的辛勤付出，同时感谢中国海洋大学水产动物病害与免疫学研究室绳秀珍教授、唐小千教授以及在读博士研究生的参与。该书的翻译出版得到国家出版基金和江苏省海洋科学与技术优势学科建设经费的资助。

由于译者水平所限，对原文的理解和把握可能会存在不完全准确的地方，敬请读者批评指正。

译 者

2020年11月

前　言

任何一种集约化的生物产业（水产养殖），无论是在陆地上还是在海上，都会遭受疾病的困扰。野生鱼类出现的零星疾病感染，若发生在池塘或网箱养殖中就有可能造成鱼类严重死亡。由于传染性疾病的暴发，水产养殖业遭受严重经济损失的事件时有发生。抗菌药物的使用可能给人类健康、水环境甚至是水产品安全带来负面影响。因此，水产养殖应更加重视生物安全措施，控制疾病的发生，才能实现可持续发展。

疫苗接种既经济，又环保。疫苗的使用能够预防疾病，减少养殖鱼类的发病率和死亡率，降低使用药物治疗的成本，提高水产品质量，给水产养殖业及其从业者带来显著效益。疫苗接种还能减轻患病鱼的痛苦，提高动物福利。最重要的是，疫苗接种可减少具有潜在生态危害的化学药品的使用，对水产养殖业的进一步发展具有重要的意义。

本书旨在对鱼类疫苗学给予全面的介绍。总论部分，介绍了疫苗学的基础知识，包括鱼类的免疫系统、疫苗生产与质控、疫苗接种策略与程序、疫苗副作用及其表观特征与作用机理等。各论部分，介绍了主要养殖鱼类的重要细菌性疾病和病毒性疾病的疫苗接种。最后一章，介绍了甲壳动物的免疫系统及免疫激活。

在各论部分的各章节中，其内容的组成结构是类似的，主要包括疾病的危害性、发生流行、病原、重要的抗原、特异性的疫苗产品、疫苗接种程序、疫苗接种的效果以及可能的副作用等与疫苗接种相关的内容。

鱼类疫苗学方面的研究成果积淀丰厚，特别是近二十年尤为丰硕。尽管我们鼓励各章节的作者尽量提供综述性文献以减少参考文献的数量，但本书所列出的参考文献依然偏多。

鱼类疫苗学是涉及多学科门类的综合性科学。本书各章节均由该领域的权威专家撰写，在确保科学性的同时，我们也充分考虑了作者的地域代表性。

出版一部有助于全球水产养殖业可持续发展的专业书是我们的目标。鱼类养殖在世界许多地方都是一个快速发展的行业。可以预见，水产业将在淡水养殖和海水养殖方向得到进一步发展，并为生产健康安全的食品、发展地方经济，乃至为一些国家减轻贫困做出积极贡献。

［挪］冉·古廷（Roar Gudding）
［挪］爱特·林汉格（Atle Lillehaug）
［挪］奥斯坦·伊温森（Øystein Evensen）

目录

第三章 非复制型疫苗

第四章 可复制型疫苗

第五章 DNA 疫苗

第六章 鱼类黏膜疫苗接种

第七章 鱼类疫苗佐剂

第八章 鱼类先天性免疫反应

第九章　鱼类获得性免疫反应

第十章　鱼类疫苗的研发、生产和质控

第十一章　鱼类疫苗的法规和授权

第十二章 疫苗接种策略和程序

第十三章 疫苗接种的副作用

第十四章 鱼类疫苗学的发展趋势

第十五章 弧菌病的疫苗接种

第十六章 疖疮病的疫苗接种

第十七章 发光杆菌病的疫苗接种

第十八章 鮰肠道败血症的疫苗接种

第十九章 耶尔森菌病的疫苗接种

第二十章 链球菌病和乳球菌病的疫苗接种

第二十一章 鱼立克次氏体病的疫苗接种

第二十二章 细菌性肾病的疫苗接种

第二十三章 黄杆菌病的疫苗接种

第二十四章 病毒性出血性败血症和传染性造血器官坏死症的疫苗接种

第二十五章 传染性胰脏坏死症的疫苗接种

第二十六章 传染性鲑贫血症的疫苗接种

第二十七章 锦鲤疱疹病毒病的疫苗接种

第二十八章 鲑甲病毒病的疫苗接种

第二十九章 野田病毒病的疫苗接种

第三十章 甲壳动物的免疫系统和免疫刺激

第一章　鱼类疫苗接种简史

Roar Gudding and Thomas Goodrich

本章概要　SUMMARY

利用疫苗开展持续有效的疾病预防，是保障水产养殖业蓬勃发展的一个重要举措。1976 年，鲑鲁克氏耶尔森菌疫苗作为第一个水产疫苗，在美国获批上市。此后，随着水产养殖业的发展，疫苗的应用范围扩大到其他鱼类和其他国家。本章回顾了水产疫苗的发展历史，包括鱼类疫苗的研究成果和疫苗行业的一些先驱。

第一节　引　言

人类对免疫的认知已有数百年的历史。公元前 400 年，古希腊历史学家——修西底德(Thucydides) 在其第二部著作《伯罗奔尼撒战争史》(*The Peloponnesian War*) 中对瘟疫进行了描述。他注意到患瘟疫康复的人再次遭遇瘟疫时并不会发生感染（Humphrey 和 White，1970)。

疫苗接种用于预防传染性疾病已有 200 多年历史。英国医生爱德华·詹纳（Edward Jenner）注意到，经常接触牛痘的挤奶工对天花产生了抵抗力（Jenner，1798)。1796 年，他给一个男孩接种了牛痘，以诱导其产生对天花病毒的免疫力。詹纳对此事用“Vaccine inoculation”予以描述，之后逐渐演变为“Vaccination”(疫苗接种)。“Vaccine”是从意为奶牛的拉丁词“*vacca*”衍生出来的。

路易·巴斯德（Louis Pasteur）和他的同事们证明，动物接种减毒活微生物后可保护其免受致病微生物的感染（Fenner 等，1997)。他们在预防狂犬病和大量病原微生物方面的研究成果，对人类和动物健康做出了重要贡献（Humphrey 和 White，1970)。

1881 年，巴斯德在一次演讲中提议将预防接种用“Vaccination”一词表述，以表达对詹纳杰出工作的敬意（Fenner 等，1997)。“Vaccinology”（疫苗学）一词出现时间不长，由索尔克(Salk）于 1977 年提出，其含义为：基于微生物刺激免疫系统，使个体和群体获得预防传染性疾病发生的多学科交叉的科学。

使用疫苗接种预防疾病是现代医学的重要里程碑之一。疫苗应用正在得到进一步的全面发展。如今，疫苗接种被认为是一种安全有效预防人类和动物疾病的方法。

第二节 水产养殖

鱼类养殖，较之畜牧，在许多国家是一种较新的生物生产方式。当养殖鲑发生疾病时，常用抗生素及化学药物进行治疗，甚至用于疾病预防。然而，早在 1930 年代末，鲑鱼苗孵化场对有效免疫预防已经凸显出了需求。

1938 年，Snieszko 等发表了（可能是第一篇）关于使用点状气单胞菌（*Aeromonas punctata*）免疫鲤后获得免疫保护力的报道。由于文章是用波兰语写的，因而知晓范围受到了影响。最早的关于鱼类免疫的英文文章似乎是由 Duff 发表的，他的研究表明，经注射和口服免疫后，鳟获得了对杀鲑气单胞菌（*Aeromonas salmonicida*）的免疫保护（Duff，1939，1942）。

第二次世界大战（以下简称“二战”）后，抗菌药物的有效作用，或许是人们对免疫预防失去兴趣的重要原因，并且 Snieszko 和 Friddle（1949）的研究结论是，在控制鱼类疖疮病上，磺胺甲基嘧啶优于口服疫苗的效果。在这一时期，有关使用疫苗预防鱼类疾病的报道，即使在科学文献中也是寥寥无几（Newman，1993），因此，二战后的前 30 年，水产养殖业被称为水产养殖的药物时代（Evelyn，1997）。直到 20 世纪 70 年代，疫苗才在商业化的水产养殖业中开始应用。

第三节 免疫学

疫苗学包含多门学科，免疫学是其中之一。Van Muiswinkel（2008）总结了过去 150 年间与疫苗学相关的大量鱼类免疫学和生物学研究资料，并发现发表这些成果的科学家从事的专业包括生物学、解剖学、血液学、生理学、鱼类学、微生物学、病理学、鱼病学等。关于免疫诱导鱼体产生体液抗体的研究（Nybelin，1935；Pliszka，1939）在二战前就已有报道，科学家们继而研究了免疫球蛋白、补体以及细胞免疫应答等（Ridgway 等，1966；Cushing，1970；Corbel，1975；Press，1998）。为更好地理解鱼类免疫系统的进化发展，科学家们对鱼类比较免疫学和发育免疫学也开展了研究（Ambrosius，1967）。

科学家们在包括鲤和鲑等多种鱼类上开展了免疫学研究，同时也研究了温度和其他环境因子在疫苗学中的重要作用（Snieszko，1974）。佐剂对鱼类免疫学和疫苗学的发展起到了重要的作用，Ambrosius 和 Lehmann（1965）发现氢氧化铝和弗氏佐剂等提高了鱼体免疫球蛋白的总量，且弗氏佐剂的效果更显著。

同时，科学家们还研究了疫苗的接种途径。在给鳟注射杀鲑气单胞菌抗原后，鱼体能产生较好的免疫反应（Krantz 等，1963），然而注射法并非既方便又经济的方式，而且还会对鱼体造成伤害。因此，口服免疫被认为是疾病预防的唯一实际可行方法（Snieszko，1970）。

对鱼类疫苗的评价，也受到了在人和陆地动物上疫苗研究结果的影响。人的脊髓灰质炎和家禽的新城疫口服疫苗的成功，影响了从事鱼类免疫学和疫苗学的科学家们的研究重点和方向

(Stone 等，1969；Sabin 和 Boulger，1973)。工厂化养殖家禽和鱼类，有共同之处，因此研究不同脊椎动物的科学家们可共享信息，相互借鉴。

家禽业的疫苗喷雾接种经验，也被借鉴作为鱼类疫苗接种的方式。鱼类喷雾疫苗接种技术被授予美国专利（Garrison 等，1980)，但该方法在实际生产中并没有得到广泛应用。

第四节 疾病预防

疾病是病原、宿主和环境三者互相影响的结果（Snieszko，1974）（图 1-1)。在野生鱼群中，疾病通常被认为是正常生物学过程的一部分，而在养殖鱼类中，由于养殖规模和密度增加，这一情形逐渐发生了改变，在野生鱼群中这种正常疾病现象，在养殖鱼类中有时就成了一个棘手问题。

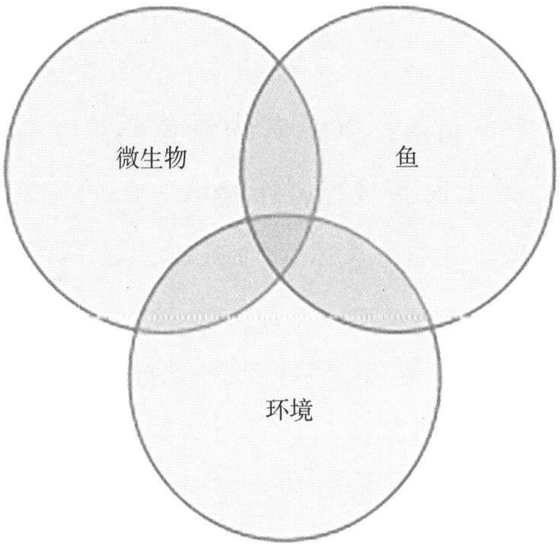

图 1-1 疾病发生与病原、宿主、环境的相互关系
（参照 Snieszko，1974）

疾病预防应从影响疾病发生的三个因素入手，与陆生动物养殖相比，水产养殖业存在一些明显的劣势。首先，致病微生物可以通过水体进行传播。其次，消毒是非常困难的，如果在海上，消毒几乎是不可能实现的。再次，药物的副作用和风险也影响了抗生素在治疗上的使用（当然，应该限制抗生素的使用)。有些环境因素是可以调控的，但有些环境因素，如水温，控制就较为困难。因此，鱼类的天然免疫力和获得性免疫力是养殖鱼类健康的决定性因素。

在现代水产养殖业的短暂的历史中，尽管鱼类疫苗接种有成功也有失败，但通过激活鱼类免疫系统预防疾病发生的举措，得到普遍认可，成为水产养殖生产不可分割的组成部分。由于科学发展和科研成果在实际生产中得到快速转化，疫苗接种已成为全球水产养殖业发展的重要环节。

第五节 科学产出——成果与学术会议

鱼类疫苗学及其相关领域的科学产出取得了巨大的成就，在科学数据库中对“Fish vaccination”（鱼类疫苗接种）的检索显示，约有 10 000 篇相关论文（Science Direct，Blackwell)，这些论文大多是在过去近 20 年里发表的。到 20 世纪 80 年代末，关于疫苗的出版物总数已超过 100 部（Newman，1993)。

针对鱼类疫苗的综述文章，就发表了多篇（如 Snieszko，1970；Harrell，1979；Leong 和 Fryer，1993；Newman，1993；Press 和 Lillehaug，1995；Ellis，1997；Gudding 等，1999；Vinitnantharat 等，1999；Sommerset 等，2005；Toranzo 等，2009)。同时，关于鱼类疫苗学的国际会议及其会议论文集也促进了鱼类疫苗学的发展（如 Anderson 和 Hennesen，1981；de Kinkelin，1984；Gudding 等，1997；Midtlyng，2005)。20 世纪末，许多学位论文的选题是鱼类

疫苗领域（如 Lamers，1985；Erdal，1990；Lillehaug，1993；Kolb，1994；Hoel，1997；Joosten，1997；Midtlyng，1998），还有的学位论文涉及鱼类免疫学、鱼类细菌学和养殖鱼类的疾病预防等多方面。另外，利用疫苗预防养殖鱼类疾病的相关内容也出现在教科书以及专著的章节中（如 Ellis，1988；Tatner，1993；Adams 等，1997；Midtlyng，1997）。

一、鲑鳟鱼类

尽管在多种鱼上开展了免疫学的研究，但鱼类疫苗学的研究主要集中在鲑鳟鱼类上，Bullock 等（1971）在其所著的疖疮病一章中，总结了 20 世纪 60 年代末的实际状况，在化学药物治疗一节中，明确表明："虽然投喂低剂量的药物能够有效地控制流行性疖疮病，但是必须小心谨慎，以免药物对鱼类产生毒性，并且要警惕可能出现杀鲑气单胞菌的耐药菌株。"关于疫苗接种，作者认为在饲料中加入抗原的口服免疫，是对疖疮病唯一可行的常规免疫接种方法。此外，作者还预言水产养殖的药物时代即将结束，疫苗接种用于疾病预防将大有可为。

Hayashi 等（1964）很早就报道了预防弧菌病的疫苗接种效果，其研究结果表明，注射一定量的浓缩弧菌疫苗，可以有效地控制虹鳟弧菌病发生。一年后，Ross 和 Klontz（1965）研究结果显示，给虹鳟幼鱼投喂含有鲁克氏耶尔森菌（*Yersinia ruckeri*）细胞成分的颗粒饵料可以预防虹鳟红嘴病（耶尔森菌病）。

1976 年，应用免疫方法预防弧菌属（*Vibrio*）和耶尔森菌属（*Yersinia*）引起的鱼类疾病，在水产养殖业中得到普遍认可，由 Wildlife Vaccine Inc. 生产的红嘴病疫苗获得生产许可（1976 年），第一个鱼类疫苗诞生了（Tebbit 等，1981）。科学实验和生产应用都证明疫苗可明显减少耶尔森菌病的发病率以及死亡率，并且通过注射或者浸泡灭活疫苗，即使不混合佐剂使用，鱼体均获得了免疫保护力（Bullock 和 Anderson，1984；Evelyn，1984）。

20 世纪 80 年代，挪威的养殖鲑中出现了一种危害极为严重的疾病，最初称为"Hitra disease"（特拉岛病），现在称为冷水弧菌病。虽然刚开始对这种传染性疾病的病因有一些争议，但很快该病被证明是由一种新的致病菌引起，病原被鉴定为杀鲑弧菌（*Vibrio salmonicida*）（Egidius 等，1986）。1988 年以来，挪威大多数的大西洋鲑和虹鳟都通过接种疫苗（最初用浸泡法）预防这种疾病。

由杀鲑气单胞菌引起的疖疮病，是鲑养殖中疾病预防面临的巨大挑战。基于浸泡免疫预防弧菌感染的成功经验，人们对疖疮病疫苗的效果充满信心，寄予厚望。然而，浸泡疖疮病疫苗产生的保护效果不佳，注射杀鲑气单胞菌全菌疫苗虽可激活鱼体的免疫保护，但免疫保护力和持续时间并不理想。

在陆生动物疫苗中，将佐剂加入疫苗中使用，已有几十年的时间了。用于人和某些动物疫苗的铝盐佐剂，当用于鱼类疫苗时，也较好地增强了疫苗的免疫保护效应（Erdal 和 Reitan，1992；Lillehaug，1993）。进而，为了获得持久的免疫效果，人们发现油类佐剂，如矿物油，无论是攻毒实验测试还是养殖生产中的结果，都能降低大西洋鲑的发病率和死亡率，效果比较理想（Midtlyng，1998）。佐剂的严重副作用是引发局部炎症反应，不过政府、从业者和消费者认为有效的疾病预防比动物福利更为重要。

添加矿物油佐剂的鳗弧菌（*V. anguillarum*）、杀鲑弧菌（*V. salmonicida*）和杀鲑气单胞菌（*A. salmonicida*）疫苗，为疾病的预防发挥了重要作用，如果没有疫苗，这些疾病将会给水产养殖业带来巨大的经济损失。在疫苗使用以前，水产养殖业中使用大量的抗菌药物进行疾病治疗，这保障不了该行业的可持续发展。疫苗接种使挪威水产养殖得以成功发展，关于其意义，一位业内资深人士发出这样的感慨："这个产业或许在鱼类发生高死亡率，蒙受经济损失时，尚可生存，但大量使用抗生素产生的副作用将使它走向末日。"因此，疫苗接种是保障鲑养殖产业可持续发展的关键因素之一。挪威水产养殖的抗生素使用量减少，是疫苗学史上的一个成功典范（Gudding，2014）。

挪威在鱼类疫苗方面取得成功的原因有很多。首先是公立和私营机构以及企业中的科学家的贡献及其富有的创新精神，其次是科学界、政府以及行业间的良好合作，还有鱼类疫苗研发、生产、养殖试验等环节的高效运作等。

在挪威，鱼类疫苗生产许可证获批的官方过程还是较快的。鱼类疫苗的早期（1993 年以前）管理，监管许可由农业部负责。考虑到处于积累和探索阶段，当时的法规要求不多，注重的是安全性和有效性。1993 年，疫苗监管和审批许可移交到挪威药品局，这表明，按照国际规则，向更全面的监管框架转变。回想起来，可以说，过去简单但有效的监管工作（在挪威水产养殖业的最初几年），对减少抗生素使用作出了重大贡献，并没有造成严重的负面效应（Midtlyng 等，2011）。

在水产养殖的早期阶段，主要的病毒性疾病包括传染性胰腺坏死、病毒性出血性败血症和传染性造血器官坏死等。对于后两种病毒疾病，并没有优先考虑研制疫苗，而是采取扑杀发病群体的措施。无毒或减毒的活疫苗，已实验性地获得对这些疾病预防的效果（Fryer，1976；Vestergaard-Jørgensen，1976；Hill 等，1980）。在实验条件下，这些活疫苗具有良好的免疫保护率，由于有些活疫苗在受免鱼体内仍保留有毒性，出于安全性考虑，没有进行更多活疫苗的研究工作；也由于担心活疫苗残留会导致水环境中其他鱼类感染发病，因此限制了活疫苗在养殖生产中的应用。

灭活病毒疫苗在实验条件下，可为受免鱼提供一些保护，但在实际应用时与大多数细菌疫苗相比，其免疫保护效果相对较低。因此，水产养殖业对灭活病毒疫苗的效果并不满意（Biering 等，2005）。

二、非鲑鳟鱼类

鲑养殖的成功刺激了其他海水鱼类养殖的发展。在地中海周边的一些国家，海鲈、海鲷和大菱鲆的养殖产量大幅增加，与此同时，疾病的确也是最突出的问题。在美国，斑点叉尾鮰产业遭受细菌性疾病的困扰。为预防疾病，许多国家都使用了疫苗（Håstein 等，2005），这些疫苗，有些是具有国家或国际许可证的商业疫苗，有些则是少数鱼类养殖场使用的自体疫苗或实验性疫苗。

弧菌病和巴斯德菌病是地中海地区养殖鱼类的主要细菌性疾病。研究发现，具有与致病微生物相同抗原成分的灭活疫苗，可为鱼体提供良好的免疫保护（Santos 等，1991；Gravningen 等，1998）。在许多养殖海鲈、海鲷和大菱鲆的地区，疫苗接种是预防这些疾病的重要措施。

细菌性疾病是影响美国斑点叉尾鮰产业的首要问题。斑点叉尾鮰的重要细菌病原有鮰爱德华菌（*Edwardsiella ictaluri*）、柱状黄杆菌（*Flavobacterium columnare*）和嗜水气单胞菌（*Aeromonas hydrophila*）。由于粗放型的养殖与管理限制了疫苗的使用，目前可用于预防鮰爱德华菌和柱状黄杆菌引起感染的疫苗是减毒的浸泡活疫苗（Shoemaker 等，2009）。

世界上，水产养殖主要在亚洲。目前，疫苗接种尚未成为该地区鱼类疾病预防的重要手段。然而，行业中已经开始重视疫苗在预防罗非鱼、鲶等鱼类疾病的作用。2011 年，越南批准了首个可用于预防一种鲶疾病的商品疫苗。

第六节 成功与失败

总体来说，鱼类疫苗接种史是一部成功史。一条小鱼在稀释的疫苗中浸泡几秒后，就可以预防某种传染病，事实说明，生物体具有足够的能力应对感染。

在鱼类疫苗学的研究历程中，基于鱼类免疫系统预防疾病，曾遇到过一些难题。对不同鱼类的研究表明，鱼类、鸟类和哺乳动物的基本免疫机制是相似的，但物种间免疫差异对免疫预防的策略和方法有很大影响。

母源免疫是保护哺乳动物和鸟类的新生个体免受疾病侵染的基本组成部分，因而可推断鱼类中也有类似机制，这在卵胎生的孔雀鱼上得以证实：研究发现母源免疫可以从免疫过的母体转移到后代中（Takahashi 和 Kawahara，1987）。但在鲑中，转移的母源抗体似乎处于非常低的水平，不足以保护后代免受感染（Lillehaug 等，1996）。

免疫系统的早熟能够弥补母源免疫的不足。免疫持续和免疫记忆对疫苗学至关重要，两者均随着鲑的生长而增加（Johnson 等，1982a）。体重为 2g 的鲑在浸泡疫苗后可以获得免疫保护（Johnson 等，1982b），4g 的虹鳟浸泡疫苗后可以获得更长时间的免疫保护。在水产养殖中，免疫持续时间应满足生产周期（接种疫苗至收获）的需求。

疫苗接种预防鱼类的一些细菌感染效果显著，但对细胞内寄生微生物，如病毒和胞内细菌，所引起的感染效果似乎并不明显，希望减毒疫苗和/或 DNA 疫苗能解决这一问题。

第七节 先　驱

鱼类疫苗接种史上的成就，是由科学文献承载的，由许多大学和研究机构科学家的研究成果组成。

疫苗也是商品，由私营公司开发、生产和销售。鱼类疫苗的制造商，通常是小型企业。那些热衷于鱼类疫苗的开拓者，他们提出想法并参与疫苗生产过程的各个环节，他们既有研发方面的知识，又了解疫苗的正确使用，为鱼类疫苗学的早期发展做出了重要贡献。

许多早期的开拓者都是优秀的科学家，他们把疫苗预防疾病的理论知识和实践经验与商业利益结合在一起。以下列举了一些鱼类疫苗公司和开拓者，但并不能囊括行业中所有公司和贡献者。

科罗拉多野生动物疫苗公司（The Colorado Company Wildlife Vaccines）是第一个获得鱼类疫苗生产许可的企业，生产的细菌疫苗销往国内外，公司拥有 Guy Tebbit、John Rohovec 和 Thomas Goodrich 专家团队。Tavolek 股份有限公司（Tavolek Inc.），是强生公司（Johnson & Johnson）的子公司，是第二个具有鱼类疫苗生产许可的公司，Keith Johnson 和 Don Amend 是公司技术专家，他们参与了耶尔森菌和其他细菌疫苗开发和生产的大部分工作。然而，两家公司都经历过因当时市场小而无盈利的经营阶段。

1980 年代，西雅图的生物医学研究实验室（Biomed Research Laboratories），组建了由分别从事水产养殖、疫苗学、孵化场疾病预防和菌苗生产领域的 Stephen Newman、Tony Novotny、James Nelson 和 Robert Busch 组成的高素质精干专家团队，进入疫苗市场。

在鱼类疫苗历史上还有两家值得提及的小公司，Aqua Health Inc. 和 Aquaculture Vaccines Ltd.（最初是 Wildlife Vaccines 的子公司），在鱼类疫苗接种的早期阶段发挥了重要作用。这两家公司的科学技术专家分别是 William Patterson 和 Patrick Smith。

在欧洲，鲑产量的增长是面向市场创立疫苗公司的基础。Apothekernes 公司，即后来的雅来（Alpharma）公司，是一个总部在奥斯陆（挪威首都）的国际性制药公司。该公司发现了水产养殖的商业潜力和鱼类疫苗的商机，起初在特罗姆索（TromsØ），后来在奥沃尔哈拉（Overhalla）开展鱼类疫苗的生产。在 20 世纪 90 年代早期，该公司与生物医学研究实验室（Biomed Research Laboratories）合并，成为雅来集团（Alpharma group）的一部分；后来，水产养殖板块从总公司分离出来，创建了法玛克（Pharmaq）公司。

在卑尔根（Bergen），以大学的科学家为专家团队的 Norbio 公司成立。后来，Norbio 被荷兰英特威（Intervet）公司收购，其后英特威与先灵宝雅（Shering-Plough）合并，同时兼并了 Aquaculture Vaccines Ltd.，2011 年起，鱼类疫苗成为默沙东公司（Merck Sharp Dome）的一部分。

公司交易在疫苗行业并不罕见，一些制药公司和投资者看到了鱼类疫苗的经济前景，开始涉足此行业。如国际制药公司诺华公司（Novartis）与有实力的加拿大水产健康有限公司（Aqua Health Inc）联合开展鱼类疫苗业务。

在水产养殖业蓬勃发展的国家，考虑到鱼类疫苗较好的经济前景，相继建立了疫苗公司。如智利的 Centrovet 和以色列的 Kovax 等公司就是区域性鱼类疫苗生产企业，其目标是为水产养殖业提供有效、安全的疫苗。

第八节　结　语

鱼类疫苗学的历史记载了如何利用疫苗激活鱼体的免疫系统，从而预防致病微生物的影响。短短几年，大学的科学家，研究机构和企业界已经掌握了微生物学和养殖鱼类免疫系统的基础知识，具备了水产养殖疫苗的研发、生产、销售和应用的能力。

鱼类疫苗学尚处于起步阶段，大多数产品还是第一代疫苗。但在科研成果、实践经验上的积累，为疫苗的进一步研发奠定了良好基础，这对全球水产养殖在环境、社会和经济方面的可持续发展有重大意义。

参考文献

Adams, A., Thompson, K. D. and Roberts, R. J. 1997. Fish vaccines, in *Vaccine Manual. The Production and Quality Control of Veterinary Vaccines for Use in Developing Countries* (eds N. Mowat and M. Rweyemanu). FAO Animal Production and Health Services No. 35. Rome, FAO, 127-142.

Ambrosius, H. 1967. Untersuchungen über die Immunglobuline niederer Wirbeltiere. *Allerg Asthma* 13: 111-119.

Ambrosius, H. and Lehmann, R. 1965. Beiträge zur Immunbiologie poikilothermer Wirbeltiere. III Der Einfluss von Adjuvanten auf de Antikörper-produktion. *Acta Biol Med Ger* 14: 830-844.

Anderson, D. P. and Hennesen, W. (eds) 1981. Fish biologics: serodiagnostics and vaccines. *Dev Biol Stand* 49: 496 pp.

Biering, E., Villoing, S., Sommerset, I. and Christie, K. E. 2005. Update on viral vaccines for fish. *Dev Biol (Basel)* 121: 97-113.

Bullock, G. L. and Anderson, D. P. 1984. Immunization against *Yersinia ruckeri*, cause of enteric red mouth, in *Symposium on Fish Vaccination* (ed. P. de Kinkelin). Paris, OIE, 151-166.

Bullock, G. L., Conroy, D. A. and Sniezko, S. F. 1971. Bacterial disease of fishes, in Diseases of Fishes (eds S. F. Snieszko and H. R. Axelrod). Jersey City, TFH Publications 10, 139.

Corbel, M. J. 1975. The immune response of fish: A review. *J Fish Biol.* 7: 539-563.

Cushing, J. E. 1970. Immunology of fish, in *Fish Physiology*, Vol. 4. (eds W. S. Hoar and D. J. Randall). New York, Academic Press, 465-500.

de Kinkelin, P. (ed.) 1984. *Symposium on fish vaccination: theoretical background and practical results on immunization against infectious diseases*. Paris, OIE.

Duff, D. C. B. 1939. Some serological relationships of the S, R, and G phases of *Bacillus salmonicida*. *J Bacteriol* 38: 91-100.

Duff, D. C. B. 1942. The oral immunization of trout against *Bacterium salmonicida*. *J Immunol* 44: 87-94.

Egidius, E., Wiik, R., Andersen, K., *et al.*, 1986. *Vibrio salmonicida* sp. nov., a new fish pathogen. *Int J Syst Evol Microbiol* 36: 518-520.

Ellis, A. E. (ed.) 1988. *Fish Vaccination*. London, Academic Press.

Ellis, A. 1997. Vaccines for farmed fish, in *Veterinary Vaccinology* (eds P. P. Pastoret, J. Blancou, P. Vannier and C. Verschueren). Amsterdam, Elsevier, 411-417.

Erdal, J. E. 1990. *Immune responses and protection after vaccination of farmed Atlantic salmon against bacterial diseases*. PhD thesis, Norwegian College of Veterinary Medicine, Oslo.

Erdal, J. E. and Reitan, L. J. 1992. Immune response and protective immunity after vaccination of Atlantic salmon (*Salmo salar* L) against furunculosis. *Fish Shellfish Immunol* 2: 99-108.

Evelyn, T. P. T. 1997. A historical review of fish vaccinology. *Dev Biol Stand* 90: 3-12.

Evelyn, T. P. T. 1984. Immunization against pathogenic Vibrios, in *Symposium on Fish Vaccination* (ed. P. de Kinkelin). Paris, OIE, 121-150.

Fenner, F., Pastoret, P. P., Blancou, J. and Terre, J. 1997. Historical introduction, in *Veterinary Vaccinology* (eds P. P. Pastoret, J. Blancou, P. Vannier and C. Verschueren). Amsterdam, Elsevier, 3-19.

Fryer, J. L., Rohovec, J. S., Tebbit, G. L., *et al.*, 1976. Vaccination for control of infectious diseases in Pacific salmon. *Fish Pathol* 10: 155-164.

Garrison, R. L., Gould, R. W., O'Leary, P. J. and Fryer, J. L. 1980. Spray immunization of fish. *US Patent No*. 4, 223, 014.

Gravningen, K., Thorarinsson, R., Johansen, L. H., *et al*., 1998. Bivalent vaccines for sea bass (*Dicentrachus labrax*) against vibriosis and pasteurellosis. *J Appl Ichthyol* 14: 159-162.

Gudding, R. 2014. Vaccination as a preventive measure, in *Fish Vaccination* (eds R. Gudding A. Lillehaug and Ø. Evensen). Chichester, John Wiley & Sons Ltd., 12-21.

Gudding, R., Lillehaug, A. and Evensen, Ø. 1999. Recent developments in fish vaccinology. *Vet Immunol Immunpathol* 72: 203-212.

Gudding, R., Lillehaug, A., Midtlyng, P. J. and Brown, F. (eds) 1997. Fish vaccinology. *Dev Biol Stand* 90: 484 pp.

Harrell, L. W. 1979. Immunization of fishes in world mariculture: A review. *Proc World Maricult Soc* 10: 534-544.

Hayashi, K., Kobayashi, S., Kamata, T. and Ozaki, H. 1964. Studies on the vibrio disease of rainbow trout. II. Prophylactic vaccination against the vibrio disease. *J Fac Fish Prefect Univ Mie-Tsu* 6: 181-191.

Hill, B. J., Dorson, D. M. and Dixon, P. F. 1980. Studies on the immunization of trout against IPN, in *Fish Diseases* (ed. W. Ahne). Berlin, Springer Verlag, 29-36.

Hoel, K. 1997. *Adjuvant effects of bacterial components in fish vaccines*. PhD thesis, Norwegian College of Veterinary Medicine, Oslo.

Humphrey, J. H. and White, R. G. 1970. *Immunology for Students of Medicine*, 3rd edn. Oxford, Blackwell Scientific Publications, 1-34.

Håstein, T., Gudding, R. and Evensen, Ø. 2005. Bacterial vaccines for fish-an update of the current situation worldwide. *Dev Biol* (*Basel*) 121: 55-74.

Kolb, C. 1994. *Die Vakzination von Fischen*. PhD thesis, Justus-Liebig-Universität, Giessen.

Krantz, G. E., Reddecliff, J. M. and Heist, C. E. 1963. Development of antibodies against *Aeromonas salmonicida* in trout. *J Immunol* 91: 757-760.

Jenner, E. 1798. *An inquiry into the causes and effects of the variolae vaccinae, a disease discovered in some of the western counties of England particularly Gloucestershire, and known by the name of cow pox*. London, Sampson Low.

Johnson, K. A., Flynn, J. K. and Amend, D. F. 1982a. Duration of immunity in salmonids vaccinated by direct immersion with *Yersinia ruckeri* and *Vibrio anguillarum* bacterins. *J Fish Dis* 5: 207-213.

Johnson, K. A., Flynn, J. K. and Amend, D. F. 1982b. Onset of immunity in salmonid fry vaccinated by direct immersion in *Vibrio anguillarum* and *Yersinia ruckeri* bacterins. *J Fish Dis* 5: 197-205.

Joosten, E. 1997. *Immunological aspects of oral vaccination in fish*. PhD thesis, University of Wageningen.

Lamers, C. H. J. 1985. *The reaction of the immune system of fish to vaccination*. PhD thesis, University of Wageningen.

Leong, J. C. and Fryer, J. L. 1993. Viral vaccines for aquaculture. *Annu Rev Fish Dis* 3: 225-240.

Lillehaug, A. 1993. *Effects and side-effects of vaccination of salmonid fish*. PhD thesis, Norwegian College of Veterinary Medicine, Oslo.

Lillehaug, A., Sevatdal, S. and Endal, T. 1996. Passive transfer of specific maternal immunity does not protect Atlantic salmon (*Salmo salar* L.) fry against yersiniosis. *Fish Shellfish Immunol* 6: 521-535.

Midtlyng, P. J. 1997. Vaccination against furunculosis, in *Furunculosis-Multidisciplinary Fish Disease Research* (eds E. M. Bernoth, A. E. Ellis, P. J. Midtlyng, G. Olivier and P. Smith). San Diego, Academic Press,

382-404.

Midtlyng, P. J. 1998. *Evaluation of furunculosis vaccines in Atlantic salmon*. PhD thesis, Norwegian College of Veterinary Medicine, Oslo.

Midtlyng, P. J. 2005. Progress in fish vaccinology. *Dev Biol* (*Basel*) 121: 335 pp.

Midtlyng, P. J., Grave, K. and Horsberg, T. E. 2011. What has been done to minimize the use of antibacterial and antiparasitic drugs in Norwegian aquaculture. *Aquac Res* 42: 28-34.

Newman, S. G. 1993. Bacterial vaccines for fish. *Annu Rev Fish Dis* 3: 145-185.

Nybelin, O. 1935. Über Agglutininbildung bei Fischen. *Z Immunitätsforsch* 84: 74-79.

Pliszka, F. 1939. Untersuchungen über die Agglutininbildung bei Fischen. *Zbl Bakt Abt* 1 143: 262-264.

Press, CMcL. 1998. Immunology of fishes, in *Handbook of Vertebrate Immunology* (eds P. P. Pastoret, P. Griebel, H. Bazin and A. Govaerts). San Diego, Academic Press, 3-62.

Press, C. and Lillehaug, A. 1995. Vaccination in European salmonid aquaculture: a review of practices and prospects. *Br Vet J* 151: 46-69.

Ridgway, G. J., Hodgins, H. O. and Klontz, G. W. 1966. The immune response in teleosts, in Phylogeny of Immunity (eds R. T. Smith, P. A. Miescher and R. A. Good). Gainesville, FL, University of Florida Press, 199-208.

Ross, A. J. andKlontz, G. W. 1965. Oral immunization of rainbow trout (*Salmo gairdneri*) against an etiologic agent of "red mouth disease". *J Fish Res Bd Can* 22: 713-719.

Sabin, A. B. and Boulger, L. R. 1973. History of Sabin attenuated poliovirus oral live vaccine strains. *J Biol Stand* 1: 115-118.

Salk, J. and Salk, D. 1977. Control of influenza and poliomyelitis with killed virus vaccines. *Science* 195: 834-847.

Santos, Y., Bandin, I., Nunez, S., *et al.*, 1991. Protection of turbot *Scophthalmus maximus* (L.), and rainbow trout, *Oncorhynchus mykiss* (Richardson), against vibriosis using two different vaccines. *J Fish Dis* 14: 407-411.

Shoemaker, C. A., Klesius, P. H., Evans, J. J. and Aria, C. R. 2009. Use of modified live vaccines in aquaculture. *J World Aquac Soc* 40: 573-585.

Snieszko, S. F. 1970. Immunization of fishes: a review. *J Wildl Dis* 6: 24-30.

Snieszko, S. F. 1974. The effects of environmental stress on outbreaks of infectious diseases of fishes. *J Fish Biol* 6: 197-208.

Snieszko, S. F. and Friddle, S. B. 1949. Prophylaxis of furunculosis in brook trout (*Salvelinus fontinalis*) by oral immunization and sulfamerazine. *Progress Fish Cult* 11: 161-168.

Snieszko, S., Piotrowska, W., Kocylowski, B. and Marek, K. 1938. Badania bakteriologiczne i serogiczne nad bakteriami posocznicy karpi. Memoires de l'Institut d'Ichtyobiologie et Pisciculture de la Station de Pisciculture Experimentale a Mydlniki de l'Universite Jagiellonienne a Cracovie Nr 38.

Sommerset, I., Krossøy, B., Biering, E. and Frost, P. 2005. Vaccines for fish in aquaculture. *Exp Rev Vaccines* 4: 89-101.

Stone, H. D., Ritchie, A. E. and Boney, W. A. 1969. Immunization of chickens against Newcastle disease with betapropiolactone-killed virus antigen administered in drinking water. *Avian Dis* 13: 568-578.

Takahashi, Y. and Kawahara, E. 1987. Maternal immunity in newborn fry of ovoviviparous guppy. *Nippon Suis Gakk* 53: 721-735.

Tatner, M. F. 1993. Fish vaccines, in *Vaccines for Veterinary Applications* (ed. A. R. Peters). Oxford, Butterworth-Heinemann, 199-224.

Tebbit, G. L., Erickson, J. D. and Van de Water, R. B. 1981. Development and use of *Yersinia ruckeri* bacterins to control enteric redmouth disease. *Dev Biol Stand* 49: 395-401.

Toranzo, A. E., Romalde, J. L., Magariños, B. and Barja, J. L. 2009. Present and future aquaculture vaccines against bacterial fish diseases. *Op Méditerr* 86: 155-176.

Van Muiswinkel, W. B. 2008. A history of fish immunology and vaccination I. The early days. *Fish Shellfish Immunol* 25: 397-408.

Vestergaard-Jørgensen, P. E. 1976. Partial resistance of rainbow trout (*Salmo gairdneri*) to viral haemorrhagic septicaemia (VHS) following exposure to non-virulent Egtved-virus. *Nord Vet Med* 28: 570-571.

Vinitnantharat, S., Gravningen, K. and Greger, E. 1999. Fish vaccines. *Adv Vet Med* 41: 539-550.

第二章　预防疫苗接种

Roar Gudding

本章概要　SUMMARY

预防措施作为农业和水产养殖业生物安全的一部分，已变得越来越重要。在水产养殖中，如果处于患病风险鱼的数量较多，从实用、技术以及经济的角度来看，则治疗就会困难，或者不可能，因此，应用疫苗接种进行疾病预防尤为重要。疫苗接种可减少抗生素的使用，有利于水产养殖的可持续发展。在水产养殖，疫苗接种的另一益处是，接种疫苗和未接种疫苗的个体都能提高抗病力，即所谓的群体免疫。疫苗接种被认为是一种安全且经济的预防措施，因此，基于激发鱼体免疫系统的疾病预防，成为现代化水产养殖发展的基础。鱼类疫苗接种在养殖鲑中的应用已取得成功，此外，在其他鱼类，如海鲈、海鲷、鲤和斑点叉尾鮰中的应用也越来越广泛。

第一节　引　言

水产养殖是世界上增长最快的食品生产行业（联合国粮食及农业组织，2012）。全球范围内，捕捞渔业的产量已趋于平稳，随着世界人口的持续增长，水产养殖将为满足人类对水产品不断增长的需求做出重大的贡献。近几十年来，以集约化和新品种养殖为特征的水产养殖业取得了重大进展。

在繁育和饲料方面取得的科学成果，为推动水产养殖的发展做出了贡献，同时，确保养殖鱼类的健康也是提高产量的决定因素。人们越来越深刻地认识到，养殖对象良好的健康状况对水产养殖的成功至关重要。这一认知是从病原微生物造成的高发病率和死亡率给一些企业甚至某些国家的水产养殖造成了巨大的经济损失（Barton 和 Fløysand，2010）的经历中得到的。

可持续发展意味着既满足当代人的需要，而又不损害后代人的需求（Brundtland，1987）。任何像水产养殖这样的生物生产都必须是可持续的。这意味着水产养殖业应采取对环境造成的负面影响是可接受的和可恢复的方式进行生产经营。可持续性的指标有许多：疾病的流行和危害范围的监控是其中之一；用抗生素或抗菌药物来治疗鱼类的细菌性感染，其使用总量是评估可持续性的另一个指标；水产品的低药物残留和患病鱼中微生物的低耐药，也是可持续水产养殖的指标（Gudding，2012）。

水产养殖可能会改变养殖水体的生态环境，如当地鱼类种群及其病原等。在野生鱼类上出现的疾病感染现象，若发生在高密度网箱养殖的鱼中，可能成为大问题。鲑养殖改变了鲑行业的发展态势，引起了养殖和垂钓利益相关者的紧张关系甚至是冲突。甚至有人推测，鱼类养殖会导致进化出致病力更强的病原体（Pulkkinen 等，2010）。

疾病造成产量减少，是对行业和社会的直接经济影响，但也存在疾病对水产品质量的间接影响。因此，越来越多的消费者和利益相关者都关注疾病的预防，这不仅有利于食品供给和食品安全，同时能减少流行病、降低发病率，也提高了水产动物的福利（Turnbull 和 Kadri，2007）。

和其他食品生产一样，水产养殖既具有社会、经济效益，也有风险。人们对养殖鱼类的认可度，对水产养殖的未来发展至关重要（Schlag，2010）。因此，生物安全有助于改善鱼类健康、提高动物福利和提高经济效益，这需要长期的努力，是实现可持续水产养殖的重要工作。

第二节 生物安全与疫苗接种

生物安全是包括有助于维持健康群体的各种措施。随着行业的发展，生物安全开始成为水产养殖管理日益重要的一部分。预防、控制和扑杀都是为了消除疾病或减少疾病的发生频率以及降低对产品影响而采取的生物安全措施。

预防胜于治疗，这在预防人类和畜牧动物疾病中是众所周知的。在水产养殖中，如果处于患病风险鱼的数量较多，从实用、技术以及经济的角度来看，治疗就会困难，或者不可能，因此预防措施尤其重要。

鱼类疫苗接种，作为水产养殖业生物安全的重要组成部分，对预防和控制传染性疾病正变得日益重要。接种过疫苗的动物发病风险降低，即使未接种疫苗的动物也可能因为群体免疫而得到保护，同时免疫预防措施也减少了抗生素的使用。在挪威，多年来，鲑在转移到海水养殖前都进行了几种疾病的预防接种。1990 年以来（尤其是 1995 年以后），挪威每年在鲑养殖中使用的抗生素还不到 1t（图 2-1）。

预防、控制和扑杀是生物安全的三大基石。扑杀水生动物是困难的，有时是不可能的。与畜牧业相比，水产养殖系统中的病原体很难通过消毒手段彻底杀灭，而疫苗通常能预防疾病，但并不一定能阻止感染。因此，既然在水生环境中根除传染病有难度，那么，最现实的选择是预防控制疾病发生，也就是做到“有病不发病”。

宿主、病原和环境三者间的联系（Sniezko，1974）是生物安全的预防控制基础。免疫预防主要是基于微生物和宿主之间的相互作用，但有利的环境也有助于增强宿主的抵抗力。为了达到最佳效果，应该增加疫苗接种这一生物安全措施，同时结合降低感染压力、改善对抗病力有影响的环境因素等措施。

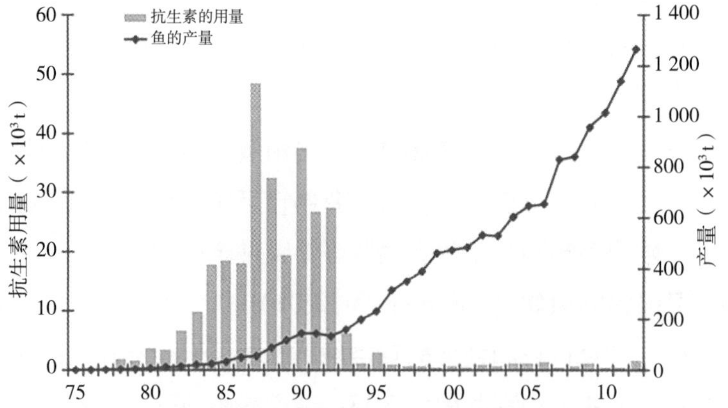

图 2-1　挪威鲑养殖抗生素的年使用量

通过疫苗接种已实现了对几种鱼类细菌性疾病的预防（Press 和 Lillehaug，1995；Gudding 等，1999；Håstein 等，2005；Sommerset 等，2005；Bravo 和 Midtlyng，2007）。大多数的鱼类疫苗是灭活疫苗，且免疫保护效果良好。然而，疫苗不能保护鱼类免受感染，因此疫苗不能提供无菌的免疫。

接种疫苗群体，与未接种疫苗群体相比，疾病的流行和发生率较低，致病性微生物也明显减少。然而，由于疾病不同，预防效果的程度也会不同，因而使用疫苗预防疾病的效果存在差异。

第三节　疫苗接种在水产养殖中的应用

鱼类疫苗主要用于预防和控制鱼类细菌性传染病，也用于预防和控制一些病毒病（Biering 等，2005），但到目前为止，还没有针对真菌或寄生虫病的鱼类疫苗。

多数疫苗用于预防鲑等洄游性鱼类的疾病。但也有用于预防其他养殖鱼类，如海鲈、海鲷、鳕、罗非鱼、斑点叉尾鮰、黄尾鰤和大菱鲆等疾病的疫苗（Håstein 等，2005；Samuelsen 等，2006；Toranzo 等，2009）。近年来，在亚洲一些国家养殖的鲶和石斑鱼中也已使用疫苗或在研制疫苗（Harikrishnan 等，2011）。

通过浸泡或腹腔注射灭活疫苗的方式可实现鱼体对弧菌病、耶尔森菌病和巴斯德菌病等免疫预防。其他疾病如疖疮病则需要使用混合佐剂的灭活疫苗。目前，混合佐剂疫苗的接种途径只能是注射方式。

灭活疫苗作为首选的主要原因是安全。减毒活疫苗虽然不会引起某一鱼类生病，但可能会通过水体造成其他易感鱼类发生疾病。由胞内微生物引起的疾病，灭活疫苗往往是无效的，因此减毒活疫苗就有了市场需求。目前，活疫苗已在实验中使用，甚至有的活疫苗已获得国家许可，如鮰爱德华菌活疫苗（Shoemaker 等，2009）。

微生物学和免疫学是疫苗学的发展基础。近年来，分子生物学在疫苗生产和控制方面得到广泛应用，如用分子生物学方法在载体菌中插入致病微生物毒力因子基因，获得抗原生产的疫苗，称重组疫苗（亚单位疫苗），传染性胰腺坏死（IPN）疫苗属于这一类（Ramstad 等，2007）。

转基因疫苗通常是改变了某些基因的活疫苗。构建的实验性杀鲑气单胞菌减毒活疫苗，就是芳香基因突变的转基因疫苗（Marsden等，1996），实验表明，疫苗的免疫保护效果较好。由于灭活疫苗已达到满意的免疫保护效果，因此这种转基因疫苗仅作为灭活疫苗的安全替代。

DNA疫苗是指含有编码毒力因子或其他因子的基因，可用于预防致病物生物感染的基因疫苗。DNA疫苗诱发机体产生抗体反应以及细胞免疫。目前已有许多针对鱼类疾病的实验性DNA疫苗，2010年，传染性造血器官坏死病DNA疫苗在加拿大和美国获得许可。尽管DNA疫苗的安全性一直受到质疑，但获准许可表明，监管部门对DNA疫苗的安全证明材料是认可的。

哺乳动物、鸟类和鱼类的免疫系统有很多相似之处，但在预防疾病上的重要生物学差异是母源免疫。在陆生动物中，母体接种疫苗获得的免疫可转移给后代，从而保护后代免受感染。在某些鱼类中似乎存在母源抗体的转移，但到目前为止，在亲鱼中应用疫苗实现保护仔鱼的目的尚未成功（Lillehaug等，1996）。

在水产养殖中，激活免疫系统只是用于预防传染病，即免疫预防。然而，在陆生动物中，疫苗也被用于内分泌、新陈代谢和生殖的调节。其中一个例子是，通过给公猪接种疫苗，使其产生抗某些激素或某些具有活性化合物的抗体，以防止公猪产生膻味（Fuchs等，2009）。在水产养殖中，不育鱼可减少与野生鱼的基因交流。为了获得不育鱼，可用鸡卵的表面蛋白抗原免疫成鱼，母源抗体转移给后代，可能会阻止后代生殖细胞发育，结果产生一尾不育的鱼。

第四节 预防不同疾病的疫苗接种

鱼类疫苗多数是用来预防由环境中无法根除的病原微生物引起的鱼类传染病，如弧菌属（*Vibrio*）、气单胞菌属（*Aeromonas*）、耶尔森菌属（*Yersinia*）等微生物引起的疾病可以通过疫苗接种进行预防和控制。

疫苗在预防和控制外来鱼类疾病上的应用极为罕见，外来疾病主要是病毒性的，与预防细菌感染的疫苗相比，灭活病毒疫苗的保护效果一般较差。

另一个原因是，疫苗接种可能成为某些地区或国家活动物交易或产品贸易的障碍。这是由于很难区分开接种过疫苗的动物和受感染的动物，即使接种了疫苗，也很难说清楚动物有没有感染某种疾病。例如，欧盟关于传染性鲑贫血症（ISA）的规定，在传染性鲑贫血症流行的地区或国家允许接种ISA疫苗，而在没有该疾病的地区，禁止接种ISA疫苗。

在畜牧养殖中，这一问题是通过扑杀和免疫两个途径相结合解决的，即区分感染动物和免疫动物（DIVA）疫苗措施（Meeusen等，2007）。猪的伪狂犬病就是通过疫苗接种与实施扑杀相结合控制的疾病之一。

到目前为止，水产养殖中还没有实施DIVA疫苗措施。鱼类疫苗学现在的首要任务是研发有效的病毒疫苗，下一个目标可能会在传染性鲑贫血症（ISA）、病毒性出血性败血症（VHS）和传染性造血器官坏死症（IHN）上实施DIVA疫苗计划。

第五节 群体免疫

疫苗接种作为预防措施的依据是，接种过疫苗和未接种疫苗的个体的抗病能力都会提高。当群体中的大部分个体接种了疫苗，可为没有免疫力或免疫力低下的个体提供保护，呈现群体免疫现象，这种效应称为“群体免疫”（herd immunity）（Anderson 和 May，1985）。

养殖场中养殖鱼的数量很多，有些个体可能漏接疫苗，有些个体可能疫苗接种效果不佳。因此，群体免疫是水产养殖中疫苗接种效果显著的一个重要因素（Gudding 等，2013）。

当具有防止感染传播（即免疫或抗病）的个体数达到一定比例时，才能实现群体免疫。群体免疫的效果与引起疾病的微生物和用于预防疾病的疫苗有关。设疫苗的效力为 φ，病原的基本再生数为 R_0，群体免疫可用以下数学公式估计：

$$\frac{1}{\varphi}\times\left(1-\frac{1}{R_0}\right)$$

疫苗效力是指攻毒或疾病暴发期间，接种疫苗和未接种疫苗的群体死亡数之间的比例关系。“疫苗效力”与“相对保护率”（RPS）含义相近。相对保护率描述的是疫苗功效的百分比（Amend，1981）。相对保护率可以由下面的公式计算：

$$\mathrm{RPS}=\left(1-\frac{\text{免疫鱼的死亡率}}{\text{未免疫鱼的死亡率}}\right)\times 100\%$$

病原的基本再生数 R_0是决定传染病传播速率和感染比例的关键因素。R_0表示在易感群体中，一个初发感染直接传染健康个体数的平均值（Anderson 和 May，1991）。如果 R_0大于 1，说明平均每一个病例会导致一个以上个体感染发病，从而导致疾病流行。另一方面，当 R_0小于 1 时，很可能不会导致疾病暴发流行。

水产养殖中的 R_0与陆生动物疾病的 R_0相同，都取决于疾病的特性，如传播力和传染期。R_0还与宿主密度有关，因为密度决定了健康个体与发病个体间的接触频率。关于鱼类疾病的 R_0数据很少，据估计，传染性鲑贫血症的 R_0为 1.3～2.5（Mardones 等，2011）。

用群体免疫方程模拟传染性鲑贫血症的免疫效果，如果传染性鲑贫血症的 R_0是 2.5，疫苗的相对保护率是 80%，那么群体免疫的阈值就是 75%，这意味着群体中 75% 的个体应接种疫苗并且 80% 以上的个体产生了免疫保护，才能达到有效的群体免疫效果。疫苗接种效果，取决于受免鱼的应答率和相对保护率（RPS）。

第六节 经济利益

鱼类养殖者，总是要从利益和成本的经济角度考虑是否接种疫苗，与经济有关的事项很多，其中包括：疫苗与其他方法的比较，疫苗是否纳入接种规划，疫苗的接种途径如何，疫苗类型，疫苗接种时间、周期，等等。

有时，尽管给鱼接种了疫苗，但是攻毒实验的相对保护率很低，甚至暴发了疾病，因此会产生对疫苗接种效果的质疑。然而，有更多的数据、事例表明，疫苗接种的效果是显著的。一项关于鲑甲病毒引起的胰腺病（PD）的疫苗接种效果的对比研究表明，疫苗对几个重要的经济性生产参数产生了积极影响（Bang Jensen 等，2012），未接种疫苗的鲑的胰腺病发病率比接种疫苗的鲑高 3 倍。此外，接种疫苗的鲑的累计死亡率降低，生长速度加快。

由于养殖场及周边一带可能使用相同的水源，因此具体使用疫苗时也要考虑养殖场所在区域的环境条件。拒绝接种疫苗可能会造成养殖区的感染压力，从而影响到养殖鱼类和野生鱼类。

有一些经济模型可以用来提高疫苗接种决策的可预测性（Lillehaug，1989；Thorarinsson 和 Powell，2006）。利用这些模型以及死亡率数据、疫苗接种成本和鱼的市场价格进行了预测，结果表明即使在低疾病风险和低鱼价情况下，鲑的疫苗接种在经济上也是有益的。然而，海鲈养殖需要市场价格和/或疾病风险高到一定程度，疫苗接种在经济上才有意义。

第七节　风险评估

接种疫苗通常是一种安全的预防措施。然而，人们越来越担心接种疫苗给鱼、接种员甚至是环境带来副作用，因此，应对接种疫苗的积极作用以及免疫预防的负面影响进行评估（Wooldridge，2007）。

在疾病防控管理中，风险评估已经成为量化利益和风险的有用工具。在一些国家，风险评估是医药用品评价的重要一环。欧洲医药管理局对疫苗评估的最后一项是“使用与风险平衡”评价。一些国家的动物卫生局，收益与风险评估是疫苗评估的最后一步。如果许可的产品与其他的疾病预防和控制措施相冲突或危及其他疾病预防和控制措施时，该产品可能会被禁止使用。

活疫苗可能在环境中传播，是不可控的潜在微生物来源，具有不可预测的生态效应。因此，对活疫苗的使用采取限制性政策是有道理的。

对转基因疫苗的态度和活疫苗类似。然而，使用分子生物学技术生产的疫苗风险似乎与新技术知识呈负相关。现代基因工程技术可以精确地构建出具有明确衰减基因的改造生物。与传统的减毒疫苗相比，通过基因删除或修饰的减毒疫苗具有更高的环境安全性（Frey，2007）。

每项风险评估主要针对的都是特定国家或地区，但有时也适用于一个养殖场。养殖场在使用自体疫苗或非许可的疫苗产品之前，进行效益和风险评估尤为重要。

接种外来抗原，会引发受免动物一系列生物过程。引起的副作用可能是系统性的，也可能是局部的。

第八节　鱼类疫苗市场

过去，接种疫苗预防疾病的方法主要在鲑鳟鱼类上应用，这可能与疫苗接种的费用有关。近年来，免疫预防已越来越多地应用于非鲑鳟鱼类，如海鲈、海鲷、鲤和斑点叉尾鮰。但与畜牧业相比，水产养殖的疫苗接种应用仍处于初级阶段。在大多数国家，鱼类疫苗是由商业疫苗公司生产，是经国家和国际权威机构许可的，他们为全球水产养殖业提供多种鱼类、多种类的疫苗产品。目前还没有关于鱼类疫苗市场价值的可靠数据。

在某些情况下，鱼类疫苗市场可能会受到限制，因为有的养殖场可能在使用自体（自主研发）疫苗。大多数自体疫苗是在生产条件可控的疫苗企业生产的，因此，有理由相信自体疫苗的需求将会增加。随着新的养殖品种的不断出现，新的或变异的病原也会随之而来，如果已有的疫苗不能发挥作用，那么自体疫苗可能是一个解决办法。对自体疫苗的一般要求是，按照监管部门的指导原则进行生产和管理。

鱼类疫苗接种具有辉煌的前景。人们在对环境、社会和经济等多方面可持续发展的水产养殖业的需求可能会与日俱增，加之鱼类疫苗学的科学成就，将有助于全球鱼类疫苗市场的进一步发展。

参考文献

Amend，D. F. 1981. Potency testing of fish vaccines. *Dev Biol Stand* 49：447-454.

Anderson，R. M. and May，R. M. 1985. Vaccination and herd immunity to infectious diseases. *Nature*：323-329.

Anderson，R. M. and May，R. M. 1991. *Infectious Diseases of Humans*：*Dynamics and Control*. Oxford，Oxford University Press.

Bang Jensen，B.，Kristoffersen，A. B.，Myr，C. and Brun，E. 2012. Cohort study of effect of vaccination on pancreas disease in Norwegian salmon aquaculture. *Dis Aquat Org* 102：23-31.

Barton，J. R. and Fløysand，A. 2010. The political ecology of Chilean salmon aquaculture，1982-2010：a trajectory from economic development to global sustainability. *Global Environ Change* 20：739-752.

Biering，E.，Villoing，S.，Sommerset，I. and Christie，K. E. 2005. Update on viral vaccines for fish. *Dev Biol* (*Basel*) 121：97-113.

Bravo，S. and Midtlyng，P. J. 2007. The use of fish vaccines in Chilean salmon industry 1999-2003. *Aquaculture* 270：36-42.

Brundtland，G. H. 1987. Address at closing ceremony of the World Commission on Environment and Development，Tokyo，27 February 1987. Available at：http：//www. regjeringen. no/upload/SMK/Vedlegg/Taler% 20og% 20artikler% 20av% 20tidligere% 20statsministre/Gro% 20Harlem% 20Brundtland/1987/Address _ at _ Eighth _ WCED _ Meeting. pdf [accessed 3 May 2013].

Food and Agriculture Organization (FAO) Fisheries & Aquaculture. 2012. State of world aquaculture. Available at：http：//www. fao. org/fishery/topic/13540/en [accessed 3 May 2013].

Frey，J. 2007. Biological safety concepts of genetically modified live bacterial vaccines. *Vaccine* 25：5598-5605.

Fuchs，T.，Thun，R.，Parvizi，N.，*et al.*，2009. Effect of a gonadotropin-releasing factor vaccine on follicle-stimulating hormone and luteinizing hormone concentrations and on the development of testicles and the expression of boar taint in male pigs. *Theriogenology* 72：672-680.

Gudding，R. 2012. Disease prevention as a basis for sustainable aquaculture，in *Improving Biosecurity through Prudent and Responsible Use of Veterinary Medicines in Aquatic Food Production* (eds M. G. Bondad-Reantaso，J. R. Arthur and R. P. Subasinghe). FAO Fisheries and Aquaculture Technical Paper Vol. 547. Rome，Food and Agriculture Organization，141-146.

Gudding，R.，Lillehaug，A. and Evensen，Ø. 1999. Recent developments in fish vaccinology. *Vet Immunol Immunopathol* 72：203-212.

Gudding, R., Lillehaug, A. and Tavornpanich, S. 2013. Immunoprophylaxis in biosecurity programs. *World Aquaculture* 44: 60-63.

Harikrishnan, R., Balasundaram, C. and Heo, M.-S. 2011. Fish health aspects in grouper aquaculture. *Aquaculture* 320: 1-21

Håstein, T., Gudding, R. and Evensen, Ø. 2005. Bacterial vaccines for fish-An update of the current situation worldwide. *Dev Biol* (*Basel*) 121: 55-74.

Lillehaug, A. 1989. A cost-effectiveness study of three different methods of vaccination against vibriosis in salmonids. *Aquaculture* 83: 227-236.

Lillehaug, A., Sevatdal, S. and Endal, T. 1996. Passive transfer of specific maternal immunity does not protect Atlantic salmon (*Salmo salar* L.) fry against yersiniosis. *Fish Shellfish Immunol* 6: 521-535.

Mardones, F. O., Perez, A. M., Valdes-Donoso, P. and Carpenter, T. E. 2011. Farm-level reproduction number during an epidemic of infectious salmon anemia virus in southern Chile in 2007-2009. *Prev Vet Med* 102: 175-184.

Marsden, M. J., Vaughan, L. M., Foster, T. J. and Secombes, C. J. 1996. A live (Δ*aroA*) *Aeromonas salmonicida* vaccine for furunculosis preferentially stimulates T-cell responses relative to B-cell responses in rainbow trout (*Onchorhyncus mykiss*). *Infect Immun* 64: 3863-3869.

Meeusen, E. N. T., Walker, J., Peters, A., *et al.*, 2007. Current status of veterinary vaccines. *Clin Microbiol Rev* 20: 489-510.

Press, CMcL. and Lillehaug, A. 1995. Vaccination in European salmonid aquaculture. A review of practices and prospects. *Br Vet J* 151: 45-69.

Pulkkinen, K., Suomalainen, L.-R., Read, A. F., *et al.*, 2010. Intensive fish farming and the evolution of pathogen virulence: the case of columnaris disease in Finland. *Proc Biol Sci* 277: 593-600.

Ramstad, A., Romstad, A. B., Knappskog, D. H. and Midtlyng, P. J. 2007. Field validation of experimental challenge models for IPN vaccines. *J Fish Dis* 30: 723-731.

Samuelsen, O. B., Nerland, A. H., Jørgensen, T., *et al.*, 2006. Viral and bacterial diseases of Atlantic cod *Gadus morhua*, their prophylaxis and treatment: a review. *Dis Aquat Org* 71: 239-254.

Schlag, A. K. 2010. Aquaculture: an emerging issue for public concern. *J Risk Res* 13: 829-844.

Shoemaker, C. A., Klesius, P. H., Evans, J. J. and Arias, C. R. 2009. Use of modified live vaccines in aquaculture. *J World Aquac Soc* 40: 573-585.

Sniezko, S. F. 1974. The effects of environmental stress on outbreaks of infectious diseases of fishes. *J Fish Biol* 6: 197-208.

Sommerset, I., Krossøy, B., Biering, E. and Frost, P. 2005. Vaccines for fish in aquaculture. *Exp Rev Vaccines* 4: 89-101.

Thorarinsson, R. and Powell, D. B. 2006. Effects of disease risk, vaccine efficacy, and market price on the economics of fish vaccination. *Aquaculture* 256: 42-49.

Toranzo, A. E., Romalde, J. L., Magariños, B. and Barja, J. L. 2009. Present and future aquaculture vaccines against fish bacterial diseases. *Op Méditerr* 86: 155-176.

Turnbull, J. F. and Kadri, S. 2007. Safeguarding the many guises of farmed fish welfare. *Dis Aquat Org* 75: 173-183.

Wooldridge, M. 2007. Risk modelling for vaccination: a risk assessment perspective. *Dev Biol* (*Basel*) 130: 87-97.

第三章　非复制型疫苗

Hetron Mweemba Munang'andu，Stephen Mutoloki and Øystein

本章概要　SUMMARY

非复制型疫苗主要有灭活疫苗、天然或合成的具有病原结构的亚单位疫苗、病原产物制备的疫苗等。非复制型疫苗主要是能诱发 B 细胞产生循环抗体的免疫反应，由于组成它们的抗原不具有致病性，所以它们的主要优势是安全。 随着分子生物学研究方法的进步和对保护性抗原认知的提升，通过注射途径，有许多亚单位疫苗被证实能够诱发机体产生保护性免疫反应。非复制型疫苗，在灭活免疫原的过程中所用的螯合或交联的化学物质会使抗原失去部分免疫原性，因此，鱼类疫苗通常配用佐剂，这样可以弥补这类疫苗中的抗原免疫原性低的缺陷。可以预期，在未来的几年里，鱼类疫苗将以体外方法取代现在批量性的体内生产方法。体外方法可能是，基于抗体或基于抗原，通过建立的诱导抗体应答水平或疫苗抗原量与免疫保护耦合关系生产疫苗。

第一节　引　言

非复制型疫苗，顾名思义，就是灭活的，天然或合成的具有病原结构的，或病原产物制备的疫苗。由于病原失活，不可能引起疾病，因此安全性是这类疫苗的主要优点，但仍能激发机体的免疫反应。

第二节　类　型

非复制型疫苗一般可分为以下四类。

一、灭活疫苗

当提到这类病毒疫苗时，通常倾向于用“Inactivated”（灭活）一词，而不用“Killed”（杀死），因为此类疫苗中不存在活病毒。在挪威水产养殖中，灭活疫苗是使细菌性疾病减少的主要预防措施。路易·巴斯德研发了第一代灭活疫苗（Pasteur，1880），制备流程有四个基本步骤：①分离病原体；②培养病原体；③灭活；④对宿主接种灭活微生物。尽管这些步骤仍然是现代灭活疫苗研发中不可或缺的内容，但为了优化疫苗的性能，这些步骤已经变得更加精细。例如，第一步不仅是病原的分离和鉴定，已扩展为鉴定参与疫苗保护的免疫原性成分。第二步，由病原体的大量繁殖培养，精细到使用不同的系统培养，如生物反应器可以大量生产疫苗，从而满足水产养殖中疫苗使用持续增长的需求。至于第三步灭活，已有多种不改变微生物抗原结构的灭活程序方法。

灭活疫苗中病原体全部失活。目前在水产养殖应用的商品疫苗中，灭活疫苗占绝大多数。它们有容易制备、生态安全且价格低廉等优势。市场上的疫苗产品中，灭活细菌疫苗最多，其次是病毒疫苗（表3-1和表3-2）。现在，还没有针对寄生虫和真菌的商业疫苗。

表3-1 用于水产养殖的灭活细菌疫苗（许可和实验）

疾病	病原	免疫的鱼类
经典弧菌病	利斯顿菌/鳗弧菌（*Listonella/Vibrio anguillarum*）	鲑科鱼类、尖吻鲈、石斑鱼、海鲈、海鲷、黄尾鰤
冷水弧菌病	杀鲑弧菌（*Vibrio salmonicida*）	鲑科鱼类
冬季溃疡病	粘摩替亚菌（*Moritella viscosa*）	鲑科鱼类
温水弧菌病	溶藻弧菌（*Vibrio alginolyticus*）、创伤弧菌（*Vibrio vulnificus*）	尖吻鲈、石斑鱼、海鲈、海鲷、金鲷
爱德华菌病	鮰爱德华菌（*Edwardsiella ictaluri*）	斑点叉尾鮰、巨鲶
疖疮病	杀鲑气单胞菌（*Aeromonas salmonicida*）	鲑科鱼类
耶尔森菌病	鲁克氏耶尔森菌（*Yersinia ruckeri*）	鲑科鱼类
巴斯德菌病	美人鱼发光杆菌杀鱼亚种（*Photobacterium damsela* subsp. *piscicida*）	海鲈、海鲷
黄杆菌病	柱状黄杆菌（*Flavobacterium columnare*）	鲑科鱼类
曲挠杆菌病	海曲挠杆菌（*Flexibacter maritimus*）	鲑科鱼类
链球菌病	海豚链球菌（*Streptococcus iniae*）	鲑科鱼类、罗非鱼、大菱鲆、黄尾鰤
	无乳链球菌（*Streptococcus agalactiae*）	罗非鱼
乳球菌病	格氏乳球菌（*Lactococcus garviae*）	黄带鰤、紫鰤、海鲈、海鲷、罗非鱼
鱼立克次氏体病	鲑鱼立克次氏体（*Piscirickettsia salmonis*）	鲑科鱼类
细菌性肾病	鲑肾杆菌（*Renibacterium salmoninarum*）	鲑科鱼类

表3-2 用于水产养殖的灭活病毒疫苗（许可和实验）

病毒病/病原	主要免疫的鱼类	免疫原蛋白
传染性胰脏坏死 / IPNV	鲑科鱼类	可变区蛋白-2（VP-2）
传染性鲑贫血症 / ISAV	鲑科鱼类、石斑鱼	血凝素酯酶（HE）
虹彩病毒病（Iridoviral disease）	红鲷，黄尾鰤	全病毒
草鱼出血症 / GCHDV	草鱼	糖蛋白（G）
鲑胰腺病 / SPDV	鲑科鱼类	融合蛋白（E2）

二、亚单位疫苗

亚单位疫苗是利用重组载体表达病原体的特异免疫原性蛋白，在体外大量表达，并提纯重组蛋白，然后制成的疫苗。表 3-3 列出了一些应用于水产养殖中病毒和细菌的免疫原性蛋白疫苗。载体和细胞系的选择应符合三个方面的条件：①能够在体外便于外源蛋白的大量生产；②容易操作；③能表达抗原蛋白原有的构象结构。现在，已有应用不同原核细胞和真核细胞系统制备生产的针对鱼类疾病的亚单位疫苗。一般来说，完成初期研发之后，期望这类疫苗的价格会更便宜一些。

表 3-3　用于鱼类亚单位疫苗的免疫原蛋白和载体（许可和实验）

分类	病原	免疫原蛋白	载体	参考文献
细菌性	海豚链球菌（*Streptococcus iniae*）	铁结合蛋白-Sip11	大肠杆菌（*Escherichia coli*）	Cheng 等，2010
	嗜水气单胞菌（*Aeromonas hydrophila*）	外膜蛋白 Omp48	大肠杆菌	Khushiramani 等，2012
	迟缓爱德华菌（*Edwardsiella tarda*）	鞭毛蛋白	大肠杆菌	Zhang 等，2012
	杀鲑气单胞菌（*Aeromonas salmonicida*）	膜孔蛋白	大肠杆菌	Lutwyche 等，1995
	粘摩替亚菌（*Moritella viscosa*）	粘摩替亚菌外膜蛋白 1，MvOmp1	大肠杆菌	Bjornsson 等，2011
	溶藻弧菌（*Vibrio alginolyticus*）	鞭毛蛋白（flaA 和 flaB）	大肠杆菌	Liang 等，2012
	鲑鱼立克次氏体（*Piscirickettsia salmonis*）	可变蛋白	大肠杆菌	Wilhelm 等，2006
病毒性	传染性胰脏坏死病毒（Infectious pancreatic necrosis virus）	可变蛋白-2（VP-2）	大肠杆菌	Munang'andu 等，2012
			铜绿假单胞菌（*Pseudomonas aeruginosa*）	Munang'andu 等，2012
			塞姆利基森林病毒（Semliki forest virus）	McKenna 等，2001
			酵母	Allnutt 等，2007
			杆状病毒（*Baculovirus*）	Shivappa 等，2005
	传染性造血器官坏死病毒（Infectious hematopoietic necrosis virus）	糖蛋白（G）	杀鲑柄杆菌（*Caulobacter salmonicida*）	Simon 等，2001
		糖蛋白（G）	大肠杆菌	Xu 等，1991
	病毒性出血性败血症病毒（Viral hemorrhagic septicemia virus）	糖蛋白（G）	大肠杆菌	Estepa 等，1994；Lorenzen 等，1993
		糖蛋白（G）	杀鲑气单胞菌	Noonan 等，1995
		糖蛋白（G）	杆状病毒	Koener 和 Leong，1990
	鲤春病毒血症病毒（Spring viremia carp of virus，SVCV）	糖蛋白（G）	杆状病毒	Koener 和 Leong，1990
	病毒性神经坏死病毒（Viral nervous necrosis virus）	衣壳蛋白（C）	大肠杆菌	Yuasa 等，2002

三、合成肽疫苗

肽疫苗是以氨基酸的短序列为抗原制备的疫苗。由于肽分子太小，所以需要耦合到载体上，使其有足够的免疫原性。Emmenegger 和同事们于 1995 年研发了由传染性造血器官坏死病毒（IHNV）的合成肽抗原与牛血清白蛋白耦合制成的水产养殖中最早的合成肽疫苗。该疫苗与弗氏完全佐剂混合后通过腹腔注射使用。近来，Lin 和同事们于 2002 年利用聚合酶链式反应技术合成的糖基化磷脂酰肌醇锚定膜蛋白多肽，又称抑动抗原，用来预防淡水鱼白点病，实验表明多肽对多子小瓜虫有抑制作用。

四、颗粒疫苗

细菌和寄生虫裂解物、声波处理的细胞和其他亚细胞物质如脂多糖，都含有制备疫苗的某些抗原。然而，这类疫苗尚未在水产养殖中应用。

第三节 疫苗的灭活方法

灭活的主要目的是使病原体失去致病力，同时保有其构象结构，以激活保护性免疫反应。灭活可作用于：①微生物的胞外或胞外表面组分，如细胞膜或病毒囊膜；②胞内或内部的成分，如核酸。灭活的目的是使病原体丧失复制能力。一般而言，灭活分为物理方法和化学方法两大类。

一、物理灭活

微生物的物理灭活有紫外线方法、高温方法和声波方法。当微生物的核酸被紫外线照射，就会发生二聚化，从而阻碍微生物的复制能力。热失活的原理是：在特定温度下，大多数微生物的传染性和免疫原性两个基本特性，都会以相同的速率丧失，而在其他温度下，二者的丧失速率不同。因此，这样就能找到既使病原微生物失去传染性，又能保留其免疫原性的温度。大多数微生物，在 56～60℃ 的温度范围内，持续 10～30min，可以破坏其传染性而保留其免疫原性。热灭活已被用于细菌疫苗的制备。在实验条件下，有研究表明，热灭活的效果优于化学灭活（Antipa，1976）。鱼类病原菌的热灭活可追溯到 1935 年，Nybelin（1935）将热灭活的鳗弧菌接种鳟和鲤后，诱导产生了较高的抗体水平。另一个用于细菌和寄生虫的灭活方法是声波灭活法，声波破坏了微生物的形态结构，同时保留了其免疫原性，这对受免鱼能产生获得性免疫反应至关重要。在鱼类疫苗学中，这种方法主要用于实验条件下的细菌疫苗（表 3-4），然而获得生产许可的商品疫苗，均未使用这种方法。

表 3-4 用于制备病毒和细菌疫苗的灭活方法举例

疫苗类型	灭活类型	灭活方法	病原	参考文献
病毒疫苗	物理法	加热	传染性造血器官坏死病毒（Infectious hematopoietic necrosis virus）	Anderson 等，2008
			病毒性出血性败血症病毒（Viral hemorrhagic septicemia virus）	Tafalla 等，2008
			传染性鲑贫血症病毒（Infectious salmon anemia virus）	Workenhe 等，2008
	化学法	双乙烯二胺（BEL）	传染性造血器官坏死病毒	Anderson 等，2008
			β诺达病毒（野田病毒）（Betanodavirus）	Kai and Chi，2008
		β-丙内酯（BPL）	传染性造血器官坏死病毒	Anderson 等，2008

（续）

疫苗类型	灭活类型	灭活方法	病原	参考文献
		甲醛	传染性造血器官坏死病毒	Anderson 等，2008
			传染性胰脏坏死病毒（Infectious pancreatic necrosis virus）	（Munang'andu 等，2013a
			鲫造血器官坏死病毒（Crucian carp hematopoietic necrosis virus）	Sato and Okamoto，2010
			条石斑虹彩病毒（Rock bream iridovirus）	Seo 等，2013
			病毒性出血性败血症病毒	Adelmann 等，2008
			白斑症病毒（White spot syndrome virus）	Singh 等，2005
细菌疫苗	物理法	加热	嗜水气单胞菌（*Aeromonas hydrophila*）	Song 等，1976
			杀鲑气单胞菌（*Aeromonas salmonicida*）	Snieszko 和 Friddle，1949
			迟缓爱德华菌（*Edwardsiella tarda*）	Song 和 Kou，1981
			鳗弧菌（*Vibrio anguillarum*）	Antipa，1976a
		声波	嗜水气单胞菌	Thune 和 Plumb，1982
			杀鲑气单胞菌	Smith 等，1980
			鳗弧菌	Evelyn，1984
	化学法	三氯甲烷	杀鲑气单胞菌	Duff，1942
			鲁克氏耶尔森菌（*Yersinia ruckeri*）	Amend 等，1983
		甲醛	嗜水气单胞菌	Nayak 等，2004
			杀鲑气单胞菌	Santos 等，2005
			鳗弧菌	Palm 等，1998
		苯酚	鲁克氏耶尔森菌	Anderson 和 Ross，1972

二、化学灭活

在哺乳动物和鱼类灭活疫苗中，最常用的化学试剂是甲醛。与β-丙内酯（BPL）和双乙烯二胺（BEL）不同，低浓度的甲醛能够改变不同微生物胞内和胞外物质的特性，但仍然能够保持其免疫原性（Hiatt，1964）。甲醛具有易获得、价格低廉、使用方便和相对安全等优点，使其成为一个有吸引力的选择。除了极少数例外情况，目前甲醛灭活的方法几乎应用到所有的水产养殖灭活疫苗中。

化学灭活可通过低 pH、溶剂、洗涤剂使微生物变性，或破坏微生物表面脂质双层膜来实现。当化合物分子与脂质双层相互作用时，使外膜功能失调，从而阻止病原体的复制。这些化合物杀死了病原体，但并不会改变影响微生物免疫特性的形态结构。除了使外膜变性，像 BPL 和 BEL 这样的化合物，通过使病毒基因组发生不可逆的改变，阻断病毒在细胞内复制，但没有改变病毒的外表形态（Bahnemann，1990；Budowsky，1991；Budowsky 等，1993）。在实验条件下，用于鱼类病毒灭活的化合物如表 3-4 所示。

第四节 灭活效果评价

优化灭活步骤在疫苗的商品化开发中至关重要，其中主要的优化步骤包括：

（1）确定用于灭活的化合物浓度。

（2）灭活时长，采集不同时长灭活的微生物样本，确定最佳灭活所需的时间。

(3) 活性测试，检测微生物已完全失活，在适宜繁殖的条件下不会复活。

(4) 确保灭活方法不会破坏微生物的免疫原性。

(5) 确定研制的疫苗在受免鱼体内能诱发预期的免疫应答。这需要功效实验，不仅评估疫苗降低感染率和死亡率的效果，还需评估其诱导的免疫反应。

(6) 由于鱼类是供人类食用的，因此确保灭活过程不会对公众健康构成威胁至关重要，这就需要检测用于灭活微生物的药物（化学试剂）的残留，判定其公共卫生风险。

第五节　灭活疫苗效果的测定

为了优化疫苗性能，重要的是建立与免疫保护作用相关的阈值，作为疫苗生产的基准。这就需要有识别疫苗保护与获得性免疫反应关联的生物标记，它可以作为疫苗具有保护作用的临界值。保护关联阈值不达标的被认为是非优疫苗，原则上，这种疫苗不允许用于商业用途。因此，确定与保护相关的关联阈值对新疫苗认证至关重要（Plotkin，2008；FDA，2012）。

相对保护率

传统的疫苗功效评价依赖于攻毒实验。保护功效通过以接种疫苗组的鱼与未接种疫苗的对照组鱼攻毒后的相对保护率（RPS）来计算（Amend，1981），但接种疫苗组的鱼，用研发疫苗的活病原攻毒后，应出现少量死亡。这种方法最大的问题是，攻毒模型需要对攻毒剂量进行筛选和优化，包括：①使用能够引起高死亡率的已知致病性毒株；②使用标准化的攻毒剂量；③易感的实验鱼；④优化实验条件，避免应激过大而影响攻毒实验的结果；⑤实验设计要考虑到统计分析，从而能够显示出免疫组和对照组之间的显著差异。对于大多数鱼类疫病，攻毒模型没有经过优化，因而很难重复，使得不同批次疫苗的功效数据难以进行比较分析。另外，这种方法并不适用于已知病原对易感鱼类不会造成高死亡率的情形。

第六节　疫苗的保护机制

值得指出的是，在水产养殖中，并非所有的疾病都会引起死亡。显然，有些疾病只是使感染动物日渐衰弱，导致产量下降，并不会引起大量死亡。因此，衡量疫苗对这类疾病的效果最合理的方法是，确定疫苗接种诱导的保护性免疫机制。例如，众所周知，胰腺病（PD）引起鱼体病理状态，死亡率并不高，但减缓患病鱼的生长，使产量降低，给水产养殖造成重大损失。在哺乳动物中，疫苗研发采取了一种更为实用的方法，在对其发病机理和保护机制充分明晰的基础上，疫苗可获得许可。例如，脊髓灰质炎病毒和麻疹疫苗是预防病毒血症（Fox，1984；Chen 等，1990）；狂犬病疫苗是防止病毒附着于乙酰胆碱上，乙酰胆碱是神经元连接处的神经受体（Hanham 等，1993）。因此，这类疾病疫苗接种的特质是诱导获得机理保护的免疫应答。同样，阻断感染性鲑贫血病毒与唾液酸受体的结合（Hellebo 等，2004；Aamelfot 等，2012），可以预防感染性鲑贫血病的发生。对 PD 而言，阻止胰腺和心脏的病理变化，可起到保护作用。关于大西洋鲑的传染性胰腺坏死病毒（IPNV）病，研究表明，阻断病毒对靶器官的感染和阻止病毒增殖，使病毒的量低于 10^7（病理学关联剂量，靶器官组织损伤的关键决定因素）可以预防疾病的发生（Munang'andu 等，2013a）。可预期，随着对水生动物病原感染生物学研究的深入，疫苗效力研究将会进一步发展并精准化，并建立相关的阈值，达到保护性免疫能够防止接种鱼由感染转变为发病的目的。

第七节 抗体与免疫保护的关系

传统上，抗体是大多数疫苗保护功效评价的常用元素。已有研究表明，当抗体水平达到阈值以上，则动物就受到了免疫保护（Pulendran 等，2010）。因此，抗体水平被认为是保护性免疫的标志。由此，许多哺乳动物疫苗，是基于抗体效价获得许可的。但这种方法尚未在水产养殖商业疫苗生产许可上运用。然而，在鱼类疫苗学抗体水平的研究上已取得了巨大进展，例如，在大西洋鲑中开展的关于 IPNV 疫苗的两项独立功效研究（Munang'andu 等，2013b）表明，大西洋鲑攻毒前产生的抗体水平与高致病性挪威 Sp 株 NVI-015 攻毒后的存活率相一致。同样，Romstad 和同事们于 2013 年发现，大西洋鲑接种疖疮病疫苗，其抗体水平的升高与攻毒后死亡率的降低之间存在高度的相关性。因此可以设想，未来将用建立的抗体滴度与保护性免疫关联的阈值，作为水产养殖疫苗生产许可的依据。

第八节 抗原剂量与免疫保护的关系

出于伦理原因，人类疫苗功效无法通过攻毒实验进行评估，与之不同的是，在鱼类疫苗学中已经证明了某些疾病的保护性免疫与抗原剂量相关。因此，可以通过剂量与免疫保护的关联，确定抗原剂量的阈值。有研究证明，大西洋鲑接种传染性胰脏坏死疫苗的剂量为保护性阈值 10%的次优剂量时，其免疫保护显著降低（Munang'andu 等，2013）。因此，建立一个与免疫保护相关联的抗原剂量阈值是非常重要的，并且还可作为疫苗生产的基准。

参考文献

Aamelfot, M., Dale, O. B., Weli, S. C., *et al.*, 2012. Expression of the infectious salmon anemia virus receptor on Atlantic salmon endothelial cells correlates with the cell tropism of the virus. *J Virol* 86: 10571-10578.

Adelmann, M., Kollner, B., Bergmann, S. M., *et al.*, 2008. Development of an oral vaccine for immunisation of rainbow trout (*Oncorhynchus mykiss*) against viral haemorrhagic septicaemia. *Vaccine* 26: 837-844.

Allnutt, F. C., Bowers, R. M., Rowe, C. G., *et al.*, 2007. Antigenicity of infectious pancreatic necrosis virus VP2 subviral particles expressed in yeast. *Vaccine* 25: 4880-88.

Amend, D. F. 1981. Potency testing of fish vaccines. *Dev Biol Stand* 49: 447-454.

Amend, D. F., Johnson, K. A., Croy, T. R. and McCarthy, D. H. 1983. Some factors affecting the potency of *Yersinia ruckeri* bacterins. *J Fish Dis* 6: 337-344.

Anderson, D. F. and Ross, A. J. 1972. Comparative study of Hagerman redmouth disease oral bacterins. *Prog Fish Cult* 34: 226.

Anderson, E., Clouthier, S., Shewmaker, W., *et al.*, 2008. Inactivated infectious haematopoietic necrosis virus (IHNV) vaccines. *J Fish Dis* 31: 729-745.

Antipa, R. 1976. Field testing of injected *Vibrio anguillarum* bacterins in pen-reared Pacific salmon. *J Fish Res Board Can* 33: 1291-1296.

Bahnemann, H. G. 1990. Inactivation of viral antigens for vaccine preparation with particular reference to the application of binary ethylenimine. *Vaccine* 8: 299-303.

Bjornsson, H., Marteinsson, V. P., Friojonsson, O. H., *et al.*, 2011. Isolation and characterization of an antigen from the fish pathogen *Moritella viscosa*. *J Appl Microbiol* 111: 17-25.

Budowsky, E. I. 1991. Problems and prospects for preparation of killed antiviral vaccines. *Adv Virus Res* 39: 255-290.

Budowsky, E. I., Smirnov, Y. and Shenderovich, S. F. 1993. Principles of selective inactivation of viral genome. VIII. The influence of β-propiolactone on immunogenic and protective activities of influenza virus. *Vaccine* 11: 343-348.

Chen, R. T., Markowitz, L. E., Albrecht, P., *et al.*, 1990. Measles antibody-reevaluation of protective titers. *J Infect Dis* 162: 1036-1042.

Cheng, S., Hu, Y. H., Jiao, X. D. and Sun, L. 2010. Identification and immunoprotective analysis of a *Streptococcus iniae* subunit vaccine candidate. *Vaccine* 28: 2636-2641.

Duff, D. C. B. 1942. The oral immunization of trout against *Bacterium salmonicida*. *J Immunol* 44: 87-94.

Emmenegger, E., Huang, C., Landolt, M., *et al.*, 1995. Immune response to synthetic peptides representing antigenic sites on the glycoprotein of infectious hematopoietic necrosis virus. *Vet Res* 26: 374-378.

Estepa, A., Thiry, M. and Coll, J. M. 1994. Recombinant protein fragments from hemorrhagic septicemia rhabdovirus stimulate trout leukocyte anamnestic responses *in vitro*. *J Gen Virol* 75: 1329-1338.

Evelyn, T. P. T. 1984. Immunization against pathogenic vibrios, in *Symposium on Fish Vaccination* (ed. P. de Kinkelin). Paris, OIE, 121-150.

Fox, J. P. 1984. Modes of action of poliovirus vaccines and relation to resulting immunity. *Rev Infect Dis* 6: S352-355.

Hanham, C. A., Zhao, F. and Tignor, G. H. 1993. Evidence from the antiidiotypic network that the acetylcholine receptor is a rabies virus receptor. *J Virol* 67: 530-542.

Hellebo, A., Vilas, U., Falk, K. and Vlasak, R. 2004. Infectious salmon anemia virus specifically binds to and hydrolyzes 4-O-acetylated sialic acids. *J Virol* 78: 3055-3062.

Hiatt, C. W. 1964. Kinetics of the inactivation of viruses. *Bacteriol Rev* 28: 150-163.

Kai, Y. H. and Chi, S. C. 2008. Efficacies of inactivated vaccines against betanodavirus in grouper larvae (*Epinephelus coioides*) by bath immunization. *Vaccine* 26: 1450-1457.

Khushiramani, R. M., Maiti, B., Shekar, M., *et al.*, 2012. Recombinant *Aeromonas hydrophila* outer membrane protein 48 (Omp48) induces a protective immune response against *Aeromonas hydrophila* and *Edwardsiella tarda*. *Res Microbiol* 163: 286-291.

Koener, J. F. and Leong, J. A. C. 1990. Expression of the glycoprotein gene from a fish rhabdovirus by using baculovirus vectors. *J Virol* 64: 428-430.

Liang, H. Y., Wu, Z. H., Jian, J. C. and Liu, Z. H. 2012. Construction of a fusion flagellin complex and evaluation of the protective immunity of it in red snapper (*Lutjanus sanguineus*). *Lett Appl Microbiol* 55: 115-121.

Lin, Y., Cheng, G., Wang, X. and Clark, T. G. 2002. The use of synthetic genes for the expression of ciliate proteins in heterologous systems. *Gene* 288: 85-94.

Lorenzen, N., Olesen, N. J., Vestergård-Jørgensen, P. E., *et al.*, 1993. Molecular cloning and expression in

Escherichia coli of the glycoprotein gene of VHS virus, and immunization of rainbow trout with the recombinant protein. *J Gen Virol* 74: 623-630.

Lutwyche, P., Exner, M. M., Hancock, R. E. and Trust, T. J. 1995. A conserved *Aeromonas salmonicida* porin provides protective immunity to rainbow trout. *Infect Immun* 63: 3137-3142.

McKenna, B. M., Fitzpatrick, R. M., Phenix, K. V., *et al.*, 2001. Formation of infectious pancreatic necrosis virus-like particles following expression of segment A by recombinant semliki forest virus. *Mar Biotechnol* (*NY*) 3: 103-110.

Munang'andu, H. M., Fredriksen, B. N., Mutoloki, S., *et al.*, 2012. Comparison of vaccine efficacy for different antigen delivery systems for infectious pancreatic necrosis virus vaccines in Atlantic salmon (*Salmo salar* L.) in a cohabitation challenge model. *Vaccine* 30: 4007-4016.

Munang'andu, H. M., Fredriksen, B. N., Mutoloki, S., *et al.*, 2013a. Antigen dose and humoral immune response correspond with protection for inactivated infectious pancreatic necrosis virus vaccines in Atlantic salmon (*Salmo salar* L). *Vet Res* 44: 7.

Munang'andu, H. M., Sandtro, A., Mutoloki, S., *et al.*, 2013b. Immunogenicity and cross protective ability of the central VP2 amino acids of infectious pancreatic necrosis virus in Atlantic salmon (*Salmo salar* L.). *PLoS One* 8: e54263.

Nayak, D. K., Asha, A., Shankar, K. M. and Mohan, C. V. 2004. Evaluation of biofilm of *Aeromonas hydrophila* for oral vaccination of *Clarias batrachus* -a carnivore model. *Fish Shellfish Immunol* 16: 613-619.

Noonan, B., Enzmann, P. J. and Trust, T. J. 1995. Recombinant infectious hematopoietic necrosis virus and viral hemorrhagic septicemia virus glycoprotein epitopes expressed in *Aeromonas salmonicida* induce protective immunity in rainbow trout (*Oncorhynchus mykiss*). *Appl Environ Microbiol* 61: 3586-3591.

Nybelin, O. 1935. Über Agglutininbildung bei Fischen. *Z Immunitätsforschung* 84: 74-79.

Palm, R. C., Jr.,, Landolt, M. L. and Busch, R. A. 1998. Route of vaccine administration: effects on the specific humoral response in rainbow trout *Oncorhynchus mykiss*. *Dis Aquat Org* 33: 157-166.

Pasteur, L. 1880. De l'attenuation du virus du cholera des poules. *CR Acad Sci Paris* 91: 673-680.

Plotkin, S. A. 2008. Correlates of vaccine-induced immunity. *Clin Infect Dis* 47: 401-409.

Pulendran, B., Li, S. Z. and Nakaya, H. I. 2010. Systems vaccinology. *Immunity* 33: 516-529.

Romstad, A. B., Reitan, L. J, . Midtlyng, P., *et al.*, 2013. Antibody responses correlate with antigen dose and *in vivo* protection for oil-adjuvanted, experimental furunculosis (*Aeromonas salmonicida* subsp *salmonicida*) vaccines in Atlantic salmon (*Salmo salar* L.) and can be used for batch potency testing of vaccines. *Vaccine* 31: 791-796.

Santos, Y., Garcia-Marquez, S., Pereira, P. G., *et al.*, 2005. Efficacy of furunculosis vaccines in turbot, *Scophthalmus maximus* (L.): evaluation of immersion, oral and injection delivery. *J Fish Dis* 28: 165-172.

Sato, A. and Okamoto, N. 2010. Induction of virus-specific cell-mediated cytotoxic responses of isogeneic ginbuna crucian carp, after oral immunization with inactivated virus. *Fish Shellfish Immunol* 29: 414-421.

Seo, J. Y., Chung, H. J. and Kim, T. J. 2013. Codon-optimized expression of fish iridovirus capsid protein in yeast and its application as an oral vaccine candidate. *J Fish Dis* 36: 763-768.

Shivappa, R. B., McAllister, P. E., Edwards, G. H., *et al.*, 2005. Development of a subunit vaccine for infectious pancreatic necrosis virus using a baculovirus insect/larvae system. *Dev Biol* (*Basel*) 121: 165-174.

Simon, B., Nomellini, J., Chiou, P., *et al.*, 2001. Recombinant vaccines against infectious hematopoietic necrosis virus: production by the *Caulobacter crescentus* S-layer protein secretion system and evaluation in laboratory trials. *Dis Aquatic Org* 44：17-27.

Singh，I. S. B.，Manjusha，M.，Pai，S. S. and Philip，R. 2005. *Fenneropenaeus indicus* is protected from white spot disease by oral administration of inactivated white spot syndrome virus. *Dis Aquatic Org* 66：265-270.

Smith，P. D.，McCarthy，D. H. and Paterson，W. D. 1980. Further studies on furunculosis vaccination，in *Fish Diseases*，*Third COPRAQ Session*（ed. W. Ahne）. Berlin，Springer-Verlag，119.

Snieszko，S. and Friddle，S. B. 1949. Prophylaxis of furunculosis in brook trout（*Salvelinus fontinalis*）by oral immunization and sulphamerazine. *Prog Fish Cult* 11：161.

Song，Y. L. and Kou，G. H. 1981. Immune response of eels（*Anguilla japonica*）against *Aeromonas hydrophila* and *Edwardsiella anguillimortiferum*（*E. tarda*）infection. *Proc Rep Chin US Coop ScL Seminar on Fish Dis*，*Nat-ScL Conn Set* 3：107.

Song，Y. L.，Chen，S. N. and Kou，G. H. 1976. Agglutinating antibodies production and protection in eel（*Anguilla japonica*）inoculated with *Aeromonas hydrophila*（*A liquefaciens*）antigens. *J Fish Soc Taiwan* 4（2）：25-29.

Tafalla，C.，Sanchez，E.，Lorenzen，N.，*et al.*，2008. Effects of viral hemorrhagic septicemia virus（VHSV）on the rainbow trout（*Oncorhynchus mykiss*）monocyte cell line RTS-11. *Mol Immunol* 45：1439-1448.

Thune，R. L. and Plumb，J. A. 1982. Effect of delivery method and antigen preparation on the production of antibodies against *Aeromonas hydrophila* in channel catfish. *Prog Fish Cult* 44：53.

United States Food and Drug Administration. 2012. *Complete list of vaccines licensed for immunization and distribution in the United States*. US Food and Drug Administration. New Hampshire Avenue，Silver Spring，MD 20993.

Wilhelm，V.，Miquel，A.，Burzio，L. O.，*et al.*，2006. A vaccine against the salmonid pathogen *Piscirickettsia salmonis* based on recombinant proteins. *Vaccine* 24：5083-5091.

Workenhe，S. T.，Kibenge，M. J.，Wright，G. M.，*et al.*，2008. Infectious salmon anaemia virus replication and induction of alpha interferon in Atlantic salmon erythrocytes. *Virol J* 5：36.

Xu，L.，Mourich，D. V.，Engelking，H. M.，*et al.*，1991. Epitope mapping and characterization of the infectious hematopoietic necrosis virus glycoprotein，using fusion proteins synthesized in *Escherichia coli*. *J Virol* 65：1611-1615.

Yuasa，K.，Koesharyani，I.，Roza，D.，*et al.*，2002. Immune response of humpback grouper，*Cromileptes altivelis*（Valenciennes）injected with the recombinant coat protein of betanodavirus. *J Fish Dis* 25：53-56.

Zhang，M.，Wu，H. Z.，Li，X. Y.，*et al.*，2012. *Edwardsiella tarda* flagellar protein FlgD：A protective immunogen against edwardsiellosis. *Vaccine* 30：3849-3856.

第四章　可复制型疫苗

Craig A. Shoemaker and Phillip H. Klesius

本章概要　SUMMARY

鱼类免疫学及鱼类抗病力的早期研究表明，感染后存活下来的鱼能抵御该病原的再次感染，这是研发可复制型疫苗（或活体疫苗）的基础。可复制型疫苗在人类医学、兽医学和水生动物医学领域，已成功地用于疾病的预防。可复制型疫苗所采用的减毒策略主要包括：实验室传代、抗原拟态、物理或化学诱变以及基因工程的分子技术等。实验室研究已证实活疫苗的有效性，其能诱导鱼类的黏膜免疫、细胞免疫和体液免疫反应，且适用于浸泡免疫接种方式。但是在欧盟以及亚洲和南美洲国家，监管部门对可复制型疫苗的许可持谨慎态度。目前，鱼类有三个细菌性活疫苗和一个病毒性活疫苗获批使用，即细菌性肾病活疫苗（加拿大、智利和美国）、鮰肠型败血症活疫苗、鮰柱形病活疫苗（美国）以及病毒性出血性败血症病毒（VHSV）活疫苗（德国）。在水产养殖中使用可复制型疫苗是一个很好的选择，但是前提是其对鱼、环境和人类造成的风险比较低或者几乎不存在。所以，监管部门必须根据疫苗的安全性、毒力恢复可能性、质粒的缺失和/或者不含编码耐药性基因等科学数据，对疫苗进行审批。疫苗的使用是所有集约化动物生产过程中，保障动物健康和生物安全管理的重要方式，只有这样才会获得最大效益。

第一节　引　言

鱼类免疫学及鱼类抗病力的早期研究表明，鱼类像动物和人类一样，在感染中存活下来的个体通常具有抵御同一病原再次感染的能力（Fujihara 和 Nakatani，1971；Fijan 等，1977；Shoemaker 和 Klesius，1997；Ahne 等，2002；Panangala 等，2009；Lorenzen 等，2010），这是研发鱼类可复制型疫苗（或活体疫苗）的基础（Hartman 和 Noga，1980；Norqvist 等，1989；Vaughan 等，1993；Thornton 等，1994；Lawrence 等，1997；Hernanz Moral 等，1998；Marsden 等，1998；Klesius 和 Shoemaker，1999；Thune 等，1999；Perelberg 等，2005）。以上文献中提到的鱼类活疫

苗均已在实验室研究阶段证明了其有效性，但欧盟以及亚洲和南美洲国家的监管部门对活疫苗的使用许可持谨慎态度。可复制型疫苗能诱导鱼类的细胞免疫和体液免疫，且适用于浸泡免疫接种方式，这就更适合应用于由于注射免疫费用不划算的价格较低的鱼类。到目前为止，在加拿大和智利有一种细菌性活疫苗获得销售许可（仅许可 Renogen 公司），可在水产养殖中使用。在美国有三种细菌性活疫苗获得许可（Shoemaker 等，2009）：细菌性肾病、鮰肠型败血症、鮰柱形病活疫苗，分别由 Renogen、AQUAVAC-ESC、AQUAVAC-COL 公司生产。在德国，病毒性出血性败血症病毒（VHSV）活疫苗获得使用许可（Gomez-Casado 等，2011）。本章主要介绍水产养殖中可复制型疫苗研发所采用的减毒策略，包括最初在人类医学和兽医学上应用的策略（Linde 等，1990；Schurig 等，1991；Schnell 等，1994；Tizard，1999）。本章还会涉及活疫苗的安全性、免疫接种和免疫反应的产生等方面的内容。

第二节　细菌性疫苗的减毒策略

一、实验室传代

比较早的减毒策略是采用培养基和/或组织培养进行实验室传代培养，以减弱致病菌的毒性。Daly（2001）用实验室减毒的鲑肾杆菌（*Renibacterium salmoninarum*）菌株免疫大西洋鲑（*Salmo salar*）。一般鲑肾杆菌只在含血清和 L-半胱氨酸的 KDM-2 琼脂培养基上生长。研究分离获得的可在胰蛋白酶大豆琼脂培养基上（TSA）生长和可在脑心浸液琼脂培养基（BHIA）上生长的两株鲑肾杆菌菌株，经体外传代培养 25 代后，TSA 株和 BHIA 株的致病力分别仅为 8%和 0%，注射免疫大西洋鲑后保护期可达 60～74d。

Swain（2010）有关嗜水气单胞菌（*Aeromonas hydrophila*）减毒疫苗的研究发现，经实验室连续传代 8 年培养出的两株减毒嗜水气单胞菌菌株缺少野生型母本菌株具有的脂多糖，腹腔注射该减毒株，其致病力是野生型菌株的十万分之一。腹腔注射免疫印度野鲮（*Labeo rohita*）可获得免疫保护（Swain 等，2010）。

二、环境菌替代

第二种方法主要包括采用对病原不易感的宿主进行病原替代来培养病原，分离或使用天然无致病性的菌株，或使用与病原抗原性相似的环境菌株作为替代菌株。迄今为止，应用此策略研制可复制型疫苗最成功的是 Renogen 公司，通过免疫注射关节杆菌（*Arthrobacter davidanieli*）预防鲑肾杆菌引起的细菌性肾病（Griffiths 等，1998；Salonius 等，2005）。关节杆菌是与鲑鱼肾杆菌在分类地位和抗原性上相近的环境菌。Griffiths（1998）发现抗关节杆菌的血清能与鲑肾杆菌的菌体表面糖类发生交叉反应，说明鲑免疫接种关节杆菌可复制型疫苗产生了抗体，因此产生了免疫保护力。Salonius（2005）报道，用关节杆菌可复制型疫苗注射免疫大西洋鲑后，相对保护率可达 80%以上（Amend，1981）。

Itano（2006）应用系统发生相近性和抗原交叉反应性确认了诺卡菌（*Nocardia seriolae*）候选疫苗株，从环境诺卡菌中筛选分离出诺卡菌 *N. soli* 和 *N. fluminea* 菌株，用其注射免疫黄尾鰤（*Seriola quinqueradiata*），研究其产生的对诺卡菌（*N. seriolae*）感染的免疫保护力，但攻毒实验结果显示保护作用很小。

Cheng（2010）从牙鲆（*Paralichthys olivaceus*）中分离出一株天然无致病力的迟缓爱德华菌（*Edwardsiella tarda*）。研究发现，从人粪便分离的迟缓爱德华菌分离株（ATCC 15947）对毛腹鱼（*Trichogaster trichopterus*）无致病性（Tan 等，2002）。腹腔注射或口服 ATCC 15947 菌株感染牙鲆，仅短时感染，在牙鲆体内 8～12d 被清除。免疫结果表明，腹腔注射免疫牙鲆相对保护率（RPS）可达 79%～81%，口服或浸泡免疫牙鲆 RPS 为 21%～56%。

三、物理或化学诱变

第三种方法主要是应用物理或化学方法使病原体突变。疫苗用菌株必须对宿主没有致病力。Ishiguro（1981）应用物理诱变法，将培养温度升高至 30℃，分离出杀鲑气单胞菌（*A. salmonicida*）外膜蛋白 A 缺失菌株。Thornton（1994）发现了外膜蛋白 A 缺失菌株和/或外膜脂多糖（LPS）O 抗原缺失株。两菌株致病力显著降低，可作为候选疫苗株。进一步研究表明，外膜蛋白 A 和 LPS 抗原缺失株可作为有效的疫苗，因为该疫苗发生了两种突变，加强了安全性。

使用抗生素是最常用也是最有效的化学诱变法。利福平能抑制细菌依赖 DNA 的 RNA 多聚酶活性，阻断 RNA 转录过程（Wehrli 等，1968），使 LPS 合成受损（Arias 等，2003；Zhang 等，2006），从而减弱革兰氏阴性细菌的致病性（Norqvist 等，1989；Schurig 等，1991；Klesius 和 Shoemaker，1999）。Norqvist（1989）应用利福平和链霉素诱导出耐药的鳗弧菌（*Vibrio anguillarum*），其毒力显著降低，作为活疫苗可对虹鳟（*Oncorhynchus mykiss*）产生免疫保护。Klesius 和 Shoemaker（1999）应用类似的方法研制出第一个获得许可的斑点叉尾鮰（*Ictalurus punctatus*）活疫苗（见第八章）。鮰爱德华菌（*Edwardsiella ictaluri*）为胞内寄生菌，将其培养在含有不同浓度的利福平的培养基中（利福平浓度最高为 320μg/mL）。对突变菌株引起斑点叉尾鮰肠败血症的致病力检测表明，浸泡感染突变株对斑点叉尾鮰无致病性，去除利福平后体外培养或者在鱼体内传代培养 5 代，突变株也未恢复致病性。接种该突变株 14d 后，斑点叉尾鮰幼鱼产生免疫保护力（Klesiu 和 Shoemaker，1999）。该活疫苗（AQUAVAC-ESC©）经过美国农业部动植物卫生检验局兽医生物制品中心的严格审批，授权给英特威/先灵宝雅公司（Intervet/Schering Plough）（现默克动物保健公司，Merck Animal Health）。该疫苗在斑点叉尾鮰的发眼卵期（Klesius 等，2000；Shoemaker 等，2002，2007）、鱼苗期（Shoemaker 等，1999；Wise 和 Terhune，2001；Carrias 等，2008）和幼鱼期（Wise 等，2000；Lawrence 和 Banes，2005；Karsi 等，2009）应用均证明有效。据 Bebak 和 Wagner（2012）报道，2009 年美国斑点叉尾鮰鱼苗业孵化 15d 的鱼苗接种鮰爱德华菌活疫苗的约占 12.3%，约为 1.35×10^8 尾。

斑点叉尾鮰肠败血症活疫苗获批后，应用利福平对病原减毒从而筛选活疫苗候选株的手段在其他鱼类疾病病原中被推广应用。Shoemaker 团队研发出减毒和保护效果良好的柱状黄杆菌（*F. columnare*）活疫苗（AQUAVAC-COL©）（Shoemaker，2007，2011；Sundell 等，2014）。疫苗 AQUAVAC-COL©的实验室（Shoemaker 等，2011）和野外（Bebak 等，2009）研究结果均表明其对大口黑鲈（*Micropterus salmoides floridanus*）鱼苗有效。Bebak 和 Wagner（2012）报道，2009 年美国孵化 20d 左右的斑点叉尾鮰和蓝尾鮰以及两种鱼的杂交鱼苗接种柱状黄杆菌活疫苗的数量约为 2.04×10^8 尾，占总鱼苗量的 17.0%。LaFrentz（2008）使用利福平对嗜冷黄杆菌（*F. psychrophilum*）减毒，研制出既安全又有效的活疫苗，成功免疫虹鳟，目前正在美国进行生产试验（B. R. LaFrentz）。这种减毒方法还成功地在鱼类嗜水气单胞菌（*Aeromonas hydrophila*）（Klesius 等，2011）、迟缓爱德华菌（*E. tarda*）（Evans 等，2006；Sun 等，2010）和鳗弧菌（*V. anguillarum*）（Yu 等，2012）中应用，均获得较好的活疫苗候选株。目前为止，这是研制有效的兽用疫苗产品最实用的方法之一（Linde 等，1990；Schurig 等，1991；Bhatnagar 等，1994；Klesius 和 Shoemaker，1999；Gantois 等，2006；LaFrentz 等，2008；Shoemaker 等，2011）。

已有应用新霉素对海豚链球菌（*Streptococcus iniae*）（Pridgeon 和 Klesius，2011a）（见第二十章）和鮰爱德华菌（Pridgeon 和 Klesius，2011b）进行减毒的研究。也有新霉素和利福平两种抗生素联合使用对嗜水气单胞菌进行减毒的研究（Pridgeon 和 Klesius，2011c）。这些疫苗候选株实验室研究均显示出良好的安全性和免疫保护效果。

四、基因工程

第四种方法是应用分子生物学技术研制候选可复制型疫苗。基因工程是通过破坏病原的代谢途径或毒力基因从而实现毒力减弱。早期的方法是制备营养缺陷型的突变体（Hoiseth 和 Stocker，1981；Smith 等，1984）。通常，这种方法是在自杀质粒中，通过等位基因替换，将细菌芳香氨基酸代谢途径的合成基因 *aroA* 替换成包含抗生素耐药标记基因的 DNA 序列。*aroA* 基因缺失使细菌不能产生叶酸合成所必需的对氨基苯甲酸，进而无法在宿主体内存活。Vaughan（1993）首次将这种方法应用在鱼类病原的减毒中，研究表明 *aroA* 缺失的杀鲑气单胞菌（*A. salmonicida*）对褐鳟（*Salmo trutta*）无致病性，具有免疫保护效果。随后，一系列针对鮰爱德华菌（*E. ictaluri*）（Thune 等，1999）、嗜水气单胞菌（*A. hydrophila*）（Hernanz Moral 等，1998；Vivas 等，2004）和鲁克氏耶尔森菌（*Yersinia ruckeri* 01）（Temprano 等，2005）*aroA* 缺失的减毒候选疫苗被研制出来，均通过实验室试验证实了疫苗的有效性，但使用高剂量的突变株免疫可能会导致宿主动物死亡（Lawrence 等，1997；Thune 等，1999；Grove 等，2003）。虽然基因工程可研制出减毒疫苗分离株，但疫苗的免疫持续性较差，仅 24～72h，且不能在幼鱼体内产生足够的免疫力。

以其他代谢途径基因为靶基因进行减毒的有环腺苷 3′，5′-磷酸腺苷（cAMP）受体蛋白（crp）和腺苷酸环化酶（cya）。Santander（2011）根据曾应用在其他肠杆菌的类似方法（Curtiss 和 Kelly，1987）构建出鮰爱德华菌的 *crp* 框内缺失突变体，*crp* 突变体对斑点叉尾鮰幼鱼无致病力，浸泡免疫后能产生免疫保护作用。

还有以毒力基因或毒力调节基因为靶基因的基因修饰（Karsi 等，2009）。Cooper 等（1996）应用微型转座子技术干扰软骨素硫酸酯酶的产生，该酶可能是鮰爱德华菌的毒力因子。斑点叉尾鮰免疫这种突变体后，对致病性爱德华菌产生了免疫保护。还有研究利用转座子技术构建爱德华菌的脂多糖 O 抗原缺失株研制活疫苗（Lawrence 等，2001；Lawrence 和

Banes，2005），研究者虽然成功构建出鮰爱德华菌的脂多糖 O 抗原缺失株，但通过浸泡免疫斑点叉尾鮰没有产生保护效果（Lawrence 和 Banes，2005）。Igarashi 和 Iida（2002）应用相似技术，通过转座子诱变技术产生低嗜铁素的突变株，研制出减毒的迟缓爱德华菌（*E. tarda*）活疫苗，免疫罗非鱼（*Oreochromis niloticus*）后，再以致死浓度的病原菌攻毒，获得保护效果。Leung（1997）还构建出含 *mini-Tn5*（转座子突变体）生长不良和蛋白酶缺失的嗜水气单胞菌（*A. hydrophila*）突变株，研制出用于免疫毛足鲈（*Trichogaster trichopterus*）的活疫苗，但该疫苗株并未完全减毒。

Buchanan（2005）利用随机转座子（Tn917）诱变技术，在杂交条纹鲈（*Morone chrysops* × *M. saxatilis*）体内培养筛选，构建出海豚链球菌（*Streptococcus iniae*）磷酸葡萄糖变位酶基因缺失株，磷酸葡萄糖变位酶主要影响细菌多糖荚膜的形成，荚膜的有无决定细菌的致病性（Barnes 等，2003；Buchanan 等，2005）。该分离株对杂交条纹鲈的致病力减弱，腹腔注射活疫苗，再以致死浓度的海豚链球菌感染，条纹鲈对感染产生免疫保护（Buchanan 等，2005）。Locke（2010）研究了 3 株海豚链球菌突变株活疫苗的浸泡免疫效果，一株是磷酸葡萄糖变位酶的突变株，另外两个分别是荚膜多糖（Locke 等，2007）和 M 样蛋白（Locke 等，2008）的突变株，以灭活海豚链球菌疫苗为对照。结果表明，M 样蛋白突变株免疫组产生的免疫保护率竟达 100%，但造成了免疫鱼轻度患病。荚膜多糖突变体和磷酸葡聚糖变位酶突变体无致病性，但只产生了较低的免疫保护。

弗朗西斯菌病是重要的新发鱼类疾病，病原是一种胞内生长的革兰氏阴性菌（Birkbeck 等，2011）。Soto 等（2010）构建了 *iglC* 突变的七星弗朗西斯菌（*Francisella asiatica*）株。*iglC* 是一种胞内生长毒力岛基因，可使弗朗西斯菌在巨噬细胞中存活。罗非鱼在浸泡免疫突变菌株后产生了免疫保护，实验室条件下 RPS 达 68%～87%。

五、活菌表达抗原疫苗

目前，比较安全的、用于鱼类细菌性病原的减毒策略是使用酵母菌或细菌表达致病菌抗原，例如酿酒酵母菌（*Saccharomyces cerevisiae*）或乳酸乳球菌（*Lactococcus lactis*）。Zhu（2006）制备出细胞表面表达哈维氏弧菌（*Vibrio harveyi*）溶血素的酵母细胞，注射实验表明，表达溶血素的酵母细胞无致病性，注射免疫活疫苗，大菱鲆产生免疫保护，但是对牙鲆没有免疫保护作用。Anuradha（2010）应用活的乳酸乳球菌重组表达嗜水气单胞菌（*Aeromonas hydrophila*）气溶素基因 *D1* 和 *D4*，实验表明，注射或口服免疫重组乳酸乳球菌，罗非鱼对嗜水气单胞菌的感染产生免疫保护。

第三节　病毒性疫苗的减毒策略

一、实验室传代

最早用实验室传代法研究鱼类病毒减毒的报道之一是一项专利，其描述了根据脊髓灰质炎病毒突变体的分离方法（Sabin 等，1954），对斑点叉尾鮰病毒（CCV）进行实验室传代培养减毒（Hartman 和 Noga，1980）。将 CCV 接种于胡子鲶（*Clarius batrachus*）的细胞系中，进行传代培养 62 代，获得了减毒的 CCV 株（V62）。减毒的 CCV 株感染斑点叉尾鮰细胞系形成的噬斑显著变小，以 1×10^6pfu*/mL 的减毒病毒株免疫接种鱼体，第 45～60 天加强免疫 1 次，加强免疫后存活率可达 97%，对照组存活率仅为 22%。

Kölbl（1989）就鲤春病毒（SVCV）和病毒性出血性败血症病毒（VHSV）的减毒研究申请了专利。该方法以鸡成纤维细胞作为替代宿主细胞系传代培养 SVCV，将培养温度由 20℃ 升至 30℃，培养约 350 代获得减毒病毒株。免疫接种鲤（*Cyprinus carpio*）后，无致病性，能产生免疫保护。采用相同的减毒方法，VHSV 经过约 250 代培养筛选出减毒病毒株。德国批准生产的 VHSV 活疫苗就是采用这种研制方法（Gomez-Casado 等，2011），但专利中并没有免疫虹鳟后有关疫苗有效性的数据。Enzmann（1998）实验室的实验结果表明，疫苗的免疫保护期较好，免疫保护力能够维持 6 个月。有意思的是，20 世纪 80 年代，就已有应用细胞培养传代构建 VHSV 减毒株，免疫虹鳟幼鱼能产生显著保护效果的报道（De Kinkelin 和 Béarzotti-Le Beer，1981；Vestergård-Jørgensen，1982）。

在锦鲤疱疹病毒（KHV）中也采用了相似的减毒策略。将强毒分离株在锦鲤鳍细胞系中传代培养 26 代，分离出 4 个减毒病毒株。腹腔注射方式免疫减毒病毒，与对照组相比，具有明显的免疫保护效果。Ronen（2003）的研究表明，鲤在发病温度 18～25℃ 下养殖才可达到有效的疫苗免疫效果；水温高于 27℃，鲤并不能产生保护性免疫力（Perelberg 等，2005）。在 KHV 疫苗的免疫应答研究中检测出抗 KHV 的特异性抗体，抗体对免疫保护力的产生发挥作用。为进一步使疫苗减毒并降低毒性恢复的可能性，Perelberg（2005）将分离出的减毒株进行紫外线（UV）照射，选取暴露于 UV 中 30s 后的病毒分离株作为疫苗。在发病温度下，鲤免疫接种 10 pfu/mL 和 100pfu/mL 的减毒病毒，攻毒后免疫组死亡率为 10%～20%，对照组的死亡率为 98%。

二、分子生物学技术

Zhang 和 Hanson（1995）通过敲除胸苷激酶（TK）基因研制出减毒重组 CCV。这一策略在其他大型的 DNA 病毒如 α-疱疹病毒上的应用表明，不含 *TK* 基因的病毒突变株复制能力降低（Wilcox 等，1992）。Zhang 和 Hanson（1995）的研究显示，减毒的 CCV 对斑点叉尾鮰具有感染性，但同时浸泡免疫能产生免疫保护力。

研发 VHSV 和传染性造血器官坏死病毒（IHNV）的可复制型疫苗采用了反向遗传学方法，这是一种有发展前景的方法。这种方法最初应用于负链 RNA 的狂犬病毒减毒疫苗的研究中（Schnell 等，1994）。Biacchesi 等（2000）详细描述了构建可表达外源基因的活体重组（r）IHNV。使用逆转录聚合酶链反应（RT-PCR）扩增 IHNV 基因组的全长 cDNA，将其克隆到转录质粒中 T7 RNA 聚合酶启动子和自催化型肝炎病毒核糖体中间。将该质粒转染到表达 T7 RNA 聚合酶的重组牛痘病毒感染的鱼类细胞中，再从表达 IHNV 基因的质粒转染的鱼类细胞获得活 rIHNV。这个实验首次验证了应用反向遗传学方法构建活疫苗是可行的。rIHNV 活病毒与野生

* pfu 为噬菌斑形成单位。——编者注

型病毒，在产生细胞病变效果和对虹鳟的致死率方面都是相同的。因此该策略可以用于研制新型活病毒疫苗株（Biacchesi 等，2000）。Biacchesi 等（2002）研制出 rIHNV，该重组病毒中 IHNV 的外膜糖蛋白 G 蛋白基因被 VHSV 的 G 蛋白取代，并且敲除了 *NV* 基因，代之以绿色荧光蛋白（GFP）基因。这个重组病毒被定义为 rIHNV-Gvhsv-GFP（GPF 标记的 G 蛋白重组 IHNV 活病毒）（Biacchesi 等，2002）。首次开展 rIHNV 作为活疫苗候选株的研究表明，*NV* 基因的缺失可导致病毒毒力不可逆的衰减，并可诱导虹鳟产生较高的免疫保护力（Thoulouze 等，2004）。Novoa 等（2006）用斑马鱼（*Danio rerio*）为模型研究了 rIHNV-Gvhsv-GFP 对 VHSV 感染的免疫效果，首次证明了免疫接种重组活病毒后，机体对 VHSV 感染有免疫保护效果。Romero（2008）的研究也表明活的重组病毒缺乏毒力，并对野生型的 IHNV 和 VHSV 感染均有保护效果，RPS 分别为 62% 和 61%。Bootland 和 Leong（2011）指出，虽然重组表达活病毒在实验室研究取得了成功，但在商业化应用之前，必须在管理、安全性、规模方面满足要求。

第四节 诱导免疫力

可复制型疫苗进入宿主并持续存在于体内，会激活机体产生细胞免疫、体液（抗体）免疫和黏膜免疫（Clark 和 Cassidy-Hanley，2005）。活疫苗在体内的存活和复制使宿主产生强烈的细胞免疫应答，从而获得持续的免疫保护。$CD4^+$ 和 $CD8^+$ T 细胞免疫应答对胞内感染产生免疫保护作用（Seder 和 Hill，2000）。Laing 和 Hansen（2011）对已有鱼类 T 细胞免疫的研究进行了综述，指出基因组研究已经鉴定出与 T 细胞功能相适应的成分，在鱼类中存在 T 细胞亚群。研究表明，鱼类中存在主要组织相容性复合体（MHC）Ⅰ类（Antao 等，2001）和Ⅱ类分子（Godwin 等，2000）。MHC 分子与其匹配抗原被相应的 T 细胞和 B 细胞亚群识别，并产生免疫应答。Seder 和 Hill（2000）发现，活疫苗能够诱导 Th1 型和 $CD8^+$ T 淋巴细胞的免疫应答。Marsden（1996）证明，可复制型的 *aroA* 缺失杀鲑气单胞菌（*A. salmonicida*）活疫苗能增强虹鳟的 T 细胞和 B 细胞免疫反应，且对 T 细胞的增强效果要高于 B 细胞。最近，有关鲶免疫细菌性活疫苗后肝组织基因表达的转录组研究表明，MHC-Ⅰ类分子基因及急性期反应基因均上调表达，这表明有活跃的抗原加工和递呈（Peatman 等，2008；Pridgeon 等，2012）。MHC-Ⅰ类分子基因的诱导表达说明，胞内细菌感染引起了 CD8 T 细胞的免疫反应。T 细胞亚群的免疫反应随后可诱导干扰素-γ 的产生（Milev-Milovanovic 等，2006），增强了胞内病原的清除（Seder 和 Hill，2000）。免疫可复制型细菌疫苗（Russo 等，2009）或病毒疫苗（Romero 等，2011）引起的巨噬细胞介导的免疫反应在产生免疫保护效果中发挥重要作用。据此，可以预测鱼类对活疫苗免疫反应的一般特性。

Seder 和 Hill（2000）认为，灭活疫苗或亚单位疫苗在激活细胞免疫反应方面是无效的。灭活疫苗或提纯的重组抗原主要诱导机体的体液（抗体介导）免疫反应。早期用于水产养殖的甲醛灭活鮰爱德华菌（*Edwardsiella ictaluri*）疫苗仅有少量可被摄入鱼体内（Nusbaum 和 Morrison，1996）。甲醛灭活可能改变病原重要的表面抗原（Bader 等，1997）。以上两种情况会导致灭活疫苗不起作用，使幼鱼不能对胞内病原感染产生免疫保护（Thune 等，1997）。大多数灭活疫苗与佐剂一起使用并通过注射免疫达到理想效果。早期弧菌疫苗通过浸泡途径免疫(Sommerset 等，2005)，而注射免疫接种才能获得最佳保护效果（Norqvist 等，1989)。一般来说，灭活疫苗产生的免疫保护期少于 4 个月，并且只对胞外病原或产毒素的病原有免疫保护效果。若延长灭活疫苗的免疫保护期，通常采取多次免疫和/或使用佐剂的方式。有研究表明，免疫接种细菌性活疫苗或病毒性活疫苗，免疫保护期分别超过 4 个月（Klesius 和 Shoemaker，1997）或 6 个月（Enzmann 等，1998)。活疫苗还能诱导免疫系统尚不成熟的鱼产生免疫保护，这是因为活疫苗能持续存活在机体中，可待鱼体免疫系统成熟后再产生免疫应答；同时/或者疫苗抗原被宿主免疫细胞上相应的 MHC 分子递呈表达，由此激活了与禽类类似的免疫活性细胞，产生免疫应答。因此，活疫苗诱导幼鱼产生免疫保护是可信的（Mast 和 Goddeeris，1999)。

第五节　疫苗接种

可复制型疫苗具有接种方式简便的优点，可通过饲料/水口服免疫接种（Frey，2007)，或者浸泡免疫接种（Norqvist 等，1989；Klesius 等，2004；Perelberg 等，2005)，用于温水养殖鱼类的可复制型疫苗就是在稚鱼放入池塘前进行浸泡免疫。活疫苗菌株能够定植并感染宿主，通过浸泡产生免疫保护效果。研究表明，鱼体浸泡低剂量的鳗弧菌（*Vibrio anguillarum*）活疫苗，可获得较高水平的免疫保护效果（Norqvist 等，1989)。Norqvist 等（1989）认为，活疫苗在宿主体内复制可增强机体对疫苗的免疫应答。

第六节　疫苗的安全性

可复制型疫苗的安全性是指，疫苗株既不能使被接种的动物发病（疫苗的毒力恢复)，又不能传播到环境中造成污染。Frey（2007）全面概述了兽用转基因活菌疫苗的生物安全内涵，其中的基本原则同样适用于水产养殖用可复制型疫苗。在美国，农业部（USDA）动植物卫生检验局（APHIS）兽医生物制品中心负责兽医生物制品的许可和注册（www. aphis. usda. gov)。安全性研究包括，以 10 倍的免疫剂量免疫鱼体，监测鱼与鱼的子代评估疫苗恢复毒力的可能；可复制型疫苗释放对水环境影响的风险；病原在自然环境中存在对人感染的可能；可复制型疫苗包含的质粒和/或含有编码耐药性基因的可能。以上方面均须重点评估。LaFrentz 等（2011）最早开展鱼类可复制型疫苗中多重耐药质粒的研究。研究表明，鮰爱德华菌（*E. ictaluri*）活疫苗不存在对水产养殖常用抗菌药物的耐药性，并且不含 IncA/C 或其他质粒不相容群。随着分子生物学技术的日益发展，将会研究开发出更安全的可复制型疫苗（Wang 等，2013）用于防治使用灭活疫苗无效的胞内感染性疾病。

第七节　致　谢

作者由衷感谢 Scott LaPatra、Dehai Xu 和 Ben LaFrentz 对手稿进行的严格审校，并提出了有益的建议。该工作得到了美国农业部农业研究服务处 CRIS 项目编号 6420-32000-024D 的支持。在本章中提及的商标或商品仅是为了提供具体的信息，并非由美国农业部认可或推荐。

参考文献

Ahne, W., Bjorklund, H. V., Essbauer, S., *et al.*, 2002. Spring viremia of carp (SVC). *Dis Aquat Org* 52: 261-272.

Amend, D. F. 1981. Potency testing of fish vaccines. *Dev Biol Stand* 49: 447-454.

Antao, A. B., Wilson, M., Wang, J., *et al.*, 2001. Genomic organization and differential expression of channel catfish MHC class I genes. *Dev Comp Immunol* 25: 579-595.

Anuradha, K., Foo, H. L., Mariana, N. S., *et al.*, 2010. Live recombinant *Lactococcus lactis* vaccine expressing aerolysin genes D1 and D4 for protection against *Aeromonas hydrophila* in tilapia (*Oreochromis niloticus*). *J Appl Microbiol* 109: 1632-1642.

Arias, C. R., Shoemaker, C. A., Evans, J. J. and Klesius, P. H. 2003. A comparative study of *Edwardsiella ictaluri* parent (EILO) and *E. ictaluri* rifampicin-mutant (RE-33) isolates using lipopolysaccharides, outer membrane proteins, fatty acids, Biolog, API 20E and genomic analyses. *J Fish Dis* 26: 415-421.

Bader, J. A., Vinitnantharat, S. and Klesius, P. H. 1997. Comparison of whole-cell antigens of pressure- and formalin-killed *Flexibacter columnaris* from channel catfish (*Ictalurus punctatus*). *Am J Vet Res* 58: 985-988.

Barnes, A. C., Young, F. M., Horne, M. T. and Ellis, A. E. 2003. *Streptococcus iniae*: serological differences, presence of capsule and resistance to immune serum killing. *Dis Aquat Org* 53: 241-247.

Bebak, J. and Wagner, B. 2012. Use of vaccination against enteric septicemia of catfish and columnaris disease by the US catfish industry. *J Aquat Anim Health* 24: 30-36.

Bebak, J., Matthews, M. and Shoemaker, C. 2009. Survival of vaccinated, feed-trained largemouth bass fry (*Micropterus salmoides floridanus*) during natural exposure to *Flavobacterium columnare*. *Vaccine* 27: 4297-4301.

Bhatnagar, N., Getachew, E., Straley, S., *et al.*, 1994. Reduced virulence of rifampicin-resistant mutants of *Francisella tularensis*. *J Infect Dis* 170: 841-847.

Biacchesi, S., Thoulouze, M.-I., Béarzotti, M., *et al.*, 2000. Recovery of NV knockout infectious hematopoietic necrosis virus expressing foreign genes. *J Virol* 74: 11247-11253.

Biacchesi, S., Béarzotti, M., Bouguyon, E. and Brémont, M. 2002. Heterologous exchanges of the glycoprotein and the matrix protein in a *Novirhabdovirus*. *J Virol* 76: 2881-2889.

Birkbeck, T. H., Feist, S. W. and Verner-Jeffreys, D. W. 2011. *Francisella* infections in fish and shellfish. *J Fish Dis* 34: 173-187.

Bootland, L. M. and Leong, J. A. C. 2011. Infectious haematopoietic necrosis virus, in *Viral, Bacterial and Fungal Infections*, 2nd edn (eds P. T. K. Woo and D. W. Bruno). Wallingford, CAB, 66-109.

Buchanan, J. T., Stannard, J. A., Lauth, X., *et al.*, 2005. *Streptococcus iniae* phosphoglucomutase is a virulence factor and target for vaccine development. *Infect Immun* 73: 6935-6944.

Carrias, A. A., Terhune, J. S., Sayles, C. A. and Chappel, J. A. 2008. Effects of an extended hatchery phase and vaccination against enteric septicemia of catfish on the production of channel catfish, *Ictalurus punctatus*, fingerlings. *J World Aquac Soc* 39: 259-266.

Cheng, S., Hu, Y., Zhang, M. and Sun, L. 2010. Analysis of the vaccine potential of a natural avirulent *Edwardsiella tarda* isolate. *Vaccine* 28: 2716-2721.

Clark, T. G. and Cassidy-Hanley, D. 2005. Recombinant subunit vaccines: potentials and constraints. *Dev Biol (Basel)* 121: 153-163.

Cooper II,, R. K., Shotts Jr., E. B. and Nolan, L. K. 1996. Use of a mini-transposon to study chondroitinase activity associated with *Edwardsiella ictaluri*. *J Aquat Anim Health* 8: 319-324.

Curtiss, III,, R. and Kelly, S. M. 1987. *Salmonella typhimurium* deletion mutants lacking adenylate cyclase and cyclic AMP receptor protein are avirulent and immunogentic. *Infect Immun* 55: 3035-3043.

Daly, J. G., Griffiths, S. G., Kew, A. K., *et al.*, 2001. Characterization of attenuated *Renibacterium salmoninarum* strains and their use as live vaccines. *Dis Aquat Org* 44: 121-126.

De Kinkelin, P. and Béarzotti-Le Beer, M. 1981. Immunization of rainbow trout against viral haemorrhagic septicemia (VHS) with a thermoresistant variant of the virus. *Dev Biol Stand* 49: 431-439.

Enzmann, P. J., Fichtner, D., Schütze, H. and Walliser, G. 1998. Development of vaccines against VHS and IHN: oral application, molecular marker and discrimination of vaccinated fish from infected populations. *J Appl Ichthyol* 14: 179-183.

Evans, J. J., Klesius, P. H. and Shoemaker, C. A. 2006. A modified live *Edwardsiella tarda* vaccine for aquatic animals. US Patent No. 7, 067, 122.

Fijan, N., Petrinec, Z., Štancl, Z., *et al.*, 1977. Vaccination of carp against spring viremia: comparison of intraperitoneal and peroral application of live virus to fish kept in ponds. *Bull Off Int Épizoot* 87: 441-442.

Frey, J. 2007. Biological safety concepts of genetically modified live bacterial vaccines. *Vaccine* 25: 5598-5605.

Fujihara, M. P. and Nakatani, R. E. 1971. Antibody production and immune responses of rainbow trout and coho salmon to *Chondrococcus columnaris*. *J Fish Res Board Can* 28: 1253-1258.

Gantois, I., Ducatelle, R., Timbermount, L., *et al.*, 2006. Oral immunization of laying hens with live vaccine strains of TAD *Salmonella* vac® E and TAD *Salmonella* vac® T reduces internal egg contamination with *Salmonella enteritidis*. *Vaccine* 24: 6250-6255.

Godwin, U. B., Flores, M., Quiniou, S., *et al.*, 2000. MHC class II *A* genes in the channel catfish (*Ictalurus punctatus*). *Dev Comp Immunol* 24: 609-622.

Gomez-Casado, E., Estepa, A. and Coll, J. M. 2011. A comparative review on European-farmed finfish RNA viruses and their vaccines. *Vaccine* 29: 2657-2671.

Griffiths, S. G., Melville, K. J. and Salonius, K. 1998. Reduction of *Renibacterium salmoninarum* culture activity in Atlantic salmon following vaccination with avirulent strains. *Fish Shellfish Immunol* 8: 607-619.

Grove, S., Høie, S. and Evensen, Ø. 2003. Distribution and retention of antigens of *Aeromonas salmonicida* in Atlantic salmon (*Salmo salar* L.) vaccinated with a Δ*aro*A mutant or formalin-inactivated bacteria in oil-adjuvant. *Fish Shellfish Immunol* 15: 349-358.

Hartman, J. X. and Noga, E. J. 1980. Channel catfish virus disease vaccine and method of preparation thereof and method of immunization therewith. US Patent No. 4, 219, 543.

Hernanz Moral, C., Flaño Del Castillo, E., López Fierro, P., *et al.*, 1998. Molecular characterization of the *Aeromonas hydrophila aro*A gene and potential use of an auxotrophic *aro*A mutant as a live attenuated vaccine. *Infect Immun* 66: 1813-1821.

Hoiseth, S. K. and Stocker, B. A. D. 1981. Aromatic dependent *Salmonella typhimurium* are non-virulent and effective as live vaccines. *Nature* 291: 238-239.

Igarashi, A. and Iida, T. 2002. A vaccination trial using live cells of *Edwardsiella tarda* in tilapia. *Fish Pathol* 37: 145-148.

Ishiguro, E. E., Kay, W. W., Ainsworth, T., *et al.*, 1981. Loss of virulence during culture of *Aeromonas salmonicida at high temperature. J Bacteriol* 148: 333-340.

Itano, T., Kawakami, H., Kono, T. and Sakai, M. 2006. Live vaccine trials against nocardiosis in yellowtail *Seriola quinqueradiata*. *Aquaculture* 261: 1175-1180.

Karsi, A., Gülsoy, N., Corb, E., *et al.*, 2009. High-throughput bioluminescence-based mutant screening strategy for identification of bacterial virulence genes. *Appl Environ Microbiol* 75: 2166-2175.

Klesius, P. H. and Shoemaker, C. A. 1997. Heterologous isolates challenge of channel catfish, *Ictalurus punctatus*, immune to *Edwardsiella ictaluri*. *Aquaculture* 157: 147-155.

Klesius, P. H. and Shoemaker, C. A. 1999. Development and use of modified live *Edwardsiella ictaluri* vaccine against enteric septicemia of catfish. *Adv Vet Med* 41: 523-537.

Klesius, P. H., Shoemaker, C. A. and Evans, J. J. 2000. *In ovo* methods for utilizing live *Edwardsiella ictaluri* against enteric septicemia in channel catfish. US Patent No. 6, 153, 202.

Klesius, P. H., Evans, J. J. and Shoemaker, C. A. 2004. Warmwater fish vaccinology in catfish production. *Anim Health Res Rev* 5: 305-311.

Klesius, P. H., Shoemaker, C. A. and Evans, J. J. 2011. Modified live *Aeromonas hydrophila* vaccine for aquatic animals. US Patent No. 7, 988, 977 B2.

Kölbl, O. 1989. *Process for producing an avirulent cold blood virus*. World Intellectual Property Organization, WO/1989/005154.

LaFrentz, B. R., LaPatra, S. E., Call, D. R. and Cain, K. D. 2008. Isolation of rifampicin resistant *Flavobacterium psychrophilum* strains and their potential as live attenuated vaccine candidates. *Vaccine* 26: 5582-5589.

LaFrentz, B. R., Welch, T. J., Shoemaker, C. A., *et al.*, 2011. Modified live *Edwardsiella ictaluri* vaccine, AQUAVAC-ESC, lacks multidrug resistance plasmids. *J Aquat Anim Health* 23: 195-199.

Laing, K. J. and Hansen, J. D. 2011. Fish T-cells: recent advances through genomics. *Dev Comp Immunol* 35: 1282-1295.

Lawrence, M. L. and Banes, M. M. 2005. Tissue persistence and vaccine efficacy of an O polysaccharide mutant strain of *Edwardsiella ictaluri*. *J Aquat Anim Health* 17: 228-232.

Lawrence, M. L., Cooper, R. K. and Thune, R. L. 1997. Attenuation, persistence, and vaccine potential of an *Edwardsiella ictaluri pur*A mutant. *Infect Immun* 65: 4642-4651.

Lawrence, M. L., Banes, M. M. and Williams, M. L. 2001. Phenotype and virulence of a transposon derived lipopolysaccharide O side-chain mutant strain of*Edwardsiella ictaluri*. *J Aquat Anim Health* 13: 291-299.

Leung, K. Y., Wong, L. S., Low, K. W. and Sin, Y. M. 1997. Mini-Tn5 induced growth- and protease-deficient mutants of *Aeromonas hydrophila* as live vaccines for blue gourami, *Trichogaster trichopterus* (Pallas). *Aquaculture* 158: 11-22.

Linde, K., Beer, J. and Bondarenko, V. 1990. Stable *Salmonella* live vaccine strains with two or more attenuating mutations and any desired level of attenuation. *Vaccine* 8: 278-282.

Locke, J. B., Colvin, K. M., Datta, A. K., *et al.*, 2007. *Streptococcus iniae* capsule impairs phagocytic clearance and contributes to virulence in fish. *J Bacteriol* 189: 1279-1287.

Locke, J. B., Aziz, R. K., Vicknair, M. R., *et al.*, 2008. *Streptococcus iniae* M-like protein contributes to virulence in fish and is a target for live attenuated vaccine development. *PloS ONE* 3: e2824.

Locke, J. B., Vicknair, M. R., Ostland, V. E., *et al.*, 2010. Evaluation of *Streptococcus iniae* killed bacterin and live attenuated vaccines in hybrid striped bass through injection and bath immersion. *Dis Aquat Org* 89:

117-123.

Lorenzen, E., Brudeseth, B. E., Wiklund, T. and Lorenzen, N. 2010. Immersion exposure of rainbow trout (*Oncorhynchus mykiss*) fry to wildtype *Flavobacterium psychrophilum* induces no mortality, but protects against later intraperitoneal challenge. *Fish Shellfish Immunol* 28: 440-444.

Marsden, M. J., Vaughan, L. M., Fitzpatrick, R. M., *et al.*, 1998. Potency testing of a live genetically attenuated vaccine for salmonids. *Vaccine* 16: 1087-1094.

Marsden, M. J., Vaughan, L. M., Foster, T. J. and Secombes, C. J. 1996. A live (Δ*aro*A) *Aeromonas salmonicida* vaccine for furunculosis preferentially stimulates T-cell responses relative to B-cell responses in rainbow trout (*Oncorhynchus mykiss*). *Infect Immun* 64: 3863-3869.

Mast, J. and Goddeeris, B. M. 1999. Development of immunocompetence of broiler chickens. *Vet Immunol Immunopathol* 70: 245-256.

Milev-Milovanovic, I., Long, S., Wilson, M., *et al.*, 2006. Identification and expression analysis of interferon gamma genes in channel catfish. *Immunogenetics* 58: 70-80.

Norqvist, A., Hagström, A. and Wolf-Watz, H. 1989. Protection of rainbow trout against vibriosis and furunculosis by the use of attenuated strains of *Vibrio anguillarum*. *Appl Environ Microbiol* 55: 1400-1405.

Novoa, B., Romero, A., Mulero, V., *et al.*, 2006. Zebrafish (*Danio rerio*) as a model for the study of vaccination against viral hemorrhagic septicemia virus (VHSV). *Vaccine* 24: 5806-5816.

Nusbaum, K. E. and Morrison, E. E. 1996. Entry of 35 S-labeled *Edwardsiella ictaluri* into channel catfish (*Ictalutus punctatus*). *J Aquat Anim Health* 8: 146-149.

Panangala, V. S., Shoemaker, C. A., Klesius, P. H., *et al.*, 2009. Cross-protection elicited in channel catfish (*Ictalurus punctatus* Rafinesque) immunized with a low dose of virulent *Edwardsiella ictaluri* strains. *Aquacult Res* 40: 915-926.

Peatman, E., Terhune, J., Baoprasertkul, P., *et al.*, 2008. Microarray analysis of gene expression in the blue catfish liver reveals early activation of the MHC class I pathway after infection with *Edwardsiella ictaluri*. *Mol Immunol* 45: 553-566.

Perelberg, A., Ronen, A., Hutoran, M., *et al.*, 2005. Protection of cultured *Cyprinus carpio* against a lethal viral disease by an attenuated virus vaccine. *Vaccine* 23: 3396-3403.

Pridgeon, J. W. and Klesius, P. H. 2011a. Development and efficacy of a novobiocin-resistant *Streptococcus iniae* as novel vaccine in Nile tilapia (*Oreochromis niloticus*). *Vaccine* 29: 5986-5993.

Pridgeon, J. W. and Klesius, P. H. 2011b. Development of a novobiocin-resistant *Edwardsiella ictaluri* as a novel vaccine in channel catfish (*Ictalurus punctatus*). *Vaccine* 29: 5631-5637.

Pridgeon, J. W. and Klesius, P. H. 2011c. Development and efficacy of novobiocin and rifampicin resistant *Aeromonas hydrophila* as novel vaccines in channel catfish and Nile tilapia. *Vaccine* 29: 7896-7904.

Pridgeon, J. W., Yeh, H.-Y., Shoemaker, C. A., *et al.*, 2012. Global gene expression in channel catfish after vaccination with an attenuated *Edwardsiella ictaluri*. *Fish Shellfish Immunol* 32: 524-533.

Romero, A., Figueras, A., Thoulouze, M. I., *et al.*, 2008. Infectious hematopoietic necrosis recombinant viruses induced protection for rainbow trout (*Oncorhynchus mykiss*). *Dis Aquat Org* 80: 123-135.

Romero, A., Dios, S., Bremont, M., *et al.*, 2011. Interaction of the attenuated recombinant rIHNV-Gvhsv GFP virus with macrophages from rainbow trout (*Oncorhynchus mykiss*). *Vet Immunol Immunopathol* 140: 119-129.

Ronen, A., Perleberg, A., Abramowitz, J., *et al.*, 2003. Efficient vaccine against the virus causing a lethal disease in cultured *Cyprinus carpio*. *Vaccine* 21: 4677-4684.

Russo, R., Shoemaker, C. A., Panangala, V. S. and Klesius, P. H. 2009. *In vitro* and *in vivo* interaction of macrophages from vaccinated and non-vaccinated channel catfish (*Ictalurus punctatus*) to *Edwardsiella ictaluri*. *Fish Shellfish Immunol* 26: 543-552.

Sabin, A. B., Hennessen, W. A. and Winsser, J. 1954. Studies on variants of poliomyelitis virus I. Experimental segregation and properties of avirulent variants of three immunologic types. *J Exp Med* 99: 551-576.

Salonius, K., Siderakis, C., MacKinnon, A. M. and Griffiths, S. G. 2005. Use of *Arthrobacter davidanieli* as a live vaccine against *Renibacterium salmoninarum* and *Piscirickettsia salmonis* in salmonids. *Dev Biol* (*Basel*) 121: 189-197.

Santander, J., Mitra, A. and Curtiss, III,, R. 2011. Phenotype, virulence and immunogenicity of *Edwardsiella ictaluri* cyclic adenosine 3′, 5′-monophosphate receptor protein (Crp) mutants in catfish host. *Fish Shellfish Immunol* 31: 1142-1153.

Schnell, M. J., Mebatsion, T. and Conzelmann, K. K. 1994. Infectious rabies viruses from cloned cDNA. *EMBO J* 13: 4195-4203.

Schurig, G. G., Roop, II,, R. M., Bagchi, T., *et al.*, 1991. Biological properties of RB51; a stable rough strain of *Brucella abortus*. *Vet Microbiol* 28: 171-188.

Seder, R. A. and Hill, A. V. S. 2000. Vaccines against intracellular infections requiring cellular immunity. *Nature* 406: 793-798.

Shoemaker, C. A. and Klesius, P. H. 1997. Protective immunity against enteric septicemia in channel catfish, *Ictalurus punctatus* (Rafinesque), following controlled exposure to *Edwardsiella ictaluri*. *J Fish Dis* 20: 361-368.

Shoemaker, C. A., Klesius, P. H. and Bricker, J. M. 1999. Efficacy of a modified live *Edwardsiella ictaluri* vaccine in channel catfish as young as seven days post hatch. *Aquaculture* 176: 189-193.

Shoemaker, C. A., Klesius, P. H. and Evans, J. J. 2002. *In ovo* method for utilizing the modified live *Edwardsiella ictaluri* vaccine against enteric septicemia in channel catfish. *Aquaculture* 203: 221-227.

Shoemaker, C. A., Klesius, P. H. and Evans, J. J. 2007. Immunization of eyed channel catfish, *Ictalurus punctatus*, eggs with monovalent *Flavobacterium columnare* vaccine and bivalent *F. columnare* and *Edwardsiella ictaluri* vaccine. *Vaccine* 25: 1126-1131.

Shoemaker, C. A., Klesius, P. H., Evans, J. J. and Arias, C. R. 2009. Use of modified live vaccines in aquaculture. *J World Aquac Soc* 40: 573-585.

Shoemaker, C. A., Klesius, P. H., Drennan, J. D. and Evans, J. J. 2011. Efficacy of a modified live *Flavobacterium columnare* vaccine in fish. *Fish Shellfish Immunol* 30: 304-308.

Smith, B. P., Reina-Guerra, M., Stocker, B. A., *et al.*, 1984. Aromatic-dependent *Salmonella dublin* as a parenteral modified live vaccine for calves. *Am J Vet Res* 45: 2231-2235.

Sommerset, I., Krossøy, B., Biering, E. and Frost, P. 2005. Vaccines for fish in aquaculture. *Exp Rev Vaccines* 4: 89-101.

Soto, E., Wiles, J., Elzer, P., *et al.*, 2011. Attenuated *Francisella asiatica iglC* mutant induces protective immunity to francisellosis in tilapia. *Vaccine* 29: 593-598.

Sun, Y., Liu, C.-S. and Sun, L. 2010. Isolation and analysis of the vaccine potential of an attenuated *Edwardsiella tarda* strain. *Vaccine* 28: 6344-6350.

Sundell, K., Högfors-Rönnholm, E. and Wiklund, T. 2014. Vaccination against diseases caused by *Flavobacteriaceae* species, in *Fish Vaccination* (eds R. Gudding, A. Lillehaug and Ø. Evensen). Oxford, John Wiley & Sons Ltd.,

273-288.

Swain, P., Behera, T., Mohapatra, D., *et al.*, 2010. Derivation of rough attenuated variants from smooth virulent *Aeromonas hydrophila* and their immunogenicity in fish. *Vaccine* 28: 4626-4631.

Tan, Y. P., Lin, Q., Wang, X. H., *et al.*, 2002. Comparative proteomic analysis of extracellular proteins of *Edwardsiella tarda*. *Infect Immun* 70: 6475-6480.

Temprano, A., Riaño, J., Yugueros, J., *et al.*, 2005. Potential use of a *Yersinia ruckeri* O1 auxotrophic *aro*A mutant as a live attenuated vaccine. *J Fish Dis* 28: 419-427.

Thornton, J. C., Garduño, R. A. and Kay, W. W. 1994. The development of live vaccines for furunculosis lacking the A-layer and O-antigen of *Aeromonas salmonicida*. *J Fish Dis* 17: 195-204.

Thoulouze, M. I., Bouguyon, E., Carpentier, C. and Brémont, M. 2004. Essential role of the NV protein of *Novirhabdovirus* for pathogenicity in rainbow trout. *J Virol* 78: 4098-4107.

Thune, R. L., Collins, L. A. and Pena, M. P. 1997. A comparison of immersion, immersion/oral combination and injection methods for the vaccination of channel catfish *Ictalurus punctatus* against *Edwardsiella ictaluri*. *J World Aquac Soc* 28: 193-201.

Thune, R. L., Fernandez, D. H. and Battista, J. R. 1999. An *aro*A mutant of *Edwardsiella ictaluri* is safe and efficacious as a live, attenuated vaccine. *J Aquat Anim Health* 11: 358-372.

Tizard, I. 1999. Grease, anthraxgate, and kennel cough: a revisionist history of early veterinary vaccines. *Adv Vet Med* 41: 7-24.

Vaughan, L. M., Smith, P. R. and Foster, T. J. 1993. An aromatic-dependent mutant of the fish pathogen *Aeromonas salmonicida* is attenuated in fish and is effective as a live vaccine against the salmonid disease furunculosis. *Infect Immun* 61: 2172-2181.

Vestergård-Jørgensen, P. E. 1982. Egtved virus: temperature dependent immune response of trout to infection with low virulence virus. *J Fish Dis* 5: 47-55.

Vivas, J., Riaño, J., Carracedo, B., *et al.*, 2004. The auxotrophic *aroA* mutant of *Aeromonas hydrophila* as a live attenuated vaccine against *A. salmonicida* infections in rainbow trout (*Oncorhynchus mykiss*). *Fish Shellfish Immunol* 16: 193-206.

Wang, Y., Wang, Q., Yang, W., *et al.*, 2013. Functional characterization of *Edwardsiella tarda* twin-arginine translocation system and its potential use as a biological containment in live attenuated vaccine of marine fish. *Appl Microbiol Biotechnol* 97: 3545-3557.

Wehrli, W., Knusel, P., Schmid, K. and Staehelin, M. 1968. Interaction of rifamycin with bacterial RNA polymerase. *Proc Natl Acad Sci* 61: 667-673.

Wilcox, C. L., Crnic, L. S. and Pizer, L. I. 1992. Replication, latent infection, and reactivation in neuronal culture with a herpes simplex virus thymidine kinase-negative mutant. *Virology* 187: 348-352.

Wise, D. J. and Terhune, J. 2001. The relationship between vaccine dose and efficacy in channel catfish *Ictalurus punctatus* vaccinated as fry with a live attenuated strain of *Edwardsiella ictaluri* (RE-33). *J World Aquac Soc* 32: 177-183.

Wise, D. J., Klesius, P. H., Shoemaker, C. A. and Wolters, W. R. (2000) Vaccination of mixed and full-sib families of channel catfish *Ictalurus punctatus* against enteric septicemia of catfish with a live attenuated *Edwardsiella ictaluri* isolate (RE-33). *J World Aquac Soc* 31: 206-212.

Yu, L.-P., Hu, Y.-H., Sun, B.-G. and Sun, L. 2012. C312M: an attenuated *Vibrio anguillarum* strain that induces immunoprotection as an oral and immersion vaccine. *Dis Aquat Org* 102: 33-42.

Zhang, G. H. and Hanson, L. A. 1995. Deletion of thymidine kinase gene attenuates channel catfish herpesvirus while maintaining infectivity. *Virology* 209: 658-663.

Zhang, Y., Arias, C. R., Shoemaker, C. A. and Klesius, P. H. 2006. Comparison of lipopolysaccharide and protein profiles between *Flavobacterium columnare* strains from different genomovars. *J Fish Dis* 29: 657-663.

Zhu, K., Chi, Z., Li, J., *et al.*, 2006. The surface display of haemolysin from *Vibrio harveyi* on yeast cells and their potential applications as live vaccine in marine fish. *Vaccine* 24: 6046-6052.

第五章　DNA 疫苗

Eirik Biering and Kira Salonius

本章概要　SUMMARY

DNA 疫苗是指编码保护性抗原的一个或多个基因（不是抗原本身）的核酸疫苗。它是由病原的一个或多个基因构成，目的是使这些基因能够在体内以较短时间产生抗原蛋白，从而诱导机体产生免疫保护反应。DNA 疫苗通常以纯化的可复制的环形细菌质粒形式存在。DNA 疫苗的结构含有调控元件，既确保质粒在细菌宿主中的复制，又保证在接种动物体内基因能够表达。核酸疫苗或基因疫苗不仅指 DNA 疫苗，还包括 RNA 疫苗。

第一节　引　言

大多数水产养殖用疫苗是与佐剂配合使用，通过注射接种的灭活疫苗，主要通过诱导机体产生保护性抗体实现免疫保护。这对预防细菌性疾病是行之有效的（Sommerset 等，2005），因为疫苗诱导机体产生的中和抗体能有效地预防控制细菌的感染。对于病毒性疾病，还需要刺激机体产生细胞免疫才能获得足够的保护（Biering 等，2005）。事实上，目前水产养殖中使用的大多数病毒疫苗的效果并不理想（Gomez-Casado 等，2011），说明灭活疫苗对预防许多病毒性疾病并不适用。

DNA 疫苗是指编码保护性抗原的一个或多个基因（不是抗原本身）的核酸疫苗。它是由病原的一个或多个基因构成，目的是使这些基因能够在体内以较短时间产生抗原蛋白，从而诱导机体产生免疫保护反应。DNA 疫苗通常以纯化的可复制的环形细菌质粒形式存在。DNA 疫苗的结构含有调控元件，既确保质粒在细菌宿主中的复制，又保证在接种动物体内基因能够表达（图 5-1）。核酸疫苗或基因疫苗不仅指 DNA 疫苗，还包括 RNA 疫苗。

DNA 疫苗一般不含佐剂，通过肌肉注射接种。质粒转染细胞，主要转染肌细胞，也转染像树突状细胞之类的专职性抗原递呈细胞等（Liu，2011）。转染后的细胞表达质粒的抗原蛋白并将其递呈在 MHC-Ⅰ类分子上，这种方式与病毒感染相似，从而激活机体获得性免疫系统的抗体、辅助性 T 细胞和细胞毒性 T 淋巴细胞（CTLs）的三个免疫应答途径。关于鱼类

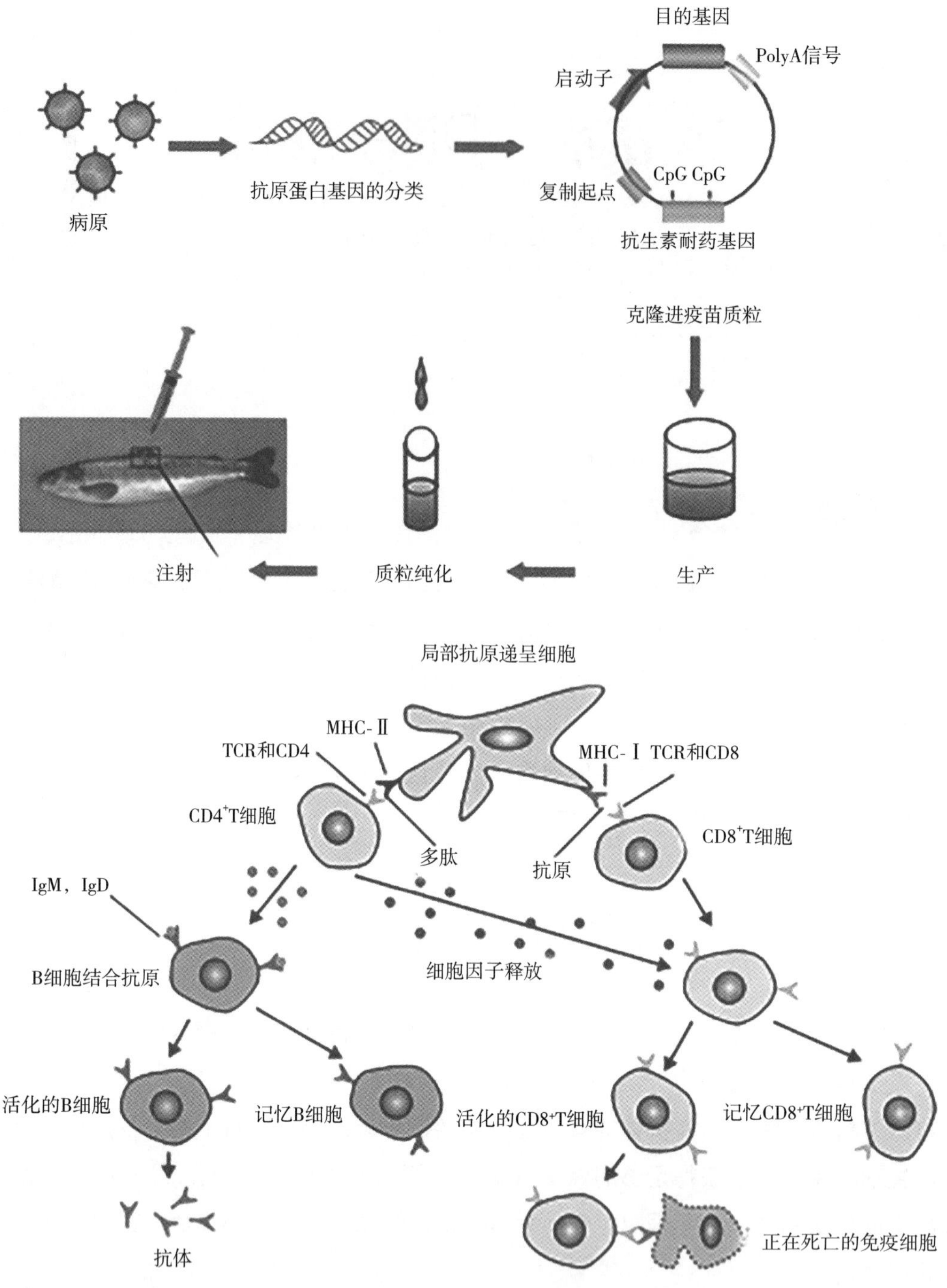

图 5-1　获得性免疫反应概述

DNA 疫苗被肌肉细胞捕获后降解或被肌细胞表达。转染的细胞产生并释放质粒编码的抗原，这些抗原被抗原递呈细胞（APCs）摄取。APCs 在摄取内源性和外源性抗原后，将 MHC-Ⅰ和 MHC-Ⅱ上的抗原肽分别递呈给 CD8+ 和 CD4+ 细胞。活化的幼稚 CD4+ 细胞将启动细胞和体液免疫反应，如活化 B 细胞分泌针对该抗原的特异性抗体，CD8+ T 细胞受体与 MHC-Ⅰ递呈的抗原肽相结合，在 Th1 细胞产生的细胞因子帮助下，启动细胞毒性 T 细胞反应，消除产生和表达病毒抗原的感染细胞。获得性免疫反应的特征是产生记忆细胞，当再次感染同一病原，记忆细胞可在短时间被激活

CD4，分化抗原簇 4 的共同受体；CD8，分化抗原簇 8 的共同受体；MHC-Ⅰ，主要组织相容性复合体Ⅰ类蛋白；MHC-Ⅱ，主要组织相容性复合体Ⅱ类蛋白；IgM，免疫球蛋白 M；IgD，免疫球蛋白 D；TCR，T 细胞受体；Th1，辅助性 T 细胞

（引自 Dufour，2001 年）

DNA 疫苗诱导的免疫应答的研究表明，免疫反应主要分为三个阶段（Kurath，2008；Tonheim 等，2008）：早期是非特异性的抗病毒反应，例如病毒性出血性败血症病毒（VHSV）疫苗也能抵御其他病毒的感染（Sommerset 等，2003）；之后是特异性的抗病毒应答，表现为特异

性抗体的产生和保护力增强；最后是长期的抗病毒应答，这个阶段保护力和抗体滴度虽然降低，但保护效果仍很显著。由于 DNA 疫苗能够诱导免疫系统细胞的免疫应答，能有效杀死胞内潜在的病毒；还因为 DNA 疫苗比灭活疫苗能在更大程度上模拟病毒感染，诱导产生更有效的保护性免疫反应。因此，与传统灭活疫苗相比，DNA 疫苗在防治已知病毒性疾病方面具有特殊的优势。

第二节 DNA 疫苗与传统灭活疫苗的优缺点对比

一、试验疫苗的研发

对于细菌性疾病，一旦病原被鉴定出来，疫苗的研发相对简单。致病菌可通过培养、灭活、与（或无）佐剂混合免疫鱼体，再进行攻毒实验验证疫苗的免疫效果。但这个过程对于研发病毒疫苗就较为困难。制备细胞培养的病毒灭活疫苗，必须有敏感的细胞系和最优的培养条件。如果制备重组亚单位疫苗，必须鉴定出保护性抗原，将其克隆到表达系统中，获得活性恢复的抗原，才能制成试验疫苗，通常需要尝试多个抗原和多种表达系统才能成功。相比之下，构建 DNA 疫苗要快得多，唯一的前提是要掌握病原基因组的详细信息，将选择的序列克隆到质粒中即可构建成试验性疫苗。如今，基因序列的数据很容易获得，尤其是病毒的序列数据，即使像鲑鱼虱等更复杂的生物序列数据也易获得（Skern-Mauritzen 等，2013）。DNA 疫苗的研发，与病毒细胞培养或重组表达相比，过程更加快捷，省去了在容器内培养病原的过程，还可同时构建多个不同基因和基因组合的试验性疫苗，甚至还可以很容易研发出多价疫苗（Einer-Jensen 等，2009），以及由保守基因构建的，能对相关病毒株具有交叉保护的 DNA 疫苗（Ulmer 等，1993）。事实上，采用 DNA 方法或者最近使用越来越多的 RNA 方法研发疫苗具有很多优势，在许多情况下这也是筛选抗原的标准方法。构建 DNA 疫苗较为容易，但最终获得有效的 DNA 疫苗较为困难。在大西洋鲑中，发现 DNA 疫苗在许多病毒性疾病中并未展现出高效的免疫保护作用，这是否与疫苗的结构设计、转染、蛋白产量或细胞转运有关，还有待进一步研究。

二、生产与储存

获得较理想的试验性疫苗后，接下来就是小规模生产试验。大多数鱼类细菌疫苗的生产相对简单，包括病原菌的培养、灭活、加工，与佐剂配伍或直接浸泡免疫。病毒疫苗的生产较为复杂，需要通过细胞培养病毒获得足量抗原或者是在异源表达系统中克隆，获得重组表达的保护性抗原。细胞培养生产病毒比较昂贵且复杂，尤其是出现新的病毒时，找到易感可进行病毒大规模生产的细胞系并非易事。异源性表达也有其不足，较难获得具有免疫活性的抗原，并且产量并不一定能满足生产需要，这些情况必须在生产过程中逐一优化。相比而言，尽管 DNA 疫苗产量会受到嵌入基因大小的影响，但质粒的生产是相对通用的，由大肠杆菌（*E. coli*）大规模生产质粒较容易实现。但纯化过程需要的缓冲液量与生产传统细菌抗原的发酵量相同，因而没有规模效益。由于尚未制定出 DNA 疫苗质粒质量和纯度的标准，制造商就必须证明其制定的质检标准对于接种动物是安全和有效的。质粒在各种条件下相对稳定，使其储存和运输更容易。相比之下，油类佐剂乳剂疫苗在说明书注明的储存条件下（如室温），可能会不稳定或被分解。

三、制剂与接种

除了活疫苗和浸泡用的细菌疫苗外，大多数的鱼类疫苗均与油类佐剂混合使用，以获得有效和持久的免疫保护，这类疫苗（通常是多价的）可预防多种细菌和病毒性疾病（Sommerset 等，2005）。这类疫苗采用腹腔注射免疫，在鲑养殖业中已实现注射接种自动化。Vaxfectin 等佐剂可以和 DNA 疫苗联合使用（Hartikka 等，2001；Sullivan 等，2010）。研究发现，在质粒中插入细胞因子、共刺激因子序列和靶向分子可增强免疫效果，还可提升鱼体对质粒的摄入和递呈（Liu，2011）。在水产养殖中，常以盐溶液“裸”DNA 疫苗进行肌肉注射方式接种。目前为止，通过腹腔注射 DNA 疫苗效果甚微，达不到肌肉注射免疫效果（Corbeil 等，2000；McLauchlan 等，2003）。因此，目前很难将 DNA 疫苗纳入鲑养殖业所使用的配套产品中。

第三节　兽用 DNA 疫苗

目前只有三个 DNA 疫苗获得许可（Kutzler 和 Weiner，2008；Liu，2011），且均为兽用疫苗，其中两个是针对病毒性疾病，即 West Nile-Innovator© 公司生产的马的西尼罗河病毒疫苗（Davis 等，2001）和 Apex-IHN© 公司生产的鲑传染性造血器官坏死病毒（IHNV）疫苗（Salonius 等，2007）。DNA 疫苗也可用于预防非传染性疾病，如癌症。第三个 DNA 疫苗是 Oncept™ 公司生产的狗的黑色素瘤疫苗（Bergman 等，2003）。DNA 疫苗也可用于接种像激素等活性分子，并非用于引起免疫反应。第四个已注册的 DNA-质粒产品——LifeTide© 公司的 SW5，它能提高仔猪成活率（Khan 等，2010）。截至目前，还没有任何 DNA 疫苗被许可用于人类。

开发鱼类核酸疫苗或 DNA 疫苗的初衷是希望通过简单、高度纯化的质粒在活体组织中转染，诱导机体产生细胞和体液免疫反应。因此，该技术最适用于病毒疫苗的开发。目前已开展研发弹状病毒 VHSV 和 IHNV 的 DNA 疫苗（Anderson 等，1996；Lorenzen 等，1998；Corbeil 等，1999）。疾病种类和基于糖蛋白构建的疫苗的效果会随着剂量（Garver 等，2005）、早期的先天性免疫反应（Kim 等，2000）、跨病毒物种的异质性保护（Lorenzen 等，1998）等效果的不同而改变，最终影响鱼类 DNA 疫苗的效果（Salonius 等，2007）。在非鲑鳟鱼类中，已有鲤春病毒血症的 DNA 疫苗的研发研究，免疫后鲤获得显著的免疫保护（RPS 为 48%）（Kanellos 等，2006）；免疫观赏鱼锦鲤也获得显著的保护效果，免疫 28d 后的攻毒实验结果表明，RPS 为 50%～80%（Emmenegger 和 Kurath，2008）。

除了弹状病毒科，DNA 疫苗在其他鱼类疾病模型中的功效并不如意。这可能与 DNA 疫苗接种后鱼体细胞中蛋白的组装、递呈或运输的效率和形式有关。有研究表明，表达传染性鲑贫血病毒（ISAV）血凝集素-酯酶的 DNA 疫苗在大西洋鲑中的 RPS 达 39.5%～60.5%（Mikalsen

等，2005)。编码传染性胰腺坏死病毒（IPNV）多肽蛋白开放阅读框的质粒疫苗显示出对大西洋鲑的良好免疫保护，RPS 为 84%，而对照组死亡率为 34%（Mikalsen 等，2004），但未见关于 IPNV 和 ISAV DNA 疫苗的进一步研究报道。在欧洲，鉴于 DNA 疫苗的备案要求以及接受度的不确定性，行业只对优于传统疫苗的 DNA 疫苗，以及对动物福利有严重影响又没有有效疫苗的疾病研发出的 DNA 疫苗进行备案登记。最近，有两篇关于鲑甲病毒亚型Ⅲ质粒疫苗的效果评估报告。Xu 和同事们在 2012 年比较研究了 E1 和 E2 刺突蛋白的质粒 DNA 疫苗，E1 和 E2 亚单位疫苗与传统细胞培养灭活 SAV 疫苗的效果。结果表明，大西洋鲑免疫 4 周后攻毒，与空质粒组相比，注射灭活疫苗组病毒载量的相对表达量减至 1/49。采取加强免疫的接种，初次免疫后的第 6 周再次免疫，仅加强免疫一次，即先接种 20μg E1 DNA 疫苗再接种 50μg E1 亚单位疫苗，病毒载量显著减少（85%）。根据胰腺和心脏组织坏死程度的组织病理，以及 R-qPCR检测的病毒 RNA 载量，对 SAV 核酸疫苗进行了评价（Simard 等，2009）。大西洋鲑接种疫苗后的第 10 周和第 28 周的保护水平显示，10μg 和 20μg 的免疫剂量能够产生较强和持久的保护反应。大西洋鲑接种 DNA 疫苗与接种商用油类佐剂型灭活 SAV 疫苗相比，体内病毒载量和发病情况显示，在接种 SAV 核酸疫苗（NAV）的第 10 周和第 28 周内均有较明显的效果。在北欧，尽管使用了其他 SAV 疫苗，鉴于鲑胰腺疾病仍造成严重的经济损失并危及动物福利（Gomez-Casado 等，2011），SAV DNA 疫苗有望成为第一个在欧洲水产养殖中被授权的核酸疫苗。

第四节 生物安全与监管

安全性是所有药物产品获得授权许可的核心问题，对于像鱼类与环境密切相关生产食品型动物，疫苗安全问题包括：①对动物的安全性；②对环境的安全性；③对用户和消费者的安全性。Holm（2004）全面论述了生产食品型动物 DNA 疫苗免疫的安全性；Lorenzen 和 LaPatra（2005）论述了用于鱼类 DNA 疫苗免疫的安全问题。在许多方面，DNA 疫苗比同类的疫苗技术更安全。由于不使用油类佐剂，避免了传统疫苗接种引起的腹膜炎问题（Lillehaug 等，1992；Midtlyng 等，1996；Garver 等，2005；Kurath 等，2006）。研究表明，鲑接种油类佐剂疫苗后产生的严重副作用，可能是受免鱼体发生了系统性自身免疫（Koppang 等，2008；Haugarvoll 等，2010），这对鱼的健康和福利将会产生负面影响。DNA 疫苗不存在活疫苗毒性恢复的问题，也没有灭活疫苗不完全灭活的风险，正如 Holm（2004）所述，接种疫苗会产生免疫耐受和自身免疫的问题，但与传统疫苗产生的副作用相比，DNA 疫苗的风险相对较小。DNA 疫苗主要的安全问题是存在疫苗组分整合到接种者的基因组中，以及随疫苗扩散到环境、鱼类后代或消费者体内的风险。尽管电穿孔传递质粒 DNA 后可检测到染色体整合（Wang 等，2004），但现有数据表明这种情况非常罕见（Holm，2004；Tonheim 等，2008），鲑肌肉注射 DNA 疫苗后的整合基因概率是自发突变的 1/43（Salonius 等，2007）。由此可见，疫苗组分整合到接种者染色体中概率很低，对消费者的风险可以忽略不计。如果 DNA 疫苗是需要整合的设计，那么风险肯定会增加，但风险评估总是会针对特定质粒的具体特性进行全面的评估。值得关注的是，在任何情况下，鱼肉中的残余疫苗 DNA 的含量与食物中的核酸总量相比都是微不足道的。

对食品中DNA的安全性分析发现，与未修饰的核酸造成的风险相比，食品中重组DNA不会带来新的风险（Jonas等，2001）。

在本书的其他章节已介绍了DNA疫苗的监管，在这里将简要讨论欧洲转基因生物（GMOs）立法的一些相关问题。在欧洲，与转基因生物相关的监管框架（指令2001/18/EC）是否适用于接种DNA疫苗的动物仍存在一些不确定性。DNA疫苗仅由遗传物质组成，并不会被认为是一种有机体（指令条款2的定义），因为DNA疫苗本身无法将遗传物质复制或转移到鱼的基因组中。根据第2001/18/EC指令，转基因产品必须征询民意并标注转基因生物标签，还要求转基因产品需申请上市许可、进行环境风险评估和售后监督。在欧洲，接种DNA疫苗鱼类的现状仍处于评估状态。在欧洲，GMO的许可程序是相当复杂的，堪比疫苗许可的类似程序。然而，由于获得疫苗许可证的责任方是药品生产者，如果接种DNA疫苗的鱼被认定为转基因鱼，目前尚不清楚谁是申请销售许可的责任者。此外，由于2001/18/EEC指令适用于具有稳定和明确的基因修饰的转基因生物，须对其进行风险评估和监管，但如何对接种DNA疫苗的鱼类进行评估和监管还不十分明确。目前，评估或鉴定的基因修饰还没有明确，只有疫苗整合到接种者基因组的风险评估。然而，关于转基因食品和饲料1829/2003条例第16条规定：用转基因饲料饲喂的动物或用转基因药品处理的动物来源的产品既不受本规定中授权的要求，也不受标签的限制。

基于此，如果DNA疫苗属于转基因药品，那么鱼就不属于转基因生物。Foss和Rogne在两篇论文（Foss和Rogne，2003，2007）中详细地讨论了此问题和可能的解决方案。显然，只要这些监管问题得不到解决，疫苗生产商就不会将DNA疫苗引入欧洲市场。

总之，DNA疫苗免疫在某些传统技术不能提供有效产品的情况下，是一种有效的方法。随着研发深入，DNA疫苗将会有新用途和新发现，但在水产养殖中，DNA疫苗不可能是开发疫苗产品的通用解决方案，最有可能成为传统疫苗的有力补充产品。疫苗的应用使鲑养殖发展迅速，欧洲是世界鲑的主产区和主要市场，又由于DNA疫苗技术对鱼类的潜在价值与作用，因而有必要在欧盟内确立DNA疫苗的监管规范。

参考文献

Anderson，E. D.，Mourich，D. V.，Fahrenkrug，S. C.，*et al.*，1996. Genetic immunization of rainbow trout（*Oncorhynchus mykiss*）against infectious hematopoietic necrosis virus. *Mol Mar Biol Biotechnol* 5：114-122.

Bergman，P. J.，McKnight，J.，Novosad，A.，*et al.*，2003. Long-term survival of dogs with advanced malignant melanoma after DNA vaccination with xenogeneic human tyrosinase：a phase I trial. *Clin Cancer Res* 9：1284-1290.

Biering，E.，Villoing，S.，Sommerset，I. and Christie，K. E. 2005. Update on viral vaccines for fish. *Dev Biol*（*Basel*）121：97-113.

Corbeil，S.，LaPatra，S. E.，Anderson，E. D.，*et al.*，1999. Evaluation of the protective immunogenicity of the N，P，M，NV and G proteins of infectious hematopoietic necrosis virus in rainbow trout *Oncorhynchus mykiss* using DNA vaccines. *Dis Aquat Org* 39：29-36.

Corbeil，S.，Kurath，G. and LaPatra，S. E. 2000. Fish DNA vaccine against infectious hematopoietic necrosis virus：

efficacy of various routes of immunisation. *Fish Shellfish Immunol* 10：711-723.

Davis，B. S.，Chang，G. J.，Cropp，B.，*et al.*，2001. West Nile virus recombinant DNA vaccine protects mouse and horse from virus challenge and expresses *in vitro* a noninfectious recombinant antigen that can be used in enzyme-linked immunosorbent assays. *J Virol* 75：4040-4047.

Einer-Jensen，K.，Delgado，L.，Lorenzen，E.，*et al.*，2009. Dual DNA vaccination of rainbow trout（*Oncorhynchus mykiss*）against two different rhabdoviruses，VHSV and IHNV，induces specific divalent protection. *Vaccine* 27：1248-1253.

Emmenegger，E. J. and Kurath，G. 2008. DNA vaccine protects ornamental koi（*Cyprinus carpio koi*）against North American spring viremia of carp virus. *Vaccine* 26：6415-6421.

Foss，G. S. and Rogne，S. 2003. Gene medication or genetic modification? The devil is in the details. *Nat Biotechnol* 21：1280-1281.

Foss，G. S. and Rogne，S. 2007. When gene medication is also genetic modification-regulating DNA treatment. *Vaccine* 25：5613-5618.

Garver，K. A.，Conway，C. M.，Elliott，D. G. and Kurath，G. 2005. Analysis of DNA-vaccinated fish reveals viral antigen in muscle，kidney and thymus，and transient histopathologic changes. *Mar Biotechnol*（*NY*）7：540-553.

Gomez-Casado，E.，Estepa，A. and Coll，J. M. 2011. A comparative review on European-farmed finfish RNA viruses and their vaccines. *Vaccine* 29：2657-2671.

Hartikka，J.，Bozoukova，V.，Ferrari，M.，*et al.*，2001. Vaxfectin enhances the humoral immune response to plasmid DNA-encoded antigens. *Vaccine* 19：1911-1923.

Haugarvoll，E.，Bjerkas，I.，Szabo，N. J.，*et al.*，2010. Manifestations of systemic autoimmunity in vaccinated salmon. *Vaccine* 28：4961-4969.

Holm A. 2004. *DNA-vaccines for food-producing animals. Technical review and discussion of safety issues.* Ministry of Food，Agriculture and Fisheries and the Danish Veterinary Institute，Copenhagen.

Jonas，D. A.，Elmadfa，I.，Engel，K. H.，*et al.*，2001. Safety considerations of DNA in food. *Ann Nutr Metab* 45：235-254.

Kanellos，T.，Sylvester，I. D.，D'Mello，F.，*et al.*，2006. DNA vaccination can protect *Cyprinus carpio* against spring viraemia of carp virus. *Vaccine* 24：4927-4933.

Khan，A. S.，Bodles-Brakhop，A. M.，Fiorotto，M. L. and Draghia-Akli，R. 2010. Effects of maternal plasmid GHRH treatment on offspring growth. *Vaccine* 28：1905-1910.

Kim，C. H.，Johnson，M. C.，Drennan，J. D.，*et al.*，2000. DNA vaccines encoding viral glycoproteins induce nonspecific immunity and Mx protein synthesis in fish. *J Virol* 74：7048-7054.

Koppang，E. O.，Bjerkås，I.，Haugarvoll，E.，*et al.*，2008. Vaccination-induced systemic autoimmunity in farmed Atlantic salmon. *J Immunol* 181：4807-4814.

Kurath，G. 2008. Biotechnology and DNA vaccines for aquatic animals. *Rev Sci Tech* 27：175-196.

Kurath，G.，Garver，K. A.，Corbeil，S.，*et al.*，2006. Protective immunity and lack of histopathological damage two years after DNA vaccination against infectious hematopoietic necrosis virus in trout. *Vaccine* 24：345-354.

Kutzler，M. A. and Weiner，D. B. 2008. DNA vaccines：ready for prime time? *Nat Rev Genet* 9：776-788.

Lillehaug，A.，Lunder，T. and Poppe，T. T. 1992. Field testing of adjuvanted furunculosis vaccines in Atlantic salmon，*Salmo salar* L. *J Fish Dis* 15：485-496.

Liu，M. A. 2011. DNA vaccines：an historical perspective and view to the future. *Immunol Rev* 239：62-84.

Lorenzen，N. and LaPatra，S. E. 2005. DNA vaccines for aquacultured fish. *Rev Sci Tech* 24：201-213.

Lorenzen, N., Lorenzen, E., Einer-Jensen, K., *et al.*, 1998. Protective immunity to VHS in rainbow trout (*Oncorhynchus mykiss*, Walbaum) following DNA vaccination. *Fish Shellfish Immunol* 8: 261-270.

McLauchlan, P. E., Collet, B., Ingerslev, E., *et al.*, 2003. DNA vaccination against viral haemorrhagic septicaemia (VHS) in rainbow trout: size, dose, route of injection and duration of protection-early protection correlates with Mx expression. *Fish Shellfish Immunol* 15: 39-50.

Midtlyng, P. J., Reitan, L. J. and Speilberg, L. 1996. Experimental studies on the efficacy and side-effects of intraperitoneal vaccination of Atlantic salmon (*Salmo salar* L.) against furunculosis. *Fish Shellfish Immunol* 6: 335-350.

Mikalsen, A. B., Torgersen, J., Aleström, P., *et al.*, 2004. Protection of Atlantic salmon *Salmo salar* against infectious pancreatic necrosis after DNA vaccination. *Dis Aquat Org* 60: 11-20.

Mikalsen, A. B., Sindre, H., Torgersen, J. and Rimstad, E. 2005. Protective effects of a DNA vaccine expressing the infectious salmon anemia virus hemagglutinin-esterase in Atlantic salmon. *Vaccine* 23: 4895-4905.

Salonius, K., Simard, N., Harland, R. and Ulmer, J. B. 2007. The road to licensure of a DNA vaccine. *Curr Opin Investig Drugs* 8: 635-641.

Simard, N. C., Hay, V., Aarflot, L., *et al.*, 2009. Evaluation of a DNA vaccine prototype for the control of salmon pancreas disease virus (SPDV) in Atlantic salmon, *Salmo salar*. Oral presentation and abstract, 14th EAFP International Conference, Prague.

Skern-Mauritzen R, Malde K, Besnier F, *et al.*, 2013. How does sequence variability affect *de novo* assembly quality? *J Nat Hist* 47: 901-910.

Sommerset, I., Krossoy, B., Biering, E. and Frost, P. 2005. Vaccines for fish in aquaculture. *Exp Rev Vaccines* 4: 89-101.

Sommerset, I., Lorenzen, E., Lorenzen, N., *et al.*, 2003. A DNA vaccine directed against a rainbow trout rhabdovirus induces early protection against a nodavirus challenge in turbot. *Vaccine* 21: 4661-4667.

Sullivan, S. M., Doukas, J., Hartikka, J., *et al.*, 2010. Vaxfectin: a versatile adjuvant for plasmid DNA- and protein-based vaccines. *Exp Opin Drug Deliv* 7: 1433-1446.

Tonheim, T. C., Bøgwald, J. and Dalmo, R. A. 2008. What happens to the DNA vaccine in fish? A review of current knowledge. *Fish Shellfish Immunol* 25: 1-18.

Ulmer, J. B., Donnelly, J. J., Parker, S. E., *et al.*, 1993. Heterologous protection against influenza by injection of DNA encoding a viral protein. *Science* 259: 1745-1749.

Wang, Z., Troilo, P. J., Wang, X., *et al.*, 2004. Detection of integration of plasmid DNA into host genomic DNA following intramuscular injection and electroporation. *Gene Ther* 11: 711-721.

Xu, C., Mutoloki, S. and Evensen, Ø. 2012. Superior protection conferred by inactivated whole virus vaccine over subunit and DNA vaccines against salmonid alphavirus infection in Atlantic salmon (*Salmo salar* L.). *Vaccine* 30: 3918-3928.

第六章　鱼类黏膜疫苗接种

Jan H. W. M. Rombout and Viswanath Kiron

本章概要　SUMMARY

疾病的预防对水产养殖业的可持续发展至关重要，疫苗接种似乎已成为预防鱼类疾病最适宜的方法。虽然商业化疫苗的数量一直在增长，但大多数疫苗必须采用注射方式免疫。注射或许是最有效的免疫方式，但劳动强度大、成本高，且不适用于小鱼。由于大多数病原体是通过黏膜表面进入鱼体，因此黏膜相关的免疫保护也十分必要。此外，通过饲料口服免疫可以解决注射免疫的某些弊端，特别是若低成本生产疫苗，疫苗用量相当于或者低于浸泡免疫，则口服免疫更具优势。自 1942 年 Duff 首次报道口服免疫至今，很多研究都表明了黏膜疫苗的前景。然而，在大多数情况下，黏膜疫苗尚未进入商业化生产。本章介绍黏膜疫苗接种方法，着重疫苗的摄入、免疫激活等，探讨其作为未来疫苗的可能。

第一节　引　言

通过疫苗免疫预防疾病是控制病原传播最合适的方法，既经济、环保又符合伦理道德。然而，仍需要投入大量研究提升现有的疫苗并研发针对病毒性和细菌性疾病的新疫苗，尤其是还应重点关注疫苗的有效性、安全性以及免疫接种途径。同时，新型疫苗接种方法必须节约成本，易于操作，对动物的刺激要尽可能小。

在新方法中，黏膜疫苗接种似乎符合所有预期标准。本章介绍关于此类疫苗接种最新的研究进展，以及该方法尚未广泛使用的原因。内容包括黏膜免疫系统、适宜黏膜的抗原和佐剂，以及摄入机制。最后介绍几种重要养殖鱼类的黏膜系统在结构和功能上的差异。

第二节　黏膜疫苗接种历程

最早记载黏膜疫苗免疫的是 Duff（1942）在《免疫学杂志》上发表的文章，研究表明，鳟持续投喂（大于 64d）含杀鲑气单胞菌（*Aeromonas salmonicida*）的饲料后，浸泡感染，死亡率由对照组的75%下降到口服免疫组的25%，保护效果与抗体产生呈现出对应关系。此后，由于普遍采用药物防治鱼类疾病，很少关注鱼类疫苗免疫。1976年，鲁克氏耶尔森菌（*Yersinia ruckeri*）口服疫苗在美国第一个获得许可，不久后，鳗弧菌/奥达弧菌（*Vibrio anguillarum/ordalii*）浸泡免疫疫苗获得许可（Plant和LaPatra，2011）。然而，之后研发出的大多数疫苗均是采用腹腔注射（IP）和浸泡/浸浴免疫。浸泡方式被认为是黏膜免疫接种，因为抗原通过皮肤、鳃和肠等器官的黏膜上皮细胞摄入，可能引起上述器官局部的免疫反应。大多数研究只监测免疫后的死亡率，很少关注抗原摄取机制和引发的局部或系统免疫反应。更详细的研究历程可参读Van Muiswinkel（2008）以及Plant和LaPatra（2011）的文献。表6-1列出了1977—1983年部分文献，比较了注射免疫和黏膜免疫后存活率或死亡率的研究结果。

表6-1 大西洋鲑或虹鳟注射、浸泡、口服、灌肠接种细菌疫苗后攻毒死亡率比较（1977—1983年）

文献/病原	免疫接种方式				
	对照*	注射	浸泡	口服	灌肠
鳗弧菌（*Vibrio anguillarum*）					
Håstein等（1977）	46	19	42	76	—
Baudin-Laurencin和Tangtrongpiros（1980）	34	1.5	2	32	—
Evelyn和Ketchetson（1980）	42	1	2	18	—
Amend和Johnson（1981）	52	0	4	27	—
Horne等（1982）	100	7	53	94	—
Johnson和Amend（1983b）	55	—	—	15～45	0
鲁克氏耶尔森菌（*Yersinia ruckeri*）					
Johnson和Amend（1983a，1983b）	65	—	18	24	3

注：*鱼未被免疫，作为攻毒对照。

表6-1中死亡率数据表明，口服免疫鳗弧菌疫苗和鲁克氏耶尔森菌疫苗的鲑科鱼可获得免疫保护，但注射和浸泡免疫通常效果更好。然而，Johnson和Amend（1983a，1983b）的研究表明，如果抗原不被降解进入后肠，则肠道疫苗接种会很有希望。因此，必须以某种方式保护抗原，使它们在多数鱼类的前肠中不被消化（Rombout等，2011），在后肠中可以被高效地摄入和递呈处理（Rombout等，1985；Rombout和van den Berg，1989）。尽管有人声称口服（生物）胶囊疫苗可以起到保护作用，但并没有商业应用成功的报道，原因可能是其经济效益较注射或者浸泡免疫低。

第三节 鱼类黏膜免疫与系统免疫

在哺乳动物上已证明，黏膜表面暴露于抗原可引起局部反应，最终产生特异性抗体IgA分泌于黏膜表面。哺乳动物具有共同黏膜免疫系统，即对一个上皮细胞的刺激可导致其他黏膜器官产生特异性IgA反应（Brandtzaeg和Pabst，2004）。这种现象是基于系统免疫和黏膜免疫细胞上存在所谓的归巢受体。在黏膜器官中，免疫细胞一旦被致敏（归巢受体），将会通过识别内皮静脉窦的上皮细胞的黏膜“地址分子”（归巢受体的特异性配体），特异性地返回至黏膜组织中。通过这种方式，机体表面可以受到产生于另一处黏膜表面的特异性抗体保护。一般而言，黏膜免疫接种是一种有效的保护机体的方式，能够阻止病原进入机体。但必须说明的是，哺乳动物黏膜疫苗的数量非常有限，部分原因是在医疗和兽医领域用注射疫苗更划算。而且，系统免疫反应的亲和力和记忆免疫均优于黏膜免疫反应。此外，几乎所有的疫苗都含佐剂以减少抗原的用量，而目前有多种多样的系统免疫佐剂均比黏膜免疫佐剂效果好。

与哺乳动物常规做法不同，黏膜疫苗在鱼类中可能更适用，因为与注射免疫相比，黏膜疫苗可以大规模使用，劳动强度较低。此外，黏膜免疫相对无应激，可用于个体太小而无法注射免疫的幼鱼。目前，浸泡免疫是黏膜疫苗接种最广泛使用的方法，但如表 6-1 所示，浸泡免疫效果不如注射免疫。浸泡时，抗原能进入皮肤、鳃，甚至肠道，但尚未阐明抗原进入机体的机制。黏膜疫苗研发是经验性的，以鱼的死亡率判断效果，对黏膜疫苗诱导机体免疫反应的类型，或使机体获得有效保护的免疫反应知之甚少。在鱼类中，必须小心地对单一黏膜表面刺激，因为鱼类的黏膜免疫系统的功能可能与哺乳动物不同。研究表明，鲤黏膜中的 B 细胞和 T 细胞（Rombout，未发表）或虹鳟肠道黏膜上皮的 T 细胞（Bernard 等，2006）上未发现特异性归巢受体。另一方面，鲤经口服或灌肠免疫后在皮肤黏液中检测到抗体，但香鱼（Kawai 等，1981）和鲤（Rombout 等，1986）在口服免疫后，血清中均检测不出抗体。目前还无法给出定论，因为研究发现，鲤的黏液抗体与血清抗体是不同的（Rombout 等，1993）。最近在鲤（IgZ；Savan 等，2005a）、虹鳟（IgT；Hansen 等，2005）和河鲀（IgT；Savan 等，2005b）中报道了一种新型的免疫球蛋白 H 链，这种 IgT 样分子可能参与鳟的黏膜免疫（Zhang 等，2010）。鲤中可能部分保留了该功能，研究表明，在鲤黏膜组织中同时有 μ 和 ζ 域分子的嵌合体形式（Ryo 等，2010），这解释了为何抗鲤血清 IgM 的单克隆抗体（mAb：WCI12）可以检测鲤黏膜免疫反应，而抗鳟血清 IgM 单抗不能检测黏膜免疫反应（Zhang 等，2010）。由于存在物种差异，对鱼类作出通用结论应当谨慎。多聚免疫球蛋白受体（pIgR）是一种免疫球蛋白结合分子，通过转胞吞作用介导免疫球蛋白向黏液转运。在许多鱼类中都发现了 pIgR（Rombout 等，2011），鳟 pIgR 可能与 IgT 结合（Zhang 等，2010）。在肠上皮细胞或肝细胞等极性细胞中，IgM 和 IgT 或 IgZ 可转运到黏液中，但推测皮肤上皮细胞不具备这个功能。免疫球蛋白与皮肤上皮细胞结合即使不转运对皮肤也有免疫保护。尚未发表的关于鲤 pIgR 的研究表明，在可能是器官依赖性的分子裂解区存在一个氨基酸差异。可推测，这种差异意味着 pIgR 不同的转运途径。由于对鱼类黏膜免疫所知有限，应该投入更多的研究增加对鱼类，特别是对主要养殖鱼类黏膜免疫的认知，以期研发出激活黏膜和系统免疫的经济有效的疫苗。

第四节　浸泡免疫

根据 Plant 和 LaPatra（2011）关于鱼类疫苗接种方式的归纳，浸泡可能是最简单的疫苗接种方法，但并非适合所有的养殖情况。各种浸泡免疫方法具体如下：

（1）直接浸泡：将鱼放入含疫苗的水里较短时间（约 30s），然后再放回养殖池（Lillehaug，2014）。这种方法不需要任何预处理，根据其保护效果，目前是使用最广泛的方法。不过，表 6-1 的数据表明，直接浸泡获得的保护效果不如注射免疫。

（2）高渗浸泡：先将鱼在高浓度氯化物溶液中浸泡几分钟，再到疫苗溶液中浸泡接种。这种方法对鱼体有刺激应激，但有研究发现鱼体只是短期出现了皮质醇水平升高，皮肤和鳃也只是一过性的损伤（Huising 等，2003），这种暂时性损伤可能有佐剂功能，因为可使皮肤和鳃更好地吸收抗原，增强免疫反应；但也有研究得出相反的结果（Plant 和 LaPatra，2011）。在鲤中，高渗浸泡免疫后第 6～8 周才引起皮肤黏膜免疫反应，而在免疫后的第 4 周就已检测到系统免疫反应（Huising 等，2003）。这可能是由于 pIgR 结合黏液 Ig 转运到皮肤上皮细胞的时间较长。

（3）喷雾接种：最初是针对较大的鱼（>20g）接种鳗弧菌疫苗而研发的，考虑到对鱼体刺激较强（Ellis，1988），之后很少使用。

（4）超声波浸泡法：是一种较新的方法，应用高频声波（20kHz）增强细胞的渗透性。已有超声波浸泡法与腹腔注射法在抗原摄入和保护水平方面的对比研究（Plant 和 LaPatra，2011）。该方法已应用于虹鳟病毒性出血性败血症病毒（VHSV）DNA 疫苗接种，其相对存活率达到 50%（Fernandez-Alonso 等，2001）。这种方法较正常的浸泡免疫节省 80% 抗原用量，但对于细菌性抗原或质粒的接种则用量非常大，这可能限制了其在水产养殖中的广泛应用。

（5）穿刺浸泡法：在浸泡免疫之前，使用多点穿刺工具在鱼体表造成小的创伤。使用这种方法对虹鳟接种海豚链球菌（*Streptococcus iniae*）菌苗，获得的免疫保护与腹腔注射达到相同的水平（Nakanishi 等，2002）。

实际上，高渗浸泡效果与超声波浸泡和多点穿刺浸泡效果相当。一般而言，皮肤和鳃损伤破坏了机体的保护屏障，从而能更好地传递抗原，增强免疫反应和增加免疫保护。其他的接种技术对正常的水产养殖并不太适用，主要原因是引起急性应激、需要反复接种和增加设备的成本。然而，如同肠灌注递呈抗原一样，浸泡免疫可诱发局部或黏膜免疫反应，其优势长期被忽略。

第五节 口服免疫

口服疫苗接种操作容易，没有接种应激，适用于所有规格鱼的大规模接种，理论上是最适宜的免疫方式（Vandenberg，2004；Rombout 等，2011；Plant 和 LaPatra，2011）。然而，它也有缺点，例如，每个动物摄取的疫苗量不确定，疫苗到达可递呈抗原的后肠时，未被降解的疫苗量不确定（Rombout 等，2011）。虽然早就有口服疫苗的研究报道（Duff，1942），但一直以来口服疫苗的效果并不一致（Plant 和 LaPatra，2011），部分原因可能是上述的局限性。不同养殖鱼类消化道的差异会影响疫苗的效果，有胃和无胃、消化道的长短，甚至有的鱼类无中肠（Inami 等，2009）等，这些造成难以显著提升口服疫苗的免疫效果。下面归纳了口服疫苗接种的方法，每种方法其目的都是能安全地递送抗原至后肠，诱导局部或系统免疫反应，最重要的是诱导免疫记忆以及实现长期的免疫保护。

一、抗原未包被的口服免疫

研究表明，抗原不经包被的疫苗口服免疫比注射或浸泡免疫产生的免疫保护水平要低得多（表6-1），即使每日投喂持续更长时间，效果也如此。例如，在首次口服疫苗研究（Duff，1942）中，鳟饲喂疫苗64～70d，死亡率才从75%降至25%。延长投喂时间，就会需要更多疫苗，这会增加成本。尽管这样，研究人员仍进行了疫苗反复饲喂摄入实验，一方面是要保证所有的动物都能摄取足够的抗原；另一方面，在哺乳动物中发现蛋白质抗原的反复服用可以诱导免疫抑制和/或耐受性（Brandtzaeg和Pabst，2004）。据报道，鲤及其他鱼类有类同的基因依赖免疫抑制现象（Joosten等，1997a）。若要提高这种方法的效率，就必须减少抗原在胃和前肠中的降解，且使饲喂频率降至最低。

二、生物包被

许多鱼类最初都是用如轮虫、卤虫和水蚤等活饵料饲喂。这些活体饵料是滤食性生物，可以摄取像细菌一样的颗粒抗原，如果能将抗原带到鱼类消化道后部释放，它们可以作为疫苗的生物包被。在香鱼（Kawai等，1989）、鲤和海鲷（Joosten等，1995）的早期尝试表明，生物包被效果良好，这是由于幼鱼产生了免疫记忆，应用该方法免疫斑马鱼，也获得了良好的保护效果（Lin等，2005）。另外，还有生物包被协调效应的报道，在接种（投喂）卤虫包被的鳗弧菌疫苗后，欧洲鲈生长加快，饲料转化率提高，并且机体应激反应较低（Chair等，1994）。这种方法只有当鱼体免疫系统发育完善时才能应用。过早接种疫苗会导致鱼在自我/非自我识别阶段，无法识别抗原，从而减弱免疫应答，在受精后1～2个月的鲤中发现过这种情况（Joosten等，1995）。这一现象在哺乳动物中被称为新生儿免疫耐受。

另一种生物包被的方式是利用微藻。例如，莱茵衣藻（*Chlamydomonas reinhardtii*）被用作鱼类抗原传递系统（Siripornadulsil等，2007）。应用莱茵衣藻表达鲑肾杆菌（*Renibacterium salmoninarum*）蛋白57，通过浸泡或饲喂该藻实施免疫。虽然这两种方式都诱导了特异的抗体应答，但只在浸泡免疫鱼的皮肤黏液中检测到特异性抗体。

三、口服疫苗

科学研究的进步使得植物基因工程得以发展，可在转基因植物叶片、果实或鳞茎中表达产生相关抗原（Richter和Kipp，1999）。这些植物产品可以大量种植并可直接加工到鱼饲料中。采用这种方式可能研发出生物包被疫苗，能防止抗原在鱼的前肠被降解。尽管食用疫苗可能具有成本效益，但大多数情况下，生产的抗原量太低不足以使接种鱼产生免疫反应。到目前为止，仅在鲤上发现了使用这种疫苗接种的研究报道（Companjen等，2005，2006）。研究将鲤春病毒（SVCV）G蛋白基因与大肠杆菌不耐热肠毒素基因（LTB）连接，或将绿色荧光蛋白（GFP）基因与LTB基因连接，分别插入质粒转染马铃薯细胞，结出的LTB-GFP马铃薯块茎产生了一定量的适用于研究鱼体抗原摄入和免疫反应的GFP。研究显示，LTB-GFP较仅GFP的摄取效果好得多，表明LTB是一种较好的肠道佐剂，可增强鱼体抗原的吸收，进而又研究了针对 GFP 的免疫记忆。尽管在马铃薯块茎中形成了非常稳定的 LTB-G 结构，但 LTB-G 含量过低，不适用于口服免疫。只有当植物块茎或其他适合于鱼饲料的植物中积累足够的 LTB-G 时，这种方法才可能非常有用。

四、微生物膜

生物膜也可用于鱼类的口服免疫。细菌在微小的几丁质颗粒上培养并聚集在一起，形成生物膜。当以这样的生物膜颗粒饲喂鱼时，细菌聚集形成的外膜可保护内部细菌不被消化。将热灭活的嗜水气单胞菌（*Aeromonas hydrophila*）几丁质颗粒生物膜疫苗经饲料投喂印度鲤、野鲮和鲤后，用嗜水气单胞菌攻毒感染表明，鱼体获得了显著的免疫保护，并对生物膜能有效地利用；同时，与对照相比，其产生了较强抗体免疫反应（Azad 等，1999，2000）。使用这种疫苗在胡子鲶（*Clarias batrachus*）中也获得了相同的结果（Nayak 等，2004）。为了检测黏膜免疫反应所起的作用，在鲤（40g）上进行了研究。鲤经 10d 的口服免疫，分别在 0d、10d、20d 和 30d 时间点（0d 即第一次口服免疫后的第 10 天）取 5 尾鱼，每尾鱼采集血液和肠道黏膜，用 ELISA 检测血清（图 6-1a 左）和肠道黏液（图 6-1b 右）中抗嗜水气单胞菌的抗体水平。

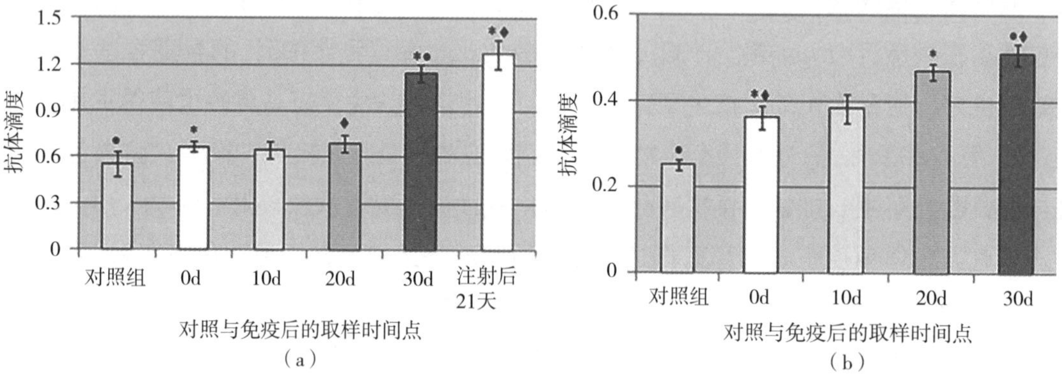

图 6-1　不同免疫方式对鲤的免疫效果比较

（a）对照组、口服组和注射组的鲤血清中抗嗜水气单胞菌的抗体滴度（平均值 ± 标准差，OD=405nm）。单因素方差的 *Dunnett* 多重比较免疫组 0d 的取样和其他取样点的数值。成组设计 *t* 检验，用于对照组与免疫后的 30d 取样组，与 20d 取样组和注射免疫组（注射后 21d）的比较。（b）对照组和免疫组的鲤肠道黏液中抗嗜水气单胞菌抗体滴度（平均值 ± 标准差，OD=405 nm）。相同符号表示两组间显著差异（$P<0.05$）

（Rombout 等，未发表）

图 6-1 表明，口服生物膜疫苗后，鲤体内可检测到系统免疫和黏膜免疫反应。在鲤上的另一研究也得到了一致的免疫保护效果（Azad 等，1999，2000）。从图 6-1 可以看出，免疫后的第 30 天采样点，血清中抗体滴度与注射免疫的第 21 天相当。口服免疫后第 10 天，肠道黏液中抗嗜水气单胞菌的抗体逐渐升高，到免疫后的第 30 天采样时达到最高水平。必须说明的是，由于黏液 Ig 是 μ 域和 ζ 域嵌合分子（Ryo 等，2010），实验中使用的抗血清 IgM 的单克隆抗体（WCI12）与黏液抗体的亲和力较低，因此利用抗血清 IgM 的单克隆抗体检测黏膜免疫应答不是最敏感的方法。即便如此，这些结果仍然是首次获得的有关鱼类系统免疫和黏膜免疫的动力学数据。

五、微型胶囊

在一些鱼类中，已有关于采用多种天然和人工合成的聚合物研制纳米和微米颗粒，用其经口递送抗原的研究。抗原通常封装在颗粒胶囊内，也可与颗粒共价连接。一般来说，均采用无毒的、可生物降解的颗粒，以满足口服要求。多数研究发现，以下颗粒能够在鱼肠末端吸收良好、增加特异性抗体产生和/或得到更好的免疫保护效果：海藻酸微球（Joosten 等，1997b；Romalde 等，2004；Rodrigues 等，2006；Tian 等，2008a）、脂质体（Yasumoto 等，2006a，2006b）、聚乳酸-羟基乙酸（PLGA）微粒（O'Donnell 等，1996；Lavelle 等，1997；Ellis，1998；Benoit 等，2001；Tian 等，2008b）和壳聚糖微粒（Kumar 等，2008；Tian 等，2008c）。

海藻酸盐微颗粒的制备方法应根据不同鱼类而确定，使其难被鱼类消化降解。例如，应用在鲤（无胃）与鳟（有胃）的海藻酸盐颗粒的配制方法一定是不同的，这样才能既引起鱼体的系统免疫反应，又能诱发黏膜免疫反应（Joosten 等，1997b）。在提到的研究中，发现海藻酸盐颗粒可诱导鱼体的免疫记忆，但是没有检测到对弧菌病的保护作用。在牙鲆中，联合使用海藻酸盐颗粒与表达淋巴囊肿病毒（LCDV）主要衣壳蛋白质粒的 DNA 疫苗，通过牙鲆不同组织中的病毒蛋白基因和标记基因（GFP）的表达结果发现，口服 DNA 疫苗也是一种有效的免疫策略（Tian 等，2008a）。但在其他鱼上使用海藻酸盐颗粒的研究结果却表明其效果不佳（金鱼，Maurice 等，2004；虹鳟，Altun 等，2010；尼罗罗非鱼，Leal 等，2010）。对褐鳟和虹鳟口服免疫（移液管灌喂）传染性胰腺坏死病毒（IPNV）DNA 疫苗，获得了较高的相对免疫保护率，说明该疫苗接种方式可行（de Las Heras 等，2010）。如前所述（Joosten 等，1997b），海藻酸盐颗粒的配制必须因鱼而异，避免抗原在鱼的前肠降解。应注意的是，颗粒并不太稳定，应在合适的肠段部位释放。目前尚不清楚这些研究所用颗粒是否符合要求，也不清楚这些颗粒在水产养殖业是否经济实用。

聚乳酸-羟基乙酸和壳聚糖颗粒被认为是可生物降解的、黏膜黏附的微球（Plant 和 LaPatra，2011），非常适合用于 DNA 质粒疫苗的包被（Tian 等，2008b，2008c）。研究表明，载有质粒的 PLGA 颗粒可以通过肠道运输，质粒在体内表达后，机体会产生特异性抗体（Tian 等，2008b）。尽管该方法适合口服 DNA 疫苗，但尚不清楚肠道上皮细胞的确切运输机制，以及将其应用于水产养殖中的成本问题。另一研究（Altun 等，2010）显示，与海藻酸盐颗粒相比，PLGA 颗粒在免疫保护效果方面略胜一筹。

壳聚糖颗粒在鱼类疫苗抗原传递能力方面的研究表明，其具有很好的适用性（Ramos 等，2005；Kumar 等，2008；Tian 等，2008c）。在鲈中的研究表明，亚洲鲈口服编码鳗弧菌外膜蛋白 38 的质粒 DNA 后，获得 46%的相对保护率（Kumar 等，2008）。

聚乙二醇（PEG）也可用于疫苗胶囊，在虹鳟 VHSV 疫苗上使用，获得 60%以上的相对保护率（Adelmann 等，2008）。

六、其他方法

研究表明，减毒细菌疫苗也可以通过浸泡或口服方式接种。哈维氏弧菌（*Vibrio harveyi*）的外膜蛋白被转化到共生菌荧光假单胞菌（*Pseudomonas fluorescens*）中，以海藻酸盐微颗粒包被，经口服免疫牙鲆，相对保护率可以达到 61%（Sun 等，2009）。

颗粒饲料的保护涂层中不管是否包含如维生素 E 或十二烷基硫酸钠等吸收增强剂，其保护都较好。在上述保护涂层颗粒中加入弧菌疫苗，口服免疫非洲鲶，产生的抗体水平与腹腔注射效果相同（Vervarcke 等，2004）。

在鲑科鱼疫苗中添加抗蛋白酶和膜渗透增强剂有助于保护抗原不被消化。这种口服佐剂获得了专利（Oraljectᵀᴹ），将其应用于尼罗罗非鱼海豚链球菌（*Streptococcus iniae*）疫苗，获得63%的相对保护率，腹腔注射免疫的相对保护率为 100%（Shoemaker 等，2006）。

第六节 展 望

在以上介绍的黏膜免疫技术中，微囊口服疫苗最适合应用于鱼类。尽管如此，市场上几乎没有口服疫苗，这可能是由于其在水产饲料中的稳定性较差，以及获得的免疫保护作用有限（Komar 等，2004），最主要原因是其经济成本较高。事实上，在水产养殖中研发“食用疫苗”应该是广泛应用口服疫苗的解决方法。例如，在植物中生产病毒三聚体 G 蛋白结合的 LTB 就非常稳定，既可作为疫苗又具有黏膜佐剂功能，添加到饲料中可作为有效的口服疫苗。然而，植物产生的转基因蛋白质的量要足够多，才可省去添加到饲料前的浓缩过程。同时，应深入开展鱼类免疫研究，以期达到使用少量的抗原即可获得最佳保护效果的目的。另外，与哺乳动物一样，还需要使用合适的肠佐剂才能实现理想的免疫效果。

浸泡免疫结合超声波或皮肤/鳃穿刺等方式，免疫可能会更加有效，可增强鱼类对抗原的摄入以及免疫反应。尽管这种方法前景较好，但也需要进一步探讨这些方法的效果以及受免鱼福利方面的问题。

鱼类 pIgR 和新 Ig 类型（IgT 或 IgZ）的最新研究结果或许会有助于阐明鱼类黏膜免疫应答，特别是黏膜免疫反应的保护作用机制。近十年来，应用分子生物学方法明晰了一系列的鱼类免疫分子特性。有些免疫分子可作为候选佐剂（Secombes，2011），能够增强免疫应答并诱发免疫保护反应。

总之，在水产养殖中，鱼类的黏膜疫苗接种尚未得到充分开发利用。鱼类防御机制方面的最新进展将有助于进一步探讨更新的疫苗传递模式和接种技术，研发出适用于养殖鱼类的黏膜疫苗。

参考文献

Adelmann，M.，Köllner，B.，Bergmann，S. M.，*et al.*，2008. Development of an oral vaccine for immunisation of rainbow trout（*Oncorhynchus mykiss*）against viral haemorrhagic septicaemia. *Vaccine* 26：837-844.

Altun，S.，Kubilay，A.，Ekici，S.，*et al.*，2010. Oral vaccination against lactococosis in rainbow trout（*Oncorhynchus mykiss*）using sodium alginate and poly（lactide-co-glycolide）carrier. *Kafkas Univ Vet Fak Derg* 16：S211-217.

Amend，D. F. and Johnson，K. A. 1981. Current status and future need of *Vibrio anguillarum* bacterins. *Dev Biol Stand* 49：403-417.

Azad，I. S.，Shankar，K. M.，Mohan，C. V. and Kalita，B. 1999. Biofilm vaccine of *Aeromonas hydrophila* standardization of dose and duration for oral vaccination of carps. *Fish Shellfish Immunol* 9：519-528.

Azad，I. S.，Shankar，K. M.，Mohan，C. V. and Kalita，B. 2000. Uptake and processing of biofilm and free cell vaccines of *Aeromonas hydrophila* in Indian major carps and common carp following oral vaccination-antigen localization by a monoclonal antibody. *Dis Aquat Org* 43：103-108.

Baudin-Laurencin，F. and Tangtrongpiros，J. 1980. Some results of vaccination against vibriosis in Brittany，in *Fish Diseases*（ed. W. Ahne）. Third COPRAQ Session. Berlin，Springer，60-68.

Benoit，M.-E.，Ribet，C.，Distexhe，J.，*et al.*，2001. Studies on the potential of microparticles EnTropping pDNA-Poly（aminoacids）complexes as vaccine delivery systems. *J Drug Target* 9：253-266.

Bernard，D.，Sox，A.，Rigottier-Gois，L.，*et al.*，2006. Phenotypic and functional similarity of gut intraepithelial and systemic T cells in a teleost fish. *J Immunol* 176：3942-3949.

Brandtzæg，P. and Pabst，R. 2004. Let's go mucosal：communication on slippery ground. *Trends Immunol* 25：570-577.

Chair，M.，Gapasin，R. S. J.，Dehasque，M. and Sorgeloos，P. 1994. Vaccination of European sea bass fry through bioencapsulation of *Artemia nauplii*. *Aquac Int* 2：254-261.

Companjen，A. R.，Florack，D. E. A.，Bastiaans，C. I.，*et al.*，2005. Development of a cost-effective oral vaccination method against viral diseases in fish. *Dev Biol*（*Basel*）121：143-150.

Companjen，A. R.，Florack，D. E. A.，Slootweg，T.，*et al.*，2006. Improved uptake of plant derived LTB-linked proteins in carp gut and induction of specific humoral immune responses upon infeed delivery. *Fish Shellfish Immunol* 21：251-260.

De Las Heras，A. I.，Saint-Jean，S. R. and Perez-Prieto，S. I. 2010. Immunogenic and protective effects of an oral DNA vaccine against infectious pancreatic necrosis virus in fish. *Fish Shellfish Immunol* 28：562-570.

Duff，D. C. B. 1942. The oral immunization of trout against *Bacterium salmonicida*. *J Immunol* 44：87-94.

Ellis，A. E. 1988. General principles of fish vaccination，in *Fish Vaccination*（ed. A. E. Ellis）. London，Academic Press，1-19.

Ellis，A. E. 1998. Meeting the requirements for delayed release of oral vaccines for fish. *J Appl Ichthyol* 14：149-152.

Evelyn，T. P. T. and Ketchetson，J. E. 1980. Laboratory and field observations on anti-vibriosis vaccines，in *Fish Diseases*（ed. W. Ahne）. Third COPRAQ-Session. Berlin，Springer，60-68.

Fernandez-Alonso，M.，Rocha，A. and Coll，J. M. 2001. DNA vaccination by immersion and ultrasound to trout haemorrhagic septicaemia virus. *Vaccine* 19：3067-3075.

Hansen，J. D.，Landis，E. D. and Phillips，R. B. 2005. Discovery of a unique Ig heavy chain isotype（IgT）in rainbow trout：implications for a distinct B cell developmental pathway in teleost fish. *Proc Natl Acad Sci USA* 102：6919-6924.

Håstein，T.，Refsti，T. and Gjedrem，T. 1977. Preliminary results of vaccination trials against vibriosis in rainbow trout. *Bull Off Int Epizoot* 87：487-488.

Horne，M. T.，Tatner，M.，McDerment，S.，*et al.*，1982. Vaccination of rainbow trout，*Salmo gairdneri* Richardson，at low temperatures and the long persistence of protection. *J Fish Dis* 5：343-345.

Huising，M. O.，Guichelaar，T.，Hoek，C.，*et al.*，2003. Increased efficacy of immersion vaccination in fish with hyperosmotic pretreatment. *Vaccine* 21：4178-4193.

Inami，M.，Taverne-Thiele，A. J.，Schrøder，M. B.，*et al.*，2009. Immunological differences in intestine and rectum of Atlantic cod（*Gadus morhua* L.）. *Fish Shellfish Immunol* 26：751-759.

Johnson，K. A. and Amend，D. F. 1983a. Comparison of efficacy of several delivery methods using *Yersinia ruckeri* bacterin on rainbow trout，*Salmo gairdneri* Richardson. *J Fish Dis* 6：331-336.

Johnson, K. A. and Amend, D. F. 1983b. Efficacy of *Vibrio anguillarum* and *Yersinia ruckeri* bacterins applied by oral and anal intubation of salmonids. *J Fish Dis* 6: 473-476.

Joosten, P. H. M. , Avilés-Trigueros, M. , Sorgeloos, P. and Rombout, J. H. W. M. 1995. Oral vaccination of juvenile carp (*Cyprinus carpio*) and gilthead seabream (*Sparus aurata*) with bioencapsulated *Vibrio anguillarum* bacterin. *Fish Shellfish Immunol* 5: 289-299.

Joosten, P. H. M. , Engelsma, M. Y. , van der Zee, M. D. and Rombout, J. H. W. M. 1997a. Induction of oral tolerance in carp (*Cyprinus carpio* L.) after feeding protein antigens. *Vet Immunol Immunopathol* 60: 187-196.

Joosten, P. H. M. , Tiemersma, E. , Threels, A. , *et al.* , 1997b. Oral immunization of fish against *Vibrio anguillarum* using alginate microparticles. *Fish Shellfish Immunol* 7: 471-485.

Kawai, K. , Kusuda, R. and Itami, T. 1981. Mechanisms of protection in ayu orally vaccinated for vibriosis. *Fish Pathol* 15: 257-262.

Kawai, K. , Yamamoto, S. and Kusuda, R. 1989. Plankton-mediated oral delivery of *Vibrio anguillarum* vaccine to juvenile ayu. *Nippon Suisan Gakkaishi* 55: 35-40.

Komar, C. , Enright, W. J. , Grisez, L. and Tan, Z. 2004. Understanding fish vaccination. *Aqua Culture Asia Pacific Magazine* Nov/Dec: 27-29.

Kumar, S. R. , Ishaq Ahmed, V. P. , Parameswaran, V. , *et al.* , 2008. Potential use of chitosan nanoparticles for oral delivery of DNA vaccine in Asian sea bass (*Lates clacrifer*) to protect from *Vibrio* (*Listonella*) *anguillarum*. *Fish Shellfish Immunol* 25: 47-56.

Lavelle, E. C. , Jenkins, P. G. and Harris, J. E. 1997. Oral immunization of rainbow trout with antigen microencapsulated in poly (DL-lactide-co-glycolide) microparticles. *Vaccine* 15: 1070-1078.

Leal, C. A. G. , Carvalho-Castro, G. A. , Sacchetin, P. S. C. , *et al.* , 2010. Oral and parental vaccines against *Flavobacterium columnare*: evaluation of humoral immune response by ELISA and *in vivo* efficiency in Nile tilapia (*Oreochromis niloticus*) . *Aquac Int* 18: 657-666.

Lillehaug, A. 2014. Vaccination strategies and procedures, in *Fish Vaccination* (eds R. Gudding, A. Lillehaug and Ø. Evensen) . Oxford, John Wiley & Sons, Ltd. , 140-152.

Lin, J. H. , Yu, C. C. , Lin, C. C. and Yang, H. L. 2005. An oral delivery system for recombinant subunit vaccine to fish. *Dev Biol* (*Basel*) 121: 175-180.

Maurice S, Nussinovitch A, Jaffe N, *et al.* , 2004. Oral immunization of *Carassius auratus* with modified recombinant A-layer proteins entrapped in alginate beads. *Vaccine* 23: 450-459.

Nakanishi, T. , Kiryu, I. and Ototake, M. 2002. Development of a new vaccine delivery method for fish: percutaneous administration by immersion with application of a multiple puncture instrument. *Vaccine* 20: 3764-3769.

Nayak, D. K. , Asha, A. , Shankar, K. M. and Mohan, C. 2004. Evaluation of biofilm of *Aeromonas hydrophila* for oral vaccination of *Clarias batrachus* -a carnivore model. *Fish Shellfish Immunol* 16: 613-619.

O'Donnell, G. B. , Reilly, P. , Davidson, G. A. and Ellis, A. E. 1996. The uptake of gamma globulin incorporated into poly (D, L-lactide-co-glycolide) microparticles following oral intubation in Atlantic salmon, *Salmo salar* L. *Fish Shellfish Immunol* 6: 507-520.

Plant, K. P. and LaPatra, S. E. 2011. Advances in fish vaccine delivery. *Dev Comp Immunol* 35: 1256-1262.

Ramos, E. A. , Relucio, J. L. V. and Torres-Villanueva, C. A. T. 2005. Gene expression in tilapia following oral delivery of chitosan-encapsulated plasmid DNA incorporated into fish feeds. *Mar Biotechnol* (*NY*) 7: 89-94.

Richter, L. and Kipp, P. B. 1999. Transgenic plants as edible vaccines. *Curr Top Microbiol Immunol* 240: 159-176.

Rodrigues, A. P. , Hirsch, D. , Figueiredo, H. C. P. , *et al.* , 2006. Production and characterisation of alginate microparticles incorporating *Aeromonas hydrophila* designed for fish oral vaccination. *Process Biochem* 41: 638-643.

Romalde, J. L., Luzardo-Alvárez, A., Ravelo, C., *et al.*, 2004. Oral immunization using alginate microparticles as a useful strategy for booster vaccination against lactococcosis. *Aquaculture* 236: 119-129.

Rombout, J. H. W. M. and van den Berg, A. A. 1989. Immunological importance of the second gut segment of carp. I. Uptake and processing of antigens by epithelial cells and macrophages. *J Fish Biol* 35: 13-22.

Rombout, J. H. W. M., Lamers, C. H. J., Helfrich, M. H., *et al.*, 1985. Uptake and transport of intact macromolecules in the intestinal epithelium of carp (*Cyprinus carpio* L.). *Cell Tiss Res* 239: 519-530.

Rombout, J. H. W. M., Blok, L. J., Lamers, C. H. J. and Egberts, E. 1986. Immunization of carp (*Cyprinus carpio* L.) with a *Vibrio anguillarum* bacterin: indications for a common mucosal immune system. *Dev Comp Immunol* 10: 341-351.

Rombout, J. H. W. M., Taverne. N., van de Kamp, M. and Taverne-Thiele, A. J. 1993. Differences in mucus and serum immunoglobulin of carp (*Cyprinus carpio* L.). *Dev Comp Immunol* 17: 309-317.

Rombout, J. H. W. M., Abelli, L., Picchietti, S., *et al.*, 2011. Teleost intestinal immunology. *Fish Shellfish Immunol* 31: 616-626.

Ryo, S., Wijdeven, R., Tyagi, A., *et al.*, 2010. Common carp have two subclasses of bony fish specific IgZ showing differential expression in response to infection. *Dev Comp Immunol* 34: 1183-1190.

Savan, R., Aman, A., Nakao, M., *et al.*, 2005a. Discovery of a novel immunoglobulin heavy chain gene chimera from common carp (*Cyprinus carpio* L.). *Immunogenetics* 57: 458-463.

Savan, R., Aman, A., Sato, K., *et al.*, 2005b. Discovery of a new class of immunoglobulin heavy chain from fugu. *Eur J Immunol* 35: 3320-3331.

Secombes, C. J. 2011. Fish immunity: the potential impact on vaccine development and performance. *Aquac Res* 42 (suppl. 1): 90-92.

Shoemaker, C. A., Vandenberg, G. W., Desormeaux, A., *et al.*, 2006. Efficacy of *Streptococcus iniae* modified bacterin delivered using OraljectTM technology in Nile tilapia (*Oreochromis niloticus*). *Aquaculture* 255: 151-156.

Siripornadulsil, S., Dabrowski, K. and Sayre, R. 2007. Microalgal vaccines. *Adv Exp Med Biol* 616: 122-128.

Sun, K., Zhang, W., Hou, J. and Sun, L. 2009. Immunoprotective analysis of VhhP2, a *Vibrio harveyi* vaccine candidate. *Vaccine* 27: 2733-2740.

Tian, J., Sun, X. Q., Chen, X., *et al.*, 2008a. Formation and oral administration of alginate microspheres loaded with pDNA codon for lymphocystis disease virus (LCDV) to Japanese flounder. *Fish Shellfish Immunol* 24: 592-599.

Tian, J., Sun, X. Q., Chen, X., *et al.*, 2008b. The formulation and immunization of oral poly (DL-lactide-co-glycolide) microparticles containing a plasmid vaccine against lymphocystis disease virus in Japanese flounder (*Paralichthys olivaceus*). *Int Immunopharmacol* 8: 900-908.

Tian, J., Yu, J. and Sun, X. Q. 2008c. Chitosan microspheres as candidate plasmid vaccine carrier for oral immunization of Japanese flounder (*Paralichthys olivaceus*). *Vet Immunol Immunopathol* 126: 220-229.

Vandenberg, G. W. 2004. Oral vaccines for finfish: academic theory or commercial reality. *Annu Health Res Rev* 5: 302-304.

Van Muiswinkel, W. B. 2008. A history of fish immunology and vaccination I. The early days. *Fish Shellfish Immunol* 25: 397-408.

Vervarcke, S., Ollevier, F., Kinget, R. and Michoel A. 2004. Oral vaccination of African catfish with *Vibrio an-*

guillarum O2：effect on antigen uptake and immune response by absorption enhancers in lag time coated pellets. *Fish Shellfish Immunol* 16：407-414.

Yasumoto，S.，Yoshimura，T. and Miyazaki，T. 2006a. Oral immunization of common carp with a liposome vaccine containing *Aeromonas hydrophila* antigen. *Fish Pathol* 41：45-49.

Yasumoto，S.，Kuzuya，Y.，Yasuda，M.，*et al.*，2006b. Oral immunization of common carp with a liposome vaccine fusing herpesvirus antigen. *Fish Pathol* 41：141-145

Zhang，Y.-A.，Salinas，I.，Li，J.，*et al.*，2010. IgT，a primitive immunoglobulin class specialized in mucosal immunity. *Nat Immunol* 11：827-835.

第七章　鱼类疫苗佐剂

Carolina Tafalla，Jarl Bøgwald，Roy A. Dalmo，
Hetron Mweemba Munang'andu and Øystein Evensen

本章概要　SUMMARY

在全球范围内，疫苗接种是水产养殖业控制传染性疾病的首选方法。然而通常仅使用疫苗不能保障良好的免疫保护效果，特别是重组蛋白疫苗和灭活疫苗。因此，常使用佐剂提升疫苗的效果。各类商品化的细菌及病毒疫苗常使用矿物油与植物油等传统佐剂，但这些疫苗仅能采用注射免疫方式，并会产生不良的副作用。目前针对鱼类胞内病原体的疫苗仍有待完善，本章将介绍一些可选择的技术方法。口服免疫是最理想的鱼类免疫接种方式，但到目前为止还没有专门的口服疫苗佐剂实现商品化。可以设想，抗原-佐剂组合制品，直接接种激发免疫应答，同时对鱼体无或很少有副作用的疫苗，有可能在未来投入使用。本章总结了鱼用疫苗佐剂的现状，介绍了一些可能突破鱼类疫苗学局限、有前途的新一代佐剂的研究进展。

第一节　引　言

从经济、环境和道德伦理方面讲，鱼类疫苗免疫预防疾病是控制病原传播的最佳方法，也是目前水产养殖业的有效方法。传统疫苗包括活疫苗、减毒疫苗、灭活疫苗、亚单位疫苗等。在许多情况下，灭活或亚单位疫苗的免疫原性较弱，疫苗效力低，而活疫苗在许多国家的水产养殖中尚未获得许可。因此，需要在现有或新型疫苗中应用佐剂，以提升疫苗的保护效果。

在过去，鱼类疫苗通过反复试验法（常规的疫苗设计）研制，包括病原鉴定和培养，然后将全细胞和油剂制成疫苗。这些策略相当有效，水产养殖业使用疫苗后，鱼类的死亡率大幅降低，抗生素使用也极大减少，这就是欧美国家所谓的“高档”水产养殖。考虑到水产养殖业快速增长的前景，尤其是在亚洲国家，仍然存在对有效疫苗的需求。另外，如今许多给经济造成损失巨大的疾病是由胞内病原感染引起的，研发针对这类病原的有效疫苗非常困难。对预防某些病毒性疾病，DNA 疫苗是最具前景的疫苗类型，研究表明，肌肉注射编码抗原的 DNA 质粒，

鱼体可产生免疫保护应答。鱼体的细胞可将抗原直接表达在细胞表面，或由主要组织相容性复合体（MHC）-Ⅰ类分子和MHC-Ⅱ类分子处理和递呈，从而激活体液免疫反应和细胞免疫反应(Biering和Salonius，2014)。因此，鱼类疫苗的研发应设计合理，使用适宜的佐剂与最佳抗原组合，且对机体的副作用最小，以期制备出对特定病原提供更有效免疫反应的疫苗。

鱼类免疫学的许多方面仍处于探索阶段，目前还不清楚对某个病原起有效保护作用的是哪种免疫机制。而且，目前已知的鱼类有将近22 000种，且大部分都有各自独特的免疫特性。毫无疑问，鱼类的先天性防御系统是非常有效的，帮助机体清除病毒、细菌甚至寄生虫，以抵御许多传染性病原体。然而，许多传染性病原可逃脱先天性的防御，因此需获得性免疫反应发挥作用。脊椎动物的获得性免疫反应能够识别和记忆特定的病原，每当机体再次遇到同一病原时，能够增强再次免疫反应。在高等的脊椎动物中，针对胞外病原的获得性免疫通常是由体液免疫反应（抗体）介导的，而针对胞内病原（包括病毒）的获得性免疫则通常依赖细胞免疫反应，包括细胞毒性T细胞。而在鱼类中，免疫系统的调控与哺乳动物有很大不同，甚至在不同的鱼类之间也有很大的差别。在哺乳动物中，先天性与获得性免疫系统之间存在着紧密的功能连接，而在鱼类中则知之甚少。最新的免疫学研究表明，疫苗免疫后，鱼体先天性免疫细胞所识别的信号大小和特异性决定了其后的获得性免疫反应。因此，鱼类的先天性免疫和获得性免疫之间可能是由细胞因子和转录因子等先天性受体和信号分子所控制，这可为未来疫苗佐剂的合理设计提供关键思路。

第二节 疫苗制剂

一般来说，疫苗由两部分组成，一部分是能激活获得性免疫反应的特异性成分，另一部分是能诱导先天性免疫反应的非特异性成分。特异性成分是用于疫苗生产的微生物抗原，而非特异性成分是佐剂。疫苗配方的选择取决于疫苗所引起免疫反应的性质，还应避免不良副作用并确保疫苗的稳定性等。疫苗制剂的核心是疫苗产品的安全性，必须权衡疫苗获得免疫保护效果与使用佐剂可能产生的副作用之间的关系。实际上，疫苗制剂生产应相对容易、成本效益好，且具有较长的保质期。

第三节 佐剂作用原理

佐剂，来源于“adjuvare”，意为“帮助”，是能够增强机体对免疫原引发免疫反应的辅助制剂。佐剂帮助激活早期免疫应答，诱导长期的效应反应，如抗体产生或细胞毒性T细胞形成，并且无需加强免疫接种（Singh和O'Hagan，2003)。佐剂的“辅助性”机制很复杂。佐剂能使抗原的递呈（至下一级淋巴器官）有充足的时间（Singh和O'Hagan，2003)。目前，佐剂普遍被认为分两类：“信号1”促进剂和“信号2”促进剂（Schijns，2001)。根据这两种信号模型，抗原（即信号1）及其他次级信号（信号2）均是激活T细胞和B细胞必要的(Ribeiro和Schijns，2010)。疫苗抗原的特性决定了信号1最终对提高免疫细胞的作用，而抗原的特性受疫苗存放时间地点、注射部位、抗原浓度，以及抗原降解、乳化持续效果等影响。信号 1 型佐剂的作用形式不依赖于受体，而信号 2 型佐剂主要作用于先天性免疫受体。当 MHC 和抗原肽复合物被 T 细胞受体（TCRs）识别时，信号 2 型佐剂可增强先天性免疫反应，释放细胞因子和损伤相关分子模式（DAMPS），进而调节免疫受体的表达以及 T 细胞分化和活化相关的免疫反应。如果佐剂效果最佳，可为机体有效的抗原特异性免疫反应提供良好的环境。总之，在辅助机体对特定病原体产生获得性免疫应答方面，佐剂将变得更加重要。佐剂已被定义为能够调节抗原固有免疫原性的一类异质化合物（Guy，2007），可根据其化学性质或物理性质进行分类。然而，即使是同类的化合物，在免疫调节活性上也有显著差异。尽管目前还不清楚大多数佐剂的具体作用机制，新的分类方法将会以佐剂诱发的免疫反应特性为依据。

第四节　抗原成分

抗原可由纯化的蛋白质、糖蛋白、糖类或是病原微生物具有免疫原性的多肽组成，是疫苗的特定成分，主要作用是诱导接种疫苗的鱼体产生获得性免疫保护反应。疫苗的优化制剂可通过选择最具保护性的疫苗株，用基因工程设计表达免疫原性蛋白，以及确定免疫保护效果最佳剂量等实现。

第五节　佐　剂

佐剂的作用主要有以下几个方面：

(1) 增强机体对抗原的免疫反应，延长免疫反应的持续时间；

(2) 调节抗体反应的广度、亲和性和特异性；

(3) 激活有效的细胞介导的免疫反应；

(4) 提高抗原的免疫原性和血清抗体转换；

(5) 促进诱导黏膜免疫；

(6) 有利于减少抗原剂量且不需加强免疫；

(7) 减少多价疫苗中的抗原竞争。

第六节　抗原传递系统

抗原传递系统是指抗原在宿主中被递呈的途径。递呈到细胞内的抗原会引起细胞免疫反应，在细胞外的抗原引起体液中和免疫反应。因此，选择传递系统以专门诱导细胞免疫反应或体液免疫反应。到目前为止，一般采用抗原的乳化悬浮液、生物降解微粒以及与细菌毒素融合等手段，探究鱼类疫苗抗原的外源性或内源传递系统。

第七节　传递载体

佐剂作为“传递载体”的主要作用与信号 1 型佐剂的概念有关，包括：①携带抗原和作为免疫调节剂；②缓慢释放抗原，从而减少加强免疫次数；③摄入过程中保护抗原；④可生物降解；⑤作为免疫刺激物，吸引抗原由递呈位点到达作用位点。这里必须指出，抗原传递和免疫刺激是两回事，但大多数佐剂的设计都会考虑其既作为抗原运载工具又作为免疫刺激剂。为提高抗原传递的效率，同时提高免疫刺激，更复杂的传递系统正在研发。例如，聚乳酸-羟基乙酸（PLGA）纳米粒子能刺激机体产生强烈的先天性免疫反应，增强对包被抗原的吸收，PLGA纳米粒子还能控制抗原的释放，可作为单剂疫苗，不必加强免疫（Fredriksen 和 Grip，2012）。然而，还需要更多深入的研究，以阐明鱼类疫苗学中这些创新性工作。

第八节 乳化型疫苗

“乳化”是指一种液体（以极小液滴）均匀地分散在不相溶的另一种液体中，第一种液体称为分散相，第二种称为连续相。鱼类疫苗中最常用的乳化方式是油包水佐剂，含抗原的水相形成微小水球，油相作为稳定剂。一般来说，大多数乳化疫苗都是采用注射接种，将抗原储存在细胞外部，通过 MHC-Ⅱ类途径协同 Th2 反应，产生体液免疫应答。目前大部分水产养殖用商业疫苗为乳剂，尤其以油类佐剂疫苗为主。

一、油类佐剂疫苗

大西洋鲑接种疖疮病和弧菌病（冷水弧菌病）油类佐剂疫苗能产生长期持久的免疫保护。就保护水平而言，使用疖疮病油类佐剂疫苗，大西洋鲑转移到海水养殖后，不会出现疖疮病暴发。然而，其缺点是导致鱼体注射部位形成可见的病变，会持续到收获期（Midtlyng 等，1996）。在某些情况下，这种病变可能会使鱼肉加工处理后品质降级。腹腔内出现的病变，主要是形成黑化以及器官之间和/或腹膜之间的粘连（Mutoloki 等，2004；Poppe 和 Koppang，2014）。

在注射部位滞留抗原是产生长期保护的必须条件，也称为贮藏效应。凝胶型佐剂（如氢氧化铝）和乳化佐剂具有这种特性。乳剂是液体的分散体，分为分散相和连续相，油包水乳剂抗原主要在微小水滴中，位于不混溶的分散相。大西洋鲑腹腔注射磷酸铝佐剂的单价和三价灭活杀鲑气单胞菌疫苗乳剂后 16 周，在头肾中可以检测出滞留的杀鲑气单胞菌亚种（*Aeromonas salmonicida* subsp. *salmonicida*）的脂多糖（LPS）和外膜蛋白 A，在脾脏中可以检测到滞留的 LPS。研究表明，大西洋鲑腹腔注射福尔马林灭活菌，16 周后应用 PCR 方法能在鱼体内检测到杀鲑气单胞菌的 16S rDNA（Høie 等，1996），免疫组化证明，杀鲑气单胞菌的 LPS 在肾中保留长达 18 个月（Evensen 等，2005）。

二、微粒疫苗

微粒能通过抗原持续的贮藏（缓慢释放）、在注射部位的定植和有效的递呈来改善信号 1 型佐剂反应途径。微粒是除油乳剂外的另一极有发展应用前景的疫苗载体，因其许多优点而备受关注（Sinyakov 等，2006）。抗原与微粒可以通过共价键结合，也可将抗原物理性的包裹在微粒内部，相比之下，共价偶联具有明显优势：需要的抗原更少，抗原递呈细胞的处理和递呈效率高，抗原储存期的稳定性更高，多余的物质易回收。使用微粒后，较低抗原剂量即可诱导强烈的体液免疫反应。由于微粒的结构、性质可随着生产条件的轻微改变发生明显变化，且大小均一纳米粒子可通过物理化学再生方式制备出来，因此可研发亚微米聚合物颗粒，在哺乳动物中作为佐剂使用（Cui 和 Mumper，2003）。严格地说，微粒疫苗载体本身产生的生物效应属于信号 2 促进剂（Oyewumi 等，2010）。

PLGA 聚合物作为疫苗包被材料，具有生物相容性和可生物降解性，其研究已有 20 多年。微球中的抗原通过基质孔隙扩散和基质降解释放出来，可通过改变聚合物组成和分子量调节生物降解率。目前，已有一些关于 PLGA 颗粒在鱼体中的摄入和降解以及鱼体产生免疫反应的研究，大部分是口服免疫途径（O'Donnell 等，1996；Lavelle 等，1997；Tian 等，2008；Altun 等，2010；Tian 和 Yu，2011）。有一篇应用 PLGA 包被疫苗经腹腔注射免疫印度野鲮（*Labeo rohita*）的研究报道，结果显示，采用 PLGA 包被嗜水气单胞菌（*A. hydrophila*）的外膜蛋白与弗氏不完全佐剂混合成乳剂，或单独使用不加佐剂，腹腔注射印度野鲮，在免疫后的第 21 天和 42 天，加佐剂组的抗体滴度明显高于不加佐剂组的（Behera 等，2010）。

大西洋鲑腹腔注射含人 γ 球蛋白（HGG）的 PLGA 微粒后，鱼体抗体反应的剂量依赖性会短暂增加，并发现微粒载体较纳米微粒效果更佳。此外，当 PLGA 包被 HGG 疫苗与 β-葡聚糖或油类佐剂一起使用时，大西洋鲑的抗体水平随时间延长（最多 120d）持续增加（Fredriksen 和 Grip2012）。也有研究表明，使用微粒和纳米微粒包被传染性胰腺坏死病毒灭活抗原，免疫大西洋鲑，鱼体并未产生理想的免疫效果（Munang'andu 等，2012），微粒包被的疫苗效果远低于相同抗原的油包水乳剂疫苗效果。

关于 PLGA 包被的口服疫苗，在牙鲆（Tian 等，2008；Tian 和 Yu，2011）、虹鳟（Lavelle 等，1997；Altun 等，2010；Adomako 等，2012）和大西洋鲑（O'Donnell 等，1996）中也有相关研究。用 PLGA 包被构建的编码淋巴囊肿病毒（LCDV）主要衣壳蛋白的质粒，对照组是空质粒和空 PLGA 粒子。经口腔插管灌喂免疫牙鲆，28d 后，肌肉注射 LCDV 攻毒，以观察淋巴细胞囊肿的有无并量化评价免疫效果。结果显示，PLGA 颗粒包被编码 LCDV 蛋白质粒免疫组，免疫后 15～120d，出现囊肿的牙鲆数量显著减少（Tian 和 Yu，2011）。此外，以 PLGA 微囊疫苗免疫后，牙鲆血清中抗体水平逐渐增加直至 9 周，之后逐渐降低（Tian 等，2008）。

将乳球菌病的口服疫苗包被在 PLGA 微囊中，混入饲料投喂虹鳟的研究（Altun 等，2010）显示，相对保护率达到 63%；经口服加强免疫，在第一次免疫后 120d，其相对保护率超过 60%。同样，在虹鳟中人 γ 球蛋白（HGG）包被在 PLGA 微囊中的研究（Lavelle 等，1997）显示，用微囊抗原免疫虹鳟用可溶 HGG 加强免疫后，在肠道黏液中检测到特异性抗体，但在仅用可溶抗原免疫虹鳟的肠道黏液中却未检测到特异性抗体。还是在虹鳟，投喂含有编码 IHNV G 蛋白 DNA 质粒 PLGA 包被的饲料，研究发现抗 IHNV 的中和抗体水平稍有升高，但 IHNV 攻毒后，虹鳟的存活率并没有增加（Adomako 等，2012）。

第九节 可生物降解的颗粒传递系统

颗粒传递系统是通过包裹或共价方式将抗原结合到可生物降解的载体粒子中。迄今为止，用于鱼类疫苗的颗粒传递系统主要有脂质体、壳聚糖、海藻酸微球和 PLGA 纳米颗粒。颗粒传递系统的优点是：

(1) 在摄入过程中保护抗原不被降解；

(2) 同时将抗原和共刺激成分递呈至相同的抗原递呈细胞（APC）；

(3) 可使抗原持续释放；

(4) 有可能通过黏膜系统传递抗原；

(5) 吞噬体释放出的微粒抗原能够通过 MHC-Ⅰ或 MHC-Ⅱ类途径进入（Burgdorf 和 Kurts, 2008）。

抗原包囊颗粒直径为 100nm 时，很容易被抗原递呈细胞（APC）吸收，若颗粒直径大于 0.5μm，会通过吞噬作用被内化（De Temmerman 等，2011）。尽管已有研究显示，微粒疫苗能诱导鱼体的体液免疫反应，但是否能诱导细胞免疫反应尚未明晰。目前，微粒疫苗对抗原的黏膜传递和肠外传递引起极大关注，正进行的大量研究将不断优化这些疫苗的性能，以期实现微粒疫苗商品化。

第十节 融合蛋白传递系统

活疫苗和 DNA 疫苗的抗原主要进入细胞质，只有这种类型的疫苗参与 MHC-Ⅰ免疫反应。然而，目前也在探索新的方法将非复制抗原传递到细胞质中以激活 MHC-Ⅰ途径。一个有效的策略是将抗原结合到细菌毒素受体上，受体将抗原转运细胞质中，这被称为激活 MHC-Ⅰ通路的快捷方式（Smith 等，2002）。这一策略基本原理是，许多细菌毒素能够穿过细胞膜转移，因此，将异源抗原融合到感受态细菌的毒素受体上，可以作为传递系统将抗原转移到细胞内。一旦抗原与细菌毒素融合后，细菌毒素-抗原复合体通过受体介导的内吞作用被内化到细胞中，当其被酸化后，抗原被逆行转运至细胞质中。一旦抗原进入细胞质，抗原就会被加工成多肽，递呈至 MHC-Ⅰ分子。虽然在哺乳动物中已经发现了几种细菌毒素可传递抗原至内质网中（Smith 等，2002），但是在鱼类疫苗研究中，只有少量细菌毒素可用于抗原传递系统（Munang'andu 等，2012）。

第十一节 免疫调节剂

免疫调节剂诱导信号 2 佐剂效应，包括激活相关的细胞因子以及上调共刺激分子，能吸引巨噬细胞、树突状细胞等不同类型的细胞聚集到抗原部位，进而摄入并加工处理抗原，供 B 细胞或 T 细胞识别，从而诱导获得性免疫反应。这些免疫调节剂通常被称为信号 2 型佐剂。免疫调节剂应依据诱导的获得性免疫反应的类型、避免副作用以及提升疫苗中抗原效力的能力选择。表 7-1 列出了鱼类疫苗中使用的免疫调节剂，这些物质能激活、促进巨噬细胞和树突状细胞成熟并形成有效的抗原递呈细胞（APC）。然而，实际上常规疫苗制剂倾向于依据获得性的体液免疫反应或细胞介导的免疫应答选择免疫调节剂。灭活疫苗通常诱导体液免疫反应，但有研究表明，可通过设计免疫调节剂，诱导非复制性抗原产生细胞免疫反应。由此，免疫调节剂可作为抗原传递的额外诱因，激活非复制性抗原自身无法实现的细胞免疫反应。下面列举了一些免疫调节剂（或称信号2型佐剂）。

表7-1 鱼类灭活疫苗使用的佐剂

分类	佐剂
凝胶型	氢氧化铝
	磷酸铝钾
微生物成分	CpG
	细菌多糖
	破伤风梭菌和麻疹病毒
	乳酪分枝杆菌（*Mycobacterium butyricum*）
	牛分枝杆菌（*Mycobacterium bovis*）
	龟分枝杆菌（*Mycobacterium chelonae*）
	分枝杆菌细胞壁
	鞭毛蛋白
油乳剂和乳化剂	弗氏完全佐剂（FCA）
	弗氏不完全佐剂（FIA）
	矿物油和失水山梨醇单油酸酯（Span 80）
	Montanide ISA
可生物降解的颗粒	海藻酸盐
	脂质体
	可生物降解的微球
	皂苷
	PLGA纳米粒子
	壳聚糖
	β-葡聚糖
细胞因子和趋化因子	白介素
	趋化因子
维生素和其他化合物	左旋咪唑
	维生素C
	维生素E

信号2型佐剂的一个重要特征是，具有与保守的微生物分子模式相似的模式，称为病原相关分子模式（PAMPs），可被模式识别受体（PRRs）识别，PRRs主要在先天性免疫系统的细胞中被发现，包括Toll样受体（TLRs）、NOD样受体、dectin-1及RIG样解旋酶。如今，认为这种初始的识别对诱导信号2途径以及激活辅助性T细胞亚群功能非常关键。关于佐剂的研究已经聚焦于不同的PRR配体，包括不同的PAMPs，其他还有热休克蛋白（hsp）类的内源性TLR

配体（DAMPs），研究这些配体诱导相应 Th 细胞反应的能力。例如，一旦产生和表达白介素 2（IL-2）（T 细胞生长因子）和 IL-2 受体（CD25）α 亚单位就会激活幼稚 Th 细胞成为 Th0 细胞，Th 细胞就开始增殖。根据细胞因子分泌的种类，Th0 细胞将分化成 Th1 或 Th2 细胞，其中干扰素（IFN-γ）诱导 Th1 细胞产生，而白介素 4（IL- 4）诱导 Th2 细胞生产/分化（Steinman，2007）。此外，经过许多细胞世代后，Th 细胞的原始细胞分化为效应性 Th 细胞、记忆性 Th 细胞和调节性 Th 细胞。兽医学和人类医学上在用的疫苗佐剂是辅助 Th 细胞分化成多种 T 细胞亚型，如 Th1、Th2、Th9 和 Th17（Dong，2008；Chen 等，2011）。

一、含铝佐剂

铝盐的佐剂特性在 1926 年被发现（Glenny 等，1926）。铝化合物，特别是磷酸铝和氢氧化铝，被认为是少数安全的佐剂之一，允许在人类疫苗中使用。已证明，铝盐佐剂几乎专门诱导 Th2 反应，特别是有效促进机体体液免疫反应；硫酸铝能激活 NLRP3 炎性小体，诱导释放尿酸的坏死细胞死亡（Coffman 等，2010）。研究发现，作为晶体，硫酸铝和树突状细胞膜脂质有巨大的结合力，而不与炎性小体和膜蛋白结合（Flach 等，2011），随后，脂质重组，使其失去抗原摄入的吞噬反应。这种树突状细胞，借助胞间黏附分子-1（ICAM-1）和淋巴细胞功能相关抗原-1（LFA-1），与 $CD4^+$ T 细胞具有高亲和力和黏附性。

在鱼类疫苗的优化中，有关铝盐佐剂的研究较少。1995 年 Mulvey 等曾以硫酸铝（明矾）为佐剂，研究了杀鲑气单胞菌（*A. salmonicida*）疫苗免疫大西洋鲑的效果。结果表明，疫苗中添加明矾能够提高疫苗的保护水平，但结果无统计学上的显著性差异。在另一项研究中，大肠杆菌突变体作为疫苗免疫斑点叉尾鮰（*Ictalurus punctatus*）预防鮰爱德华菌（*Edwardsiella ictaluri*）引起的肠型败血症（Tyler 和 Klesius，1994）。混合或不混合明矾的灭活大肠杆菌突变体，腹腔注射斑点叉尾鮰，用强毒的鮰爱德华菌感染。结果显示，混合明矾的大肠杆菌突变体免疫组攻毒后的存活率较高（92%），单独免疫灭活大肠杆菌突变体组（54%）和免疫生理盐水组免疫保护率（56%）较低。最近的研究（Jiao 等，2010a）表明，使用氢氧化铝佐剂的迟缓爱德华菌疫苗，腹腔注射免疫牙鲆，攻毒后，佐剂疫苗免疫组相对保护率为 69%，仅用疫苗免疫组相对保护率为 34%，而与弗氏不完全佐剂混合的疫苗免疫组相对保护率为 81%。另外，甲醛灭活的虹彩病毒与明矾混合，注射或两次浸泡免疫大菱鲆，攻毒后的相对存活率分别为 83. 3%和 90. 5%（Fan 等，2012）。

二、皂苷

皂苷是类固醇或三萜烯的天然糖苷，由于其具有同时刺激 Th1 和 Th2 细胞反应的能力，已在各种哺乳动物中作为佐剂研究开发（Sun 等，2009）。使用最广的是植物皂苷（Quil A）（也常被用于免疫激活复合物，ISCOMs）及其衍生物，然而，由于其细胞毒性高以及在水相中不稳定，其他种类的皂苷也正在被研究开发。在牙鲆中，经饲料口服甲醛灭活的迟缓爱德华菌，在饲料中添加凝胶多糖和 Quil A 以及二者均不添加的比较研究结果显示，仅添加凝胶多糖也能提高牙鲆的存活率，但凝胶多糖和 Quil A 联合使用的疫苗免疫组牙鲆存活率显著升高（Ashida 等，1999）。

三、β-葡聚糖

β-葡聚糖由其模式识别受体 dectin-1 介导，能够激活哺乳动物和鱼的非特异性免疫反应（Robertsen 等，1999，Dalmo 和 Bøgwald，2008）。β-葡聚糖与 dectin-1 结合会诱导适应性的 Th1/Th17 细胞活化（Carvalho 等，2012），但鱼类的 T 细胞分化过程是否类似尚不清楚。葡聚糖通常通过注射方式免疫以达到保护效果，有研究表明，葡聚糖的作用具有剂量依赖性并是短期的。在已发表许多论文中均报道了 β-葡聚糖的免疫增强或佐剂效果（Nikl 等，1991；Chen 和 Ainsworth，1992；Rørstad 等，1993；DeBaulney 等，1996；Midtlyng 等，1996；Figueras 等，1998；Ashida 等，1999；Selvaraj 等，2005；Kamilya 等，2006）。一般来说，使用葡聚糖能够增强抗体反应和先天性细胞免疫反应，并获得一定程度的保护效果，这些已有相关的研究论述（Bricknell 和 Dalmo，2005）。然而，最近很少有关于葡聚糖佐剂活性在鱼类上的研究报道。

四、Toll 样受体的配体

一些称为 Toll 样受体（TLR）的配体（激动剂）会诱发强烈的先天性免疫反应，这对之后的获得性免疫反应可能起决定性作用。硬骨鱼类拥有的各种 TLR，数量可能是哺乳动物的近两倍，那么，鱼类 TLR 诱导的信号通路引起的免疫反应会有何不同？Zhu 等（2013）发表了一篇关于免疫相关基因的综述，包括了鱼类的 TLR。一般来说，与配体结合后，TLR 诱导产生白介素 12（IL-12），IL-12 有利于 Th1 反应（TLR3、TLR4、TLR5、TLR7、TLR8、TLR9 和 TLR11）。此外，TLR 的激活还诱导抗原的交叉递呈，在一定条件下促进细胞毒性 T 细胞反应（Manicassamy 和 Pulendran，2009）。值得一提的是，配体与 TLR3、TLR4、TLR7 和 TLR9 结合还可通过干扰素调节因子诱导产生Ⅰ型干扰素反应。

五、鞭毛蛋白

革兰氏阴性菌鞭毛的结构蛋白被称为鞭毛蛋白，是一种有效的激活剂，可以激活先天性和获得性免疫系统的多种细胞。研究表明鞭毛蛋白以及其融合结构（抗原）具有佐剂作用（Mizel 和 Bates，2010）。已知鞭毛蛋白通过 TLR5 信号引起 Th1 和 Th2 细胞的混合免疫反应。也有研究发现，含 NLRC4/IPAF 的炎性小体可与胞质定位的鞭毛蛋白结合，从而诱导干扰素（IFN-γ）的表达（Coffman 等，2010）。鞭毛蛋白的佐剂效应研究，在过去的 10 年中，主要集中在脊椎动物，也有在鱼类上的研究报道（Wilhelm 等，2006；Jiao 等，2009，2010b）。

对牙鲆迟缓爱德华菌（*E. tarda*）疫苗，配以不同形式的鞭毛蛋白，获得的疫苗保护效果显示，最好的疫苗配方是一种编码迟缓爱德华菌蛋白 Eta6 融合鞭毛蛋白 FliC 的嵌合 DNA 疫苗（Jiao 等，2009，2010b）。尽管发现迟缓爱德华菌鞭毛蛋白 FliC 自身产生的保护性免疫力低，但 FliC 可以作为分子佐剂并增强迟缓爱德华菌抗原 Eta6 诱导的特异性免疫应答。以鞭毛蛋白为佐剂免疫大西洋鲑的研究表明，与未加佐剂的对照相比，鱼的 TNF-α、IL-6、IL-8 和 IL-1β 的免疫基因表达显著上调（Hyne 等，2011）。然而，并没有诱导大西洋鲑产生对鞭毛蛋白或模式抗原鲎血淋巴（LPH）的特异性抗体反应。

鱼立克次氏体病是鲑鳟鱼类中的一种严重疾病，已成为智利水产养殖业的重大问题。由于抗生素治疗效果较差，为了控制该病，研制出一种重组亚单位疫苗，调制了三个配方：含有两种或三种该菌的重组蛋白，与弗氏不完全佐剂乳化（Wilhelm 等，2006）。结果显示，鱼体获得最高保护性反应的配方是含有鞭毛蛋白亚单位组分和鲑鱼立克次氏体 Hsp60 和 Hsp70 伴侣蛋白的疫苗配方，这说明使用一种以上的重组蛋白抗原可获得针对该疾病的良好保护效果。

六、聚肌苷酸胞苷酸

聚肌苷酸胞苷酸（poly I:C）是一种双链多核糖核苷酸，能够模拟病毒感染与 TLR3 结合，因此在包括鱼类的很多物种中，poly I:C 被广泛用于诱发 I 型干扰素反应（Eaton，1990；Jensen 等，2002；Plant 等，2005）。干扰素是在先天性抗病毒感染中具有重要作用的细胞因子。最近在感染造血器官坏死病毒（IHNV）的虹鳟体内，检测到 poly I:C 的非特异性抗病毒活性（Kim 等，2009）。虹鳟先注射 poly I:C，2d 后注射感染 IHNV，鱼体获得对 IHNV 感染的保护，并在存活的虹鳟体内检测到 IHNV 的特异性抗体。在攻毒后的第 21 天和第 49 天再次攻毒，存活率为 100%（Kim 等，2009），这说明鱼类在病毒感染期间处于抗病毒状态。poly I:C 的使用为鱼类其后产生特异性抗体提供了重要的优势。类似的还有对七带石斑鱼（*Epinephelus septemfasciatus*）野田村石斑鱼神经坏死病毒（RGNNV）免疫研究（Nishizawa 等，2009），结果显示，每尾肌肉注射 50mg 以上 poly I:C，2d 后肌肉注射 RGNNV，七带石斑鱼的存活率超过 90%。3 周后幸存的七带石斑鱼再次注射 RGNNV，注射过 poly I:C 的七带石斑鱼没有发现死亡，因为这些鱼体内 RGNNV 的抗体水平较高，所有再次攻毒存活的鱼体内都有更高水平的特异性抗体（Nishizawa 等，2009）。其后，同一研究团队的现场试验表明，七带石斑鱼接种 RGNNV 疫苗后注射 poly I:C，获得了较好的效果，但在 RGNNV 病自然暴发 20d 后，七带石斑鱼一次注射 poly I:C 也产生了显著的保护率（93.7%），而未处理的七带石斑鱼仅为 9.8%（Oh 等，2012）。

使用 poly I:C 在预防牙鲆病毒性出血性败血症病毒（VHSV）上的研究（Takami 等，2010）显示，给牙鲆注射 poly I:C 后，以 VHSV 攻毒，牙鲆的存活率为 100%，而所有未经处理的牙鲆在攻毒后的 9d 内死亡。存活的鱼进行第二次 VHSV 攻毒，存活率为 100%；而未免疫牙鲆的存活率为零。

七、CpG 寡脱氧核苷酸

细菌 DNA 与合成的寡脱氧核苷酸（ODNs）表达未甲基化的 CpG 序列，能够引起免疫刺激级联反应，具体表现为 B 淋巴细胞、T 淋巴细胞、NK 细胞、单核细胞、巨噬细胞和树突状细胞等多种免疫细胞的成熟、分化和增殖。哺乳动物 DNA 中的 CpG 序列是微生物的 1/20，因此表达 TLR9 的细胞将其作为危险信号识别。在哺乳动物中，已经证实了 CpG 寡脱氧核苷酸（CpG ODN）的佐剂功能，当与疫苗联合使用时，能够增强机体的免疫反应（Bode 等，2011）。尽管已在鱼类上开展了较多 CpGs 免疫调节的研究（Carrington 和 Secombes，2007；Liu 等，2010a，2010b），但有关其佐剂效应的研究报道仍较少。

美国在太平洋养殖的大鳞大麻哈鱼（*O. tshawytscha*）经常发生由鲑肾杆菌

(*Renibacterium salmoninarum*) 感染引起的细菌性肾病（BKD)。使用 CpG ODN 佐剂全菌疫苗免疫的研究结果显示，与仅使用疫苗相比，CpG 佐剂提高疫苗的保护效果不显著 (Rhodes 等，2004)。

在虹鳟中，CpG ODNs 的佐剂效果研究显示，使用商业化非佐剂水溶疖疮病疫苗 (Aquavac Furovac 5，灭活杀鲑气单胞菌)，以及疫苗分别与 CpG ODN 1982、CpG ODN 2133 或 ODN 2143 混合，均采用肌肉注射免疫，7 周后，以杀鲑气单胞菌致病株腹腔注射感染，与只注射疫苗（死亡率 52%）的相比，注射疫苗加 CpG ODN 2143 佐剂的死亡率明显降低，仅为 21%（Carrington 和 Secombes，2007)。

在大菱鲆和牙鲆中也开展了 CpG 序列的保护作用研究（Liu 等，2010a，2010b)。研究合成了 16 个 CpG ODNs，分别检测其在牙鲆血液中抑制细菌传播的能力，从中筛选出 4 个具有较强烈抑菌效果的 ODNs，构建了包含这 4 个 ODNs 序列的质粒 pCN6，肌肉注射牙鲆 pCN6，以 pCN3 质粒和 PBS 为对照。第 4 周后以嗜水气单胞菌攻毒，检测 20d 内鱼的死亡率。结果显示，pCN6、pCN3 和 PBS 组牙鲆累积死亡率分别是 30%、66.7% 和 63.3%（Liu 等，2010b)。以同样的方式免疫牙鲆，第 4 周以迟缓爱德华菌（*E. tarda*）攻毒，pCN6、pCN3 和 PBS 组牙鲆累积死亡率分别是 53.3%、90% 和 93.3%。所以，pCN6 质粒可提供同时针对嗜水气单胞菌和迟缓爱德华菌的非特异性保护，并可持续 4 周。在大菱鲆（Liu 等，2010a）的研究中，哈维氏弧菌（*Vibrio harveyi*）DegQ 重组亚单位疫苗，与已注射在宿主体内可产生抗菌效果的 CpG 质粒混合，腹腔注射免疫大菱鲆，设置相应的对照组，免疫后 28d，用哈维氏弧菌强毒株攻毒，记录累积死亡率。结果显示，唯一能产生显著免疫保护的疫苗配方是 DegQ + pCN5 CpG，且佐剂效应持续至少 50d。

在鱼类寄生虫感染中也观察到这种非特异性的保护作用，已发现某些 CpGs（如 CpG-ODN 1668 和 CpG-ODN 2359）对贪食迈阿密虫（*Miamiensis avidus*）的感染有保护作用（Kang 等，2012)。在鲑甲病毒疫苗中加 CpG 和 poly I:C 作为佐剂的研究显示，通过比较免疫组与对照组鲑甲病毒抗体阳性鱼的比例，发现加佐剂疫苗免疫能够显著诱导鲑产生中和抗体，可提供一定程度的免疫保护（Thim 等，2012)。研究者还指出，佐剂疫苗诱导了Ⅰ型干扰素的显著表达。

DNA 疫苗的独特之处在于，接种编码保护性抗原的细菌质粒，能够刺激机体的细胞免疫反应和体液免疫反应（Weiner 和 Kennedy，1999；Biering 和 Salonius，2014)。由于细菌质粒骨架上存在 CpG 序列，因而 DNA 疫苗具有内在的免疫刺激能力，所以在疫苗质粒中插入额外的 CpG 序列会在同一载体中提供内在佐剂，这也是提高免疫原性的简便方法。有研究报道，CpG 可提高 VHSV DNA 疫苗的免疫原性（Martinez-Alonso 等，2011)。研究将已知具有免疫刺激效应的 CpG 序列片段（2 个或 4 个）加入 DNA 疫苗中，结果显著增加了鱼体血清的中和抗体滴度，并提高了免疫基因（如 *Mx* 和 *MHC-Ⅰ*）的转录水平，首次证明添加 CpG 序列可增加 DNA 疫苗的免疫原性。

第十二节　稳定剂

稳定剂和防腐剂是确保疫苗在保质期内性质稳定，免疫后鱼在抗原摄取过程中保持其免疫原性的物质。例如，乳化剂是包含由脂肪酸链组成的亲油基团和亲水极性基团的化合物，作为表面活性剂添加到疫苗中用于稳定疫苗的极性。乳化剂的亲水亲油平衡值（HLB）决定疫苗产品属油包水乳剂（w/o）还是水包油（o/w）乳剂。低 HLB 值表明对油的亲和力高，即 w/o 乳剂；高 HLB 值表明对水的亲和力高，即 o/w 乳剂；若 HLB 值居中则为水包油包水（w/o/w）型复乳剂。总之，分散相在连续相中的聚集程度影响疫苗的储存，只要筛选出合适的乳化剂和 HLB 值，就可获得疫苗较长保质期的配方。此外，在疫苗配方中添加氨基酸和糖作为稳定剂，还可以保护疫苗免受冷冻干燥等不利条件的影响。

第十三节 结语及展望

今后研制用于水产养殖的有效疫苗，方向是将高抗原性的蛋白与可激活信号 1 和信号 2 型佐剂联合使用。佐剂必须能激发特定免疫反应（如病毒性病原），还能激活其配体反应原性。新型的、合理的疫苗设计将具有油类佐剂的优势，能引起既强烈又持久的免疫反应。因此，应该大力开展免疫保护的分子基础相关研究。

第十四节 致　谢

该项工作由 2011 年欧洲研究委员会的启动基金（项目编号 280469）、西班牙部长会议计划项目（AGL2011-29676）以及鱼专项（311993）资助。此外，感谢挪威研究委员会（合同编号 183204/S40）和特罗姆瑟研究基金委员会的支持。

参考文献

Adomako, M., St-Hilaire, S., Zheng, Y., *et al.*, 2012. Oral DNA vaccination of rainbow trout, *Oncorhynchus mykiss* (Walbaum), against infectious haematopoietic necrosis virus using PLGA [poly (D, L-lacticco-glycolic acid)] nanoparticles. *J Fish Dis* 35: 203-214.

Altun, S., Kubilay, A., Ekici, S., *et al.*, 2010. Oral vaccination against lactococcosis in rainbow trout (*Oncorhynchus mykiss*) using sodium alginate and poly (lactide-co-glycolide) carrier. *Kafkas Univ Vet Fak Derg* 16: S211-217.

Ashida, T., Okimasu, E., Ui, M., *et al.*, 1999. Protection of Japanese flounder *Paralichthys olivaceus* against experimental edwardsiellosis by formalin-killed *Edwardsiella tarda* in combination with oral administration of immunostimulants. *Fish Sci* 65: 527-530.

Behera, T., Nanda, P.K., Mohanty, C., *et al.*, 2010. Parenteral immunization of fish, *Labeo rohita* with Poly D, L-lactide-co-glycolic acid (PLGA) encapsulated antigen microparticles promotes innate and adaptive immune responses. *Fish Shellfish Immunol* 28: 320-325.

Biering, E. and Salonius, K. 2014. DNA vaccines, in *Fish Vaccination* (eds R. Gudding, A. Lillehaug and Ø. Evensen) . Oxford, John Wiley & Sons, Ltd., 47-55.

Bode, C., Zhao, G., Steinhagen, F., *et al.*, 2011. CpG DNA as a vaccine adjuvant. *Exp Rev Vaccines* 10: 499-511.

Bricknell，I. and Dalmo，R. A. 2005. The use of immunostimulants in fish larval aquaculture. *Fish Shellfish Immunol* 19：457-472.

Burgdorf，S. and Kurts，C. 2008. Endocytosis mechanisms and the cell biology of antigen presentation. *Curr Opinion Immunol* 20：89-95.

Carrington，A. C. and Secombes，C. J. 2007. CpG oligodeoxynucleotides up-regulate antibacterial systems and induce protection against bacterial challenge in rainbow trout（*Oncorhynchus mykiss*）. *Fish Shellfish Immunol* 23：781-792.

Carvalho，A.，Giovannini，G.，De Luca，A.，*et al.*，2012. Dectin-1 isoforms contribute to distinct Th1/Th17 cell activation in mucosal candidiasis. *Cell Mol Immunol* 9：276-286.

Chen，D. and Ainsworth，A. J. 1992. Glucan administration potentiates immune defence mechanisms of channel catfish，*Ictalurus punctatus* Rafinesque. *J Fish Dis* 15：295-304.

Chen，Z.，Lin，F.，Gao，Y.，*et al.*，2011. FOXP3 and RORγt：transcription regulation of Treg and Th17. *Int Immunopharmacol* 11：536-542.

Coffman，R. L.，Sher，A. and Seder，R. A. 2010. Vaccine adjuvants：putting innate immunity to work. *Immunity* 33：492-503.

Cui，Z. and Mumper，R. J. 2003. Microparticles and nanoparticles as delivery systems for DNA vaccines. *Crit Rev Ther Drug Carrier Syst* 20：103-137.

Dalmo，R. A. and Bøgwald，J. 2008. β-glucans as conductors of immune symphonies. *Fish Shellfish Immunol* 25：384-396.

DeBaulny MO，Quentel C，Fournier V，*et al.*，1996. Effect of long-term oral administration of β-glucan as an immunostimulant or an adjuvant on some non-specific parameters of the immune response of turbot *Scophthalmus maximus*. *Dis Aquatic Org* 26：139-147.

De Temmerman，M. L.，Rejman，J.，Demeester，J.，*et al.*，2011. Particulate vaccines：on the quest for optimal delivery and immune response. *Drug Discov Today* 16：569-582.

Dong，C. 2008. TH17 cells in development：an updated view of their molecular identity and genetic programming. *Nat Rev Immunol* 8：337-348.

Eaton，W. D. 1990. Antiviral activity in four species of salmonids following exposure to poly inosinic：cytidylic acid. *Dis Aquat Org* 9：193-198.

Evensen，Ø.，Brudeseth，B. and Mutoloki，S. 2005. The vaccine formulation and its role in inflammatory processes in fish-effects and adverse reactions. *Dev Biol*（*Basel*）121：117-125.

Fan，T. J.，Hu，X. Z.，Wang，L. Y.，*et al.*，2012. Development of an inactivated iridovirus vaccine against turbot viral reddish body syndrome. *J Ocean Univ China* 11：65-69.

Figueras，A.，Santarem，M. M. and Novoa，B. 1998. Influence of the sequence of administration of β-glucans and a *Vibrio damsela* vaccine on the immune response of turbot（*Scophthalmus maximus* L.）. *Vet Immunol Immunopathol* 64：59-68.

Flach，T. L.，Ng，G.，Hari，A.，*et al.*，2011. Alum interaction with dendritic cell membrane lipids is essential for its adjuvanticity. *Nat Med* 17：479-487.

Fredriksen，B. N. and Grip，J. 2012. PLGA/PLA micro- and nanoparticle formulations serve as antigen depots and induce elevated humoral responses after immunization of Atlantic salmon（*Salmo salar* L.）. *Vaccine* 30：656-667.

Glenny，A. T.，Pope，C. G.，Waddington，H. and Wallace，U. 1926. Immunological notes. XXIII. The antigenic value of toxoid precipitated by potassium alum. *J Pathol Bacteriol* 29：31-40.

Guy, B. 2007. The perfect mix: recent progress in adjuvant research. *Nat Microbiol* 5: 505-517.

Høie, S. H., Heum, M. and Thoresen, O. F. 1996. Detection of *Aeromonas salmonicida* by polymerase chain reaction in Atlantic salmon vaccinated against furunculosis. *Fish Shellfish Immunol* 6: 199-206

Hynes, N. A., Furnes, C., Fredriksen, B. N., *et al.*, 2011. Immune response of Atlantic salmon to recombinant flagellin. *Vaccine* 29: 7678-7687.

Jensen, I., Albuquerque, A., Sommer, A. I. and Robertsen, B. 2002. Effect of poly I: C on the expression of Mx proteins and resistance against infection by infectious salmon anaemia virus in Atlantic salmon. *Fish Shellfish Immunol* 13: 311-326.

Jiao, X. D., Zhang, M., Hu, Y. H. and Sun, L. 2009. Construction and evaluation of DNA vaccines encoding *Edwardsiella tarda* antigens. *Vaccine* 27: 5195-5202.

Jiao, X. D., Cheng, S., Hu, Y. H. and Sun, L. 2010a. Comparative study of the effects of aluminum adjuvants and Freund's incomplete adjuvant on the immune response to an *Edwardsiella tarda* major antigen. *Vaccine* 28: 1832-1837.

Jiao, X. D., Hu, Y. H. and Sun, L. 2010b. Dissection and localization of the immunostimulating domain of *Edwardsiella tarda* FliC. *Vaccine* 28: 5635-5640.

Kamilya, D., Maiti, T. K., Joardar, S. N. and Mal, B. C. 2006. Adjuvant effect of mushroom glucan and bovine lactoferrin upon *Aeromonas hydrophila* vaccination in catla, *Catla catla* (Hamilton). *J Fish Dis* 29: 331-337.

Kang, Y. J. and Kim, K. H. 2012. Effect of CpG-ODNs belonging to different classes on resistance of olive flounder (*Paralichthys olivaceus*) against viral hemorrhagic septicemia virus (VHSV) and *Miamiensis avidus* (Ciliata; Scuticociliata) infections. *Aquaculture* 324: 39-43.

Kim, H. J., Oseko, N., Nishizawa, T. and Yoshimizu, M. 2009. Protection of rainbow trout from infectious hematopoietic necrosis (IHN) by injection of infectious pancreatic necrosis virus (IPNV) or Poly (I: C). *Dis Aquat Org* 83: 105-113.

Lavelle, E. C., Jenkins, P. G. and Harris, J. E. 1997. Oral immunization of rainbow trout with antigen microencapsulated in poly (DL-lactide-co-glycolide) microparticles. *Vaccine* 15: 1070-1078.

Liu, C. S., Sun, Y., Hu, Y. H. and Sun, L. 2010a. Identification and analysis of a CpG motif that protects turbot (*Scophthalmus maximus*) against bacterial challenge and enhances vaccine-induced specific immunity. *Vaccine* 28: 4153-4161.

Liu, C. S., Sun, Y., Hu, Y. H. and Sun, L. 2010b. Identification and analysis of the immune effects of CpG motifs that protect Japanese flounder (*Paralichthys olivaceus*) against bacterial infection. *Fish Shellfish Immunol* 29: 279-285.

Manicassamy, S. and Pulendran, B. 2009. Modulation of adaptive immunity with Toll-like receptors. *Semin Immunol* 21: 185-193.

Martinez-Alonso, S., Martinez-Lopez, A., Estepa, A., *et al.*, 2011. The introduction of multi-copy CpG motifs into an antiviral DNA vaccine strongly upregulates its immunogenicity in fish. *Vaccine* 29: 1289-1296.

Midtlyng, P. J., Reitan, L. J. and Speilberg, L. 1996. Experimental studies on the efficacy and side-effects of intraperitoneal vaccination of Atlantic salmon (*Salmo salar* L.) against furunculosis. *Fish Shellfish Immunol* 6: 335-350.

Mizel, S. B. and Bates, J. T. 2010. Flagellin as an adjuvant: cellular mechanisms and potential. *J Immunol* 185: 5677-5682.

Mulvey, B., Landolt, M. L. and Busch, R. A. 1995. Effects of potassium aluminium sulphate (alum) used in an

Aeromonas salmonicida bacterin in Atlantic salmon, *Salmo salar* L. *J Fish Dis* 18: 495-506.

Munang'andu, H. M., Fredriksen, B. N., Mutoloki, S., *et al.*, 2012. Comparison of vaccine efficacy for different antigen delivery systems for infectious pancreatic necrosis virus vaccines in Atlantic salmon (*Salmo salar* L.) in a cohabitation challenge model. *Vaccine* 30: 4007-4016.

Mutoloki, S., Alexandersen, S. and Evensen Ø. 2004. Sequential study of antigen persistence and concomitant inflammatory reactions relative to side-effects and growth of Atlantic salmon (*Salmo salar* L.) following intraperitoneal injection with oil-adjuvanted vaccines. Fish Shellfish Immunol 16: 633-644.

Nikl, L., Albright, L. J. and Evelyn, T. P. T. 1991. Influence of seven immunostimulants on the immune response of coho salmon to *Aeromonas salmonicida*. *Dis Aquat Org* 12: 7-12.

Nishizawa, T., Takami, I., Kokawa, Y. and Yoshimizu, M. 2009. Fish immunization using a synthetic double-stranded RNA Poly (I: C), an interferon inducer, offers protection against RGNNV, a fish nodavirus. *Dis Aquat Org* 83: 115-122.

O'Donnell, G. B., Reilly, P., Davidson, G. A. and Ellis, A. E. 1996. The uptake of human gamma globulin incorporated into poly (D, L-lactide-co-glycolide) microparticles following oral intubation in Atlantic salmon, *Salmo salar* L. *Fish Shellfish Immunol* 6: 507-520.

Oh, M. J., Takami, I., Nishizawa, T., *et al.*, 2012. Field tests of Poly I: C immunization with nervous necrosis virus (NNV) in sevenband grouper, *Epinephelus septemfasciatus* (Thunberg). *J Fish Dis* 35: 187-191.

Oyewumi, M. O., Kumar, A. and Cui, Z. 2010. Nano-microparticulates as immune adjuvants: correlating particle sizes and the resultant immune responses. *Exp Rev Vaccines* 9: 1095-1107.

Plant, K. P., Harbottle, H. and Thune, R. L. 2005. Poly I: C induces an antiviral state against ictalurid herpesvirus 1 and Mx1 transcription in the channel catfish (*Ictalurus punctatus*). *Dev Comp Immunol* 29: 627-635.

Poppe, T. and Koppang, E. O. 2014. Side effects of vaccination, in *Fish Vaccination* (eds R. Gudding, A. Lillehaug and Ø. Evensen). Oxford, John Wiley & Sons, Ltd., 153-161.

Ribeiro, C. M. S. and Schijns, V. E. J. C. 2010. Immunology of vaccine adjuvants. *Methods Mol Biol* 626: 1-14. Rhodes, L. D., Rathbone, C. K., Corbett, S. C., *et al.*, 2004. Efficacy of cellular vaccines and genetic adjuvants against bacterial kidney disease in Chinook salmon (*Oncorhynchus tshawytscha*). *Fish Shellfish Immunol* 16: 461-474.

Robertsen, B. 1999. Modulation of the non-specific defence of fish by structurally conserved microbial polymers. *Fish Shellfish Immunol* 9: 269-290.

Rørstad, G., Aasjord, P. M. and Robertsen, B. 1993. Adjuvant effect of a yeast glucan in vaccines against furunculosis in Atlantic salmon (*Salmo salar* L.). *Fish Shellfish Immunol* 3: 179-190.

Schijns, V. E. J. C. 2001. Induction and direction of immune responses by vaccine adjuvants. *Crit Rev Immunol* 21: 456-463.

Selvaraj, V., Sampath, K. and Sekar, V. 2005. Administration of yeast glucan enhances survival and some non-specific and specific immune parameters in carp (*Cyprinus carpio*) infected with *Aeromonas hydrophila*. *Fish Shellfish Immunol* 19: 293-306.

Singh, M. and O'Hagan, D. T. 2003. Recent advances in veterinary vaccine adjuvants. *Int J Parasitol* 33: 469-478.

Sinyakov, M. S., Dror, M., Lublin-Tennenbaum, T., *et al.*, 2006. Nano- and microparticles as adjuvants in vaccine design: success and failure is related to host natural antibodies. *Vaccine* 24: 6534-6541.

Smith, D. C., Lord, J. M., Roberts, L. M., *et al.*, 2002. 1st class ticket to class I: protein toxins as pathfinders for antigen presentation. *Traffic* 3：697-704.

Steinman，L. 2007. A brief history of T（H）17，the first major revision in the T（H）1/T（H）2hypothesis of T cell-mediated tissue damage. *Nat Med* 13：139-145.

Sun，H. X.，Xie，Y. and Ye，Y. P. 2009. Advances in saponin-based adjuvants. *Vaccine* 27：1787-1796.

Takami，I.，Kwon，S. R.，Nishizawa，T. and Yoshimizu，M. 2010. Protection of Japanese flounder *Paralichthys olivaceus* from viral hemorrhagic septicemia（VHS）by Poly（I：C）immunization. *Dis Aquat Org* 89：109-115.

Thim，H. L.，Iliev，D. B.，Christie，K. E.，*et al.*，2012. Immunoprotective activity of a salmonid alphavirus vaccine：comparison of the immune responses induced by inactivated whole virus antigen formulations based on CpG class B oligonucleotides and poly I：C alone or combined with an oil adjuvant. *Vaccine* 30：4828-4834.

Tian，J.，Sun，X.，Chen，X.，*et al.*，2008. The formulation and immunisation of oral poly（DL-lactideco-glycolide）microcapsules containing a plasmid vaccine against lymphocystis disease virus in Japanese flounder（*Paralichthys olivaceus*）. *Int Immunopharmacol* 8：900-908.

Tian，J. and Yu，J. 2011. Poly（lactic-co-glycolic acid）nanoparticles as candidate DNA vaccine carrier for oral immunization of Japanese flounder（*Paralichthys olivaceus*）against lymphocystis disease virus. *Fish Shellfish Immunol* 30：109-117.

Tyler，J. W. and Klesius，P. H. 1994. Protection against enteric septicemia of catfish（*Ictalurus punctatus*）by immunization with the R-mutant，*Escherichia coli*（J5）. *Am J Vet Res* 55：1256-1260.

Weiner，D. B. and Kennedy，R. C. 1999. Genetic vaccines. *Sci Am* 281：50-57.

Wilhelm，V.，Miquel，A.，Burzio，L. O.，*et al.*，2006. A vaccine against the salmonid pathogen *Piscirickettsia salmonis* based on recombinant proteins. *Vaccine* 24：5083-5091.

Zhu，L.-Y.，Nie，L.，Zhu，G.，*et al.*，2013. Advances in research of fish immune-relevant genes：A comparative overview of innate and adaptive immunity in teleosts. *Dev Comp Immunol* 39：39-62.

第八章　鱼类先天性免疫反应

Jorunn B. Jørgensen

本章概要　SUMMARY

有关先天性免疫的研究发现，硬骨鱼类中既有系统发生上的保守机制，也有高等脊椎动物中不存在的特有机制。通过同源克隆和利用已知的某些硬骨鱼类的基因组和转录组数据，许多“先天性”的免疫基因，如 Toll 样受体等模式识别受体和多种细胞因子，已被鉴定出来，而且也开展了这些分子的生物活性方面的研究，但其功能活性的研究才刚刚开始。鱼类中存在重要的先天性免疫细胞，包括巨噬细胞、中性粒细胞以及一些非特异性的毒性细胞，但是因为缺乏类似白细胞分化抗原簇（CD）的特异性抗体，无法像哺乳动物一样根据 CD 标志分子鉴定白细胞及其亚群。本章主要介绍鱼类先天性免疫的最新研究进展。

第一节　引　言

先天性免疫系统是抵御感染的第一道防线，并可以直接对抗感染威胁。先天性免疫系统既存在于无脊椎动物中，也存在于脊椎动物中，对很多病原体起到免疫防御作用。在进化地位低于无颌鱼类的生物中，先天性免疫是唯一的防御机制。约 4.5 亿年前，远古的有颌类软骨鱼进化出重组激活基因（RAGs），被认为是出现了具有抗原特异性的获得性免疫（Agrawal 等，1998）。基于 RAG 基因的重排，分化发育出免疫球蛋白（Igs）和细胞受体，也就是现在熟知的 B 细胞和 T 细胞（Flajnik 和 Kasahara，2010）。最近的研究还表明，现存量很少的无颌鱼类具有抗原特异性免疫（Pancer 等，2004；Deng 等，2010）。有颌鱼类和高等脊椎动物的先天性免疫具有双重作用：控制并破坏病原体，同时诱发相应的获得性免疫应答的全面免疫反应。

第二节　先天性免疫的感应与效应

先天性免疫可大致分为感应和效应两部分。感应主要是指鱼类等所有多细胞生物如何识别感染，而效应是指细胞如何参与抗感染反应。感应和效应又可以根据来源分为细胞和体液成分。图 8-1 列出了鱼类中已知的先天性免疫成分。本章主要介绍鱼类先天性免疫的重要细胞类型，先天性免疫的细胞感应，以及细胞因子和补体成分等体液效应分子。

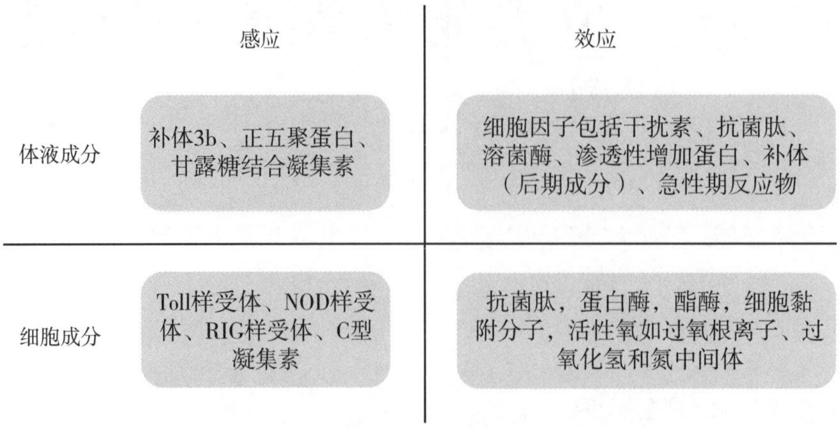

图 8-1　鱼类先天性免疫系统感应和效应阶段的细胞和体液成分

第三节　专职吞噬细胞：巨噬细胞和中性粒细胞

在脊椎动物中，先天性免疫主要依赖于髓样细胞，它是吞噬并破坏病原体的专职吞噬细胞。髓样细胞包括单核吞噬细胞和多形核吞噬细胞。单核吞噬细胞来源于单核细胞的巨噬细胞，与髓样树突状细胞（DCs）密切相关。DCs 也是来源于单核细胞，能有效地递呈抗原给获得性免疫系统的 T 细胞。在鱼类中，巨噬细胞是最早被分离出并开展研究的白细胞（Braun-Nesje 等，1981），据报道，鱼类巨噬细胞与哺乳动物一样具有多种表型（Hanington 等，2009；Forlenza 等，2011）。鱼类中，中性粒细胞是主要的多形核吞噬细胞（Ellis，2001）。硬骨鱼类没有骨髓和淋巴结，其单核细胞发生于头肾（pronephros），此外，头肾还具有抗原捕获功能（Zapata 和 Cooper，1990）。

在鱼类中，巨噬细胞和粒细胞都是重要的炎性细胞，在鱼体被感染或组织损伤时，它们就会被调动起来。巨噬细胞是腹腔内的吞噬细胞，而中性粒细胞在炎性反应过程中出现。鱼类巨噬细胞的主要鉴定特征有：形态学（豆状核和高的质/核比）、非特异性酯酶和内源性过氧化物酶活性，以及胞饮和吞噬能力（Ellis，2001；Lieschke 等，2001；Traver 等，2003）。鱼类中没有鉴定出单核细胞-巨噬细胞特异性的“分化抗原簇”（CD）抗原，然而对一些鱼类基因组测序的研究表明，鱼类与哺乳动物具有抗原同源性的候选基因越来越多。值得注意的是，由 MD2、CD14 和 TLR4 分子组成的脂多糖（LPS）-感应系统，是哺乳动物巨噬细胞的显著特征，而在鱼类中并没有完全呈现（Iliev 等，2005）。这些分子中，只有 TLR4 的同源体在部分鱼类基因组的测序中被检测到（Jault 等，2004）。中性粒细胞也是鱼类中具有强吞噬能力的细胞，可通过形态（比巨噬细胞小，且具分叶形的核）与巨噬细胞区分（Lieschke 等，2001）。鱼类中特有的是黑色素巨噬细胞，它们含有数量不等的黑色素颗粒，存在于头肾和脾脏，也能在多数鱼类的炎症组织中发现。这些细胞被认为是抗原递呈系统中的一部分（Agius 和 Roberts，2003；Koppang 等，2003）。

吞噬作用和细胞内杀伤是吞噬细胞的主要功能，微生物被吞噬形成具有膜包被的吞噬体。吞噬体与颗粒融合产生吞噬溶酶体，在吞噬溶酶体中微生物被抗菌肽（AMPs）、酶和活性氧（ROS）协同作用有效地杀死。许多研究发现，头肾的单核细胞/巨噬细胞，还有中性粒细胞可诱导抗菌反应（Neumann 等，2001；Rodriguez 等，2003；Plouffe 等，2005），包括呼吸暴发产生活性氧（ROS）（Jørgensen 和 Robertsen，1995；Rodriguez 等，2003）以及诱导型一氧化氮合成酶（iNOS）产生具有抗菌作用的含氮中间产物（Laing 等，1999；Stafford 等，2002；Forlenza 等，2008）。越来越多的研究表明，鱼类拥有抗菌肽类物质（AMPs），如防御素、自然防御相关巨噬细胞蛋白（Nramp）、自然杀伤细胞溶解酶和抗菌多肽（Falco 等，2008；Casadei 等，2009；Nam 等，2010），这些也是吞噬细胞杀伤作用的重要组分。有综述文章对 AMPs 研究进展进行了较为详细的介绍（Zhu 等，2013）。除了通过吞噬作用和分泌抗菌成分清除病原体，中性粒细胞还可以释放由 DNA 和抗菌蛋白组成的胞外结构，即中性粒胞外陷阱（NETs），捕获和杀死微生物（Brinkmann 和 Zychlinsky，2007）。利用激光共聚焦显微镜和光学显微镜技术已经证明，活化的鱼类中性粒细胞可以产生 NETs（Palic 等，2007）。

鱼类单核吞噬细胞除了具有直接的抗菌作用外，还表达 MHC-Ⅱ类分子（Iliev 等，2010），具有抗原处理和递呈作用（Vallejo 等，1991a，1991b；Wittamer 等，2011；Mutoloki 等，2014）。

第四节　自然杀伤（NK）样细胞

NK 细胞是一种淋巴细胞，能识别感染和/或应激的细胞，通过直接杀死这些细胞或分泌炎性细胞因子进行免疫防御反应。哺乳动物的 NK 细胞在抗病毒反应中起着重要的作用，可以直接杀死病毒感染的细胞，还可以产生一种介导抗病毒活性的细胞因子——γ-干扰素。与细胞毒性 T 细胞不同的是，自然杀伤样细胞是以一种非组织相容性复合体限制的方式杀死细胞，不需要致敏反应。体内和体外研究证明了鱼类具有自然杀伤样细胞。例如，鱼类白细胞未经免疫即可自发地杀死各种异种细胞（Fischer 等，1998；Somamoto 等，2000）。

Shen 等（2002）综述了斑点叉尾鮰中两种非特异的细胞毒性细胞：自然杀伤（NK）样细胞和非特异的细胞毒细胞（NCC）。两种类型的细胞都能杀死哺乳动物 NK 细胞的典型靶标细胞，如异种异体细胞和同种异体细胞。与哺乳动物 NK 细胞类似，斑点叉尾鮰 NK 样细胞是从外周血中分离出的大的颗粒细胞（Shen 等，2004），而 NCC 是在淋巴组织中发现的小的非颗粒细胞（Evans 等，1984a，1984b，1984c）。NCC 表达 34 ku 的蛋白，被命名为 NCC-受体蛋白 1（Jaso-Friedmann 等，1997a，1997b，2001）可以识别目标细胞上的一个单一且高度保守的靶标抗原。NCC 细胞被认为是哺乳动物 NK 细胞进化的前体细胞，该细胞已在多种鱼类中发现（Plouffe 等，2005）。在鱼类中发现了 NK 样细胞毒性白细胞，正在开展细胞受体与哺乳动物 NK 细胞受体在序列和结构的相关性鉴定研究（Fischer 等，2006；Yoder 和 Litman，2011）。通过对调节 NK 样细胞的细胞毒性作用受体的鉴定，才能进一步研究鱼类细胞介导的细胞毒效应。鱼的细胞毒性细胞如何被触发，以及如何以受体区别感染与未感染的细胞，均是亟待研究的课题，这将进一步阐释硬骨鱼类细胞毒性反应的机制。

第五节 先天性免疫的感应

先天性免疫系统可以直接对抗病原，无须依赖复杂的机制，与需选择细胞类型并通过增殖对外来抗原作出反应的获得性免疫系统不同。先天性免疫系统是针对存在于一类微生物或其他病原体共有结构的反应，这些共有的结构称为病原相关分子模式（PAMPS）。宿主通过可溶性的细胞遗传来的病原识别受体（PRRs，图 8-2）来识别 PAMPS。近 10 年，有关鱼类识别病毒或细菌先天性免疫受体的研究取得了很大的进展，哺乳动物中发现的大多数 PRR 家族也在鱼

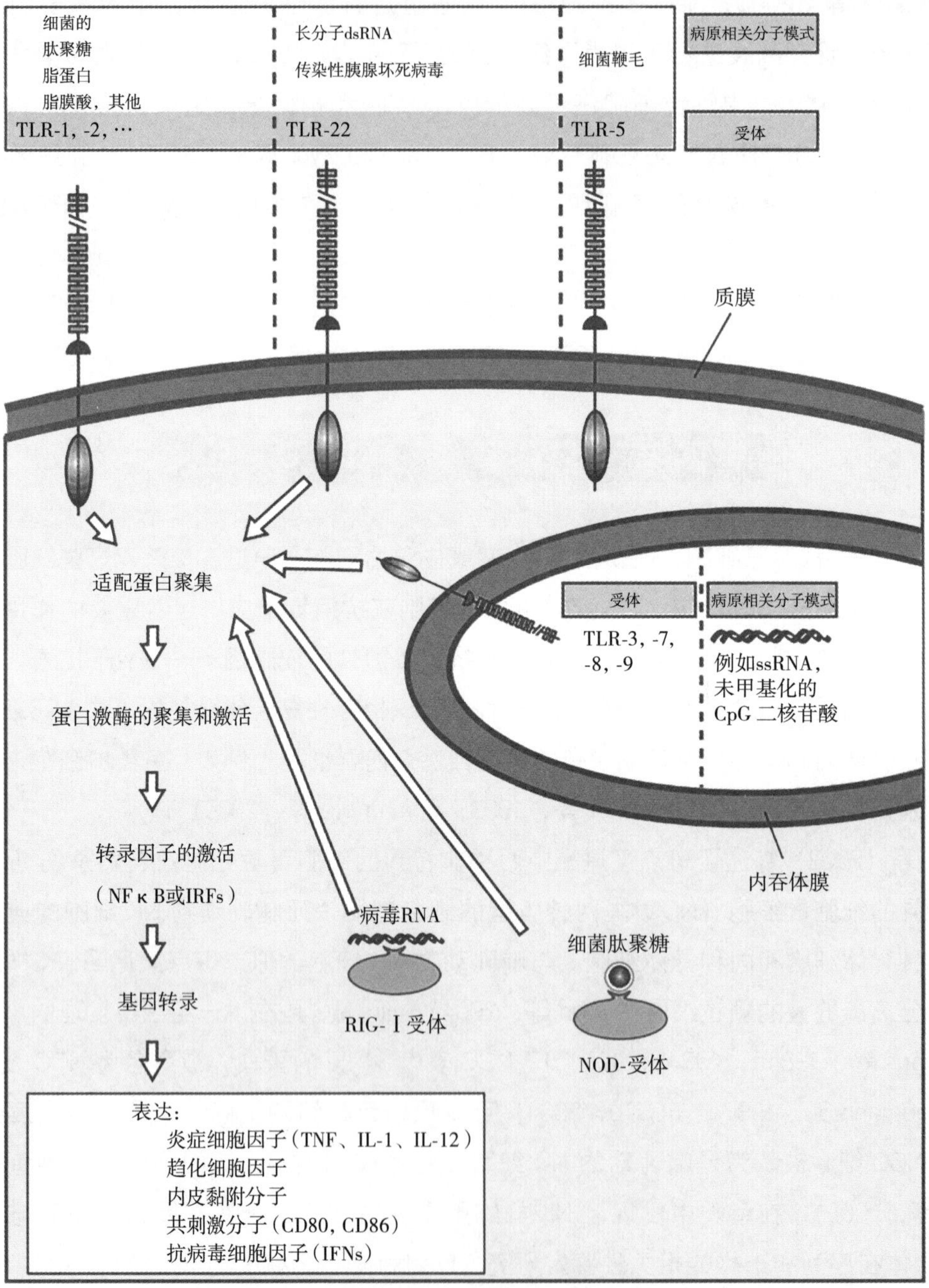

图 8-2 模式识别受体（PRRs）的特性与功能

不同的 PRRs 响应不同的微生物分子、脂蛋白和核酸。多数 PRRs 激活的细胞免疫反应具有类似的信号机制，细胞免疫反应是先天性免疫反应的主体

类中被鉴定出来，包括 Toll 样受体（TLRs）、RIG-Ⅰ样受体、NOD 样受体、C-型凝集素和补体成分（Zhu 等，2013）。然而，在大多数鱼中，PRR 与特定 PAMPS 的相互作用、受体配对后引起的信号通路机制还不完全清楚。

第六节　Toll 样受体是鱼类中研究最透彻的模式识别受体

鱼类研究得最清楚的病原识别受体（PRRs）是 Toll 样受体（TLRs），已在不同种类的鱼中鉴定出 17 种 TLRs（Palti，2011）。人类中，通过在细胞上的定位，将 TLRs 分为两种主要亚型：TLR1、TLR2、TLR4、TLR5、TLR6 和 TLR10 为Ⅰ型受体，主要分布在细胞表面，可以识别微生物脂质和来源于细菌和真菌的糖类；而 TLR3、TLR7、TLR8 和 TLR9 属于Ⅱ型受体存在于细胞核内，主要识别来源于病毒或细菌的双链或单链 RNA 和非甲基化 CpG DNA。在鱼类中这两种亚型的主要 TLR 均被发现和确认（Rebl 等，2010；Palti，2011），然而，与哺乳动物相比，鱼类 TLRs 具有更丰富的多样性，并发现 6 种哺乳动物中不存在的 TLRs，其中包括与 TLR1 和 TLR2 具有相似的序列和结构的 TLR14（Oshiumi 等，2003；Jault 等，2004；Meijer 等，2004；Hwang 等，2011；Star 等，2011），以及鱼类特有的 TLR19、TLR20、TLR22 和 TLR23（Oshiumi 等，2003；Meijer 等，2004）。在鱼类特有的 TLRs 中，河鲀的 TLR22 研究得较清楚，它可以识别细胞表面较长的 dsRNA（Matsuo 等，2008）。鱼类 TLR 具有明显的特征，例如，TLR5 在哺乳动物中是一种膜结合受体，能识别细菌的鞭毛蛋白；而在虹鳟中发现，TLR5 既是一种膜结合蛋白（TLR5m）又是一种可溶性蛋白（TLR5s）（Tsujita 等，2004）。这在其他鱼类中也有相同的发现（Oshiumi 等，2003；Hwang 等，2010）。研究发现通过激活核因子 NFκB，虹鳟膜结合型 TLR5 通路介导对鞭毛蛋白的反应，与可溶性型 TLR5（TLR5s）结合，通过正反馈方式相互作用放大对鞭毛蛋白的反应（Tsujita 等，2004）。大西洋鳕 TLR 较独特，因为缺乏识别细菌表面抗原的 TLR（TLR1、TLR2 和 TLR5），却多了识别核酸的 TLR 成员（TLR7、TLR8、TLR9 和 TLR22）（Star 等，2011）。这些结果表明，通过核酸探测系统识别病原体对大西洋鳕的免疫防御极为重要。

第七节　NOD 样受体和 RIG-Ⅰ受体在鱼类中的发现

在鱼类先天性免疫受体中，另外两种参与 TLRs 反应的重要病原识别分子是 NOD 样受体（NLRs）（Laing 等，2008）和视黄酸诱导基因Ⅰ（RIG-Ⅰ）样受体（RLRs）。两者都是胞内细胞质受体，分别参与细菌和病毒免疫防御反应。NLR 家族成员主要功能是激活 NFκB 信号（NOD1 和 NOD2）或分泌促炎症细胞因子 IL-1 和 IL-18（NALPs）（Chen 等，2009）。在哺乳动物中，NLRs 主要在免疫细胞（淋巴细胞和抗原递呈细胞，如巨噬细胞和树突状细胞）中表达，但也在上皮细胞等非免疫细胞中表达（Chen 等，2009）。硬骨鱼 NLR 家族的基因最早是在斑马鱼基因组中发现的，由三个不同的亚家族组成，包括一个独特的亚家族（Laing 等，2008）。在其他鱼类中也发现了 *NLR* 基因，包括斑点叉尾鮰（Sha 等，2009；Rajendran 等，2012a）、牙鲆（Park 等，2012）、鲤（Chen 等，2010）和虹鳟（Chang 等，2011b），并且这些基因广泛表达于各种组织和白细胞中。*NOD*1 是斑点叉尾鮰 NLR 基因家族中的一个基因，在胞内病原菌鮰爱德华菌（*Edwardsiella ictaluri*）感染后显著表达，而其他的 *NOD* 基因基本稳定表达（Sha 等，2009)。这表明斑点叉尾鮰 *NOD*1 可能参与胞内病原感染的免疫反应。鳟 *NOD*2 效应区域的过量表达可增加一些促炎症基因、Ⅰ型和Ⅱ型干扰素以及各种细胞凋亡蛋白酶的表达。这些结果表明，鱼类 *NOD*2 参与炎症反应，并可能调节细胞凋亡并诱发抗菌和抗病毒防御（Chang 等，2011b)。

RLRs 包括三个成员：RIG-Ⅰ、MDA-5 和 LGP2。它们都是 RNA 解旋酶，识别细胞质内的病毒 RNA（Kumar 等，2011)。RIG-Ⅰ在草鱼（Yang 等，2011）和斑点叉尾鮰（Rajendran 等，2012b）中已被克隆，MDA-5 和 LGP2 在虹鳟（Chang 等，2011a）和草鱼（Chang 等，2011a）中被鉴定。在虹鳟中，MDA-5 和 LGP2 均对 poly I:C 具有高亲合力（Chang 等，2011a)。此外，虹鳟（Chang 等，2011a)、牙鲆（Ohtani 等，2011）和草鱼（Wang 等，2012）的 MDA5，以及虹鳟的 LGP2 均具有抗病毒能力（Chang 等，2011a)。上述这些结果表明，类似于哺乳动物的 RLR，鱼类 RLRs 在宿主监测病毒感染方面发挥作用。几篇关于鱼类先天性免疫识别的综述，对鱼类 PRRs 的基因组学和生物学进行了更详细的介绍（Zhang 和 Gui，2012；Zhu 等，2013)。

第八节 凝集素是多功能的糖类识别分子

凝集素（lectins）是一类糖结合蛋白，能特异性结合病毒、细菌、真菌和动物的糖类结构，进而凝集各种细胞以发挥它们的生物学功能。在动物中，凝集素是先天性免疫的重要组成部分，可以诱导吞噬作用、激活血小板、启动补体级联反应和增强自然杀伤细胞的活性（Vasta 等，2004)。凝集素还可以通过介导树突状细胞（DCs）对细菌和病毒成分进行识别，调节获得性免疫反应。许多高等脊椎动物的研究表明，凝集素结合抗原后，进入溶酶体内，与结合了抗原肽的 MHC-Ⅱ类分子，递呈至 CD4 T 淋巴细胞（van Vliet 等，2007)。

C 型凝集素（CTLs）是 PRR 家族的成员。CTLs 是通过 C 型糖类识别域（CRD）或 C 型凝集素域与糖类残基相互作用的性能而命名的。CTLs 可进一步分为胶凝素、选凝素和甘露糖-巨噬细胞受体，它们直接参与免疫反应（Vasta 等，2011)。胶凝素包括甘露糖结合凝集素(MBL)，它与丝氨酸蛋白酶 MASP 结合，诱导激活补体系统（详见本章第十二节)。C 型凝集素的一些成员已在虹鳟、鲤、大菱鲆和斑马鱼等鱼类中报道（Vasta 等，2011)。所有这些 C 型凝集素在细菌入侵后，均显著被诱导高表达，并可凝集和抑制多种病原体。

在鱼类中发现的另一种凝集素是树突状细胞特异性细胞间黏附分子-3-非整合素（DC-SIGN)，也称为 CD209，是一种Ⅱ型跨膜蛋白。此外，DC-SIGN 是一种可介导树突状细胞成熟、迁移和 T 淋巴细胞活化的重要黏附受体。鱼类的首个 DC-SIGN 是在斑马鱼中发现的，与哺乳动物 DC-SIGN 家族成员特征相近，在胞外区有一个典型的 CRD 结构域（Lin 等，2009)。功能性研究表明，在体内，抗 DC-SIGN 抗体可显著抑制由外源性血蓝蛋白抗原诱导的 T 细胞活化和抗体生成。吞噬作用检测表明，DC-SIGN 蛋白并没有直接参与 APCs 的抗原摄取，但可能在 APC 迁移和聚集以及后期的 T 细胞活化中起作用（Lin 等，2009)。近年来在鱼类也发现了树突状细胞，但 DC-SIGN 是否是鱼类树突状细胞参与获得性免疫反应的功能分子需要进一步研究。

第九节 模式识别受体及其诱导的免疫反应

大多数 TLRs、NLRs 和 RLRs 在与 PAMPs 结合后，会激活不同信号通路，通过这些信号通路调节核因子 NFκB 和干扰素调节因子（IRFs）等各种转录因子的活化（Kumar 等，2011）。NFκB 的激活会增强炎性细胞因子反应，诱导树突状细胞的成熟（CD80、CD83、CD86），以及趋化因子和趋化因子受体表达增加（Janeway 和 Medzhitov，2002）。最终 PRR 的激活，特别是 TLRs 的激活，将未成熟的树突状细胞转换为能诱导获得性免疫反应的炎性表型，促进抗原特异性的 CD4 和 CD8T 细胞反应和体液反应。IRF 激活会诱导干扰素的表达，不仅产生直接抗病毒的活性，还在调节宿主其他免疫反应中具有广泛的作用（Kumar 等，2011）。在不同鱼类中，应用 PRRs 刺激剂在体外培养的白细胞以及体内实验进行了大量的研究，结果表明，细胞因子和趋化因子的产生量显著增加，共刺激分子的上调表达以及抗感染保护也显著增强（Jørgensen 等，2001a，2001b；Jensen 等，2002b；Jensen 和 Robertsen，2002；Pedersen 等，2006；Purcell 等，2006b；Strandskog 等，2007，2008a；Kileng 等，2008）。在哺乳动物和鱼类中，研究较多的 PRR 刺激剂是合成的核酸，例如 CpG -寡核苷酸和 poly I：C，在大西洋鲑中的研究表明，这些合成的刺激物可以产生强烈的Ⅰ型干扰素反应，进而增强大西洋鲑抗 IPNV、ISAV 和 SPDV 感染的免疫应答（Jensen 等，2002a；Jørgensen 等，2003；Strandskog 等，2011）。显然，先天性免疫信号通路之间的相互作用对产生应答反应也是必要的，比如多个 TLRs 的激活能产生互补或协同效应（Strandskog 等，2008b，2011；Thim 等，2012）。

第十节 参与先天性免疫的细胞因子

细胞因子是一种小分子的细胞信号蛋白，是一个庞大而多样化的蛋白家族。细胞因子由白细胞介素（ILs）和干扰素（IFNs）等免疫调节因子组成。目前，哺乳动物中已报道大约有 35 种 ILs，它们中的多数的同源体已在硬骨鱼类中发现（Secombes 等，2011）。虽然很多鱼的细胞因子基因序列已经被发现和鉴定，但是关于它们的生物学功能认知仍很有限。本章主要介绍细胞因子 IL-1β、肿瘤坏死因子 TNF-α 和Ⅰ型干扰素，它们是先天性免疫细胞中产生的重要细胞因子，也是目前特性最为明确的鱼类细胞因子。

促炎细胞因子 IL-1β 和肿瘤坏死因子 TNF-α

事实上，所有有核细胞特别是内皮细胞/上皮细胞和巨噬细胞可产生白介素 IL-1 和肿瘤坏死因子（TNF-α）。这两种细胞因子都属于多基因家族，相关成员参与各种免疫学和发育过程。

IL-1 家族在免疫调节特别是炎症反应中发挥核心作用，包括 11 个成员。在哺乳动物中，IL-1α、IL-1β 和 IL-18 已被广泛地研究（Barksby 等，2007）。目前，在鱼类中只发现了 IL-1β 和 IL-18 的同源基因（Secombes 等，2011）。许多硬骨鱼中已经克隆出 IL-1β，包括鲑（Ingerslev 等，2006）、鳟（Zou 等，1999）、鳕（Seppola 等，2008）和鲤（Engelsma 等，2003）。在大多数鱼类中，存在着不止一个 *IL-1β* 基因，例如鲑科鱼有三种不同的 *IL-1β* 基因，分别为 *IL-*

1*β*1、*IL*-1*β*2 和 *IL*-1*β*3（Husain 等，2012）。研究发现，*IL*-1*β* 表达于健康鱼几乎所有的组织中，但表达最高的是脾脏、肾和鳃等免疫器官（Ingerslev 等，2006；Husain 等，2012）。

人 IL-1β 是由无活性的 31 ku 的前体（IL-1β 前体）经过加工处理产生的分泌型的、具有生物活性的 17ku 的成熟蛋白。这个过程主要在半胱氨酸蛋白酶、天冬氨酸蛋白水解酶 1（caspase1）的作用下完成（Thornberry 等，1992）。caspase1 在激活 IL-1β 之前，于炎性小体的多蛋白复合体中，在与 PAMPs 的免疫应答过程中被激活（Martinon 等，2002）。通过对鱼类 *IL*-1*β* 基因的序列分析，未发现 caspase1 的酶切位点（Secombes 等，2011）。尽管如此，有间接证据表明，鱼类中的 *IL*-1*β* 是通过类似的半胱氨酸天冬氨酸酶进行加工修饰的，因为有研究显示重组的鳟 *IL*-1*β* 片段具有生物学活性（Hong 等，2004）且在鱼类细胞培养中出现了被加工修饰过的 *IL*-1*β*（Pelegrin 等，2004）。有研究找到了较为明确的证据，在胞内病原弗朗西斯菌（*Francisella noatunensis*）感染的斑马鱼的主要白细胞中检测到完整的激活态的 IL-1β（约 18ku）（Vojtech 等，2012）。当这些细胞预先孵育含半胱氨酸天冬氨酸酶抑制剂后，IL-1β 的激活被部分抑制。在同一研究中，斑马鱼的两种同源炎性 caspase 在激活 IL-1β 上具有不同的特异性。该研究表明，鱼类中可能存在炎性小体免疫激活途径。

在高等脊椎动物中，IL-1β 已被证明可介导多种炎症反应，包括急性反应、巨噬细胞的激活以及其后的 TNF 和 IL-6 等细胞因子的分泌，还可以活化 T 细胞、B 细胞和 NK 细胞。在感染（Husain 等，2012；Wang 等，2009）和促炎症因子刺激（Strandskog 等，2008b；Wang 等，2009；Husain 等，2012）后，鱼类白细胞 IL-1β 转录水平显著上调表达；重组表达的 IL-1β 可影响细胞增殖和免疫基因的表达（Peddie 等，2001，2002；Buonocore 等，2005）。

与 IL-1 一样，TNF-α 是一种多效细胞因子，主要调节白细胞转运和炎症反应（Ware，2003）。TNF-α 属于 TNF 配体超家族，它们是一类结构相似的参与细胞信号转导通路的蛋白质，有关鱼类的 TNF-α 超家族的描述可参考 2011 年 Wiens 和 Glenney 的报道。TNF-α 同源基因已在不同种鱼类中被鉴定出来（Laing 等，2001；Saeij 等，2003；Zou 等，2003b），与 IL-1β 相似，基因的数量（1～3）在物种间有差异（Zou 等，2002；Haugland 等，2007；Saeij 等，2003；Savan 等，2005），这可能是鱼类特异性基因/基因组重复的结果（Meyer 和 Van de Peer，2005）。

TNF-α 是Ⅱ型跨膜糖蛋白，具有一个胞外 C 端功能域和一个胞质尾区，可以是膜结合型，也可以是可溶性形式存在。可溶性形式以三聚体形式存在，由与受体结合的三个相同的亚基组成。研究显示，*TNF-α* 基因在头肾和鳃组织中稳定表达，并且在分离的头肾白细胞中可被 LPS 诱导表达（Laing 等，2001）。鱼类 TNF-α 分子间存在物种差异，跟它们激活吞噬细胞的能力直接相关。例如，在虹鳟、鲤和金鱼中的研究发现，在体外，不同浓度的 TNF-α 重组蛋白能刺激细胞吞噬活性（Zou 等，2003a；Grayfer 等，2008；Forlenza 等，2009）。在海鲷中却得到相反的结果，TNF-α 是内皮细胞的激活剂，但不能直接激活吞噬细胞（Roca 等，2008）。经 TNF-α 处理的海鲷内皮细胞上清液，能够促进海鲷白细胞的黏附、迁移和激活。这和哺乳动物 TNF 的功能相类似，即用 TNF-α 刺激内皮细胞可增加黏附分子的表达，以及 IL-8 等趋化调节因子的合成。因此，TNF-α 促进白细胞聚集和激活并促进炎症过程。与哺乳动物一样，TNF-α 和 IL-1β 这两个细胞因子的诱导可以通过激活丝裂原活化蛋白激酶 p38 来调节（Hansen 和 Jørgensen，2007）。

第十一节 干扰素

在病毒感染过程中，机体产生的最重要的防御分子为干扰素（IFNs），其因具有干扰病毒复制的能力而得名。因此，诱导干扰素的产生和调控干扰素产生的水平是病毒感染期间的关键。大多数病毒以其 dsRNA 或 ssRNA 作为病原相关分子模式，可被 PRRs 识别，并激活转录因子，然后如前所述表达Ⅰ型 IFNs。这种先天性应答在受感染的细胞及其相邻细胞中建立起“抗病毒”状态，并刺激免疫细胞感知危险信号（图 8-3）。在哺乳动物中，根据 IFN 的生物学和结构特性，将其分为三种类型：Ⅰ型（IFN-α 和 IFN-β），Ⅱ型（IFN-γ）和Ⅲ型（IFN-λ）。

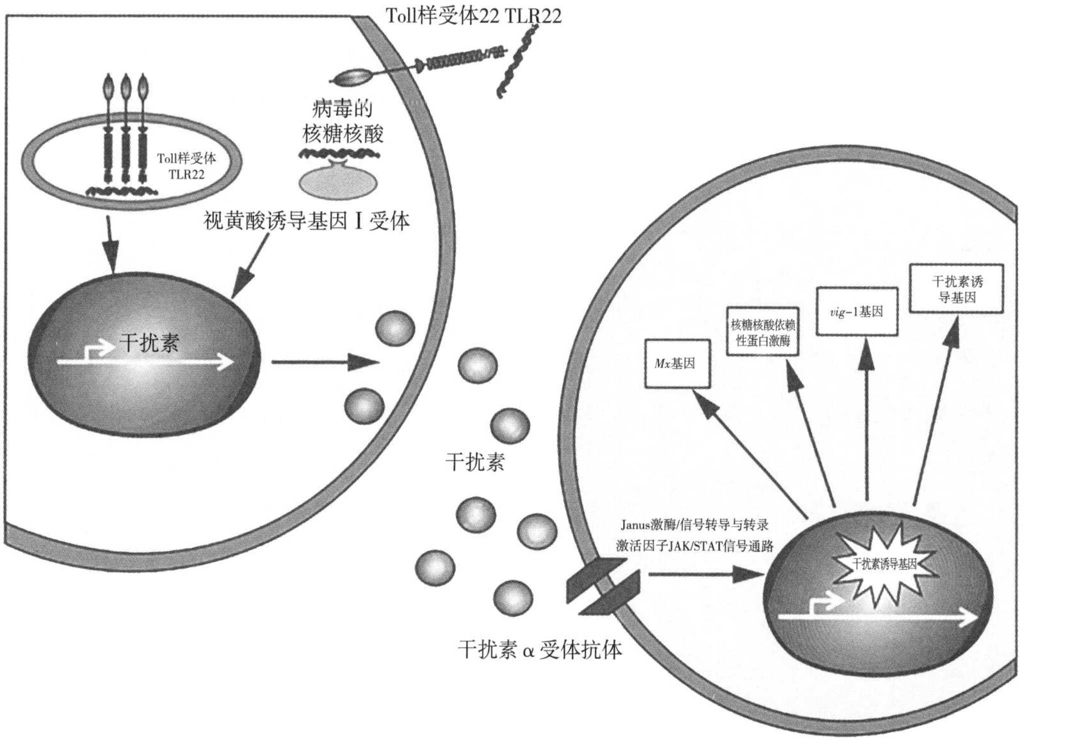

图 8-3 鱼类病毒诱导的基因和Ⅰ型干扰素信号途径

病毒进入细胞后，病毒核酸结合 RLRs 和 TLRs 等各种感应分子。此外鱼类独特的细胞表面受体 TLR22 结合 dsRNA，激活感应分子诱导干扰素的活性和分泌。干扰素有自分泌和旁分泌的活性，介导干扰素与跨膜受体（IFNAR）相互作用，激活 JAK/STAT 通路，然后激活黏病毒抵抗蛋白（Mx）、干扰素刺激基因 15 和 vig/viperin 等干扰素刺激基因（ISGs）

在鱼类中，诱导具有 IFN 特性的分泌分子的研究在 1973 年就已有报道（de Kinkelin 和 Dorson，1973），但直到 2003 年，才在三种鱼中克隆出首个Ⅰ型干扰素：斑马鱼（Altmann 等，2003）、大西洋鲑（Robertsen 等，2003）和河鲀（Lutfalla 等，2003）。和哺乳动物一样，鱼类有多个 IFN 的副本（Zou 和 Secombes，2011），例如，大西洋鲑至少有 11 个 *IFN* 基因，分为 3 个亚型（Sun 等，2009）。有研究揭示了鲑的各种 *IFN* 基因亚型的不同作用（Svingerud 等，2012）。在哺乳动物中，干扰素信号通过 JAK/STAT 通路转导，上调表达几百个干扰素刺激基因（ISGs），形成“抗病毒”状态（图 8-3）。许多 ISGs 有直接的抗病毒功能，在大西洋鲑和牙鲆中发现，Mx、ISG15 和双链 RNA 依赖性蛋白激酶（PKR）具有直接抗病毒的活性（Caipang 等，2003；Larsen 等，2004；Zhu 等，2008）。但是鱼类大多数的 ISGs 功能仍然未知。

鱼类中已发现 IFN-γ 的同系物（Zou 等，2005；Igawa 等，2006；Milev-Milovanovic 等，2006；Stolte 等，2008；Furnes 等，2009），虹鳟 IFN-γ 具有与哺乳动物类似物一样的功能特性，包括激活巨噬细胞的能力等（Zou 等，2005），并在金鱼等其他鱼类的研究中被证实（Grayfer 等，2010）。不同鱼类的Ⅰ型 IFN 和 IFN-γ 重组蛋白已成功表达，并对其生物学活性进行了研究（Zou 等，2007；Ohta 等，2011；Sun 等，2011），鱼类 IFN-γ 可诱导许多对Ⅰ型 IFN 也反应的 *ISGs* 基因的表达，这表明Ⅰ型 IFN 和Ⅱ型 IFN 的先天性反应存在交叉激活；然而，与Ⅰ型 IFN 相比，IFN-γ 诱导 ISG 的能力较弱，因此需要相对高剂量的 IFN-γ 抵抗病毒（Sun 等，2011）。此外，还发现鱼类 IFN-γ 能够诱导鲑 *TLR*8 和 *TLR*9 的表达（Skjæveland 等，2008，2009）。

许多高致病性病毒已经进化出逃避 IFN 反应的机制，且宿主免疫应答的强度与病毒拮抗机制之间的平衡决定了病毒是否能被清除。已证实，ISAV 和 IPNV 等鲑鳟鱼类病毒能干扰 IFN 应答（Garcia-Rosado 等，2008；Skjesol 等，2009）。通过鉴定 IFN 拮抗的病毒蛋白，研制该蛋白残缺或缺失的减毒病毒用作候选疫苗，诱发更显著的先天性免疫反应。

第十二节　补体系统

与哺乳动物一样，补体系统由大约 30 种不同的胞质蛋白和膜相关蛋白所组成，在先天性免疫中起着关键作用。补体激活的三种不同途径（经典途径、替代途径和凝集素途径）涉及肝脏中由肝细胞产生并分泌到血浆中的几种蛋白质，但 C1q、C7 和因子 D 等蛋白质主要在肝外组织中产生。

经典途径起始自 C1q 的识别，C1q 主要由单核细胞-巨噬细胞系的细胞合成。C1q 识别免疫复合物中的抗体。凝集素途径通过血清凝集素 MBL（甘露糖结合凝集）或纤维胶凝蛋白与入侵病原表面的糖残基结合被激活，因此使 MBL 相关丝氨酸蛋白酶（MASP）途径的补体激活。C3 分子的自发激活是替代途径激活的基础，导致大量 C3 剪切片段在激活的病原体表面上沉积。这三种途径都产生 C3 转化酶，C3 转化酶将 C3 裂解为 C3a 和 C3b。C3b 再与入侵病原体或自身结构的碳水化合物/蛋白质的羟基基团相结合。

补体的主要功能是 C3b 对微生物的调理作用，C3b 可以被吞噬细胞表达的补体受体（CR1 和 CR3）识别。补体的另一个作用是它具有细胞溶解活性，C3b 靶向结合形成蛋白酶复合体，将 C5 裂解成 C5a 和 C5b，C5a 和 C5b 作为支架与 C6～C9 共同形成攻膜复合体（MAC）。MAC 可以诱导微生物等细胞渗透溶解。小肽片段 C3a、C4a 和 C5a 对白细胞有趋化作用，刺激释放炎症调节剂，并作用于内皮细胞，以促进白细胞和血浆蛋白运动进入组织。

对鳟、鲤和斑马鱼的研究表明，哺乳动物中存在的所有补体成分的同源体在硬骨鱼中几乎都存在，并且这些补体同源体均可通过三个补体途径被激活。关于硬骨鱼补体系统的研究可参考 Boshra 等 2006 年和 Nakao 等 2011 年发表的综述文章。硬骨鱼补体系统一个有趣的特点：C3、C4、C5、C7、MBL、B 因子、I 因子等具有同型多样性，由基因复制产生（Nakao 等，2006）。鳟和鲤的 C3 异构体在蛋白质和 cDNA 水平上研究较多。C3 异构体在某些功能位点上有不同的结构，特别是在决定硫代酯与羟基或氨基基团结合特异性的硫酯催化位点上存在差异，因此其在溶血活性和与绵羊红细胞以及兔红细胞、细菌和酵母等靶标物结合的特异性上就不同，但关于决定不同结合特异性的具体功能元素目前还不清楚。这可能表明 C3 多样化代表了一种进化策略，以扩大或增强硬骨鱼补体系统的免疫识别和效应功能（Nakao 等，2006）。在硬骨鱼中，补体成分的 mRNA 不仅在肝脏中表达，在头肾、肾、肠、鳃、皮肤、脑和性腺等器官中也有表达（Gonzalez 等，2007a，2007b；Løvoll 等，2007a；Encinas 等，2010）。此外，用免疫刺激剂处理后，鳟的 C3 异构体在不同器官中表达存在差异（Løvoll 等，2007b）。

在鱼类中已证实了补体介导的吞噬作用，某些硬骨鱼的 C3b 可介导调理活性（Schraml 等，2006；Jenkins 和 Ourth，1993）。然而，该过程所涉及的特异性受体的研究仍然很少（Nakao 等，2011）。在鳟和鲤中已开展了 C3a 和 C5a 的趋化活性研究（Kato 等，2004；Rotllant 等，2004）。

第十三节　结语与展望

如上所述，在硬骨鱼免疫系统中有关先天性免疫基因同源性的探索研究已经确定了病原识别受体、补体成分和多种细胞因子等不同种类的重要分子，并开展了在感染、免疫刺激和疫苗接种等不同处理条件下的表达谱分析，以及应用微阵列技术进行全基因表达分析（Purcell 等，2006a；Krasnov 等，2011）的进一步研究。在先天性免疫的分子生物活性方面研究取得了很大进展，但对其功能活性的研究还处于起步阶段。要进一步阐明先天性免疫的作用，研究重点应转向蛋白质和细胞效应功能方面。抗原递呈细胞（APC），如树突状细胞（DCs）和巨噬细胞，在将先天性免疫反应的信息转化为获得性免疫系统的细胞可识别的信号过程中起着关键作用，因而进一步阐明这些细胞类型的特征及其与获得性免疫的关联是未来硬骨鱼研究的热点。

参考文献

Agius, C. and Roberts, R. J. 2003. Melano-macrophage centres and their role in fish pathology. *J Fish Dis* 26: 499-509.

Agrawal, A., Eastman, Q. M. and Schatz, D. G. 1998. Transposition mediated by RAG1 and RAG2 and its implications for the evolution of the immune system. *Nature* 394: 744-751.

Altmann, S. M., Mellon, M. T., Distel, D. L. and Kim, C. H. 2003. Molecular and functional analysis of an interferon gene from the zebrafish, *Danio rerio*. *J Virol* 77: 1992-2002.

Barksby, H. E., Lea, S. R., Preshaw, P. M. and Taylor, J. J. 2007. The expanding family of interleukin-1 cytokines and their role in destructive inflammatory disorders. *Clin Exp Immunol* 149: 217-225.

Boshra, H., Li, J. and Sunyer, J. O. 2006. Recent advances on the complement system of teleost fish. *Fish Shellfish Immunol* 20: 239-262.

Braun-Nesje, R., Bertheussen, K., Kaplan, G. and Seljelid, R. 1981. Salmonid macrophages: separation, *in vitro* culture and characterization. *J Fish Dis* 4: 141-151.

Brinkmann, V. and Zychlinsky, A. 2007. Beneficial suicide: why neutrophils die to make NETs. *Nat Rev Microbiol* 5: 577-582.

Buonocore F, Forlenza M, Randelli E, *et al.*, 2005. Biological activity of sea bass (*Dicentrarchus labrax* L.) recombinant interleukin-1β. *Mar Biotechnol* 7: 609-617.

Caipang, C. M. A., Hirono, I. and Aoki, T. 2003. *In vitro* inhibition of fish rhabdoviruses by Japanese flounder, *Paralichthys olivaceus* Mx. *Virology* 317: 373-382.

Casadei, E., Wang, T., Zou, J., *et al.*, 2009. Characterization of three novel β-defensin antimicrobial peptides in rainbow trout (*Oncorhynchus mykiss*). *Mol Immunol* 46: 3358-3366.

Chang, M., Collet, B., Nie, P., *et al.*, 2011a. Expression and functional characterization of the RIG-I-like receptors MDA5 and LGP2 in rainbow trout (*Oncorhynchus mykiss*). *J Virol* 85: 8403-8412.

Chang, M., Wang, T., Nie, P., *et al.*, 2011b. Cloning of two rainbow trout nucleotide-binding oligomerization domain containing 2 (NOD2) splice variants and functional characterization of the NOD2 effector domains. *Fish Shellfish Immunol* 30: 118-127.

Chen, G., Shaw, M. H., Kim, Y. G. and Nunez, G. 2009. NOD-like receptors: role in innate immunity and inflammatory disease. *Annu Rev Pathol* 4: 365-398.

Chen, W. Q., Xu, Q. Q., Chang, M. X., *et al.*, 2010. Molecular characterization and expression analysis of nuclear oligomerization domain proteins NOD1 and NOD2 in grass carp *Ctenopharyngodon idella*. *Fish Shellfish Immunol* 28: 18-29.

de Kinkelin, P. and Dorson, M. 1973. Interferon production in rainbow trout (*Salmo gairdneri*) experimentally infected with Egtvedt virus. *J Gen Virol* 19: 125-127.

Deng, L., Velikovsky, C. A., Xu, G., *et al.*, 2010. A structural basis for antigen recognition by the T cell-like lymphocytes of sea lamprey. *Proc Natl Acad Sci USA* 107: 13408-13413.

Ellis, A. E. 2001. The immunology of teleosts, in *Fish Pathology* (ed. R. J. Roberts). London, W. B. Saunders, 133.

Encinas, P., Rodriguez-Milla, M. A., Novoa, B., *et al.*, 2010. Zebrafish fin immune responses during high mortality infections with viral haemorrhagic septicemia rhabdovirus. A proteomic and transcriptomic approach. *BMC Genomics* 11: 518.

Engelsma, M. Y., Stet, R. J., Saeij, J. P. and Verburg-van Kemenade, B. M. 2003. Differential expression and haplotypic variation of two interleukin-1β genes in the common carp (*Cyprinus carpio* L.). *Cytokine* 22: 21-32.

Evans, D. L., Carlson, R. L., Graves, S. S. and Hogan, K. T. 1984a. Nonspecific cytotoxic cells in fish (*Ictalurus punctatus*). IV. Target cell binding and recycling capacity. *Dev Comp Immunol* 8: 823-833.

Evans, D. L., Graves, S. S., Cobb, D. and Dawe, D. L. 1984b. Nonspecific cytotoxic cells in fish (*Ictalurus punctatus*). II. Parameters of target cell lysis and specificity. *Dev Comp Immunol* 8: 303-312.

Evans, D. L., Hogan, K. T., Graves, S. S., *et al.*, 1984c. Nonspecific cytotoxic cells in fish (*Ictalurus punctatus*). III. Biophysical and biochemical properties affecting cytolysis. *Dev Comp Immunol* 8: 599-610.

Falco, A., Chico, V., Marroqui, L., *et al.*, 2008. Expression and antiviral activity of a β-defensin-like peptide identified in the rainbow trout (*Oncorhynchus mykiss*) EST sequences. *Mol Immunol* 45: 757-765.

Fischer, U., Ototake, M. and Nakanishi, T. 1998. *In vitro* cell-mediated cytotoxicity against allogeneic erythrocytes in ginbuna crucian carp and goldfish using a non-radioactive assay. *Dev Comp Immunol* 22: 195-206.

Fischer, U., Utke, K., Somamoto, T., *et al.*, 2006. Cytotoxic activities of fish leucocytes. *Fish Shellfish Immunol* 20: 209-226.

Flajnik, M. F. and Kasahara, M. 2010. Origin and evolution of the adaptive immune system: genetic events and selective pressures. *Nat Rev Genet* 11: 47-59.

Forlenza, M., Scharsack, J. P., Kachamakova, N. M., *et al.*, 2008. Differential contribution of neutrophilic granulocytes and macrophages to nitrosative stress in a host-parasite animal model. *Mol Immunol* 45: 3178-3189.

Forlenza, M., Magez, S., Scharsack, J. P., *et al.*, 2009. Receptor-mediated and lectin-like activities of carp (*Cyprinus carpio*) TNF-α. *J Immunol* 183: 5319-5332.

Forlenza, M., Fink, I. R., Raes, G. and Wiegertjes, G. F. 2011. Heterogeneity of macrophage activation in fish. *Dev Comp Immunol* 35: 1246-1255.

Furnes, C., Seppola, M. and Robertsen, B. 2009. Molecular characterisation and expression analysis of interferon gamma in Atlantic cod (*Gadus morhua*). *Fish Shellfish Immunol* 26: 285-292.

Garcia-Rosado, E., Markussen, T., Kileng, O., *et al.*, 2008. Molecular and functional characterization of two infectious salmon anaemia virus (ISAV) proteins with type I interferon antagonizing activity. *Virus Res* 133: 228-238.

Gonzalez, S. F., Buchmann, K. and Nielsen, M. E. 2007a. Complement expression in common carp (*Cyprinus carpio* L.) during infection with *Ichthyophthirius multifiliis*. *Dev Comp Immunol* 31: 576-586.

Gonzalez, S. F., Chatziandreou, N., Nielsen, M. E., *et al.*, 2007b. Cutaneous immune responses in the common carp detected using transcript analysis. *Mol Immunol* 44: 1664-1679.

Grayfer, L., Walsh, J. G. and Belosevic, M. 2008. Characterization and functional analysis of goldfish (*Carassius auratus* L.) tumor necrosis factor-alpha. *Dev Comp Immunol* 32: 532-543.

Grayfer, L., Garcia, E. G. and Belosevic, M. 2010. Comparison of macrophage antimicrobial responses induced by type II interferons of the goldfish (*Carassius auratus* L.). *J Biol Chem* 285: 23537-23547.

Hanington, P. C., Tam, J., Katzenback, B. A., *et al.*, 2009. Development of macrophages of cyprinid fish. *Dev Comp Immunol* 33: 411-429.

Hansen, T. E. and Jørgensen, J. B. 2007. Cloning and characterisation of p38 MAP kinase from Atlantic salmon: A kinase important for regulating salmon TNF-2 and IL-1β expression. *Mol Immunol* 44: 3137-3146.

Haugland, Ø., Mercy, I. S., Romøren, K., *et al.*, 2007. Differential expression profiles and gene structure of two tumor necrosis factor-α variants in Atlantic salmon (*Salmo salar* L.). *Mol Immunol* 44: 1652-1663.

Hong, S., Zou, J., Collet, B., *et al.*, 2004. Analysis and characterisation of IL-1 β processing in rainbow trout, *Oncorhynchus mykiss*. *Fish Shellfish Immunol* 16: 453-459.

Husain, M., Bird, S., Zwieten, R. V., *et al.*, 2012. Cloning of the IL-1β3 gene and IL-1β4 pseudogene in salmonids uncovers a second type of IL-1β gene in teleost fish. *Dev Comp Immunol* 38: 431-446.

Hwang, S. D., Asahi, T., Kondo, H., *et al.*, 2010. Molecular cloning and expression study on Toll-like receptor 5 paralogs in Japanese flounder, *Paralichthys olivaceus*. *Fish Shellfish Immunol* 29: 630-638.

Hwang, S. D., Kondo, H., Hirono, I. and Aoki, T. 2011. Molecular cloning and characterization of Toll-like receptor 14 in Japanese flounder, *Paralichthys olivaceus*. *Fish Shellfish Immunol* 30: 425-429.

Igawa, D., Sakai, M. and Savan, R. 2006. An unexpected discovery of two interferon gamma-like genes along with interleukin (IL) -22 and -26 from teleost: IL-22 and -26 genes have been described for the first time outside mammals. *Mol Immunol* 43: 999-1009.

Iliev, D. B., Jørgensen, S., Rode, M., Krasnov, A., Harneshaug, I., Jørgensen, J. B., 2010. CpG-induced secretion of MHCIIb from salmon (*Salmo salar*) APCs. *Dev Comp Immunol* 34: 29-41.

Iliev, D. B., Roach, J. C., Mackenzie, S., *et al.*, 2005. Endotoxin recognition: in fish or not in fish? *FEBS Lett* 579: 6519-6528.

Ingerslev, H. C., Cunningham, C. and Wergeland, H. I. 2006. Cloning and expression of TNF-α, IL-1β and COX-2 in an anadromous and landlocked strain of Atlantic salmon (*Salmo salar* L.) during the smolting period. *Fish Shellfish Immunol* 20: 450-461.

Janeway, C. A. , Jr. , and Medzhitov, R. 2002. Innate immune recognition. *Annu Rev Immunol* 20: 197-216.

Jaso-Friedmann, L. , Leary, J. H. III, and Evans, D. L. 1997a. NCCRP-1: a novel receptor protein sequenced from teleost nonspecific cytotoxic cells. *Mol Immunol* 34: 955-965.

Jaso-Friedmann, L. , Leary, J. H. III, , Warren, J. , *et al.* , 1997b. Molecular characterization of a protozoan parasite target antigen recognized by nonspecific cytotoxic cells. *Cell Immunol* 176: 93-102.

Jaso-Friedmann, L. , Leary, J. H. III, and Evans, D. L. 2001. The non-specific cytotoxic cell receptor (NCCRP-1): molecular organization and signaling properties. *Dev Comp Immunol* 25: 701-711.

Jault, C. , Pichon, L. and Chluba, J. 2004. Toll-like receptor gene family and TIR-domain adapters in *Danio rerio*. *Mol Immunol* 40: 759-771.

Jenkins, J. A. and Ourth, D. D. 1993. Opsonic effect of the alternative complement pathway on channel catfish peripheral blood phagocytes. *Vet Immunol Immunopathol* 39: 447-459.

Jensen, I. and Robertsen, B. 2002. Effect of double-stranded RNA and interferon on the antiviral activity of Atlantic salmon cells against infectious salmon anemia virus and infectious pancreatic necrosis virus. *Fish Shellfish Immunol* 13: 221-241.

Jensen, I. , Albuquerque, A. , Sommer, A. I. and Robertsen, B. 2002a. Effect of poly I: C on the expression of Mx proteins and resistance against infection by infectious salmon anaemia virus in Atlantic salmon. *Fish Shellfish Immunol* 13: 311-326.

Jensen, I. , Larsen, R. and Robertsen, B. 2002b. An antiviral state induced in Chinook salmon embryo cells (CHSE-214) by transfection with the double-stranded RNA poly I: C. *Fish Shellfish Immunol* 13: 367-378.

Jørgensen, J. B. and Robertsen, B. 1995. Yeast β-glucan stimulates respiratory burst activity of Atlantic salmon (*Salmo salar* L.) macrophages. *Dev Comp Immunol* 19: 43-57.

Jørgensen, J. B. , Johansen, A. , Stenersen, B. and Sommer, A. I. 2001a. CpG oligodeoxynucleotides and plasmid DNA stimulate Atlantic salmon (*Salmo salar* L.) leucocytes to produce supernatants with antiviral activity. *Dev Comp Immunol* 25: 313-321.

Jørgensen, J. B. , Zou, J. , Johansen, A. and Secombes, C. J. 2001b. Immunostimulatory CpG oligodeoxynucleotides stimulate expression of IL-1β and interferon-like cytokines in rainbow trout macrophages via a chloroquine-sensitive mechanism. *Fish Shellfish Immunol* 11: 673-682.

Jørgensen, J. B. , Johansen, L. H. , Steiro, K. and Johansen, A. 2003. CpG DNA induces protective antiviral immune responses in Atlantic salmon (Salmo salar L.) . *J Virol* 77: 11471-11479.

Kato, Y. , Nakao, M. , Shimizu, M. , *et al.* , 2004. Purification and functional assessment of C3a, C4a and C5a of the common carp (*Cyprinus carpio*) complement. *Dev Comp Immunol* 28: 901-910.

Kileng, O. , Albuquerque, A. and Robertsen, B. 2008. Induction of interferon system genes in Atlantic salmon by the imidazoquinoline S-27609, a ligand for Toll-like receptor 7. *Fish Shellfish Immunol* 24: 514-522.

Koppang, E. O. , Hordvik, I. , Bjerkås, I. , *et al.* , 2003. Production of rabbit antisera against recombinant MHC class II β chain and identification of immunoreactive cells in Atlantic salmon (*Salmo salar*) . *Fish Shellfish Immunol* 14: 115-132.

Krasnov, A. , Timmerhaus, G. , Schiotz, B. L. , *et al.* , 2011. Genomic survey of early responses to viruses in Atlantic salmon, *Salmo salar* L. *Mol Immunol* 49: 163-174.

Kumar, H. , Kawai, T. and Akira, S. 2011. Pathogen recognition by the innate immune system. *Int Rev Immunol* 30: 16-34.

Laing, K. J. , Hardie, L. J. , Aartsen, W. , *et al.* , 1999. Expression of an inducible nitric oxide synthase gene in rainbow trout *Oncorhynchus mykiss*. *Dev Comp Immunol* 23：71-85.

Laing，K. J.，Wang，T.，Zou，J.，*et al.*，2001. Cloning and expression analysis of rainbow trout *Oncorhynchus mykiss* tumour necrosis factor-α. *Eur J Biochem* 268：1315-1322.

Laing，K. J.，Purcell，M. K.，Winton，J. R. and Hansen，J. D. 2008. A genomic view of the NOD-like receptor family in teleost fish：identification of a novel NLR subfamily in zebrafish. *BMC Evol Biol* 8：42.

Larsen，R.，Røkenes，T. P. and Robertsen，B. 2004. Inhibition of infectious pancreatic necrosis virus replication by Atlantic salmon Mx1 protein. *J Virol* 78：7938-7944.

Lieschke，G. J.，Oates，A. C.，Crowhurst，M. O.，*et al.*，2001. Morphologic and functional characterization of granulocytes and macrophages in embryonic and adult zebrafish. *Blood* 98：3087-3096.

Lin，A. F.，Xiang，L. X.，Wang，Q. L.，*et al.*，2009. The DC-SIGN of zebrafish：insights into the existence of a CD209 homologue in a lower vertebrate and its involvement in adaptive immunity. *J Immunol* 183：7398-7410.

Lutfalla，G.，Crollius，H. R.，Stange-thomann，N.，*et al.*，2003. Comparative genomic analysis reveals independent expansion of a lineage-specific genefamily in vertebrates：The class II cytokine receptors and their ligands in mammals and fish. *BMC Genomics* 4：29.

Løvoll，M.，Dalmo，R. A. and Bøgwald，J. 2007a. Extrahepatic synthesis of complement components in the rainbow trout（*Oncorhynchus mykiss*）. *Fish Shellfish Immunol* 23：721-731.

Løvoll，M.，Fischer，U.，Mathisen，G. S.，*et al.*，2007b. The C3 subtypes are differentially regulated after immunostimulation in rainbow trout，but head kidney macrophages do not contribute to C3 transcription. *Vet Immunol Immunopathol* 117：284-295.

Martinon，F.，Burns，K. and Tschopp，J. 2002. The inflammasome：a molecular platform triggering activation of inflammatory caspases and processing of proIL-β. *Mol Cell* 10：417-426.

Matsuo，A.，Oshiumi，H.，Tsujita，T.，*et al.*，2008. Teleost TLR22 recognizes RNA duplex to induce IFN and protect cells from birnaviruses. *J Immunol* 181：3474-3485.

Meijer，A. H.，Gabby Krens，S. F.，Medina Rodriguez，I. A.，*et al.*，2004. Expression analysis of the Toll-like receptor and TIR domain adaptor families of zebrafish. *Mol Immunol* 40：773-783.

Meyer，A. and Van de Peer，Y. 2005. From 2R to 3R：evidence for a fish-specific genome duplication（FSGD）. *Bioessays* 27：937-945.

Milev-Milovanovic，I.，Long，S.，Wilson，M.，2006. Identification and expression analysis of interferon gamma genes in channel catfish. *Immunogenetics* 58：70-80.

Mutoloki，S.，Jørgensen，J. B. and Evensen，Ø. 2014. The adaptive immune response in fish，in *Fish Vaccination*（eds R. Gudding，A. Lillehaug and Ø. Evensen）. Oxford，John Wiley & Sons，Ltd.，104-115.

Nakao，M.，Kato-Unoki，Y.，Nakahara，M.，*et al.*，2006. Diversified components of the bony fish complement system：more genes for robuster innate defense? *Adv Exp Med Biol* 586：121-138.

Nakao，M.，Tsujikura，M.，Ichiki，S.，*et al.*，2011. The complement system in teleost fish：progress of post-homolog-hunting researches. *Dev Comp Immunol* 35：1296-1308.

Nam，B. H.，Moon，J. Y.，Kim，Y. O.，*et al.*，2010. Multiple β-defensin isoforms identified in early developmental stages of the teleost *Paralichthys olivaceus*. *Fish Shellfish Immunol* 28：267-274.

Neumann，N. F.，Stafford，J. L.，Barreda，D.，*et al.*，2001. Antimicrobial mechanisms of fish phagocytes and their role in host defense. *Dev Comp Immunol* 25：807-825.

Ohta，T.，Ueda，Y.，Ito，K.，*et al.*，2011. Anti-viral effects of interferon administration on sevenband grouper，*Epinephelus septemfasciatus*. *Fish Shellfish Immunol* 30：1064-1071.

Ohtani, M., Hikima, J., Kondo, H., *et al.*, 2011. Characterization and antiviral function of a cytosolic sensor gene, MDA5, in Japanese flounder, *Paralichthys olivaceus*. *Dev Comp Immunol* 35: 554-562.

Oshiumi, H., Tsujita, T., Shida, K., *et al.*, 2003. Prediction of the prototype of the human Toll-like receptor gene family from the pufferfish, *Fugu rubripes*, genome. *Immunogenetics* 54: 791-800.

Palic, D., Ostojic, J., Andreasen, C. B. and Roth, J. A. 2007. Fish cast NETs: neutrophil extracellular traps are released from fish neutrophils. *Dev Comp Immunol* 31: 805-816.

Palti, Y. 2011. Toll-like receptors in bony fish: from genomics to function. *Dev Comp Immunol* 35: 1263-1272.

Pancer, Z., Amemiya, C. T., Ehrhardt, G. R., *et al.*, 2004. Somatic diversification of variable lymphocyte receptors in the agnathan sea lamprey. *Nature* 430: 174-180.

Park, S. B., Hikima, J., Suzuki, Y., *et al.*, 2012. Molecular cloning and functional analysis of nucleotide-binding oligomerization domain 1 (NOD1) in olive flounder, *Paralichthys olivaceus*. *Dev Comp Immunol* 36: 680-687.

Peddie, S., Zou, J., Cunningham, C. and Secombes, C. J. 2001. Rainbow trout (*Oncorhynchus mykiss*) recombinant IL-1β and derived peptides induce migration of head-kidney leucocytes *in vitro*. *Fish Shellfish Immunol* 11: 697-709.

Peddie, S., Zou, J. and Secombes, C. J. 2002. A biologically active IL-1β derived peptide stimulates phagocytosis and bactericidal activity in rainbow trout, *Oncorhynchus mykiss* (Walbaum), head kidney leucocytes *in vitro*. *J Fish Dis* 25: 351-360.

Pedersen, G. M., Johansen, A., Olsen, R. L. and Jørgensen, J. B. 2006. Stimulation of type I IFN activity in Atlantic salmon (*Salmo salar* L.) leukocytes: synergistic effects of cationic proteins and CpG ODN. *Fish Shellfish Immunol* 20: 503-518.

Pelegrin, P., Chaves-Pozo, E., Mulero, V. and Meseguer, J. 2004. Production and mechanism of secretion of interleukin-1β from the marine fish gilthead seabream. *Dev Comp Immunol* 28: 229-237.

Plouffe, D. A., Hanington, P. C., Walsh, J. G., *et al.*, 2005. Comparison of select innate immune mechanisms of fish and mammals. *Xenotransplantation* 12: 266-277.

Purcell, M. K., Nichols, K. M., Winton, J. R., *et al.*, 2006a. Comprehensive gene expression profiling following DNA vaccination of rainbow trout against infectious hematopoietic necrosis virus. *Mol Immunol* 43: 2089-2106.

Purcell, M. K., Smith, K. D., Aderem, A., *et al.*, 2006b. Conservation of Toll-like receptor signaling pathways in teleost fish. *Comp Biochem Physiol Part D-Genomics Proteomics* 1: 77-88.

Rajendran, K. V., Zhang, J., Liu, S., *et al.*, 2012a. Pathogen recognition receptors in channel catfish: I. Identification, phylogeny and expression of NOD-like receptors. *Dev Comp Immunol* 37: 77-86.

Rajendran, K. V., Zhang, J., Liu, S., *et al.*, 2012b. Pathogen recognition receptors in channel catfish: II. Identification, phylogeny and expression of retinoic acid-inducible gene I (RIG-Ⅰ)-like receptors (RLRs). *Dev Comp Immunol* 37: 381-389.

Rebl, A., Goldammer, T. and Seyfert, H. M. 2010. Toll-like receptor signaling in bony fish. *Vet Immunol Immunopathol* 134: 139-150.

Robertsen, B., Bergan, V., Røkenes, T., *et al.*, 2003. Atlantic salmon interferon genes: cloning, sequence analysis, expression, and biological activity. *J Interferon Cytokine Res* 23: 601-612.

Roca, F. J., Mulero, I., Lopez-Munoz, A., *et al.*, 2008. Evolution of the inflammatory response in vertebrates: fish TNF-α is a powerful activator of endothelial cells but hardly activates phagocytes. *J Immunol* 181: 5071-5081.

Rodriguez, A., Esteban, M. A. and Meseguer, J. 2003. Phagocytosis and peroxidase release by seabream (*Sparus aurata* L.) leucocytes in response to yeast cells. *Anat Rec A Discov Mol Cell Evol Biol* 272: 415-423.

Rotllant, J., Parra, D., Peters, R., *et al.*, 2004. Generation, purification and functional characterization of three C3a anaphylatoxins in rainbow trout: role in leukocyte chemotaxis and respiratory burst. *Dev Comp Immunol* 28: 815-828.

Saeij, J. P., Stet, R. J., de Vries, B. J., *et al.*, 2003. Molecular and functional characterization of carp TNF: a link between TNF polymorphism and trypanotolerance? *Dev Comp Immunol* 27: 29-41.

Savan, R., Kono, T., Igawa, D. and Sakai, M. 2005. A novel tumor necrosis factor (TNF) gene present in tandem with theTNF-α gene on the same chromosome in teleosts. *Immunogenetics* 57: 140-150.

Schraml, B., Baker, M. A. and Reilly, B. D. 2006. A complement receptor for opsonized immune complexes on erythrocytes from *Oncorhynchus mykiss* but not *Ictalarus punctatus*. *Mol Immunol* 43: 1595-1603.

Secombes, C. J., Wang, T. and Bird, S. 2011. The interleukins of fish. *Dev Comp Immunol* 35: 1336-1345.

Seppola, M., Larsen, A. N., Steiro, K., *et al.*, 2008. Characterisation and expression analysis of the interleukin genes, IL-1β, IL-8 and IL-10, in Atlantic cod (*Gadus morhua* L.). *Mol Immunol* 45: 887-897.

Sha, Z., Abernathy, J. W., Wang, S., *et al.*, 2009. NOD-like subfamily of the nucleotide-binding domain and leucine-rich repeat containing family receptors and their expression in channel catfish. *Dev Comp Immunol* 33: 991-999.

Shen, L., Stuge, T. B., Zhou, H., *et al.*, 2002. Channel catfish cytotoxic cells: a mini-review. *Dev Comp Immunol* 26: 141-149.

Shen, L., Stuge, T. B., Bengten, E., *et al.*, 2004. Identification and characterization of clonal NK-like cells from channel catfish (*Ictalurus punctatus*). *Dev Comp Immunol* 28: 139-152.

Skjæveland, I., Iliev, D. B., Strandskog, G. and Jørgensen, J. B. 2009. Identification and characterization of TLR8 and MyD88 homologs in Atlantic salmon (*Salmo salar*). *Dev Comp Immunol* 33: 1011-1017.

Skjæveland, I., Iliev, D. B., Zou, J., *et al.*, 2008. A TLR9 homolog that is up-regulated by IFN-γ in Atlantic salmon (*Salmo salar*). *Dev Comp Immunol* 32: 603-607.

Skjesol, A., Aamo, T., Hegseth, M. N., *et al.*, 2009. The interplay between infectious pancreatic necrosis virus (IPNV) and the IFN system: IFN signaling is inhibited by IPNV infection. *Virus Res* 143: 53-60.

Somamoto, T., Nakanishi, T. and Okamoto, N. 2000. Specific cell-mediated cytotoxicity against a virus-infected syngeneic cell line in isogeneic ginbuna crucian carp. *Dev Comp Immunol* 24: 633-640.

Stafford, J. L., Galvez, F., Goss, G. G. and Belosevic, M. 2002. Induction of nitric oxide and respiratory burst response in activated goldfish macrophages requires potassium channel activity. *Dev Comp Immunol* 26: 445-459.

Star, B., Nederbragt, A. J., Jentoft, S., *et al.*, 2011. The genome sequence of Atlantic cod reveals a unique immune system. *Nature* 477: 207-210.

Stolte, E. H., Savelkoul, H. F., Wiegertjes, G., *et al.*, 2008. Differential expression of two interferon-γ genes in common carp (*Cyprinus carpio* L.). *Dev Comp Immunol* 32: 1467-1481.

Strandskog, G., Ellingsen, T. and Jørgensen, J. B. 2007. Characterization of three distinct CpG oligonucleotide classes which differ in ability to induce IFN α/β activity and cell proliferation in Atlantic salmon (*Salmo salar* L.) leukocytes. *Dev Comp Immunol* 31: 39-51.

Strandskog, G., Skjæveland, I., Ellingsen, T. and Jørgensen, J. B. 2008a. Double-stranded RNA-and CpG DNA-induced immune responses in Atlantic salmon: comparison and synergies. *Vaccine* 26: 4704-4715.

Strandskog, G., Villoing, S., Iliev, D. B., *et al.*, 2011. Formulations combining CpG containing oliogonucleotides and poly I: C enhance the magnitude of immune responses and protection against pancreas disease in Atlantic salmon. *Dev Comp Immunol* 35: 1116-1127.

Sun, B., Robertsen, B., Wang, Z. and Liu, B. 2009. Identification of an Atlantic salmon IFN multigene cluster encoding three IFN subtypes with very different expression properties. *Dev Comp Immunol* 33: 547-558.

Sun, B., Skjæveland, I., Svingerud, T., *et al.*, 2011. Antiviral activity of salmonid gamma interferon against infectious pancreatic necrosis virus and salmonid alphavirus and its dependency on type I interferon. *J Virol* 85: 9188-9198.

Svingerud, T., Solstad, T., Sun, B., *et al.*, 2012. Atlantic salmon type I IFN subtypes show differences in antiviral activity and cell-dependent expression: evidence for high IFNb/IFNc-producing cells in fish lymphoid tissues. *J Immunol* 189: 5912-5923.

Thim, H. L., Iliev, D. B., Christie, K. E., *et al.*, 2012. Immunoprotective activity of a salmonid alphavirus vaccine: comparison of the immune responses induced by inactivated whole virus antigen formulations based on CpG class B oligonucleotides and poly I: C alone or combined with an oil adjuvant. *Vaccine* 30: 4828-4834.

Thornberry, N. A., Bull, H. G., Calaycay, J. R., *et al.*, 1992. A novel heterodimeric cysteine protease is required for interleukin-1β processing in monocytes. *Nature* 356: 768-774.

Traver, D., Paw, B. H., Poss, K. D., *et al.*, 2003. Transplantation and *in vivo* imaging of multilineage engraftment in zebrafish bloodless mutants. *Nat Immunol* 4: 1238-1246.

Tsujita, T., Tsukada, H., Nakao, M., *et al.*, 2004. Sensing bacterial flagellin by membrane and soluble orthologs of Toll-like receptor 5 in rainbow trout (*Onchorhynchus mykiss*). *J Biol Chem* 279: 48588-48597.

Vallejo, A. N., Miller, N. W. and Clem, L. W. 1991a. Phylogeny of immune recognition: processing and presentation of structurally defined proteins in channel catfish immune responses. *Dev Immunol* 1: 137-148.

Vallejo, A. N., Miller, N. W. and Clem, L. W. 1991b. Phylogeny of immune recognition: role of alloantigens in antigen presentation in channel catfish immune responses. *Immunology* 74: 165-168.

van Vliet, S. J., den Dunnen, J., Gringhuis, S. I., *et al.*, 2007. Innate signaling and regulation of dendritic cell immunity. *Curr Opin Immunol* 19: 435-440.

Vasta, G. R., Ahmed, H. and Odom, E. W. 2004. Structural and functional diversity of lectin repertoires in invertebrates, protochordates and ectothermic vertebrates. *Curr Opin Struct Biol* 14: 617-630.

Vasta, G. R., Nita-Lazar, M., Giomarelli, B., *et al.*, 2011. Structural and functional diversity of the lectin repertoire in teleost fish: relevance to innate and adaptive immunity. *Dev Comp Immunol* 35: 1388-1399.

Vojtech, L. N., Scharping, N., Woodson, J. C. and Hansen, J. D. 2012. Roles of inflammatory caspases during processing of Zebrafish interleukin-1β in *Francisella noatunensis* infection. *Infect Immun* 80: 2878-2885.

Wang, T., Bird, S., Koussounadis, A., *et al.*, 2009. Identification of a novel IL-1 cytokine family member in teleost fish. *J Immunol* 183: 962-974.

Wang, L., Su, J., Yang, C., *et al.*, 2012. Genomic organization, promoter activity of grass carp MDA5 and the association of its polymorphisms with susceptibility/resistance to grass carp reovirus. *Mol Immunol* 50: 236-243.

Ware, C. F. 2003. The TNF superfamily. *Cytokine Growth Factor Rev* 14: 181-184.

Wiens, G. D. and Glenney, G. W. 2011. Origin and evolution of TNF and TNF receptor superfamilies. *Dev Comp Immunol* 35: 1324-1335.

Wittamer, V., Bertrand, J. Y., Gutschow, P. W. and Traver, D. 2011. Characterization of the mononuclear phagocyte system in zebrafish. *Blood* 117: 7126-7135.

Yang, C., Su, J., Huang, T., *et al.*, 2011. Identification of a retinoic acid-inducible gene I from grass carp

(*Ctenopharyngodon idella*) and expression analysis *in vivo* and *in vitro*. *Fish Shellfish Immunol* 30：936-943.

Yoder，J. A. and Litman，G. W. 2011. The phylogenetic origins of natural killer receptors and recognition：relationships，possibilities，and realities. *Immunogenetics* 63：123-141.

Zapata，A. G. and Cooper，E. L. 1990. *The Immune System*. Chichester，John Wiley & Sons，Ltd.

Zhang，Y. B. and Gui，J. F. 2012. Molecular regulation of interferon antiviral response in fish. *Dev Comp Immunol* 38：193-202.

Zhu，L. Y.，Nie，L.，Zhu，G.，*et al.*，2013. Advances in research of fish immune-relevant genes：A comparative overview of innate and adaptive immunity in teleosts. *Dev Comp Immunol* 39：39-62.

Zhu，R.，Zhang，Y. B.，Zhang，Q. -Y. and Gui，J. F. 2008. Functional domains and the antiviral effect of the double-stranded RNA-dependent protein kinase PKR from *Paralichthys olivaceus*. *J Virol* 82：6889-6901.

Zou，J.，Grabowski，P. S.，Cunningham，C. and Secombes，C. J. 1999. Molecular cloning of interleukin 1beta from rainbow trout *Oncorhynchus mykiss* reveals no evidence of an ice cut site. *Cytokine* 11：552-560.

Zou，J.，Wang，T.，Hirono，I.，*et al.*，2002. Differential expression of two tumor necrosis factor genes in rainbow trout，*Oncorhynchus mykiss*. *Dev Comp Immunol* 26：161-172.

Zou，J.，Peddie，S.，Scapigliati，G.，*et al.*，2003a. Functional characterisation of the recombinant tumor necrosis factors in rainbow trout，*Oncorhynchus mykiss*. *Dev Comp Immunol* 27：813-822.

Zou，J.，Secombes，C. J.，Long，S.，*et al.*，2003b. Molecular identification and expression analysis of tumor necrosis factor in channel catfish (*Ictalurus punctatus*). *Dev Comp Immunol* 27：845-858.

Zou，J.，Carrington，A.，Collet，B.，*et al.*，2005. Identification and bioactivities of IFN-γ in rainbow trout *Oncorhynchus mykiss*：the first Th1-type cytokine characterized functionally in fish. *J Immunol* 175：2484-2494.

Zou，J.，Tafalla，C.，Truckle，J. and Secombes，C. J. 2007. Identification of a second group of type Ⅰ IFNs in fish sheds light on IFN evolution in vertebrates. *J Immunol* 179：3859-3871.

Zou，J. and Secombes，C. J. 2011. Teleost fish interferons and their role in immunity. *Dev Comp Immunol* 35：1376-1387.

第九章　鱼类获得性免疫反应

Stephen Mutoloki, Jorunn B. Jørgensen and Øystein Evensen

本章概要　SUMMARY

获得性免疫系统是动物机体遇到病原体后针对病原感染产生保护作用的系统。由于这种反应“适应于”特定病原体，因此是特异性的。获得性免疫的主要特点是产生速度较慢，需要几天到几周时间；但随着时间的推移，它会变得精准，再次遇到相同病原后会迅速而有效地应答，因此具有记忆性。树突状细胞是专门进行抗原捕获、迁移和激活 T 细胞的细胞。 鱼类树突状细胞的鉴定已取得进展，这些研究为进一步了解硬骨鱼的抗原递呈过程奠定了基础。获得性免疫反应的主体是淋巴细胞，包括具有抗原识别多样性、特异性和记忆的细胞类型。淋巴细胞大致分为产生抗体的 B 细胞和介导细胞免疫的 T 细胞。T 细胞又进一步分为两类：细胞毒性 T 细胞，直接杀死感染的和异常的细胞；辅助性 T 细胞，通过产生细胞因子调节其他免疫细胞。根据产生的细胞因子类型，辅助性 T 细胞又分为几类亚群。与哺乳动物获得性免疫系统功能相比，已经证实，硬骨鱼具有获得性免疫所具备的必要条件，但是二者仍存在着重要差异。首先，硬骨鱼的获得性免疫反应是温度依赖型，与哺乳动物相比，硬骨鱼需更长时间形成获得性免疫反应。其次，硬骨鱼的抗体类型（IgM、IgD 和 IgT）较少，且仅有 2 种抗体能够针对抗原刺激产生反应。抗体没有类型转换，因而特异性抗体类型是由特定的 B 细胞亚群产生的。IgM 占主导地位，IgT 主要在肠道黏膜和皮肤中起作用。目前的研究表明，在大多数情况下，硬骨鱼中细胞介导的免疫应答是保守的。然而，阐释硬骨鱼获得性免疫系统的功能仍然需要更多的研究，特别是 T 细胞应答方面。硬骨鱼中参与获得性免疫反应的大多数蛋白质同源分子是保守的，尽管如此，这些蛋白质似乎存在不止一个基因拷贝。因此，不仅从比较学的角度，更要从功能性的角度研究这些基因的功能。

第一节 引　言

在前一章中，先天性免疫的作用是以保护机体为目的免疫防御的第一道防线，参与消灭病原体。激活先天性免疫不能解决所有病原对机体的侵染，有些病原体已经进化出了复杂的机制，以躲避先天性免疫系统并成功感染。这时，参与先天性免疫反应的巨噬细胞和树突状细胞等会激活获得性免疫反应（也称为特异性免疫反应）。约 4.5 亿年前，获得性免疫系统在脊椎动物进化早期的圆口鱼类和软骨鱼类的分化时出现，在所有的有颌类脊椎动物中都存在比先天性免疫系统具有更强大的应对感染的能力。表 9-1 列出了获得性免疫系统的主要特点以及与先天性免疫系统的区别。

表 9-1　哺乳动物先天性和获得性免疫系统的区别

	先天性免疫	获得性免疫
抗原识别结构	胚系编码	基因重组产生
抗原识别的特异性	一般病原体（如细菌、病毒和真菌），但可做出细微区别	很高
反应	立即的	需要时间
细胞类型	吞噬细胞、树突状细胞、自然杀伤细胞	B 细胞和 T 细胞
记忆	无，再次感染反应能力相同	有，记忆细胞对特异性的病原体有记忆，再次感染反应更有效

获得性免疫反应的形成需要时间，形成过程中会不断改善。与先天性免疫系统不同，获得性免疫系统通过基因重组产生识别抗原的结构，具有特异性和抗原识别的多样性。此外，它在感知方面是“聪明的”，因为它“学习”获得了对病原体的识别能力，并能够“记住”入侵病原，因此如果相同病原体再次入侵，可以更迅速、有效地做出应答，这是针对病原体设计疫苗的基础。

鱼类有 20 000 多种（依据 Fishbase，截至 2020 年，全世界鱼类约有34 000种，译者注），即使是非常相近的物种之间，其获得性免疫系统的特性和能力的差异仍很大。目前，获得性免疫系统如何对感染做出应答的研究主要集中在硬骨鱼，因此本章的其余部分描述仅限于硬骨鱼。总体来说，很多年前就知道鱼类中存在淋巴细胞。最初通过形态学，或者基于 IgM 试验的染色特征鉴定淋巴细胞。利用这些鉴定方法，可以通过跟踪拍摄或流式细胞术分类细胞，并进行功能试验（例如同种异体移植排斥反应）；也可用有丝分裂原（脂多糖和刀豆蛋白 A，Con A）鉴定和分类 B 细胞和 T 细胞。近年来，由于分子生物学和遗传学的发展，可以用更多的工具鉴定细胞的亚群。

第二节 淋巴细胞是获得性免疫系统的关键细胞

获得性免疫系统的关键细胞类型是淋巴细胞。这些细胞的主要作用是多样性和特异性的抗原识别，还包括抗原记忆。已鉴定出淋巴细胞主要包括 B 细胞和 T 细胞两种类型。在哺乳动物中，淋巴细胞起源于骨髓，属淋巴系细胞，与髓系的先天性免疫细胞不同。B 细胞在骨髓中成熟（鸟类的 B 细胞在法氏囊中成熟），而 T 细胞迁移到胸腺，经正向选择和成熟，然后定植于次级淋巴器官。

由于鱼类缺乏骨髓，所以认为 B 细胞和 T 细胞起源于相当于骨髓的头肾。B 细胞在头肾内发育成熟，而 T 细胞在胸腺发育成熟（Ellis 等，1988；Clem 等，1991；Salinas 等，2011）。虽然在鱼类的皮下发现了弥散的淋巴样聚集物，代表原始的肠道相关淋巴组织（GALT），但鱼类并不具有哺乳动物的生发中心、B 细胞囊泡、淋巴结和黏膜相关淋巴组织（MALTs）的淋巴结构（Zapata 和 Amemiya，2000）。最近，在鲑鳃中发现了鳃间淋巴样器官，推测可能是 T 细胞聚集的重要场所（Koppang 等，2010），但其功能有待深入研究。

第三节 抗原捕获以及淋巴细胞激活

硬骨鱼主要的抗原捕获器官是头肾和脾脏（Press 和 Evensen，1999）。研究发现，比目鱼淋巴细胞在头肾的组织中不断循环，同样鲤淋巴细胞受到抗原刺激后，嗜派洛宁碱性细胞和浆细胞形成球形聚集物。嗜派洛宁期被认为是 B 细胞的增殖阶段。另一方面，鱼类脾脏虽未达到高等脊椎动物的组织分化程度，但在高等鱼类中已有相当严密的椭球形组织结构，并参与捕获免疫复合物（Press 和 Evensen，1999）。淋巴细胞经抗原刺激后，与头肾中淋巴细胞相似，分化成浆细胞。鱼类的黑色素巨噬细胞中心（MMC）被认为是类似高等脊椎动物生发中心的原始组织，位于头肾和脾脏内，与抗原捕获结构相连，在慢性炎症的组织中也发现了 MMC 的存在。事实上，据报道 MMC 可能也是抗原保留区，小淋巴细胞会迁移到这些区域，因此认为 MMC 是免疫加工的场所。虽然有以上的认知，但抗原如何递呈以及淋巴细胞如何被激活仍需进一步的研究。

第四节 骨髓来源的抗原递呈细胞（APCs）

在先天性免疫一章中，介绍了鱼类单核吞噬细胞的直接抗菌作用（Jørgensen，2014），除此之外，这些细胞表达 MHC-Ⅱ类分子（Iliev 等，2010），有抗原加工和递呈所必需的分子基础（Vallejo 等，1991；Wittamer 等，2011）。然而，我们对鱼类抗原递呈方面了解甚少，一些基本问题还是未知，比如抗原递呈的场所以及哪类细胞激活 T 细胞的增殖。在斑点叉尾鮰中进行的鱼类 APC 功能的开创性研究表明，白细胞在与 Ig-阴性细胞（类似 T 细胞）反应时，外周血白细胞（PBL）中的单核细胞是必需的辅助细胞（Sizemore 等，1984）。在哺乳动物中，树突状细胞（DCs）被认为是免疫系统最有效和多样化的专职抗原递呈细胞，因为它们具有能激活初始 T 细胞反应的能力。DCs 形态多样是由细胞发生，也就是分化阶段所决定的（Geissmann 等，2010）。Steinman 等（1979）最先提出异质性细胞（heterogenous cell）群这个名称，他们观察到，某些可附着的脾脏细胞在体外培养后呈现出典型的星状形态。目前 DCs 被定义为一种由某些分子细胞表面的表型和功能结合的 APCs 系统（Steinman，2007）。未成熟的 DCs 位于有抗原存在的环境中，被激活的 DCs 迁移到次级淋巴器官的 T 细胞区（图 9-1）。在迁移过程中，它们失去了内化抗原的能力，将抗原递呈给初始 T 细胞，这个过程称为 DC 的成熟 (Randolph 等，2005a，2005b)。

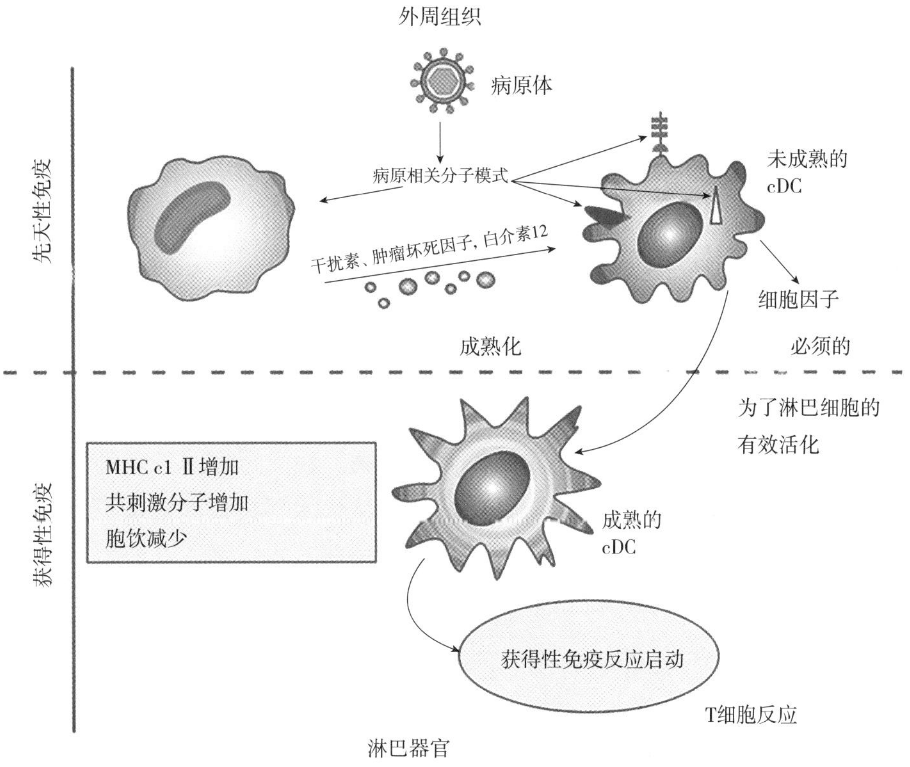

图 9-1 树突状细胞（DCs）的激活与成熟

外周组织中的未成熟 DCs 响应病原相关模式受体（PAMPs）并内化病原体。此外，被 PAMPs 激活的单核细胞/巨噬细胞等其他先天性免疫细胞分泌Ⅰ型干扰素和肿瘤坏死因子等细胞因子，随后激活 DCs。DCs 成熟后增强 MHC-Ⅱ类分子和共刺激分子的表达，迁移到淋巴器官，递呈抗原至初始 T 细胞

最近的研究中，斑马鱼（Lugo-Villarino 等，2010）和虹鳟（Bassity 和 Clark，2012）中鉴定出的专职抗原递呈细胞与哺乳动物 DCs 特征相似，包括细胞表面具有高表达的 MHC-Ⅱ类分子和 CD80、CD86 等共刺激因子。在斑马鱼中，这些细胞与植物凝集素具有高亲和力，使这些细胞富集。用电子显微镜观察，这些细胞在形态上与其他髓样细胞不同（Lugo-Villarino 等，2010)。此外，功能研究发现，这些细胞具有抗原特异性的记忆反应。虹鳟 DC 样细胞表面 MHC-Ⅱ类分子表达水平较高，与巨噬细胞和 B 细胞等专职抗原递呈细胞相比，DC 样细胞可显著刺激混合白细胞反应（Lugo-Villarino 等，2010)。虹鳟 DC 样细胞可以直接从脾脏中分离出来，也可从头肾和外周血白细胞中大量获得（Bassity 和 Clark，2012)。在大西洋鲑头肾中，应用流式细胞术检测到三类表达 MHC-Ⅱ类分子的细胞（Iliev 等，2010)，其中一类是 Ig^+ 细胞 (可能是 B 细胞，<10%)，而大多数是 Ig^- 细胞并表面高表达 MHC-Ⅱ类分子的细胞。这些表面高表达 MHC-Ⅱ类分子的髓系细胞比例较高，说明鱼类专职抗原递呈细胞很可能来源于头肾，也表明头肾可能是抗原递呈发生的重要场所。

除了鱼类专职抗原递呈细胞的形态学特征外，研究还表明，这些细胞可以被 TLR 配体 (Iliev 等，2010；Bassity 和 Clark，2012）或 CD40L 等共刺激分子（Lagos 等，2012）有效激活。激活后，这些细胞就从未成熟的表型转变为炎性表型，表现为表面 MHC-Ⅱ类分子的表达升高，内化抗原的能力降低（Iliev 等，2010)。后者内化抗原能力的降低是树突状细胞成熟的

标志（Randolph 等，2005a，2005b）。

第五节 免疫球蛋白和 B 细胞

迄今为止，在硬骨鱼中发现了三种免疫球蛋白。外周循环系统中含量最高的抗体类型是 IgM，这种抗体主要由位于头肾中的浆细胞和浆母细胞产生（Bromage 等，2004；Zwollo 等，2005），以单体、二聚体和四聚体的形式存在，不同亚型间存在功能差异。硬骨鱼中没有免疫球蛋白的类别转换现象，且 IgM 应答成熟后亲和力较差。即便这样，与高等脊椎动物的体液免疫一致，在预防疖疮病和弧菌病等胞外细菌性疾病中，这种抗体类型已被作为检测免疫保护效果的标志性分子。也有研究表明，IgM 与防护大西洋鲑传染性胰腺坏死病（Munang'andu 等，2013）和胰腺病（Xu 等，2012）等病毒性疾病感染相关。有趣的是，对于传染性胰腺坏死病疫苗来说，免疫接种后（在攻毒开始时）的抗体滴度高，但攻毒期间抗体滴度并不高，可能是因为在攻毒过程中抗体被消耗，而在没有任何抗体或抗体水平较低的情况下，大西洋鲑在攻毒早期出现死亡。大西洋鲑在免疫后对攻毒病毒产生的抗体与未接种疫苗对照组之间未观察到明显的相关性，这表明攻毒前的抗体在限制感染及其病情发展方面发挥了重要作用，可能是抗体在循环过程中和了病毒（Munang'andu 等，2013）。这些结果有力地表明了 IgM 在硬骨鱼抵抗胞内病原感染中的作用。鱼类接种疫苗后再感染是否可以诱导像哺乳动物一样的记忆应答还是待解之谜。

被动免疫实验是将受免鱼的血清抗体注射给未接种的鱼，并已证实可诱导免疫保护。例如，未接种疫苗的尼罗罗非鱼（*Oreochromis niloticus*）接种了免疫海豚链球菌（*Streptococcus iniae*）鱼的血清抗体后，产生了被动免疫保护作用（LaFrentz 等，2011）。另一研究也表明，利用接种链球菌（*S. difficile*）疫苗罗非鱼的高滴度抗体被动免疫未接种鱼，可以降低链球菌的感染。这些结果显示，抗体在未接种疫苗的鱼体中起保护作用。

另一种抗体类型 IgD，其基因虽然已经在几种硬骨鱼中被克隆出来，但仅在斑点叉尾鮰（*Ictalurus punctatus*）中有相关蛋白水平的研究（Edholm 等，2010）。斑点叉尾鮰 IgD 蛋白为单体，且分泌型 IgD 缺乏可变区，其功能尚不清楚。然而，认为其 Fc 区域具有模式识别分子的作用，此外，发现鲶具有 IgD 颗粒细胞。

还有一种抗体类型为 IgT，其基因在一些硬骨鱼中表达，但在斑点叉尾鮰尚未发现。该抗体主要在肠道黏膜中发现，以多聚体的形式存在，表明其相当于 IgA，但在血清中也观察到 IgT 单体存在形式。与 IgM 类似，IgT 是由感染肠道的病原诱导产生的（Zhang 等，2011），然而这种抗体在鱼类中研究较少，所知不多（Salinas 等，2011）。IgT 分泌细胞的位置尚不明确，抗体传送到黏膜表面的机制也不清楚。

与免疫球蛋白类型的研究一样，对产生这些抗体的 B 细胞亚群也进行了大量的研究，斑点叉尾鮰是 B 细胞研究最多的硬骨鱼之一，根据 B 细胞受体（BCRs）分成了 3 个亚群，即 IgM^+/IgD^-（IgM）、IgM^+/IgD^+ 和 IgM^-/IgD^+（IgD）B 细胞（Edholm 等，2010）。然而，在其他种类如虹鳟中，只报道了两个亚群，$IgM^+/IgD^+/IgT^-$ 和 $IgM^-/IgD^-/IgT^+$。鱼类表达 IgT 的 B 细胞而不表达 IgM，反之亦然，说明 IgM 和 IgT 产生细胞是相互独立的专门的 B 细胞类群。与高等脊椎动物相比，由于缺少抗体类别转换，硬骨鱼可能具有不同的机制来产生多种类型的B细胞亚群（Zhang 等，2011）。IgM^+ B细胞是外周循环中主要的B细胞种类，而 IgT^+ B细胞是一个小的亚群，然而，后者主要在肠黏膜中，并且也发现存在于虹鳟皮肤黏液中。以此推测，IgT是专职参与黏膜免疫的免疫球蛋白。IgM^+ 和 IgT^+ B细胞都已被证明具有较强的吞噬功能以及吞噬病原微生物的能力（Li 等，2006）。

第六节 T细胞

T细胞是协调细胞介导免疫反应的一类细胞。在哺乳动物中，T细胞可以分为 αβ 或 γδ T细胞，依据T细胞受体（结合抗原的受体，缩写为TCR）分别是包含β链的TCR-α和包含δ链的TCR-γ。αβ-T细胞在血液循环系统和淋巴器官中数量较多，被认为是“传统的T细胞”。而γδ-T细胞多在黏膜组织和上皮组织中出现，几乎无需通过MHC分子进行抗原处理和递呈便可以识别病原，因而认为属于先天性免疫系统的细胞。因此，接下来提及的T细胞，除非特别说明，均指αβ-T细胞。

成熟的T细胞具有TCR、CD3、CD28和CD45表面标记分子，参与病原体识别和信号传递。与B细胞上的BCR一样，TCR通过基因重组也具有多样性。T细胞只能识别与MHC分子结合的抗原，这意味着抗原必须是在其他细胞的表面，才能被T细胞识别。

T细胞根据细胞毒性和辅助性被进一步分成功能不同的两类。一种是细胞毒性T细胞（CTLs），其通过释放的分子诱导细胞凋亡，直接杀伤被感染的或异常的细胞。靶细胞被识别是由于其表面有与MHC分子结合的病原或异常蛋白质片段。通过这种方式，宿主能够防止胞内病原感染。另一种是辅助性T细胞，通过分泌细胞因子激活其他免疫细胞对特定病原做出反应。

在硬骨鱼中，T细胞在血液循环系统中占比例较少，虽然它们在黏膜区域数量较多。由于缺乏相关试剂产品和手段，有关它们功能的研究比较滞后。然而，在近20年，大多数实验室利用已表达的序列标签（EST）数据库和鱼类基因组数据库，应用同源基因克隆技术开展了T细胞的特性研究。例如，通过免疫扫描谱型分析技术，研究虹鳟T细胞克隆扩增和系统多样性的特征（Boudinot 等，2001），一些鱼类中*TCR*基因既具有保守性又具有新的特征（Laing 和 Hansen，2011）。同时在几种鱼类中也发现了CD3链、CD28和其他的T细胞共刺激分子和共抑制分子（Araki 等，2005；Bernard 等，2006，2007；Liu 等，2008）。CC趋化因子受体7（CC7）、CD44和CD45RA等表面标记分子可以用于区分初始T细胞、记忆T细胞和效应T细胞（Laing 和 Hansen，2011）。但是，对T细胞各亚群的免疫应答功能仍知之甚少。

第七节 细胞毒性T细胞

细胞毒性T细胞（CTLs）是细胞毒性反应的效应细胞。它们表达CD8分子和TCR共同受体与MHC-Ⅰ类分子相互作用。在某些鱼类，这些分子的抗体可用于鉴定和分选该细胞类群（Araki 等，2008；Hetland 等，2010；Shibasaki 等，2010）。

首个被证明具有特异性 CTLs 的鱼类是斑点叉尾鮰（Miller 等，1986）。此后，在鲫和虹鳟中证明了抗原特异性和病毒特异性的细胞毒作用（Nakanishi 等，2002；Fischer 等，2006；Castro 等，2011）。虹鳟接种编码 VHSV 病毒 G 蛋白的 DNA 疫苗后，其外周血可有效地溶解 VHSV 感染的组织相容性细胞，但是不能溶解传染性造血器官坏死病毒（IHNV）感染的组织相容性细胞（Utke 等，2008）。

CTLs 可释放诱导细胞凋亡的分子，在几种鱼类中已鉴定出颗粒酶、穿孔素和凋亡相关因子配体 FasL（CD95）分子的基因序列。已有研究证明，哺乳动物的 FasL 抗体与海鲷白细胞和鲶 T 细胞系有交叉反应（Cuesta 等，2003；Long 等，2004），这些结果说明了该分子从鱼类到哺乳动物高度保守。还有研究发现重组穿孔素在钙离子存在的情况下具有溶解性（Laing 和 Hansen，2011）。在一些鱼类物种中还鉴定出了 CTLs 脱颗粒标记分子 CD107 的基因序列。

第八节 辅助性 T 细胞

如上所述，辅助性 T 细胞（Th）产生细胞因子，这些细胞因子在细胞免疫反应和体液免疫反应中都有重要作用。所有 Th 细胞的一个共同特征是拥有一种表面的糖蛋白 CD4，其决定 Th 细胞与 MHC-Ⅱ类蛋白相互作用的特异性。在许多硬骨鱼中已经发现具有哺乳动物 CD4 结构保守的同源基因以及类似的 *CD4* 基因（Laing 和 Hansen，2011），这些分子也作为研究靶标，且已研制出抗鲫 CD4 单克隆抗体（Shibasaki 等，2010）。在哺乳动物中，根据产生细胞因子的类型，CD4 细胞被分为几种效应性亚群和调节性亚群（图 9-2）。Th1 细胞主要参与细胞免疫反应；Th2 细胞参与体液免疫反应；Th17 细胞增强对胞外细菌的免疫反应；而调节性 T 细胞（Tregs）参与调节或抑制免疫反应以防止免疫损伤。最近，Th22、Th9 和 Tfh 等 Th 细胞也陆续在哺乳动物中被鉴定出来（Dong，2006；Reiner，2007）。

在硬骨鱼中，已鉴定出多种辅助性 T 细胞亚群的标志性转录因子和细胞因子的同源序列（Laing 和 Hansen，2011），研究主要集中在遗传和功能相关的方面。在多种鱼类中，已测定出 Th1 细胞的标志细胞因子干扰素-γ2（IFN-γ2）序列，以及与哺乳动物 IFN-γ 没有同源性的另一个序列（鱼 IFN-γ1）（Zou 和 Secombes，2011）。在斑马鱼、虹鳟和金鱼中已发现，重组 IFN-γ 可以诱导抗病毒基因和炎症性基因的表达，与哺乳动物相似，鱼类中这种细胞因子主要由 T 细胞和 NK 细胞产生，巨噬细胞和 B 细胞也能产生 IFN-γ。T-β 是 Th1 细胞标志性转录因子，其序列也已在鱼类中被鉴定，研究发现病毒感染草鱼后，T-β 表达水平存在差异（Wang 等，2013）。

到目前为止，鱼类中还没有经典的 Th2 型细胞因子功能的研究。在基因水平上发现，*IL-4/13A* 和 *IL-4/13B* 是 *IL-4* 和 *IL-13* 共同起源的两个旁系基因（Li 等，2007），而硬骨鱼中主要的转录因子 GATA-3 在序列和发育功能上，与哺乳动物的相比是保守的。

在哺乳动物中，Th17 细胞所分泌的细胞因子是多基因家族 IL-17A 到 IL-17F，以及 IL-22，在硬骨鱼中发现了这类细胞分泌的三种细胞因子形式。已有 *IL-17A* 基因表达的研究（Mutoloki 等，2010），仍需对其功能进行探索。另一个 T 细胞亚群是 Th22 细胞，研究证明其可分泌 IL-22，与 Th17 细胞不同，它不能分泌 IL-17（Eyerich 等，2010）。虽然在硬骨鱼中有关

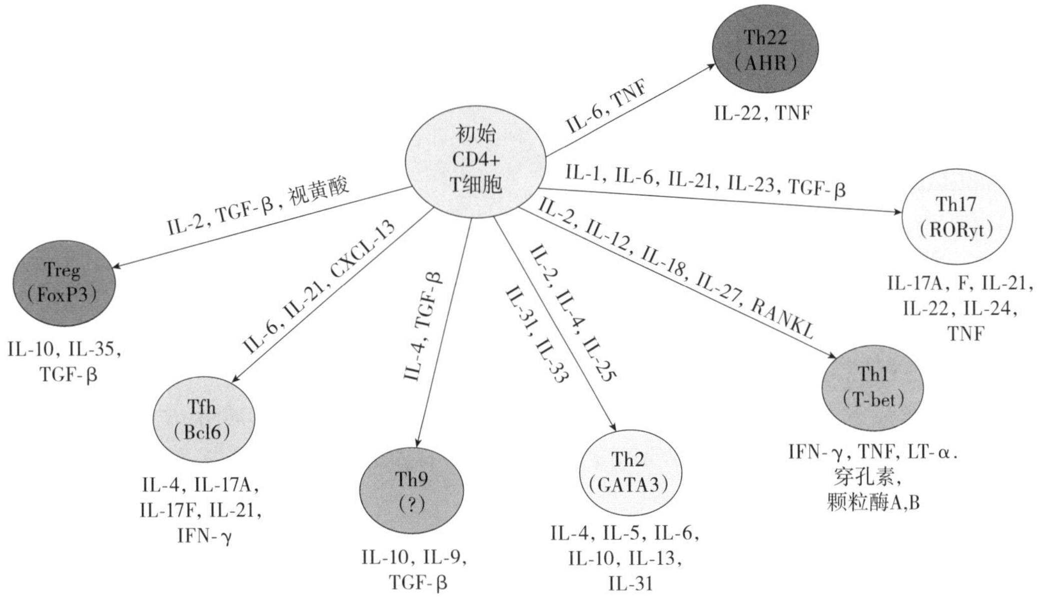

图 9-2 CD4^{+} T 细胞的分化

特异性的细胞因子决定 T 细胞的分化亚群（细胞因子标注在箭柄两侧，箭头所指为细胞亚群），关键转录因子（括号里）决定 T 细胞亚群分泌的细胞因子（在细胞亚群下面列出）。Th1 细胞亚群主要参与胞内微生物侵染的免疫反应、器官特异性自身免疫失调和慢性炎症；Th2：胞外寄生虫和过敏性疾病；Th9：功能不详；Th22：参与皮肤稳态和病理；Tfh：辅助滤泡 B 细胞；Bcl6：B 细胞淋巴瘤因子- 6；RANKL：核因子 NFκB 配体的受体激活剂

于 Th22 细胞的描述，但其还未被鉴定出来，因此需要更多的研究确定鱼类中是否存在 Th22 细胞。有关 Tfh、Th9 和 Tregs 等其他 T 细胞亚群，研究发现，Tfh 细胞特殊的作用是辅助囊泡 B 细胞变为长寿命浆细胞。在哺乳动物中，Tfh 细胞分泌 IL-21，也能分泌 IL-10，表明 Tfh 细胞对免疫应答有抑制作用。在河鲀的基因组中发现存在 IL-21，在鲑中也有 IL-21 功能的研究，但尚未证实产生 IL-21 的细胞是 Tfh 细胞。鱼类的细胞因子是否类似哺乳动物具有诱导抗体产生的功能还有待探索。B 细胞白血病/淋巴瘤- 6（Bcl-6）是该细胞亚群特异性的转录因子，已在硬骨鱼中发现，但并未有其他进一步的报道。到目前为止，在鱼类中还没有关于 Th9 细胞的报道，但对于 Treg 来说，转录因子 *Foxp3* 的功能已在许多鱼类中得到证实，其在胸腺和脾脏等 T 细胞相关组织中的表达量最高，并且在斑马鱼胚胎中，*Foxp3* 的表达量与 *IL-17* 的表达量成互反关系，这与 Foxp3 抑制 Th17 细胞的作用是一致的。

参考文献

Araki, K., Suetake, H., Kikuchi, K. and Suzuki, Y. 2005. Characterization and expression analysis of CD3varepsilon and CD3gamma/delta in fugu, *Takifugu rubripes*. *Immunogenetics* 57: 158-163.

Araki, K., Akatsu, K., Suetake, H., *et al*., 2008. Characterization of CD8^{+} leukocytes in fugu (*Takifugu rubripes*) with antiserum against fugu CD8α. Dev Comp Immunol 32: 850-858.

Bassity, E. and Clark, T. G. 2012. Functional identification of dendritic cells in the teleost model, rainbow trout (*Oncorhynchus mykiss*). *PLoS One* 7: e33196.

Bercovier, H., Ghittino, C. and Eldar, A. 1997. Immunization with bacterial antigens: infections with streptococci and related organisms. *Dev Biol Stand* 90: 153-160.

Bernard, D., Riteau, B., Hansen, J. D., *et al*., 2006. Costimulatory receptors in a teleost fish: typical CD28, elusive CTLA4. *J Immunol* 176：4191-4200.

Bernard，D.，Hansen，J. D.，Du Pasquier，L.，*et al.*，2007. Costimulatory receptors in jawed vertebrates：conserved CD28，odd CTLA4 and multiple BTLAs. *Dev Comp Immunol* 31：255-271.

Boudinot，P.，Boubekeur，S. and Benmansour，A. 2001. Rhabdovirus infection induces public and private T cell responses in teleost fish. *J Immunol* 167：6202-6209.

Bromage，E. S.，Kaattari，I. M.，Zwollo，P. and Kaattari，S. L. 2004. Plasmablast and plasma cell production and distribution in trout immune tissues. *J Immunol* 173：7317-7323.

Castro，R.，Bernard，D.，Lefranc，M. P.，*et al.*，2011. T cell diversity and TcR repertoires in teleost fish. *Fish Shellfish Immunol* 31：644-654.

Clem，L. W.，Miller，N. W. and Bly，J. E. 1991. Evolution of lymphocyte subpopulations，their interactions and temperature sensitivities，in *Phylogenesis of Immune Functions*（eds G. W. Warr and N. Cohen）. Boca Raton，FL，CRC Press，191-213.

Cuesta，A.，Esteban，M. A. and Meseguer，J. 2003. Identification of a FasL-like molecule in leucocytes of the teleost fish gilthead seabream（*Sparus aurata* L.）. *Dev Comp Immunol* 27：21-27.

Dong，C. 2006. Diversification of T-helper-cell lineages：finding the family root of IL-17-producing cells. *Nat Rev Immunol* 6：329-333.

Edholm，E. S.，Bengten，E.，Stafford，J. L.，*et al.*，2010. Identification of two IgD^{+} B cell populations in channel catfish，*Ictalurus punctatus*. *J Immunol* 185：4082-4094.

Ellis，A. E.，Stapleton，K. J. and Hastings，T. S. 1988. The humoral immune response of rainbow trout（*Salmo gairdneri*）immunised by various regimes and preparations of *Aeromonas salmonicida* antigens. *Vet Immunol Immunopathol* 19：153-164.

Eyerich，S.，Eyerich，K.，Cavani，A. and Schmidt-Weber，C. 2010. IL-17 and IL-22：siblings，not twins. *Trends Immunol* 31：354-361.

Fischer，U.，Utke，K.，Somamoto，T.，*et al.*，2006. Cytotoxic activities of fish leucocytes. *Fish Shellfish Immunol* 20：209-226.

Geissmann，F.，Manz，M. G.，Jung，S.，*et al.*，2010. Development of monocytes，macrophages，and dendritic cells. *Science* 327：656-661.

Hetland，D. L.，Jørgensen，S. M.，Skjødt，K.，*et al.*，2010. *In situ* localisation of major histocompatibility complex class I and class II and CD8 positive cells in infectious salmon anaemia virus（ISAV）-infected Atlantic salmon. *Fish Shellfish Immunol* 28：30-39.

Iliev，D. B.，Jørgensen，S. M.，Rode，M.，*et al.*，2010. CpG-induced secretion of MHCIIβ and exosomes from salmon（*Salmo salar*）APCs. *Dev Comp Immunol* 34：29-41.

Jørgensen，J. B. 2014. The innate immune response in fish，in *Fish Vaccination*（eds R. Gudding，A. Lillehaug and Ø. Evensen）. Oxford，John Wiley & Sons，Ltd.，85-103.

Koppang，E. O.，Fischer，U.，Moore，L.，*et al.*，2010. Salmonid T cells assemble in the thymus，spleen and in novel interbranchial lymphoid tissue. *J Anat* 217：728-739.

LaFrentz，B. R.，Shoemaker，C. A. and Klesius，P. H. 2011. Immunoproteomic analysis of the antibody response obtained in Nile tilapia following vaccination with a *Streptococcus iniae* vaccine. *Vet Microbiol* 152：346-352.

Lagos，L. X.，Iliev，D. B.，Helland，R.，*et al.*，2012. CD40L—a costimulatory molecule involved in the maturation of antigen presenting cells in Atlantic salmon（*Salmo salar*）. *Dev Comp Immunol* 38：416-430.

Laing，K. J. and Hansen，J. D. 2011. Fish T cells：recent advances through genomics. *Dev Comp Immunol* 35：

1282-1295.

Li, J., Barreda, D. R., Zhang, Y. A., *et al.*, 2006. B lymphocytes from early vertebrates have potent phagocytic and microbicidal abilities. *Nat Immunol* 7: 1116-1124.

Li, J. H., Shao, J. Z., Xiang, L. X. and Wen, Y. 2007. Cloning, characterization and expression analysis of pufferfish interleukin-4 cDNA: the first evidence of Th2-type cytokine in fish. *Mol Immunol* 44: 2078-2086.

Liu, Y., Moore, L., Koppang, E. O. and Hordvik, I. 2008. Characterization of the CD3ζ, CD3γδ and CD3ε subunits of the T cell receptor complex in Atlantic salmon. *Dev Comp Immunol* 32: 26-35.

Long, S., Wilson, M., Bengten, E., *et al.*, 2004. Identification of a cDNA encoding channel catfish interferon. *Dev Comp Immunol* 28: 97-111.

Lugo-Villarino, G., Balla, K. M., Stachura, D. L., *et al.*, 2010. Identification of dendritic antigen-presenting cells in the zebrafish. *Proc Natl Acad Sci USA* 107: 15850-15855.

Miller, N. W., Deuter, A. and Clem, L. W. 1986. Phylogeny of lymphocyte heterogeneity: the cellular requirements for the mixed leucocyte reaction with channel catfish. *Immunology* 59: 123-128.

Munang'andu, H. M., Fredriksen, B. N., Mutoloki, S., *et al.*, 2013. Antigen dose and humoral immune response correspond with protection for inactivated infectious pancreatic necrosis virus vaccines in Atlantic salmon (*Salmo salar* L). *Vet Res* 44: 7.

Mutoloki, S., Cooper, G. A., Marjara, I. S., *et al.*, 2010. High gene expression of inflammatory markers and IL-17A correlates with severity of injection site reactions of Atlantic salmon vaccinated with oil-adjuvanted vaccines. *BMC Genomics* 11: 336.

Nakanishi, T., Fischer, U., Dijkstra, J. M., *et al.*, 2002. Cytotoxic T cell function in fish. *Dev Comp Immunol* 26: 131-139.

Press, C. McL. and Evensen, Ø. 1999. The morphology of the immune system in teleost fish. *Fish Shellfish Immunol* 9: 309-318.

Randolph, G. J., Angeli, V. and Swartz, M. A. 2005a. Dendritic-cell trafficking to lymph nodes through lymphatic vessels. *Nat Rev Immunol* 5: 617-628.

Randolph, G. J., Sanchez-Schmitz, G. and Angeli, V. 2005b. Factors and signals that govern the migration of dendritic cells via lymphatics: recent advances. *Springer Semin Immunopathol* 26: 273-287.

Reiner, S. L. 2007. Development in motion: helper T cells at work. *Cell* 129: 33-36.

Salinas, I., Zhang, Y. A. and Sunyer, J. O. 2011. Mucosal immunoglobulins and B cells of teleost fish. *Dev Comp Immunol* 35: 1346-1365.

Shibasaki, Y., Toda, H., Kobayashi, I., *et al.*, 2010. Kinetics of $CD4^+$ and $CD8\alpha^+$ T-cell subsets in graft-versus-host reaction (GVHR) in ginbuna crucian carp *Carassius auratus langsdorfii*. *Dev Comp Immunol* 34: 1075-1081.

Sizemore, R. C., Miller, N. W., Cuchens, M. A., *et al.*, 1984. Phylogeny of lymphocyte heterogeneity: the cellular requirements for *in vitro* mitogenic responses of channel catfish leukocytes. *J Immunol* 133: 2920-2924.

Steinman, R. M. 2007. Dendritic cells: understanding immunogenicity. *Eur J Immunol* 37: S53-60.

Steinman, R. M., Kaplan, G., Witmer, M. D. and Cohn, Z. A. 1979. Identification of a novel cell type in peripheral lymphoid organs of mice. V. Purification of spleen dendritic cells, new surface markers, and maintenance *in vitro*. *J Exp Med* 149: 1-16.

Sunyer, J. O. 2012. Evolutionary and functional relationships of B cells from fish and mammals: insights into their novel roles in phagocytosis and presentation of particulate antigen. *Infect Disord Drug Targets* 12: 200-212.

Utke, K., Kock, H., Schuetze, H., *et al.*, 2008. Cell-mediated immune responses in rainbow trout after DNA immunization against the viral hemorrhagic septicemia virus. *Dev Comp Immunol* 32: 239-252.

Vallejo, A. N., Miller, N. W. and Clem, L. W. 1991. Phylogeny of immune recognition: processing and presentation of structurally defined proteins in channel catfish immune responses. *Dev Immunol* 1: 137-148.

Wang, L., Shang, N., Feng, H., *et al.*, 2013. Molecular cloning of grass carp (*Ctenopharyngodon idellus*) Tbet and GATA-3, and their expression profiles with IFN-γ in response to grass carp reovirus (GCRV) infection. *Fish Physiol Biochem* 39: 793-805.

Wittamer, V., Bertrand, J. Y., Gutschow, P. W. and Traver, D. 2011. Characterization of the mononuclear phagocyte system in zebrafish. *Blood* 117: 7126-7135.

Xu, C., Mutoloki, S. and Evensen, Ø. 2012. Superior protection conferred by inactivated whole virus vaccine over subunit and DNA vaccines against salmonid alphavirus infection in Atlantic salmon (*Salmo salar* L.). *Vaccine* 30: 3918-3928.

Zapata, A. and Amemiya, C. T. 2000. Phylogeny of lower vertebrates and their immunological structures. *Curr Top Microbiol Immunol* 248: 67-107.

Zhang, Y. A., Salinas, I. and Oriol, S. J. 2011. Recent findings on the structure and function of teleost IgT. *Fish Shellfish Immunol* 31: 627-634.

Zou, J. and Secombes, C. J. 2011. Teleost fish interferons and their role in immunity. *Dev Comp Immunol* 35: 1376-1387.

Zwollo, P., Cole, S., Bromage, E. and Kaattari, S. 2005. B cell heterogeneity in the teleost kidney: evidence for a maturation gradient from anterior to posterior kidney. *J Immunol* 174: 6608-6616.

第十章　鱼类疫苗的研发、生产和质控

Dag Knappskog, Joseph Koumans, Inger Kvitvang, Arne Marius Fiskum and Rune Wiulsrød

本章概要　SUMMARY

在全球鱼类养殖中，成熟市场和新兴市场都需要针对各种工业化养殖鱼类的安全有效疫苗。近几十年来，鱼类生物制品的生产商已由具有创业基础的当地制造商发展为专门的产业公司。通常，使用各种不同的疫苗和疫苗组合制品来预防已知的鱼类病原。鱼类疫苗主要分为三类：灭活疫苗、活疫苗和 DNA 疫苗。鱼类疫苗属于食品生产动物的生物制品，因此鱼类疫苗的所有流程步骤均由国家和国际法律准则制定的严格监管要求所管制，包括疫苗的研发和生产，生产中的转移和升级、纯化、灌装、包装，最终产品的标签和测试。

第一节　引　　言

过去的几十年里，疫苗行业历经重大的发展，已从科学技术和实践知识相结合的传统生产，发展到如今具有先进的操作系统和质量系统的高度精密、自动化的高科技产业生产。在这些高科技企业中，具有生物技术知识的科学家和具有实操专长的人员与工程师和系统控制员一起协同工作。

疫苗生产是从具有创业基础的本土公司开始，如今，由大型国际产业公司主导，其利益远远超出当地的需求。疫苗生产从最初的想法到生产和销售这一过程的复杂性，使得疫苗公司必须改善其财务基础，因此，疫苗公司间通常会发生并购。

现有的鱼类疫苗公司是专业企业或大型制药公司中的专业部，其不仅生产鱼类疫苗，还研发和生产为满足养殖鱼类治疗需求的其他医药产品，像预防海虱和真菌感染的产品以及麻醉、激素产品。

水产养殖业的疫苗由溶解在水中或水油乳剂中的活性成分（抗原）组成。疫苗含有的免疫增强剂或佐剂有多种类型，油类是其中一种（例如矿物油）。油基疫苗有三种形式：油包水、水包油、水包油包水，每一种都有其优缺点。此外，水产疫苗还需要添加乳化剂来维持其物理稳定性。油基疫苗内的抗原可以固定在水相、油相中，或固定在水油两相中。这类疫苗可含有细菌和病毒多种抗原，因此可以预防多种疾病。油基疫苗只用于腹腔注射，水基疫苗可以用于浸泡或腹腔注射，而 DNA 疫苗则用于肌肉注射。

鱼类疫苗的研发、生产和质控受国家和国际法律、法规及准则的管制（Holm 等，2014）。

第二节 生产许可证

疫苗的生产和质控是一个复杂的过程，对厂房、设备和人员的要求很高。所有商用疫苗的生产商都必须得到所在国疫苗生产管理部门的批准，这种批准是基于所有可能会影响疫苗终产品质量的相关因素的一系列文件。疫苗生产的厂房和设备必须有符合规定的标准（图 10-1），所有过程必须经过验证，并且从业人员必须具备必要的相关能力。

根据现代质量控制准则（QC）的规定，疫苗生产常规的内部控制过程必须是合理的，这些内部控制过程，在标准操作规程（SOPs）中必须进行详细描述，因此，疫苗的生产和质控必须按照药品质量管理规范（GMP）进行。

在欧盟/欧洲经济区，获得疫苗生产许可需遵从 GMP 91/412/EEC 指令，该指令也适用于兽药生产，这是管理机构现场 GMP 审计评估的重要内容。

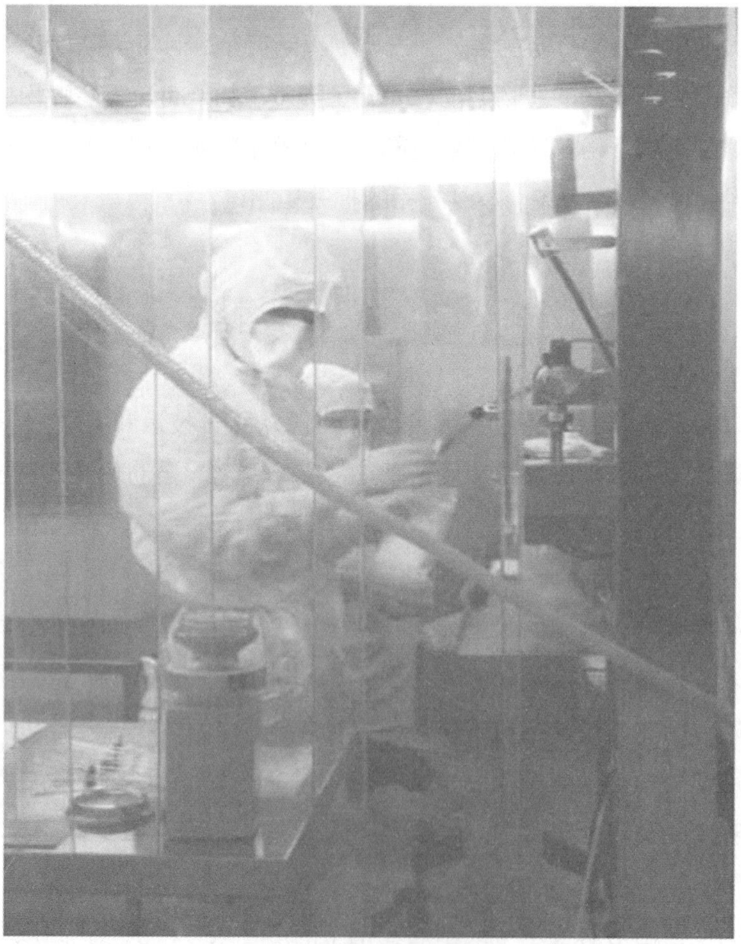

图 10-1　疫苗的无菌灌装
（经 Pharmaq 公司许可转载）

一、档案

疫苗档案有具体内容要求，档案包含疫苗的研发、安全、功效、生产和质控等相关方面的详细资料。在欧洲经济区（EEA），疫苗档案内容的共同要求可在委员会 2009/9/EC 指令的修订指令 2001/82/EC 中找到，并且在 EEA 的通用指南中有进一步的描述。

疫苗档案记载的是疫苗和疫苗生产企业。疫苗产品部分是活性成分、适应症、使用方法、禁忌等。疫苗生产企业部分是机构、厂房、产品组合和人员信息。疫苗档案还包括生产许可证的副本，并且疫苗生产企业必须声明，疫苗生产遵循 GMP 的准则。

疫苗的生产过程必须有非常详细的描述，包括制成品的质量检查。一个疫苗中的抗原是至关重要的，因此档案必须包含抗原所有相关信息资料，如来源、特性、原材料（尤其强调动物来源的）的生产和质控、存储和管理。

档案资料还应包括微生物发酵、培养基的组成以及抗原收获后下游处理的相关细节，另外还包括疫苗生产过程中质量检查相关的信息。疫苗生产过程的所有阶段均需采集样品，并且这些样品都要按照计划使用特定的方法进行检测。遵循 GMP 或药品临床试验管理规范（GCP），在实验室或现场检验时应使用经验证可靠的方法（图 10-2）。

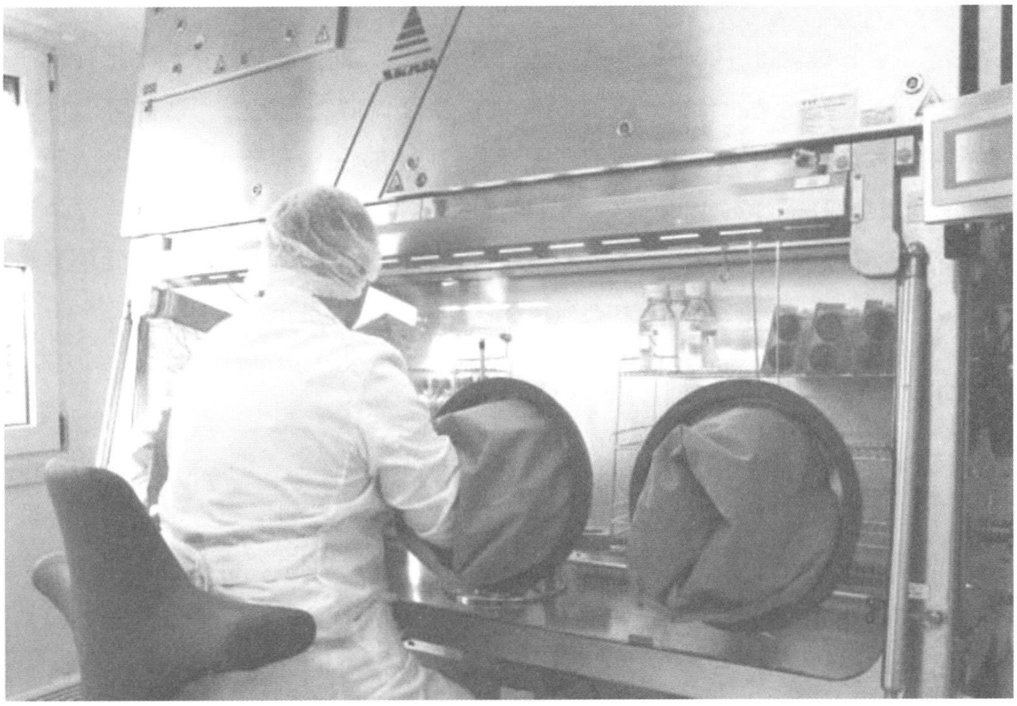

图 10-2　无菌检测隔离箱
（经 Pharmaq 公司许可转载）

二、质量体系

药品生产，必须建立有效的质量体系。质量体系的关键原则是要有良好的生产系统，这样能确保每次都能生产出同样的产品。欧盟委员会指令 91/412/EEC 和 Eudralex 的第 4 卷明确规定了兽药产品的药品质量管理规范原则和指南。

三、质量管理

质量管理（QM）涵盖了单一或多种影响产品质量的所有事项（图 10-3）。质量管理体系适用于药品生产，确保按药品生产质量管理规范和实验室管理规范执行。标准操作规程（SOPs）应该规范清楚所有的体系和程序。质量管理是管理部门规定要求的体系和程序的全部内容，这包括责任定义系统、质量文件、培训、生产控制、测试设备、原材料规格、实验室检测和管理、实验室和实地检查的质量要求、定期检修、风险管理、偏差处理、管理变更、客户投诉和召回、药物警戒、制造商/实验室/供应商的合同批准以及内部/外部审计。

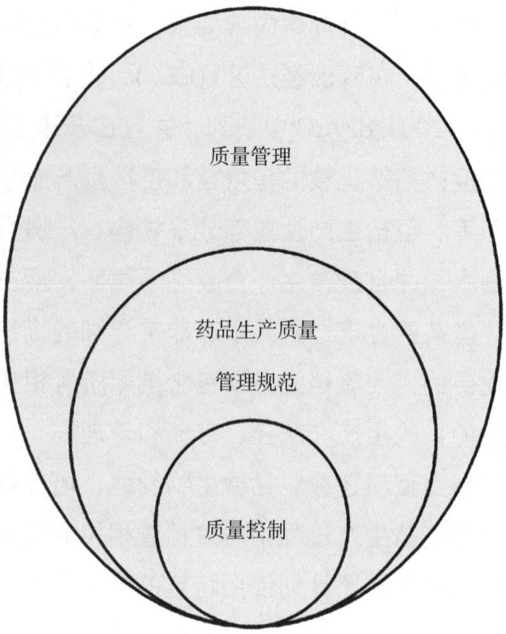

图 10-3　质量管理体系

四、药品生产质量管理规范

药品生产质量管理规范（GMP）是质量管理的一部分，确保产品始终如一地按照质量标准控制和生产，提供预期用途的合格产品。GMP 要求整个价值链可追溯，从许可的供应商处购买原材料到生产和质控在内的所有活动都必须记录，所有的关键环节都要通过双人验证（四眼原则）。产品的质量取决于多个因素，如产品生产的设施和设备必须符合产品生产的目标，并且质量达标；产品生产的程序和方案很重要且生产过程必须严格明确；从业人员的能力和态度是成功的关键因素，必须投入大量精力训练员工的能力并记入员工个人档案。

五、质量控制

质量控制（QC）是 GMP 的一部分，其内容包括规范、样品、测试、文件审查和产品发布。质控是确保产品符合既定规范的工具。通常，QC 的重点是测试最终产品，如今这种观念已被新观念取代，新观念是产品质量体现在产品生产的所有过程中，并需要在整个生产过程中取样验证产品质量。有效的质控体系保证了每一批产品的质量，并且质控是验证产品是否符合规范，而不是检验规范本身。

正如产品规范所规定的，质控用于确认产品的质量。此外，必须测试每批次的产品，验证其安全性和效力。

传统的细菌/病毒疫苗通常使用甲醛灭活，必须确认甲醛已使它们失活。灭活后，至少在不同时段两次取灭活样品，证明其已被灭活。失活过程必须经过验证，也就是说，必须证明在总失活期的 2/3 的时间内有效失活，失活动力学的速度和水平也表明甲醛对微生物失活的效果。此外，中间产品和最终产品也必须经过纯度和无菌检测 。

病毒抗原制备用的细胞种子库和病毒种子库必须按照《欧洲药典》规定的要求进行测试。这包括大量的病毒测试和无菌测试，既复杂又耗时。

质量控制的所有测试在体外进行是最理想的，但对疫苗来说，只有进行靶动物试验才能证明其安全性和有效性。疫苗的安全性是通过对靶动物使用过量的疫苗，证明疫苗对其无害。一般而言，灭活疫苗采用2倍剂量，活疫苗10倍剂量。在疫苗安全性研究期间，通常连续观察靶动物3周，记录所有的异常反应。

功效测试对证明产品的一致性十分重要，包括测定微生物的数量（活疫苗）、血清学或抗体水平。鱼类疫苗的功效测试在鱼体内进行，把免疫组的死亡率与未免疫组的死亡率进行对比，若对照组的死亡率处在符合要求的水平，则疫苗功效用相对存活率（RPS）表示，也可用参考疫苗作阳性对照。

第三节 疫苗研发

一旦确定病原体与疾病的相关性，就可以进行该病原体的疫苗研发。理想的疫苗是安全和有效的，能迅速研发和注册，能供应充足的剂量且成本低，易于管理。研发水生动物的疫苗非常具有挑战性，因为水产疫苗必须遵守许多规定（如在单个小的注射剂量中混合多种抗原组分），这些规定因靶动物不同而不同。特定的疫苗需求决定了抗原的选择和生产系统的选择，但有时这些选择也会做适当调整（最好的实现不了，但要努力做到最好）。通常，疫苗分为活疫苗、灭活疫苗和DNA疫苗，每一类还可进一步细分。

一、可复制型疫苗

经典活抗原可以是一种在特定系统中传代培养获得的减毒株，而这种特定系统和正常系统不同（如温度不同、其他物种的细胞系、缺乏特定成分的培养基等）（Shoemaker 和 Klesius，2014）。研发这类活抗原通常需要花费很长时间，并可能需要数百次的传代培养。然而，一般来说，活抗原更为有效，这是因为在较低剂量（剂量为灭活疫苗的1/1 000）情况下，活抗原仍然可以在靶动物中复制。另外，如果培养体系适应活抗原复制的话，就可以有相对较高的产出。

研发定向诱变的活抗原，需要对病原的分子生物学有广泛了解，并且需要一个可以在其基因组中引入特定突变/剪切的系统（如反向遗传学），虽然有时病原体存在明显的可操纵的候选基因，但是这些基因对诱发有效免疫反应的能力或对病原特性会产生很大的影响，因此，改造这些候选基因不太适合用于生产疫苗，除非能通过培养条件消除这种影响。这些抗原属于转基因生物（GMOs），这意味着必须遵循转基因生物的注册程序。

载体活疫苗是添加其他病原保护性抗原的载体，这种活疫苗有其优点，它是一个简化生产环节的模块化系统，但对新的、以前不存在的特性的安全问题随之而来，比如改变物种特异性或趋向性，这些特性是随着转基因的加入而引入的。这可以通过载体复制缺陷（所谓的单轮感染粒子）来抵消，但很明显，这种大规模的载体生产并不容易实现。

活抗原的另一个优点是，其自身就是有效的，并不需要佐剂，这对其使用方式（一般来说，油类佐剂疫苗只能注射）及其局部反应的安全性是有好处的。活疫苗可通过口服或浸泡接种，适用于小鱼（海洋生物）和价值相对低的鱼类。活抗原的缺点是，不能与其他的灭活抗原联合使用，并且还需要特定储存条件（冷冻或冻干）。

二、非复制型疫苗

非复制型（灭活的）疫苗是指灭活的全病原体或重组亚单位的疫苗（Munang'andu 等，2014）。重组亚单位疫苗是通过表达系统生产的，该系统可以是细菌、酵母、植物、脊椎动物细胞、重组脊椎动物或感染脊椎动物细胞的病毒载体。表达系统的类型取决于产生有效抗原需要的翻译后加工的水平。虽然灭活产品是安全的，但亚单位系统的选择也要考虑安全性。技术方面的问题也必须考虑，简单地在生产系统中培养病原体是不可能的，同时需要具有能判断诱导免疫保护的病原体成分的分子生物学知识。最后，需要评估疫苗的成本，因为有的鲑鳟鱼类病毒和培养病毒的细胞生长速度非常缓慢，这会影响疫苗的生产成本和生产能力。

一般来说，非复制型疫苗需要的免疫剂量较高，即使是结合佐剂一起使用，免疫剂量也需加倍，这就需要高产的大型生产系统和低成本。由于亚单位疫苗的免疫保护仅建立在表位很少的单一抗原上，因此，这种疫苗对野生变异株的保护能力会受到影响。同样地，载体疫苗也面临这种问题，其是含有编码病原体基因可以预防疾病的亚单位疫苗。

最后，虽然每种类型的疫苗都需要特定类型的抗原，但是选择适合每种类型疫苗的生产系统也非常重要。

三、DNA 疫苗

DNA 疫苗接种是遗传免疫，它依赖的是基因或编码保护性抗原的基因，并不是抗原本身。DNA 疫苗是一种含有一个或多个来自病原体基因的遗传结构，能瞬时促进疫苗基因产生蛋白质，引起保护性免疫反应（Biering 和 Salonius，2014）。DNA 疫苗通常是纯化的细菌质粒，一种可以在细菌中复制的环状 DNA 分子。DNA 疫苗包含调控元件，以确保质粒在细菌中的复制生产，以及疫苗基因在接种动物中表达。在某些情况下，DNA 疫苗又称核酸疫苗（NAV）或基因疫苗，RNA 疫苗也属于这个范畴。

第四节 研发试验

研发试验是为了确保产品的安全性和有效性，并早期发现生产过程中的偏差，将成本降到最低、耗时降到最短。测试内容有原材料、设备和基础设施（如环境监测）、抗原和最终产品。许多测试（如批次安全测试、批次药效测试、灭活产品的灭活控制测试、无菌测试等）是强制性的。还有一些内部测试，是为了更好地控制和监控生产过程（如批次给料生产过程中，测定生物反应器中的葡萄糖水平）。任何情况下，测试结果都必须与规范进行比对，然后才能执行预先制定的程序。在某些情况下，这可能意味着材料或抗原甚至疫苗不合格。任何时候，必须防止发生产品召回事件。

测试应满足的条件是：快速（获得结果是即时的不是滞后的）、客观（不受人为因素影响）、准确（否则会对产品的安全性、有效性和经济性产生影响）和经济（不会大幅提高产品成本）。

第五节 中试试验

疫苗从研发到生产的过程中遇到的主要挑战之一是如何扩大疫苗的实际生产规模。试想一下，一个 1mL 瓶装抗原的冷冻/解冻与 2 000 L 生物反应器抗原的冷冻/解冻的关系（图 10-4）。如果管理不当，在确保大规模生产时进行调整可能不会被认可，监管部门将把研制批次和大规模生产批次视为完全不同的产品，因此，注册登记的批次量会降到最小实际生产规模。

图 10-4 生产细菌抗原的发酵罐
（经 Pharmaq 公司许可转载）

疫苗生产过程依据总制造说明书规定执行，与生产现场无关；而批量生产说明书则是专门用于特定的生产现场。以上两个说明书涵盖了疫苗生产过程的所有阶段。

过程研发是一个非常专业的领域，因此须由过程研发团队或技术支持团队实施完成。团队应尽早参与新产品的研发，以确保所有注册实验所使用的材料与实际产品一致。同时，团队也应经常参与中试生产，以利于向规模生产转换。将产品从一个生产地点转移到另一个生产地点也被认为是传输转换，必须对这个流程进行验证和记录。新产品转移到生产之前，所有必要的质控测试也必须经过验证并移交到质控部门。产品生产过程中，过程控制进行的中间产品测试，通常在生产现场进行，这就可以避免样品从一个地点转运到另一个地点的过程中出现问题。

第六节 生　产

疫苗生产是制药行业的一个分支，GMP 认证的设施是疫苗生产的必要基础，因此，必须对疫苗的生产厂房和设备进行验证。同时，也需要对生产工艺、生产过程中的测试以及质控测试进行验证。所有这些项目都必须记录和得到认可，其理念是为了确保产品质量以及产品均一，所以验证和培训是疫苗生产的重要基石。

一、生产厂房

疫苗生产工厂必须按照严格的要求和规范进行设计、规划和建造。无菌车间是最关键的，需要在适当的位置设置公用区域以确保产品不受污染，并且人和材料的入口必须分开。操作台表面必须光滑，没有污染物可以积聚的孔隙，同时空气供应设备和通风设备必须经过高效空气过滤器处理，并且要时常调控湿度，获得最佳的相对湿度。必须为保护产品和操作人员而特别制作工作服，同时，人员和材料的流程图是 GMP 的重要组成部分。

二、生产系统

通常，有两种类型的疫苗生产系统：附着型和悬浮型。大多数细菌可以在悬浮液中生长，而生产病毒的大多数细胞基质都是附着生长。

三、细菌抗原和重组亚单位疫苗的生产

细菌抗原是常见的陆生动物疫苗和鱼类疫苗组分，通常是传统发酵生产。细菌抗原可以是全菌或是部分，或是亚单位（如毒素）。细菌抗原生产是从种子系统提供的菌种开始，它是其后生产的具体特定抗原的“母体”。

种子能在固体培养基和液体培养基中生长，可通过冷冻干燥、稳定液体、超低温冰箱或液氮保存。生产用菌种是种子的传代菌，它是发酵培养的开始，是将来数以千升计的细菌发酵液的起源。

发酵罐通常由不锈钢制作，可控制搅拌速度、溶解氧和 pH 等参数。为降低操作成本和减少验证工作，开发出了一次性发酵罐。为了减少污染风险，发酵应在一个封闭的系统中，这意味着从接种到最终产品的过程中不接触外界环境。

生产重组亚单位抗原常常也使用发酵的方法，是由基因工程菌产生特定的蛋白作为最终疫苗抗原，生产系统确立后，就能高效地大量生产抗原。

四、病毒培养

病毒具有宿主依赖性，不能用有机或无机营养培养基培养，只能在活细胞和组织中培养。在实验室或生产条件下，这些活的细胞和组织成为病毒的宿主，为此，培养病毒前，首先应培养相关的细胞或组织。

病毒颗粒大量生产的方法还有接种动物的体内培养和鸡胚培养，然而，标准方法是细胞培养。

用于生产病毒的细胞贴壁生长，无菌细胞培养瓶是贴壁细胞的标准生产系统，但近来，新的封闭系统越来越多，且培养条件较易控制（例如细胞工厂、细胞立方体和固定床生物反应器）。较早的一种替代方法是微载体生物反应器，但微载体系统的体积有限制（一般为 250～400L，而悬浮细胞可以在1 000 L 或更大体积的容器中生长）。新方法是使用“封闭系统”（例如管焊接）和一次性系统（例如单次生物反应器）。封闭系统大大降低了污染的可能性，而一次性系统可减少不同的生产运行之间的时间耗费，且具有对基础设施要求低、灵活性强的优点。

五、灭活

灭活是微生物疫苗生产过程中必不可少的一部分，灭活方法会影响疫苗的有效性和安全性。灭活疫苗需要严格的质控以确保微生物完全被灭活，因此，必须有灭活过程验证。灭活有

多种方法，如化学、温度和辐射灭活。灭活过程中的温度、浓度和持续时间是主要的参数。通常，灭活使用的化学试剂是甲醛，这是因为甲醛具有保持激发免疫反应重要抗原的结构特性的优点。

六、纯化

疫苗制备的后期流程是纯化，也就是获得或分离培养基中的细菌或病毒抗原。最常用的纯化步骤/方法有细胞溶解/裂解、均质化、微滤（0.2～0.45μm）、超滤（<0.2μm）、透析、离心和色谱等。疫苗抗原的纯化方案可能包括一个或多个步骤，这取决于每种特定抗原或疫苗的质量要求。

无论是灭活的细菌还是特定的抗原，纯化方案可能主要包括微滤、离心和超滤几个步骤。为了减少副作用，疫苗需要的纯度会更高，那么，可能需要透析和色谱等进一步的纯化步骤/方法。

微滤或离心法能充分有效地纯化获得培养液中的病毒抗原，然而，如果某些培养基成分，如胎牛血清或卵蛋白，会引发机体副作用，就需要采用微滤或超滤对抗原进一步纯化。如果病毒抗原包含在细胞内，那么纯化方案还应包括细胞裂解或均质化步骤。一般来说，抗原的提纯对生产高质量、稳定性强以及副作用低的疫苗意义重大。

七、灌装、标签和包装

注射用的油基疫苗产品可灌装在塑料瓶或静脉注射塑料袋中，通常规格为250～500mL，或5～10 000个剂量单位，每尾鱼的单次注射剂量为0.05～0.1mL。

注射用水基核酸疫苗可灌装在塑胶IV型袋中，规格为250mL，或5 000个剂量单位，每尾鱼的注射单次剂量为0.05mL。

浸泡用水基疫苗通常灌装在1 000mL的塑料瓶中，通常一瓶可以免疫20～100kg的鱼。

瓶子和袋子的规格是档案信息的一部分，未经有关部门批准不能改变。

疫苗的每个包装都应贴上批准的标签，标签内容包括疫苗产品名称、生产商、批次号、储存条件、保质期、剂量和适用鱼的大小。外包装也应贴有符合要求的标签。标签上的任何更改都必须经过授权许可。

八、质控与发布

在整个疫苗的生产过程中以及疫苗生产完成后，都要进行GMP复核确认。最后的产品要经过测试，确认化学和物理参数都在规范规定的范围，并且产品符合效力和安全性的要求。如果产品满足所有的要求，由有资质人员签发批次放行证书（指令2001/82 EC和2001/83 EC），这是疫苗产品允许销售和使用所必需的。

批次放行证书包含抗原制备、灭活、混合和配方数据以及最终产品测试的所有细节。

油基注射疫苗的最终产品测试包括但不限于以下方面：

（1）灭活，确保抗原完全灭活，除非研发的疫苗是一种活疫苗或减毒疫苗。

（2）安全，确保没有任何因疫苗导致的非正常的局部或系统反应。

(3) 最终产品的外观。

(4) 灌装体积。

(5) 无菌，确保没有外源的细菌或真菌。

(6) 疫苗中每一种抗原的效力持续稳定。

(7) 残留甲醛浓度低于安全水平（<0.05%）。

(8) 最终产物的黏稠度达厘泊级。

(9) 最终产品的颗粒大小达微米级。

(10) 可混溶性，确认产品已形成油包水状乳剂。

参考文献

Biering, E. and Salonius, K. 2014. DNA vaccines, in *Fish Vaccination* (eds R. Gudding, A. Lillehaug and Ø. Evensen). Oxford, John Wiley & Sons, Ltd., 47-55.

Commission Directive 91/412/EEC of 23 July 1991 laying down the principles and guidelines of good manufacturing practice for veterinary medicinal products. *Official Journal L* 228, 17/8/1991, p. 70-73.

EU Directive 2001/82/EC, as amended (Veterinary Medicines). European Pharmacopoeia. Available at http://www.edqm.eu/en/european-pharmacopoeia-publications-1401.html [accessed 10 April 2013].

GMP Guidelines, Eudralex vol. 4. Available at http://ec.europa.eu/health/documents/eudralex/vol-4/index_en.htm [accessed 10 April 2013].

Holm, A., Rippke, B. and Noda, K. 2014. Legal requirements and authorisation of fish vaccines, in *Fish Vaccination* (eds R. Gudding, A. Lillehaug and Ø. Evensen). Oxford, John Wiley & Sons, Ltd., 128-139.

Munang'andu, H., Mutoloki, S. and Evensen, Ø. 2014. Non-replicating vaccines, in *Fish Vaccination* (eds R. Gudding, A. Lillehaug and Ø. Evensen). Oxford, John Wiley & Sons, Ltd., 22-32.

Shoemaker, C. A. and Klesius, P. H. 2014. Replicating vaccines, in *Fish Vaccination* (eds R. Gudding, A. Lillehaug and Ø. Evensen). Oxford, John Wiley & Sons, Ltd., 33-46.

第十一章　鱼类疫苗的法规和授权

Anja Holm, Byron E. Rippke and Ken Noda

本章概要　SUMMARY

鱼类疫苗生产和销售必须获得使用疫苗的国家或地区主管部门批准，不同国家或地区的授权、立法、使用要求以及所涉及的监管机构可能有所不同，但在使用或买卖之前，应向有关国家主管部门征求建议并了解相关的法律规定。

第一节　引　言

人类和动物的疫苗被定义为药物，与治疗药物类似，必须通过销售许可的管理程序。疫苗接种可能会干预消灭和控制特定传染病的卫生计划，因此，在一些疾病已经消失或得到控制的国家会有相关疫苗的使用规则。在挪威，使用疫苗通常需要有兽医的处方或者是一个有资质并且已被授权的人，比如鱼类健康生物学家的处方。疫苗已经进行国际交易，在一些地区，如欧洲，多年以来，通用法律已对鱼类疫苗使用规则进行了统一，其中核心要素是申请销售授权的程序以及制造商审核和授权体系。上市许可申请包括产品生产和质量控制的详细说明，以及评估疫苗安全性和有效性研究的完整报告。

第二节　生产商授权

疫苗生产商必须具有由所在国家主管部门签发的有效的生产批文，该授权需要疫苗生产过程的详细资料。在欧盟，疫苗生产应符合国际标准的药品生产质量管理规范（GMP）（GMP指南，Eudralex Vol. 4），并且批文必须定期更新。主管部门可以定期或根据其他部门的要求对工厂以及所有关联的场地和生产过程进行检查。来自世界多个地区的检查机构可以开展官方合作，交流检查报告和结果。

研究设计的注意事项

（示例参照鱼类疫苗安全性和有效性评估的研究设计指南，EMA/CVMP/IWP 314550/2010）

该指南的法律依据是修订后的欧盟第 2001/82/EC 号指令，特别是其附件 1，指令 2010/63/EC（关于保护用于科研的动物）和相关的《欧洲药典》（Ph. Eur. 0062，5.2.6 和 5.2.7）专著。通用指南 EMA/CVMP/IWP/206555/2010 中在免疫兽药产品生产和质量控制要求部分概述了指导方针。

原则上，不论研究在何处进行，所有研究的结果应适用于任何条件。然而，申请人应该考虑到各种条件（例如气候、疾病情况、水温和盐度），因为这些可能影响研究结果。商品化疫苗研究使用鱼的种类、年龄、大小、重量和生理状况（如银化过程和性成熟情况）必须具有代表性，这样才能用于实际生产中。实验室研究，应使用推荐的最低接种年龄或大小的鱼，因为这些终将在产品信息中体现。此外，实验研究中使用的鱼最好是来自商业养殖场。

免疫组和对照组的鱼数量应足以获得具有统计学意义和临床可靠性的结果。然而，对于免疫后攻毒研究，要尽可能地减少未接种疫苗的对照组鱼的数量，因为这些鱼将遭受疾病及相关的痛苦。如果药典中已有该类型疫苗的记载，其相关的要求也必须酌情考虑和遵守。在所有的实验中，使用适当的方法随机分配免疫组和对照组的鱼，免疫组和对照组应采取相同的养殖方式，对照组与免疫组鱼的管理应保持一致。

如果免疫组和对照组的鱼不在同一个水槽中，为了消除不同水槽中鱼群体之间可能出现的影响，应采取一些措施；例如，每一个免疫组和对照组至少使用两个水槽。同时也应该考虑到放养密度对结果的影响。

接种疫苗过程中，每项实验室研究使用的水质参数，如温度和盐度（如淡水和海水），应与商业养殖场中的环境一致。对不同鱼类或疾病发生情况的研究可能需要同时设高温和低温条件进行。同样，其他一些研究需要遵循该种鱼类的主要生产实际规律设计研究方案。

用于安全性和有效性研究中的疫苗制剂应该与最终在市场上销售的配方一致，并且不同配方的使用需完全合理。剂量、疫苗免疫途径和用量必须和预期商业应用一致。研究所评估的参数应适用于拟定指标（如果与研究设计相关的话，应包括免疫的起始和持续时间），并且研究进行之前应该预先确定这些参数。

研究人员必须确定每个研究中使用的攻毒模型是合理的，并讨论其与自然疾病发生状况的相关性，需要考虑的因素包括：攻毒与免疫保护的相关性；攻毒机体的免疫方式（如共栖、注射等）；水温；确定攻毒的时间和评估参数指标获取的时间。

考虑到鱼是变温动物，并且鱼的免疫力是温度依赖性的，温度越高，注射部位反应的频率和强度越强，因此，鱼类疫苗的安全性和有效性的实验研究的比较数据应该以“度·日”（degree-days）为单位。度日是水温的摄氏度和天数的乘积，如 10℃ 持续 5d 即为 50℃·d。

探讨疫苗使用对受免鱼生命周期可能产生的不利影响非常重要，这一点在腹腔注射疫苗的使用中尤其重要，因为肠道粘连可能会对鱼类产卵产生不利的影响，此外，粘连/色素沉淀/损伤可能导致商品鱼的品质下降或者会使消费者拒绝购买（Evensen 等，2005）。

实验室研究是在相关模型经过严格的条件控制下产生的数据，而实地研究用于证实这些实验室研究的结果。然而，某些鱼类的疾病不存在（或很少存在）理想的攻毒模型。在这种情况下，更需要强调在反映实地疾病情况的条件下进行的实地研究。实地研究的范围要保证疫苗在地区相关鱼类的多样化水产养殖条件下，既有效又安全。

不同容器或水槽中的养殖鱼类的来源和病史必须记录保存；养殖鱼类以前的药物治疗情况、化学药品和疫苗的使用历史均要知晓；需要每天记录疾病暴发、死亡率和用药情况，以及日常管理措施，例如卫生、饲料、操作以及添加剂和化学物质使用。疫苗研发通常需要来自实验室和全面的实地研究数据。在某些情况下，申请人应对缺乏的相关资料作出解释。

研发和测试一种新型鱼类疫苗需要进行很多试验研究。实验室和实地放大研究中使用鱼的数量，以及疫苗批量生产时评估疫苗效力用鱼的数量都很多。挪威兽医研究所的一份报告表明，通过合并实验或用体外实验来替代一些鱼体内实验，实验用鱼的数量可以大大减少（Ellingsen 和 Gudding，2011），例如通过 ELISA 实验，检测与免疫保护相关的抗原（Romstad 等，2012），或者选择更为优化的统计方法。许多权威法规、指导机构或管理部门鼓励使用“3Rs”法（优化 refine、减少 reduce 和替代 replace，3Rs）（Midtlyng 等，2011），如《欧洲药典》、兽药产品注册技术要求的国际合作协调组织（VICH）和欧洲替代方法验证中心（ECVAM），但是“3Rs”法必须经过适当的验证才能确保信息数据的准确性。

在欧盟，关于保护用于科研动物的第 2010/63/EEC 法令将很快生效，并规定在科学实验中必须考虑“3Rs”法。

第三节 食品安全——最高残留限量

动物进行疫苗接种也可能影响这些动物食品的质量。在一些地区，农业或渔业部门对产品质量有规定。使用含佐剂的疫苗会造成注射部位的严重反应，从而可能导致肉质下降，也可能导致上市产品的品质降级（Aunsmo 等，2008）。动物屠宰时，可食用组织中的兽药（包括疫苗）的残留量必须低于规定的限量，这样不会对消费者产生不良影响。所有药理活性物质和辅料均已设定最高残留限量（MRLs），但免疫性抗原除外。疫苗的最终产品中通常会有辅料、佐剂或防腐剂，因此最终的水产品需要设置最高残留限量，但是也有评估结论表明，疫苗不需要设置最高残留限量。在欧盟，MRLs 由欧洲药品管理局（EMA）设定，食品法典委员会（Codex Alimentarius）领导建立全球认可的 MRLs 规程。国际上达成一致的 MRLs 标准对避免基于食品中可接受残留水平差异而设置的贸易壁垒极为重要。

大多数疫苗都没有停药期，也就是说，停药期为 0d，因为它们对消费者没有风险。但是，停药期需要由主管部门对每种产品进行评估，对于一些特别的疫苗，停药期可设置为大于 0d。

第四节 转基因生物

一些疫苗含有转基因生物（GMOs），这种疫苗在欧盟必须由欧洲药品管理局授权，而不能由国家主管机关授权（图 11-1），然而，国家主管机关可以授权转基因疫苗的临床试验。已有几种经欧盟授权销售的兽用 GMO 疫苗，例如预防马流感和猫白血病的疫苗。依照 2001/18/

EEC（谨慎投放 GMOs）法令对疫苗质量、安全性和有效性的评价，欧盟和欧洲经济区（EEA）的环境机构参与 GMOs 疫苗的授权程序。法令 2001/18/EEC 的主要目的是同意谨慎投放转基因产品，并将其投放市场的程序变得高效化和透明化，同时将此类许可的期限限制在十年（可续期），并在转基因生物投入市场后实行强制性监测。该法令还提供了一种常规方法，评估与 GMOs 投放有关个案的环境风险。根据这一法令规定，公众知情和转基因产品标签是强制性的。环境风险评估的目的是查明和评估转基因产品对人类健康和环境的潜在不利影响，包括直接和间接、短期和长期的影响。评估还包括已谨慎投放到市场的转基因产品。

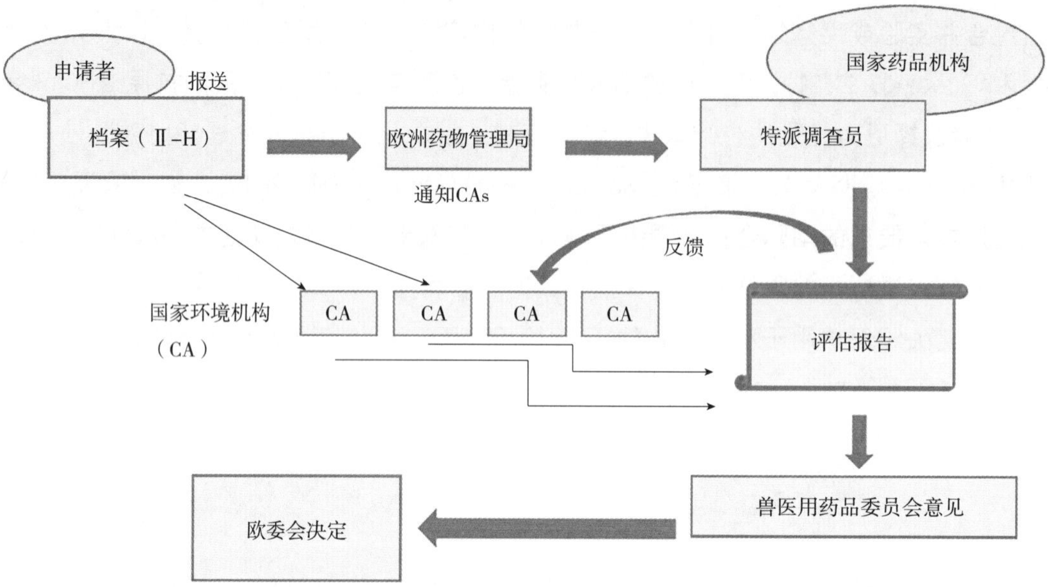

图 11-1　欧盟对含转基因生物疫苗的环境评价程序

第五节　DNA 疫苗

DNA 疫苗是将编码相关抗原的细菌质粒 DNA 注入动物体内，而 DNA 疫苗没有被定义为转基因产品，但是 DNA 疫苗也需要经过一个常规的授权过程。然而，和常规疫苗相比，DNA 疫苗需要特别考虑一些技术和安全相关的问题（Holm，2004）。此外，目前一些地区正在讨论是否必须将 DNA 疫苗接种的动物归为转基因生物。因此，迫切需要确定 DNA 疫苗的授权过程期间是否必须满足某些需求（依据 2001/18/EEC 法令），从而避免接种的动物被认为是转基因生物。

加拿大已经授权了一种预防鱼类传染性造血器官坏死病毒（IHNV）的 DNA 疫苗（加拿大食品检验局，2005；Salonius 等，2007）。

第六节　某些疫苗的禁用

许多地区对领土上没有的或已根除的疾病都有控制和监测计划。一些国家鱼类的细菌性肾病（BKD）和感染性胰腺坏死（IPN）受到监管。一旦制定了控制程序，主管当局将严格管制预防某种疾病的疫苗接种，这意味着即使疫苗在特定国家有销售许可，主管当局也可能限制或者禁止疫苗使用（例如法令 2001/82，art. 71）。因此，开始实施一项新的疫苗研发或推广应用前，一定要咨询所在国家或地区的主管部门。

第七节 未授权疫苗的使用

如果需要使用疫苗来控制一种疾病，但该国没有任何授权的预防这种疾病的疫苗，那么许多地区或国家有针对这种疫苗的特殊豁免规则或有条件的许可。许多小规模的渔业市场可能缺乏可用的控制养殖鱼类疾病的疫苗产品，因此，为了动物的福利和生产而建立一个灵活的体系非常重要，这可以避免疾病的暴发。通常负责的兽医必须向主管部门提出申请，并提交申请理由和拟使用的疫苗的信息。

第八节 自体疫苗

自体疫苗被定义为灭活的免疫兽药产品，是由从一个动物或多个动物体内获得的病原体和抗原制成，用来治疗该种动物或同区域养殖的动物。这些自体疫苗在欧盟受自己国家法律（非欧盟法律）监管，并且必须从国家主管机关获得生产或使用这种自体疫苗的许可。细菌抗原免疫已经被用来预防很多鱼类病原体的感染，例如链球菌和相关的病原体（Bercovier 等，1997）。

第九节 区域规则和主管机构

世界各地区都有自己对于药物的监管规则、程序和管理机构。药物的监管和指导方针通常在国家或地区部门的网站上公开发布。兽药注册技术要求国际协调会（VICH）及其外展服务正在努力统一药物授权的要求。VICH 是一个三边项目（欧盟-日本-美国），旨在统一兽医产品注册的技术要求，澳大利亚、加拿大和新西兰是观察员国。VICH 于 1996 年 4 月正式启动，并在 2011 年开展外展服务，鼓励非 VICH 成员的地区/国家更广泛地统一注册要求以有效地利用资源。

VICH 指南提供了统一的指导，记载了兽药产品营销授权的申请档案中需要提供的数据。VICH 也为兽药产品制定了药物警戒指南，即监测兽药产品上市后的安全性和持续有效性。然而，VICH 通常不制定如何进行数据评估或评估方法相关的指南，评估是由 VICH 成员和地区的监管当局完成。最终，一旦国家或地区通过 VICH 指南，已有的区域指导方针将被取代。

第十节 欧盟和相关欧洲经济区国家

欧洲药物管理网络由欧洲药品管理局（EMA）和来自 27 个成员的国家主管机关（NCAs）

以及3个欧洲经济区（EEA）国家（挪威、冰岛和列支敦士登）组成。

EMA负责集中、转介程序和MRL评估，并且设有中央科学委员会和CVMP（兽医用药品委员会）以及工作组。NCAs负责本国的市场，并为EMA和所有授权程序的工作推荐专家。许多国家的人药和兽药统一管理，而且通常药品和免疫产品两者都有。然而，有的国家，人药和兽药由不同的机构管理，也有的国家药品和免疫产品分开管理。

一、程序

集中审批程序使得所有成员国和EEA国家都获有合法授权（图11-2）。销售申请发送到欧洲药品管理局（EMA），EMA协调兽医用药品委员会（CVMP）对其进行科学评估。CVMP由来自欧盟（+EEA）各国的一位成员和一些具有特殊专长的其他当选成员组成，每个CVMP成员与各国的专家就具体领域进行协调。CVMP的科学意见发送给欧盟委员会，由欧盟委员会正式发布在欧盟的市场授权。然而，如果在某国已启动对相关疾病的监视和控制程序，那么在该国这种疾病的疫苗很可能会被禁止使用。

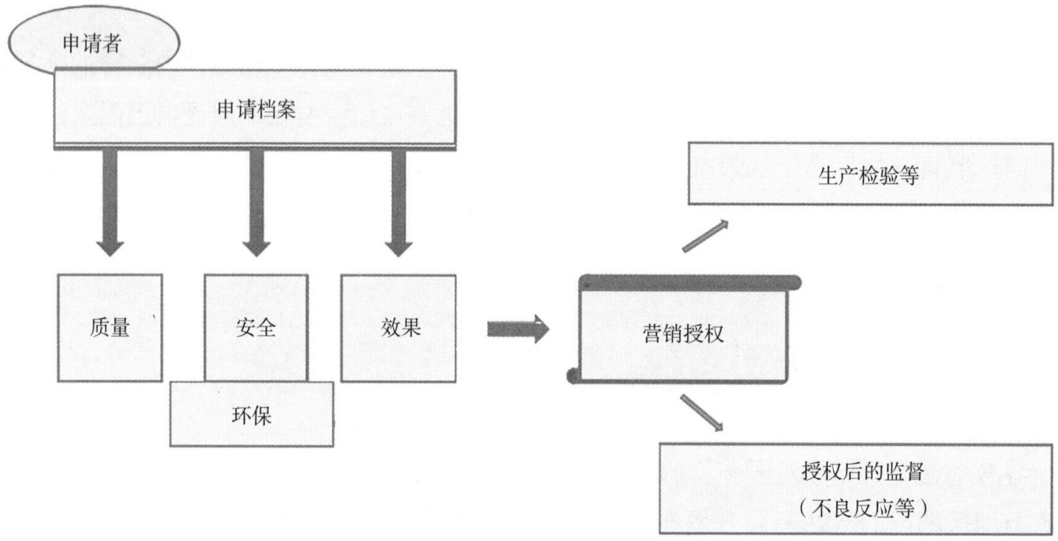

图11-2　疫苗的授权

分散程序（DCP）用于新疫苗，制造商不能或不会使用集中程序，并且DCP仅在程序所在国中有效（两个到所有国家）。一个国家作为参考成员国并进行基础评估，而其他有关成员国可提出其他的问题或条件。

互认程序（MRP）是针对EU/EEA成员国的现有授权疫苗。相关成员（一个至全部）应接受已持有疫苗授权的国家的评估，并对疫苗产品签发类似的授权。对于分散程序（DCP）和互认程序（MRP），成员通常需要在授权申请上达成协议，这样成员就可接受产品的适应症、建议用途和产品其他条款，并在国内统一发行授权。如果无法达成协议，成员则向CVMP提交就争论的问题发表科学意见的申请。欧盟委员会（European Commission）和各成员根据CVMP的意见，做出所有成员都必须遵守的具有法律约束力的决定。

只有一个国家的程序也是可能的，这样的程序将由相关国家主管机关NCA进行处理和发布。

二、欧洲药典委员会

欧洲药典委员会是欧洲理事会下属的一个有超过50个成员的组织，其主要任务是协调药品生产和药品管控的规则。欧洲药典委员会为多种的疫苗发行专著，包括许多预防严重动物疾病的疫苗，但是专著中只有几种预防鱼类疾病的疫苗。在欧盟，疫苗授权的部分法律要求应符合《欧洲药典》的相关规定。

尽管提交受试动物批次安全测试数据仍然是地区参与VICH的一项要求，2012年（2013年生效）《欧洲药典》删除了相关的试验内容。如果连续生产了足够数量的产品批次，并发现符合测试要求，监管部门可免除受试动物批次的安全测试，从而体现生产过程的一致性(VICH GL50，草案2011)。预计未来几年，VICH将会对区域要求进行修订。

三、科学咨询

EMA/CVMP成立了一个兽医科学咨询工作组（veterinary Scientific Advice Working Party，SAWP-v)，向制造商提供咨询，并根据具体要求向申请疫苗销售批文的厂商提供建议。这些建议通常与现有科学领域未发布指导意见的研究设计有关，或者针对与该领域现有的指导意见不相关的特殊产品。申请人可联络EMA，进一步了解中小企业的申报程序、费用和特殊条款等更多信息。

四、少量使用和少数动物

对于“少量使用和少数动物”（Minor Use and Minor Species，MUMS)，或称有限市场(EMA /CVMP/IWP/123243/2006）的疫苗，兽医用药品委员会（CVMP）制定了具体的指导方针。大多数的鱼类疾病需要疫苗，但是有的鱼类疫苗市场规模太小，以至于企业难以获得投资回报，因此，可以放宽这类疫苗产品的授权要求，或者在疫苗产品推向市场后，再要求提交一些科学数据，同时MUMS疫苗产品还可能获得免费的科学建议（Quigley，2012)，并且在EMA网站公布了这方面的有用文件，如最新的MUMS目录、在疾病或动物上疫苗需求的差距分析。

第十一节 美　国

在美国，兽医疫苗的监管机构是美国农业部（USDA）的兽医生物制剂中心（CVB)，CVB获得了《病毒、血清、毒素法》（VSTA）授权。VSTA是1913年3月通过并签署的一项法案，其规范了州际贸易中的兽用生物产品的销售和分销行为，VSTA在1985年进行了修订。CVB授权美国生产的所有兽用生物制剂的销售和分销批文，并且建立单一的国家监管框架。必须强调的是，兽用生物产品在美国由农业部监管，而兽药产品由食品药品监督管理局监管。

一、程序

《病毒、血清、毒素法》规定，无论是在美国境内或向美国境外销售或分销无效的、危险的、有污染的，或有害的兽用生物产品，都是违法行为。此外，该法律要求，销售或分销的产品必须是在 USDA 许可的车间设施中生产，并符合美国农业部的所有要求，联邦法规第 9 章、兽医服务备忘录和兽医生物学通告中规定了这些要求。这些法规和文件协同一致，以确保在美国生产和在美国国内销售及销往国外的产品纯净、安全、强力和有效。

CVB 收到申请和支撑材料后，将开始按照兽用疫苗的要求包括产品许可前各方面的多种评估，这包括能够很好地证明产品的纯度、安全、效能、功效的一系列步骤，通常包括以下内容：

(1) 证明用于生产免疫原的种子和细胞株的纯度和同一性。

(2) 建立一致、可控的疫苗生产流程。

(3) 证明疫苗功效，其中疫苗的有效性通常反映了对宿主动物临床疾病的预防或缓解，支持所有预期的产品标签说明。

(4) 确定佐剂安全性（如果适用的）和停用时间要求（用于食品动物）。

(5) 在多个地点和不同饲养条件下的全面现场试验中，证明对宿主动物的安全性。

(6) 多批次疫苗验证效力测定法的有效性。

(7) 成功大规模生产商品化的多批次疫苗（预审批系列）。

(8) 水产疫苗的许可遵循同样基本框架。

(9) 完成所有这些要求后，制造商就可获得生物产品许可证。

二、产品许可/许可证的类型

CVB 对不同产品颁发不同类型许可证，这些产品都有助于满足各种动物健康的需求。

完全的生物产品许可证：获得此类许可证的产品，满足 USDA 适用的纯度、安全性、效力和功效的所有要求，无论在销售上有无限制。

有条件的许可证：获得这种许可证的产品，通常都有分销限制，这些产品满足了对安全和纯度的所有要求，并已证明具有“预期疗效”。通常这些产品总体上已经被许可生产销售，并希望进一步获得完全许可证。产品的预期疗效取决于产品施用的生物体。这种类型的许可对于应对行业中新出现的疾病、利基（niche market）市场需求或特殊情况是很有用的。

自主许可证：获得这类许可证的灭活产品，用于原产地的牛群、羊群或其他群体中，在兽医或具有相关工作经验的专家指导下使用。

生物产品许可证：这类许可证是颁发给在美国境外生产并进口用于销售和分销的产品。它们满足 USDA 对许可证的所有要求，并被认为不会对美国动物健康构成威胁。该许可证颁发给美国境内检疫部门的负责官员，而不是直接颁发给外国制造商。

第十二节 日　本

在日本，人用和兽用药品均遵循《药品管理法》（第 145 号法律，1960 年）。《药品管理法》旨在提供必要的法规，以确保药品、准药品、药妆品和医疗器械在研发、制造、进口、销售、零售和使用的各个阶段的质量、有效性和安全性。农业、林业和渔业部（MAFF）制定了专门适用于兽用医疗产品（VMP）的部级条例，并根据该条例提供了相关的法规和指南。分销兽用医疗产品的厂商应当从 MAFF 那里获得每个产品的市场许可。食品安全与消费者事务局的动物产品安全部门负责制定/修订部级条例及相关通告。国家兽医分析实验室（NVAL）负责审查申请文件和制定指南。

一项新的兽用医疗产品的授权获批申请，经药品事务委员会（PASC）分会调查，确定该产品的品类（生物制剂、抗生素和普通药物等）。通常水产养殖产品委员会负责与鱼类有关的医疗产品。兽药残留委员会调查用于食品生产动物的 VMPs 在食品中的残留情况。申请还需要经过 PASC 执行委员会的审议。此外，食品安全委员会将评估食品生产动物的 VMPs 对人类健康的安全性。需要时，卫生部、劳工部和福利部必须在批准之前确定有关组织的最高残留限量。

批准鱼类疫苗所需的数据因产品的状况而异。例如，对于含有新活性成分的产品，申请人必须提交完整的档案材料，包括：产品研发的起源和过程，物理、化学和生物特性，制造工艺，适应证，效果和效力，用法与用量，稳定性，靶动物安全性，药理作用和临床试验。对含有佐剂的疫苗需进行残留物衰减的研究。靶动物安全数据应符合实验室管理规范（GLP）的部级条例（1997 年第 74 号条例），临床试验的数据建议符合药品临床试验管理规范（GCP）的部级条例（1997 年第 75 号条例）。此外，生产过程控制和质量控制应符合药品生产质量管理规范（GMP）的部级条例（1994 年第 18 号条例）。

第十三节 其他有关组织：OIE、FAO、WHO

除主管部门规定的规则外，其他有关组织也有疫苗使用和注册的建议或规则。

世界动物卫生组织（OIE）有 170 多个成员，致力于全球动物疾病控制。OIE 在协调兽医工作规则、消灭严重动物疾病、授权参考实验室以及授权兽药的使用等方面发挥非常重要的作用。世界动物卫生组织已发布了可以在线访问的《OIE 水生动物卫生法典（2011）》和《水生动物疾病诊断手册》。

世界粮食及农业组织（FAO）参与了预防动物疾病的疫苗接种计划，特别是在发展中国家。

世界卫生组织（WHO）致力于防治人畜共患病，例如禽流感疫苗的研发。世界卫生组织也与国际生物标准化协会合作，主要致力于人类疫苗研发，还在奥斯陆（1996）和卑尔根（2003）组织了鱼类疫苗免疫的国际会议。

参考文献

Aunsmo, A., Guttvik, A., Midtlyng, P.J., *et al.*, 2008. Association of spinal deformity and vaccine-induced abdominal lesions in harvest-sized Atlantic salmon, *Salmo salar* L. *J Fish Dis* 31: 515-524.

Bercovier, H., Ghittino, C. and Eldar, A. 1997. Immunization with bacterial antigens: infections with streptococci and related organisms. *Dev Biol* (*Basel*) 90: 153-160.

Canadian Food Inspection Agency. 2005. *DNA vaccine*. Available at http://epe.lac-bac.gc.ca/100/206/301/cfia-acia/2011-09-21/www.inspection.gc.ca/english/anima/vetbio/eaee/vbeaihnve.shtml [accessed 22 May 2013].

Codex alimentarius. Available at http://www.codexalimentarius.net/web/index_en.jsp [accessed 22 May 2013].

Ellingsen, K. and Gudding, R. 2011. The potential to increase use of the 3Rs in the development and validation of fish vaccines. Available at http://www.norecopa.no/norecopa/vedlegg/Rapporten---final.pdf

European Medicines Agency. http://www.ema.europa.eu/ema/ [accessed 22 May 2013].

EU Directive 2001/18/EC (Genetically Modified Organisms), EU Directive 2001/82/EC, as amended (Veterinary Medicines), EU Directive 2004/28/EC (Veterinary Medicines), EU Directive 2009/9/EC (technical requirements, = Amended annex 1 of Directive 2001/82/EC), and EU Directive 2010/63/EC (protection of test animals). Available at http://ec.europa.eu/health/documents/eudralex/vol-5/index_en.htm [accessed 22 May 2013].

European Pharmacopoeia. http://www.edqm.eu/en/european-pharmacopoeia-publications-1401.html [accessed 22 May 2013].

Evensen, Ø., Brudeseth, B. and Mutoloki, S. 2005. The vaccine formulation and its role in inflammatory processes in fish—effects and adverse effects. *Dev Biol* (*Basel*) 121: 117-125.

EudraLex, *Vol*. 4: *Good manufacturing practice* (*GMP*) *guidelines*. Available at http://ec.europa.eu/health/documents/eudralex/vol-4/index_en.htm [accessed 22 May 2013].

EMA Guideline on the design of studies to evaluate the safety and efficacy of fish vaccines (EMA/CVMP/IWP/314550/2010). Available at http://www.ema.europa.eu/ema/pages/includes/document/open_document.jsp?webContentId=WC500118226 [accessed 22 May 2013].

Food and Agriculture Organization. http://www.fao.org/fishery/en [accessed 22 May 2013].

Guideline on requirements for the production and control of immunological veterinary medicinal products (EMA/CVMP/IWP/206555/2010, consultation). Available at http://www.ema.europa.eu/ema/pages/includes/document/open_document.jsp?webContentId=WC500104492 [accessed 22 May 2013].

Holm, A. 2004. *DNA-vaccines for food-producing animals*. *Technical review and discussion of safety issues*. Available at http://www.dfvf.dk/Admin/Public/Download.aspx?file=Files%2FFiler%2FPublikationer%2FDNA-vaccines_report_-_Final1.doc [accessed 22 May 2013]. Japanese authorisation procedures: *Outline of Regulation System of Veterinary Drugs in Japan*. Available at http://www.maff.go.jp/nval/english/ [accessed 22 May 2013].

Midtlyng, P. J., Hendriksen, C., Balks, E., *et al*., 2011. Three Rs approaches in the production and quality control of fish vaccines. *Biologicals* 39: 117-128.

OIE (World Organization for Animal Health). http://www.oie.int/international-standard-setting/terrestrialmanual/access-online/ [accessed 22 May 2013].

Quigley, K. 2012. Veterinary medicines: what is the MUMS/limited markets policy? *Regulatory Rapporteur* 9: 8-9.

Romstad, A. B., Reitan, L. J., Midtlyng, P., *et al*., 2012. Development of an antibody ELISA for potency testing of furunculosis (*Aeromonas salmonicida* subsp *salmonicida*) vaccines in Atlantic salmon (*Salmo salar* L). *Biologicals* 40: 67-71.

Salonius, K., Simard, N., Harland, R. and Ulmer, J. B. 2007. The road to licensure of a DNA vaccine. *Curr Opin Invest Drugs* 8: 635-641.

VICH website and guidelines. Available at http://www.vichsec.org/ [accessed 22 May 2013].

World Health Organization. http://www.who.int/about/en/ [accessed 22 May 2013].

第十二章　疫苗接种策略和程序

Atle Lillehaug

本章概要　SUMMARY

研发具有保护效应的疫苗并正确地使用是疫苗免疫成功的关键。此外，疫苗接种成功的决定因素还包括疫苗接种预防的疾病种类、接种方法的选择、接种时间以及加强免疫。为获得最佳的保护效果，疫苗接种应在疾病可能发生前的一段时间进行，确保鱼体有足够的时间形成免疫力。另一方面，疫苗接种的时机不应过早，因为机体的免疫力会随时间延长而下降。在确定何时实施免疫接种时，水温和鱼体大小应予以考虑，因为这些是影响免疫力形成的主要因素。

在水产养殖，疫苗接种的总体目标是，减少疾病造成的损失，增加经济产出，包括权衡比较药物治疗以及其他费用与疫苗接种费用。通常，疫苗注射接种的工作量大，费用最高，但可获得较好的免疫保护效果。也可采用浸泡接种，由于浸泡免疫诱导鱼体产生的免疫力较低，免疫保护效果不够理想，因此采用浸泡接种方式一般需要进行加强免疫。当使用佐剂疫苗时，只能采用注射免疫。

如果疫苗成本、疫苗接种的劳动力和副作用的成本与疾病导致的潜在损失相当，接种疫苗鱼的死亡率减少不显著，就应权衡疫苗接种是否值得。实际上，虽然疫苗和疫苗接种程序会产生较高的成本，但是鱼体获得了较高的抗病力，通常，在经济价值较高的鱼类上实施疫苗接种，能够产生较好的经济效益。

第一节　引　言

最佳的疫苗免疫保护效果与研发高效的疫苗以及正确使用疫苗密切相关。“接种策略”是指疫苗接种预防哪种疾病，选择采用何类接种程序（Lillehaug，1997）。疫苗接种程序包括确定疫苗类型和接种方法，根据鱼类养殖周期确定疫苗接种时间，确定接种时的水温和鱼体的大小以及是否需要加强免疫，这些因素相互关联且密切相关，因此，疫苗接种程序不能仅依据某一因素，应多种因素统筹考虑。

疫苗接种策略旨在保护鱼类免受在养殖场或者某区域易发特定病原体的感染。疫苗诱导的免疫保护力应足以抵御病原的感染，并在疾病暴发风险最高时，免疫保护水平处于最佳状态，且在整个风险期，免疫保护应维持在较高水平。此外，疫苗接种策略还应考虑疫苗接种相关成本和副作用风险。总之，疫苗接种策略的目标是，从经济效益和动物福利角度，为鱼类养殖业提供一个有效的解决方案。

第二节 疫苗接种时机

接种疫苗时鱼体必须已经具有免疫能力，这样才能获得保护性的免疫应答。鱼体接种疫苗后，机体的免疫力形成一般遵循典型的变化趋势，即在免疫力达到可以保护机体的水平或机体达到的最大保护水平前，需要一段时间形成免疫力。免疫力形成一段时间后，机体的免疫水平开始下降，并最终会降到免疫保护的下限。机体免疫力形成需要的时间、达到的最大保护力以及机体免疫持续时间取决于多种因素，包括水温、疫苗接种程序和疫苗产品的特性（例如灭活疫苗、减毒疫苗、佐剂的使用情况等）。

疫苗接种策略必须与鱼类的养殖周期相适应，最好是在疾病暴发前的适当时间进行免疫接种。集约化水产养殖生产体系通常分为两个不同的阶段：第一个阶段包括孵化、开始摄食和第一个生长期，这个阶段在水箱系统中进行，用水量有限，可控制病原体进入（如消毒系统）；第二个阶段是池塘或网箱养殖，在此阶段前应考虑实施疫苗接种。对于溯河产卵的鱼类，如鲑鳟鱼类，针对预防海水环境病原体的感染，疫苗接种应在鲑鳟淡水生长阶段进行（Press 和 Lillehaug，1995）。

在水产养殖生产中，多联疫苗是疫苗接种策略的主要选择，鱼体仅接种一次疫苗就能保护鱼类免受多种疾病的感染。大西洋鲑的高产得益于，在其淡水养殖阶段，使用含有佐剂的抗多种病原体疫苗，采用一次性注射免疫的方式，诱导鱼体产生在整个海水生长周期中持续存在高水平的免疫保护力。挪威的多联疫苗包括鳗弧菌（*Vibrio anguillarum*）O1 和 O2 血清型、杀鲑弧菌（*Vibrio salmonicida*）和杀鲑气单胞菌（*Aeromonas salmonicida*），并且还可能包括粘摩替亚菌（*Moritella viscosa*）和传染性胰脏坏死病毒（IPNV）。

第三节 水 温

环境因子包括水温，影响着鱼类免疫系统的功能和免疫应答（Bowden，2008）。之前的研究建议在相对较高的温度下给鱼类包括鲑鳟鱼类接种，这是为了避免在低温环境下，鱼类免疫应答受抑制（Bly 和 Clem，1992）。然而冷水性溯河产卵鱼类的情况有所不同，如鲑鳟鱼类，春天转到海水之前，周围环境水温可能会短暂上升，并且当水温持续上升的时候，鲑的银化已经开始了，这个阶段的鲑很脆弱（Eggset 等，1999），鱼体对疫苗反应可能受到影响（Maule 等，1987）。根据这种情况，对大西洋鲑进行了水温、免疫反应、接种时间和鱼体保护力产生等相关的临床研究，在 2～10℃ 水温范围，对幼鲑接种冷水弧菌疫苗后的感染实验（Lillehaug 等，1993）表明，在水温 2～6℃，无论是浸泡或注射免疫，接种冷水弧菌疫苗鲑鱼的免疫保护力水平呈明显升高的趋势。

研究结果表明，5℃、10℃和15℃水温下，用抗病毒性出血性败血症DNA疫苗免疫虹鳟，虽然最低温度下的鱼体免疫应答较慢，但三个水温下免疫的虹鳟均产生了较强的免疫保护力(Lorenzen等，2011)。另一项研究显示，大西洋鲑接种油类佐剂多联疫苗，饲养在2～8℃不同温度下一段时间后，感染实验结果表明，所有温度实验组的大西洋鲑对冷水弧菌病均具有免疫抗性，这证实了前面的结论。然而，在2℃下，接种疫苗的大西洋鲑不能抵抗疖疮病，4℃下只有中等强度的免疫保护；而8℃时，鱼体获得较好的免疫保护效果（Sevatdal和Dalen，1997)。鱼体对杀鲑气单胞菌抗原表现出免疫应答延迟，当水温升高或使用油类佐剂疫苗时，鱼体会在免疫接种后期产生免疫保护力。斑点叉尾鮰作为温水性鱼类，免疫多子小瓜虫（*Ichthyophthirius multifiliis*）幼虫后，进行的感染实验结果表明，相比高温30℃免疫组，15℃免疫组的死亡率较高且抗体应答反应较弱（Martins等，2011)。

总之可以得出结论，接种疫苗后，水温对鱼类最终免疫保护水平的影响并不显著，至少在鱼类正常生存温度范围内影响不大。温度对不同鱼类（冷水性或者温水性）的影响不同，对不同类型的抗原的影响也不同。一般来说，疫苗公司很少标明疫苗接种温度，但是Merck动物保健公司（Merck Animal Health）建议，斑点叉尾鮰疫苗接种温度为21～29℃。

第四节 鱼体大小

鱼类免疫系统发育完全时的鱼体大小和年龄的相关数据仅在少数鱼类中有过报道。重要经济鱼类，如鲤和虹鳟2～3月龄时就具有完全免疫应答能力（Zapata等，2006)，鲑鳟鱼类在体重5g左右时免疫系统成熟，此前不适合接种疫苗（Hart等，1988)。鲑鳟鱼类体重达到20g才较为适合人工操作实施注射免疫，这也同样适用于其他鱼类。鳕的免疫系统在2月龄发育成熟（Schrøder等，1998)，大西洋庸鲽仅需要约1个月（Øvergård等，2011)。不过，亚热带淡水鱼类斑点叉尾鮰在孵化后7d，接种鮰爱德华菌（*Edwardsiella ictaluri*）活疫苗，就可产生完整的免疫应答（Shoemaker等，1999)，这说明鱼类的免疫应答方面存在巨大差异，并且不同抗原诱导的免疫反应也存在差异。

第五节 疫苗接种方法

疫苗接种方法有三种：一是肠道外注射免疫，二是浸泡免疫（通过浸浴将抗原递呈到鱼体表面，或者直接向鱼体喷洒疫苗)，三是饲喂含疫苗饲料的经口免疫。

一、注射免疫

注射免疫是产生最佳免疫保护的疫苗接种方式，而且是佐剂类疫苗的唯一选择。这种方法的缺点是耗费劳动力、技术设备要求高、不适合小鱼种（如体重小于10g)（Evensen，2009)。

注射免疫通常是通过腹腔注射的方式进行，也可通过肌肉注射。肌肉注射一般用于DNA疫苗的接种，如疫苗Apex-IHN（诺华动物保健公司，Novartis Animal Health）在加拿大许可用于预防传染性造血器官坏死病。由于腹腔注射佐剂疫苗在腹腔内会产生局部反应，为了避免腹腔注射引起卵脱落的问题，鲑亲鱼的注射接种部位多选在背部静脉窦及背鳍下方的脂肪组织，并且还可确保鱼片不受疫苗引起的炎症反应的影响（Treasurer 和 Cox，2008）。采用注射方式的主要是灭活疫苗，然而，诺华动物健康公司生产的用于预防鲑细菌性肾病的节杆菌（*Arthrobacter*）活疫苗 Renogen 也是通过腹腔注射的方式接种。

疫苗接种程序：首先捕鱼，接着浸入麻醉液中麻醉。苯佐卡因（benzocaine）常用于麻醉鲑鳟鱼类，而美他卡因（metacain）更适用于麻醉海水鱼，如鳕（Karlsson 等，2012）。麻醉期间应保持充氧，并经常更换麻醉液，以保持药效，麻醉时间与水温有关，一般需要1min。当鱼体停止扭动静躺在水中时，用网将鱼推捞到接种台上，接种台面为钢制板，边缘有挡板，可以让水和鱼停留在台面上，接种台上的水要保持流动，整个过程保持鱼体湿润。经腹腔注射疫苗后，将鱼放入流水的管道运回养殖池，不久免疫过的鱼就会逐渐苏醒。

疫苗注射使用由细管与疫苗瓶连接的自动连续注射器，在注射前，瓶内疫苗要充分摇匀，如有必要，接种过程中应反复摇晃疫苗瓶，防止抗原沉降或者乳化剂与抗原分离。疫苗的注射剂量因疫苗而异，大西洋鲑含佐剂的多联疫苗说明书以前标注的注射剂量是 0.2mL，由于副作用（腹腔内反应）问题，已减少到 0.1mL，而有些鱼用疫苗产品的注射剂量只有 0.05mL。

鲑鳟鱼类疫苗注射部位建议在腹部的中线，大约是在臀鳍基部前一个鳍长度的位置（图12-1）。通常，注射针头直径为 0.5～0.7mm，针头长 3～5mm，可穿透腹壁（图 12-2）。针头应定时更换，尤其是针头变钝时，这样也会减少鱼体间传播感染的风险。

疫苗注射通常由专业团队一起合作完成，其中 1 人进行麻醉操作并持续不断地向接种台上提供一定数量的待接种鱼，4～5 个疫苗注射员快速对鱼进行接种，避免鱼在接种桌上停留太

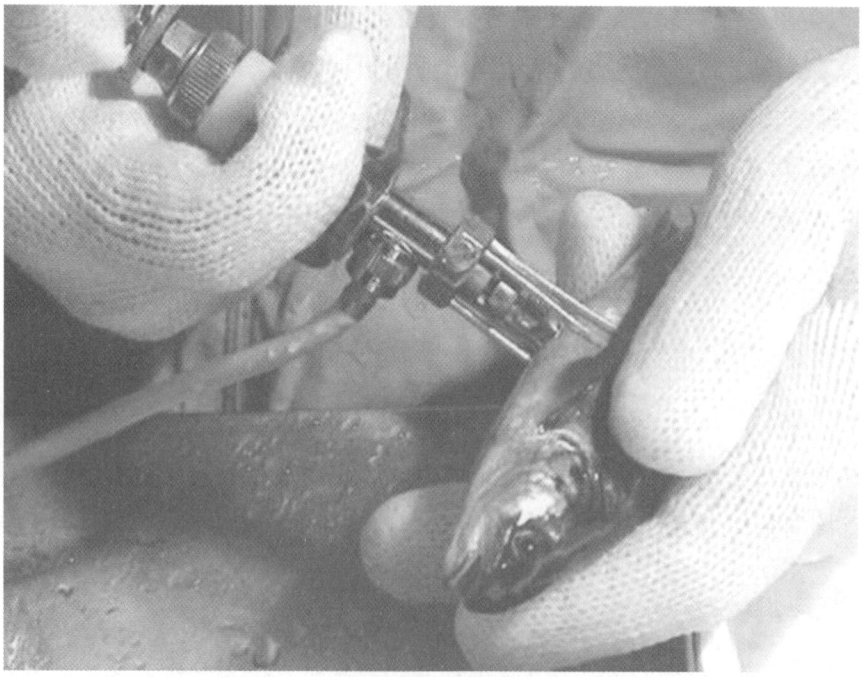

图 12-1　大西洋鲑鱼种的腹腔注射免疫

注射部位是在腹部中线，大约在臀鳍基部前一个鳍长度的位置。注射枪的两侧套有金属环，便于针头对准注射位点，防止操作员自我误伤

（引自 Pharmaq）

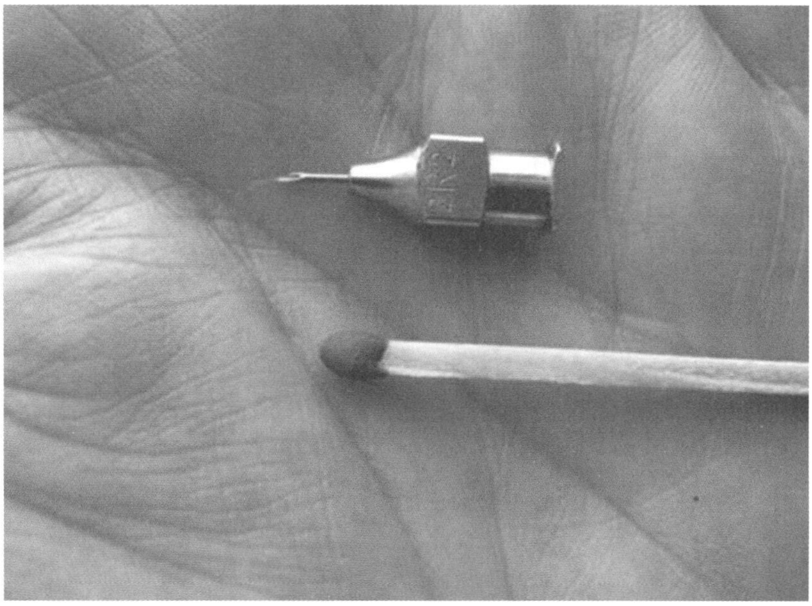

图 12-2　注射针头
应短小，一般长 3～5mm，足以穿透腹壁，直径 0.5～0.7 mm
（引自 Trygve T. Poppe）

久（图 12-3）。接种完成后，必须检查接种疫苗的鱼类的苏醒情况，保证药物麻醉的时间和浓度适当。如果鱼体苏醒过程需要的时间过长，接种过的鱼可能会在养殖池底部堆积，导致鱼体的缺氧甚至窒息。总体来说，每一接种步骤都需要优化到最快，使整个过程用时最短。

图 12-3　疫苗注射
通常，鱼类的大规模注射免疫需要专业团队合作完成。应保持运输到接种台上鱼的数量较少和台面以及管道中水的流动，以便在短时间内完成整个操作过程
（引自 Trygve T. Poppe）

鱼用疫苗如果误注射到操作员手上，可能会引起人体的严重不良反应，必须避免这种情况的发生（Leiraand Baalsrud，1997）。操作人员必须佩戴金属线制的防护手套，并在注射器上安放双层金属环，便于固定鱼体，也便于针头对准注射位点（图 12-1）。此外，在麻醉鱼的过程中，操作人员应使用橡胶手套，以保护自身的健康安全。

已研发出各式各样的鱼体注射免疫接种设备，从人工操作的自动注射器到完全自动注射的机器。Maskon 公司生产的鱼用疫苗自动注射机，每小时能完成 1 万尾鱼的快速接种（图 12-4 和图 12-5）。

图 12-4　Maskon 公司生产的鱼用疫苗自动注射机
（引自 Skala Maskon）

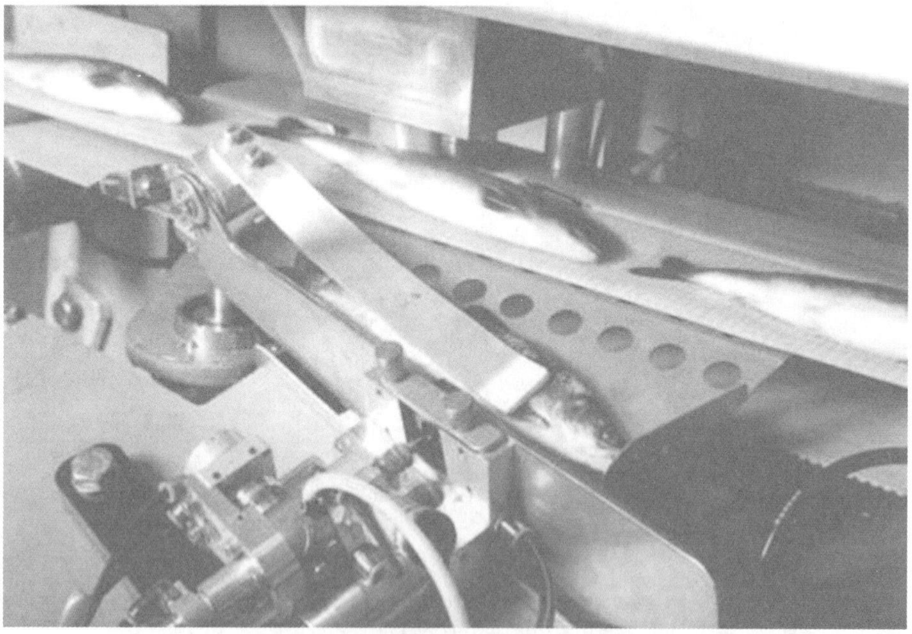

图 12-5　鱼用疫苗自动注射机（注射部分细节）
（引自 Skala Maskon）

二、浸泡免疫

使用浸泡免疫时，鱼体表面（鳃、皮肤、消化道）暴露于疫苗稀释溶液中，鱼体通过鳃、皮肤、侧线和肠道（Nakanishi 和 Ototake，1997）摄取抗原。鱼体可以在高浓度的疫苗溶液中进行短时间浸泡（30s 到 1min），或在大体积稀释的疫苗中较长时间浸泡，也可以通过向鱼体喷洒的方式免疫。浸泡免疫适用于鱼苗和幼鱼，因为小鱼不宜实施注射免疫。小鱼的疫苗消耗较少，但随着鱼的规格增大（即鱼的总生物量），疫苗消耗量会成比例地增加。

浸泡免疫在具有颗粒性抗原的福尔马林灭活弧菌和鲁克氏耶尔森菌疫苗上广泛使用。通常浸泡免疫获得的免疫保护力比注射免疫的要低，但对减毒活疫苗接种极为适宜（Evensen，2009），如柱状黄杆菌无毒活疫苗（Aquavac-Col）、鮰爱德华菌无毒活疫苗（Aquavac-Esc）（两种疫苗均由默沙东动物保健公司生产），对孵化 1 周后的鲶鱼苗就是实施浸泡免疫策略。

浸泡免疫接种疫苗溶液的稀释，通常是在容器中将疫苗与孵化用水以 1∶10 进行稀释。在每一次浸泡过程中，用大而宽松的网将鱼苗在疫苗稀释液中浸泡 30s。容器内侧壁上应有刻度，以控制接种鱼的数量。或者在每次浸泡时，将容器放在秤上称量，通常每升疫苗原液可以接种 100kg 的鱼。

浸浴免疫是一种经过改良的接种方法，该免疫方法不需要将鱼从养殖池中转运出来。浸浴免疫前，先将养殖池的进水关闭，降低水位，增加鱼群密度，然后将疫苗添加到水中，通常稀释度为 1∶1000。在浸浴接种过程中，保持充氧，停止供水 1h。这种接种方法与常规的浸泡相比，免疫鱼的数量基本相同，但诱导的免疫保护水平一般比较低（Håstein 和 Refsti，1986）。

浸泡免疫过程中，影响抗原吸收进而影响免疫保护水平的因素有很多，最重要的是抗原浓度，还有浸泡时间以及抗原的特性（颗粒的或可溶的）（Nakanishi 和 Ototake，1997）。鱼体自身特性也会影响免疫应答，如鱼的大小和对免疫接种的应激水平。为了提高鱼体对浸泡免疫的抗原摄取，试验尝试了多种浸泡免疫方法。较早的方法是鱼类在疫苗浸泡之前先进行高渗处理（Croy 和 Amend，1977），在鲤接种杀鲑气单胞菌（*A. salmonicida*）疫苗时，高渗处理增加了鲤对可溶性抗原的摄入并提高了免疫保护（Huising 等，2003）。另有研究表明，鱼体浸泡免疫接种前超声处理，可以增加金鱼和鲤对模式抗原的吸收，并增强鱼体的抗体应答（Navot 等，2005）。还有研究显示在虹鳟幼鱼接种海豚链球菌（*Streptococcus iniae*）疫苗前，采用一种多针的工具穿刺虹鳟幼鱼的皮肤，可以增强虹鳟幼鱼的抗感染能力（Nakanishi 等，2002）。然而，这些程序都需要额外的处理步骤，实用性低而且操作会使鱼体产生应激，可行性差。到目前为止，这些改进的浸泡免疫方法尚未在商业水产养殖中应用。

三、口服免疫

自从首次报道疫苗可以保护鱼类免受疾病感染以来，口服免疫方式是理想的选择，尤其适用于数量巨大、不易处理和易应激的鱼类。口服免疫已有很多报道，但是其保护效果并不明确。总体而言，由于保护效果差和抗原消耗大，口服免疫尚未应用于实际的疫苗接种中。有报道称，初次浸泡免疫后，口服加强免疫，鱼体获得了效果显著的免疫保护（Thinh 等，2009），但现在市场上商业化的口服疫苗产品很少。默沙东动物保健公司（MSD Animal Health）生产的 Aquavac ERM 口服疫苗用于早期注射接种了红嘴病（鲁克氏耶尔森菌病）疫苗虹鳟的加强免疫。智利 Centrovet 公司提供了单价和二价的鲑传染性贫血症和鲑鱼立克次氏体综合征的口服疫苗。

为了避免疫苗抗原经胃和前肠时被消化降解，进行了很多保护疫苗抗原的试验研究，以确保有足够的抗原到达后肠壁的免疫相关细胞，因为后肠是免疫应答反应的重要部位。已对耐酸涂层、缓释颗粒（如海藻酸微粒）以及微米和纳米粒子进行了研究（Plant 和 LaPatra，2011）。疫苗抗原可以先饲喂给桡足类，再添加到饵料生物如卤虫中，在水蚤和轮虫上也进行了类似的实验。近年来，蝇幼虫也被当作基因工程抗原的生产者和疫苗传递载体，通过饲喂鱼类口服免疫。然而在实际应用中，这些研究都没能在抗原剂量以及免疫保护等方面取得突破进展。

第六节 保护形成与保护持续时间

疫苗接种的时间应与预测可能发生疾病的时间相适应，使鱼体获得的最大免疫保护力与疾病可能暴发的时间一致。如果鱼体接种疫苗的时间太晚，免疫力尚未形成；如果鱼体接种疫苗的时间过早，免疫保护水平可能在疾病发生时已经下降。鲑鳟鱼类佐剂疫苗生产公司建议，鲑从疫苗接种到转移至海上至少需 500℃・d，而斑点叉尾鮰在水温 21～29℃ 时应在至少 2 周前（300～400℃・d）接种鮰爱德华菌活疫苗。对于非佐剂类疫苗，无论是浸泡接种还是注射接种，疫苗接种时间也应参考这一规律。

鲑鳟鱼类在海水养殖的周期大约为一年半，鉴于以往的养殖数据和经验，油类佐剂的注射疫苗在整个养殖过程中都具有免疫保护作用。研究证明，大西洋鲑在疫苗接种后 3～6 个月，抗杀鲑气单胞菌的特异性抗体水平以及对疖疮病感染的相对存活率（RPS）均有提高（Midtlyng 等，1996）。大西洋鲑的养殖实验还表明，在疫苗接种后的第二年，其仍然具有免疫保护力，这一结果在两次浸泡接种冷水弧菌非佐剂疫苗（Lillehaug 等，1991），以及注射接种疖疮病铝佐剂疫苗的实验中（Lillehaug 等，1992）均已被证实。但相比第一年，第二年的免疫保护水平较低。

罗非鱼注射接种实验性无乳链球菌（*Streptococcus agalactiae*）无佐剂疫苗后 6 个月（Pasnik 等，2005）和大西洋鲑注射接种杀鲑气单胞菌无佐剂疫苗后 9 个月（Bricknell 等，1999）的攻毒实验结果表明，尽管免疫保护水平随时间延长有下降的趋势，但两种疫苗对鱼均具有免疫保护作用。虹鳟肌肉注射病毒性出血性败血症病毒 DNA 疫苗（McLauchlan 等，2003）或传染性造血器官坏死病毒 DNA 疫苗（Kurath 等，2006）后，分别在接种后的第 9 个月和第 25 个月进行攻毒实验，结果表明，这些疫苗均具有显著的免疫保护作用。格氏乳球菌（*Lactococcus garvieae*）疫苗中添加不同佐剂，在虹鳟上的研究表明，添加佐剂的疫苗，不仅可以提高免疫保护水平，还可以延长免疫保护时间（Ravelo 等，2006）。

第七节 加强免疫

加强免疫能提高鱼体疫苗接种后的免疫保护水平（Tatner 和 Horne，1985），其常用于接种后没有产生足够的免疫保护，或者是小鱼在接受初次免疫时免疫力尚未完全形成时，通常在初次免疫几周后进行加强免疫。对幼鱼无论初次免疫还是加强免疫，通常采用浸泡免疫接种，例如对海鲈幼鱼的免疫，正是这样做的（Angelidis 等，2006）。

首次免疫几个月后，在鱼类的易感疾病可能再次暴发时，可采取加强免疫以恢复免疫保护力，通常这种情况适合相对较大的鱼。实际上，注射免疫比较困难，浸泡免疫需要大量的疫苗，这就对口服免疫提出了极大的需求。默沙东动物保健公司生产的 Aquavac ERM 口服疫苗，适用于虹鳟首次浸泡免疫 4～6 个月后的口服加强免疫。

第八节　疫苗接种效益

实施疫苗接种策略的目的是为鱼类养殖业提高经济效益。危害严重的传染性病原引起的疾病，能造成费用增加和严重的经济损失，包括产量减少、药物费用及其他花费增加。为了估算疫苗接种费用和疾病带来的损失之间的关系，Lillehaug（1989）建立了疫苗费用与疾病损失的盈亏平衡计算模型。疫苗接种费用除了疫苗本身，也包括接种过程的劳动力费用，以及因疫苗副作用而造成的损失，还有不同接种方法获得的保护水平差异等，这些都应在评估计算中予以考虑。据估算，在实际养殖中，大西洋鲑弧菌病的发生率造成不足 1%的经济损失，无论是注射还是浸泡免疫，均已大于疫苗接种的费用。注射免疫能产生较高的免疫保护率，如果是引起较高死亡率的疾病，疫苗注射接种会体现出显著的经济效益。

Thorarinsson 和 Powell（2006）建立了更完善的经济模型，评估引入疾病风险、疫苗效力和鱼产品的市场价格等因素，分别在大西洋鲑接种疖病疫苗和海鲈接种弧菌病疫苗中进行了评估。评估结果显示，若整个养殖周期内，未接种疫苗鱼的死亡率为 20%，接种疫苗的大西洋鲑和海鲈，每千克的净收入分别高出 0.42 欧元和 0.52 欧元。在决定使用哪种疫苗产品时，即使疫苗的价格较高，也应选择生产实践中产生最佳保护效果的疫苗产品，因为疫苗效力对经济产出影响很大。

参考文献

Angelides，P.，Karagiannis，D. and Crump，E. M. 2006. Efficacy of a *Listonella anguillarum*（syn. *Vibrio anguillarum*）vaccine for juvenile sea bass *Dicentrarchus labrax* L. *Dis Aquat Org* 71：19-24.

Bly，J. E. and Clem，L. W. 1992. Temperature and teleost immune functions. *Fish Shellfish Immunol* 2：159-171.

Bowden，T. J. 2008. Modulation of the immune system of fish by their environment. *Fish Shellfish Immunol* 25：373-383.

Bricknell，I. R.，King，J. A.，Bowden，T. J，and Ellis，A. E. 1999. Duration of protective antibodies，and the correlation with protection in Atlantic salmon（*Salmo salar* L.），following vaccination with an *Aeromonas salmonicida* vaccine containing iron-regulated outer membrane proteins and secretory polysaccharide. *Fish Shellfish Immunol* 9：139-151.

Croy，T. R. and Amend，D. F. 1977. Immunization of sockeye salmon（*Oncorhynchus nerka*）against vibriosis using hyperosmotic infiltration. *Aquaculture* 12：317-325.

Eggset，G.，Mortensen，A. and Løken，S. 1999. Vaccination of Atlantic salmon（*Salmo salar* L.）before and during smoltification：effects on smoltification and immunological protection. *Aquaculture* 170：101-112.

Evensen，Ø. 2009. Development in fish vaccinology with focus on delivery methodologies，adjuvants and formulations. *Op Méditerr Series A* 86：177-186.

Hart，S.，Wrathmell，A. B.，Harris，J. E. and Grayson，T. H. 1988. Gut immunology in fish：a review. *Dev*

Comp Immunol 12：453-480.

Håstein，T. and Refsti，T. 1986. Vaccination of rainbow trout against vibriosis by injection，dip and bath. *Bull Eur Assoc Fish Pathol* 6：45-49.

Huising，M. O.，Guichelaar，T.，Hoek，C.，*et al.*，2003. Increased efficacy of immersion vaccination in fish with hyperosmotic pretreatment. *Vaccine* 21：4178-4193.

Karlsson，A.，Rosseland，B. O.，Massabuau，J. C. and Kiessling，A. 2012. Pre-anaesthetic metomidate sedation delays the stress response after caudal artery cannulation in Atlantic cod（*Gadus morhua*）. *Fish Physiol Biochem* 38：401-411.

Kurath，G.，Garver，K. A.，Corbeil，S.，*et al.*，2006. Protective immunity and lack of histopathological damage two years after DNA vaccination against infectious hematopoietic necrosis virus in trout. *Vaccine* 24：345-354.

Leira，H. L. and Baalsrud，K. J. 1997. Operator safety during injection vaccination of fish. *Dev Biol Stand* 90：383-387.

Lillehaug，A. 1989. A cost-effectiveness study of three different methods of vaccination against vibriosis in salmonids. *Aquaculture* 83：227-236.

Lillehaug，A. 1991. Vaccination of Atlantic salmon（*Salmo salar* L.）against cold-water vibriosis duration of protection and effect on growth rate. *Aquaculture* 92：99-107.

Lillehaug，A. 1997. Vaccination strategies in seawater cage culture of salmonids. *Dev Biol Stand* 90：401-408.

Lillehaug，A.，Lunder，T. and Poppe，T. T. 1992. Field testing of adjuvanted furunculosis vaccines in Atlantic salmon，*Salmo salar* L. *J Fish Dis* 15：485-496.

Lillehaug，A.，Ramstad，A.，Bækken，K. and Reitan，L. J. 1993. Protective immunity in Atlantic salmon（*Salmo salar* L.）vaccinated at different water temperatures. *Fish Shellfish Immunol* 3：143-156.

Lorenzen，E.，Einer-Jensen，K.，Rasmussen，J. S.，*et al.*，2011. The protective mechanisms induced by a DNA vaccine in fish depend on temperature. *Scand J Immunol* 73：392.

Martins，M. L.，Xu，D. H.，Shoemaker，C. A. and Klesius，P. H. 2011. Temperature effects on immune response and hematological parameters of channel catfish *Ictalurus punctatus* vaccinated with live theronts of *Ichthyophthirius multifiliis*. *Fish Shellfish Immunol* 31：774-780.

Maule，A. G.，Schreck，C. B. and Kaattari，S. L. 1987. Changes in the immune system of coho salmon（*Oncorhynchus kisutch*）during the parr-to-smolt transformation and after implantation of cortisol. *Can J Fish Aquat Sci* 44：161-166.

McLauchlan，P. E.，Collet，B.，Ingerslev，E.，*et al.*，2003. DNA vaccination against viral haemorrhagic septicaemia（VHS）in rainbow trout：size，dose，route of injection and duration of protection—early protection correlates with Mx expression. *Fish Shellfish Immunol* 15：39-50.

Midtlyng，P. J.，Reitan，L. J. and Speilberg，L. 1996. Experimental studies on the efficacy and side-effects of intraperitoneal vaccination of Atlantic salmon（*Salmo salar* L.）against furunculosis. *Fish Shellfish Immunol* 6：335-350.

Nakanishi，T. and Ototake，M. 1997. Antigen uptake and immune responses after immersion vaccination. *Dev Biol Stand* 90：59-68.

Nakanishi，T.，Kiryu，I. and Ototake，M. 2002. Development of a new vaccine delivery method for fish：percutaneous administration by immersion with application of a multiple puncture instrument. *Vaccine* 20：3764-3769.

Navot，N.，Kimmel，E. and Avtalion，R. R. 2005. Immunisation of fish by bath immersion using ultrasound. *Dev Biol*（*Basel*）121：135-142.

Øvergård, A.-C., Fiksdal, I. U., Nerland, A. H. and Patel, S. 2011. Expression of T-cell markers during Atlantic halibut (*Hippoglossus hippoglossus* L.) ontogenesis. *Dev Comp Immunol* 36: 203-213.

Pasnik, D. J., Evans, J. J. and Klesius, P. H. 2005. Duration of protective antibodies and correlation with survival in Nile tilapia *Oreochromis niloticus* following *Streptococcus agalactiae* vaccination. *Dis Aquat Org* 66: 129-134.

Plant, K. P. and LaPatra, S. E. 2011. Advances in fish vaccine delivery. *Dev Comp Immunol* 35: 1256-1262.

Press, C. McL. and Lillehaug, A. 1995. Vaccination in European salmonid aquaculture: a review of practices and prospects. *Br Vet J* 151: 45-69.

Ravelo, C., Magarinos, B., Herrero, M. C., *et al.*, 2006. Use of adjuvanted vaccines to lengthen the protection against lactococcosis in rainbow trout (*Oncorhynchus mykiss*). *Aquaculture* 251: 153-158.

Schrøder, M. B., Villena, A. J. and Jørgensen, T. Ø. 1998. Ontogeny of lymphoid organs and immunoglobulin producing cells in Atlantic cod (*Gadus morhua* L.). *Dev Comp Immunol* 22: 507-517.

Sevatdal, S. and Dalen, V. 1997. Immunization of Atlantic salmon (*Salmo salar* L.) against furunculosis and cold water vibriosis at low temperatures. *Dev Biol Stand* 90: 472.

Shoemaker, C. A., Klesius, P. H. and Bricker, J. M. 1999. Efficacy of a modified live *Edwardsiella ictaluri* vaccine in channel catfish as young as seven days post hatch. *Aquaculture* 176: 189-193.

Tatner, M. F. and Horne, M. T. 1985. The effects of vaccine dilution, length of immersion time, and booster vaccinations on the protection levels induced by direct immersion vaccination of brown trout, *Salmo trutta*, with *Yersinia ruckeri* (ERM) vaccine. *Aquaculture* 46: 11-18.

Thinh, N. H., Kuo, T. Y., Hung, L. T., *et al.*, 2009. Combined immersion and oral vaccination of Vietnamese catfish (*Pangasianodon hypophthalmus*) confers protection against mortality caused by *Edwardsiella ictaluri*. *Fish Shellfish Immunol* 27: 773-776.

Thorarinsson, R. and Powell, D. B. 2006. Effects of disease risk, vaccine efficacy, and market price on the economics of fish vaccination. *Aquaculture* 256: 42-49.

Treasurer, J. and Cox, C. 2008. Intraperitoneal and dorsal median sinus vaccination effects on growth, immune response, and reproductive potential in farmed Atlantic salmon *Salmo salar*. *Aquaculture* 275: 51-57.

Zapata, A., Diez, B., Cejalvo, T., *et al.*, 2006. Ontogeny of the immune system of fish. *Fish Shellfish Immunol* 20: 126-136.

第十三章　疫苗接种的副作用

Trygve T. Poppe and Erling O. Koppang

本章概要　SUMMARY

疫苗接种会干扰鱼类的生物学功能，可能会产生副作用，最终对经济效益和鱼类福利造成一定的影响。虽然所有的疫苗接种方法都可能产生副作用，但大多数的副作用都比较轻微。最严重的副作用是佐剂疫苗注射免疫接种可能引起鱼类腹腔内病变，这种病变可能会影响鱼类的生长、健康和鱼肉质量，从动物福利的角度，这些影响的程度和严重性也可能超过了鱼体所能接受的极限。许多养殖的鱼类可能会受疫苗接种的影响，但在大西洋鲑中关于腹腔注射后主要的副作用的报道较多。

第一节　引　言

采用疫苗接种是当今鲑大规模养殖中的普遍做法，在其他鱼类养殖中也基本如此。在大多数国家，鱼类疫苗为多价油类佐剂疫苗，其有效地减少了因疖疮病和其他细菌病造成的损失，因此，如果没有有效的疫苗接种计划，现代化的大规模的鲑养殖和其他鱼类养殖很难持续健康地发展。

多数疫苗接种方式都可能导致急性的或慢性的副作用，但目前为止，鲑腹腔注射油类佐剂疫苗造成的副作用最为严重，因此，这是本章的主要内容。浸泡或浸浴免疫接种可能引起鱼类短暂或较长时间的食欲不振，从而影响生长，但与接种疫苗带来的巨大益处相比，这点副作用微不足道。

除塔斯马尼亚岛外，几乎所有的海水养殖鲑鳟鱼类均接种油类佐剂疫苗。基于各地的疾病流行发生状况，疫苗的成分可能会有所不同，但多数是包含几种细菌和病毒抗原的多联疫苗，并添加佐剂以提高或延长疫苗的效果，然而，多数佐剂可能产生不同程度的副作用。多年来，已研发出多种类型佐剂，包括铝盐（Lillehaug 等，1992）、硫酸铝钾（明矾）（Horne 等，1984；Mulvey 等，1995）以及烃油。目前，广泛用于鲑鳟鱼类养殖的疫苗通常含烃油（Haugarvoll 等，2010）。鲑的免疫接种是在银化期或转入海水之前，首先饥饿几日，然后麻醉，再进行腹腔注射接种。接种一般采用疫苗接种专业团队或每小时可接种20 000尾的疫苗自动接种机。注射部位在鱼体的腹部中线，从头到腹鳍基部 1.5 倍鳍长处，注射量通常为每尾0.05～0.1mL。哺乳动物初次免疫后通常会加强免疫，但与之不同的是，鲑鳟鱼类只接种一次。

自从使用疫苗以来，经常发生注射后的腹腔病变，因此，疫苗产生的副作用就备受关注。世界范围内，这种病变在大西洋鲑养殖中最为严重，而虹鳟产生的病变较少且更轻微，这可能是因为虹鳟疫苗所含抗原量比鲑的少。另外，有些虹鳟的疫苗不含佐剂，刺激较小。疫苗副作用引起的关注，一方面是涉及动物福利，另一方面是由于黑化和纤维粘连导致生长迟缓、品质下降，甚至使消费者因肉质问题而排斥商品鱼（Midtlyng 和 Lillehaug，1998）。

迄今为止，鲑养殖业中使用油类佐剂疫苗已有 20 多年，副作用仍然存在。从动物福利的角度，有些时候这些影响的程度和严重性可能超过了鱼体所能接受的程度，尽管已投入很大努力优化疫苗以及增加对动物福利的关注，但这个问题的存在仍是不争的事实。随着动物福利越来越受到重视，毫无疑问，应投入更多的努力以减少或最小化这种有效的预防措施所带来的负面影响。

第二节　急性副作用

急性副作用大多是由于操作不当、麻醉失误、疫苗污染以及真正由疫苗本身造成的副作用（Lillehaug，1989）。操作流程设计不合理、设备配置不合适以及麻醉过程等，都可能在接种疫苗期间或接种后不久导致急性副作用或鱼的死亡。

水质差、疫苗污染或接种程序不符合卫生要求都可能使细菌与疫苗同时进入腹腔，导致鱼类在接种几天后发生急性败血症，在肛门周围、胸鳍和腹鳍的基部和腹部出现广泛出血。Olsen 等（2006）报道了苏格兰和挪威大西洋鲑免疫后，腹腔感染红球菌（*Rhodococcus erythropolis*）。内部病变主要是腹水以及腹膜表面、肝脏和幽门有瘀血点；组织病理学发现腹膜病变区域，心肌外膜和骨骼肌中感染革兰氏阳性杆状细菌。有趣的是，用从病灶分离出的细菌攻毒，并不会引起同样的疾病，而分离出的细菌与油类佐剂混合注射却产生了 100% 的死亡率，这表明鱼体的免疫系统受到了某种物理或营养方面的保护。感染源一般并不确定，通常分离到的是假单胞菌和气单胞菌，也可能与接种时卫生状况不佳有关。

研究表明，大西洋鲑接种后的几小时内，与对照组相比，表现出食欲和游泳能力降低（Bjørge 等，2011）。严重的病变也伴随摄食延迟和活力下降，这类似于在其他鱼类进行的注射醋酸的疼痛反应实验，认为是受免鱼对注射疫苗表现出的通常疼痛反应。实际生产中，由于鱼体大小、水温等因素，摄食延迟和行为改变可能会持续几天到几周。在虹鳟（Rønsholdt 和 McLean，1999）、红点鲑（Pylkkö 等，2000）、白鲑（*Coregonus* sp.）（Koskela 等，2004）、狼鱼和大比目鱼（Ingilæ 等，2000）中，也出现了这种短期对生长的负面影响。对疫苗的应激会引起疼痛，扰乱正常的肠道功能，引发高耗能的机体免疫反应，造成体重降低和差的动物福利（Sørum 和 Damsgaard，2004）。

有研究报道，在大西洋鲑中，由于使用不合适的接种设备和/或接种技术将疫苗注射到了内脏器官（肝、脾、肠和鳔）或血管，造成疫苗接种后大量死亡。受损鱼的鳃苍白，尤其是鳃丝的末端。组织病理学显示，鳃丝血管和鳃小片的毛细血管内出现油滴（Skrudland 等，2002），这些栓塞性病变造成鳃丝末端坏死。

第三节 慢性副作用

所有疫苗接种方式均可能对鱼体产生慢性的影响，包括被认为是最温和的浸泡免疫方式（Lillehaug，1991）。造成鱼体的生长迟缓可能是接种操作、麻醉以及其他可能无法解释的原因。接种疫苗后生长迟缓的鱼，在后期可能会出现补偿性生长（Berg 等，2006）。

第四节 注射部位的反应

接种疫苗的鱼最常见的反应是注射部位附近的局部纤维性腹膜炎，通常在腹部中线和幽门盲囊的末端。腹壁、脾脏和幽门盲囊之间形成纤维束粘连（图 13-1），这些纤维束通常很容易撕开，对鱼肉的损害较小，对鱼体几乎没有影响。

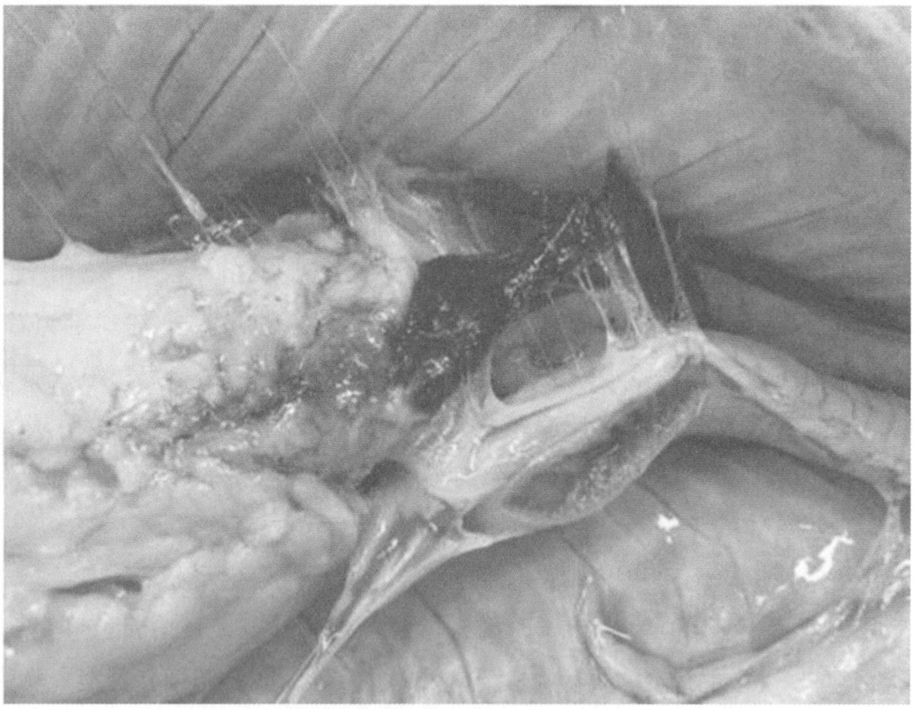

图 13-1 在注射部位形成脾脏、后肠和体壁的中度纤维束粘连
注意脾脏头部的黑化
（引自 Trygve T. Poppe）

组织学上，鲑注射部位的早期病变特点是嗜中性粒细胞浸润，随后是巨噬细胞和淋巴细胞反应增强（Mutoloki 等，2006a，2006b）。肥大细胞经常出现，特别是出现在靠近病变部位中心的圆形油滴的异常表象处。整个病变部位可能存在不同数量的黑色素巨噬细胞。几个月后，浸润炎性细胞的纤维组织形成长链的情况越来越常见。某些情况下，即便肉眼不易观察到病

变，在幽门盲囊壁和胰腺/脂肪组织之间可能已出现轻度病变和中度病变。对疫苗导致的粘连一般采用“斯皮尔伯格评分系统”进行标准化评价（Midtlyng 等，1996），“0”表示未接种疫苗的鱼或接种疫苗的鱼无可见病变，而 6 表示整个腹腔严重或广泛的病变。通常，斯皮尔伯格的分值为 2.0 或 2.5，属于动物伦理和消费者均可接受的范围。

第五节 广泛的腹部病变

鱼类接种疫苗后，时常发生严重的、大范围的病变，尤其是大西洋鲑（Poppe 和 Breck，1997）。其病变程度和病变类型多样，并可能会危害鱼体的生理功能、动物福利及产品质量。这类病变的发生不可预测，自从使用油类佐剂疫苗以来，时有发生。有些病变是由于注射不当，如疫苗注入幽门盲囊、胃壁、肝脏等腹部器官，引起大范围的肉芽肿炎症和器官粘连。还有的病变出现在腹腔的背部头侧靠近横隔处、食道、性腺、大血管和神经等部位的大面积肉芽肿粘连（图 13-2）。这些部位的病变可能严重损害性腺的正常发育以及影响多个器官系统的功能。肉芽肿组织中经常出现清晰的空腔，这是疫苗中油滴形成的异常表象（图 13-3），其显微特征是具有较厚的和大范围的典型肉芽肿性炎症结构。纤维组织、巨噬细胞、淋巴细胞和肥大细胞在异常油滴周围形成大的同心环状结构。多核巨细胞、斯普伦多雷-何博礼-样反应（Splendore-Hoeppli-like reaction）、新生血管形成和黑色素巨噬细胞可能使病变更加复杂（图 13-4）。在腹腔内膜和病变部位的大范围的黑化现象也时有发生。

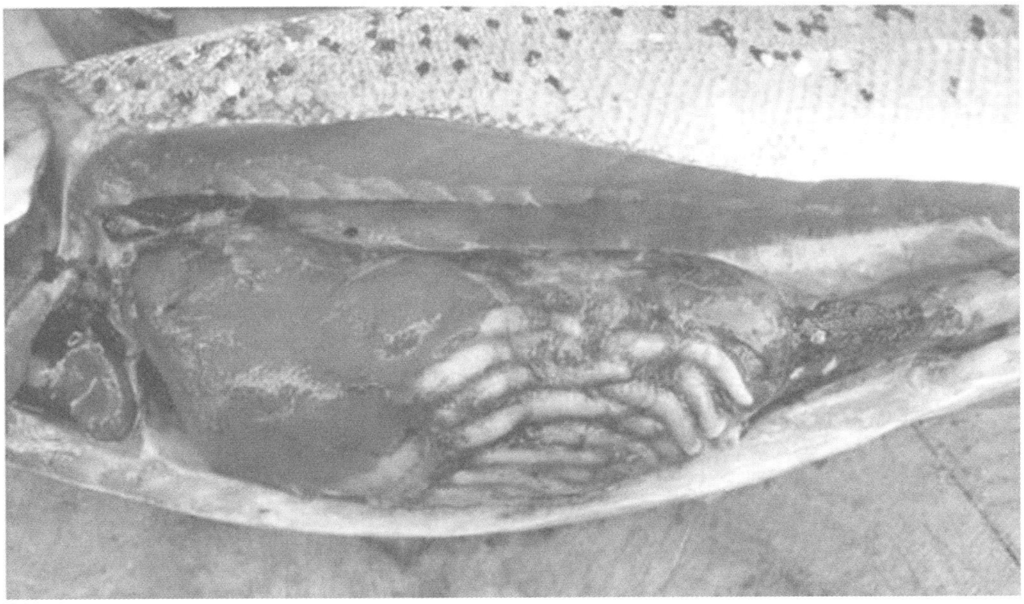

图 13-2 在幽门盲囊产生严重的肉芽肿炎症（示病变部位大面积黑化）
（引自 Trygve T. Poppe）

这种病变的病因并不明确，当然也存在争议。实验表明，传染性胰腺坏死病毒或假单胞菌（*Pseudomonas*）的感染能加重病变，因此，在许多养殖场，怀疑这种感染是发生病变的原因（Colquhoun 等，1998）。挪威的鲑养殖场广泛存在着多核微孢子虫（*Paranucleospora thereidon*/*Desmozoon salmonis*）的感染，由此考虑这可能是造成病变的原因，但尚未得到证实。

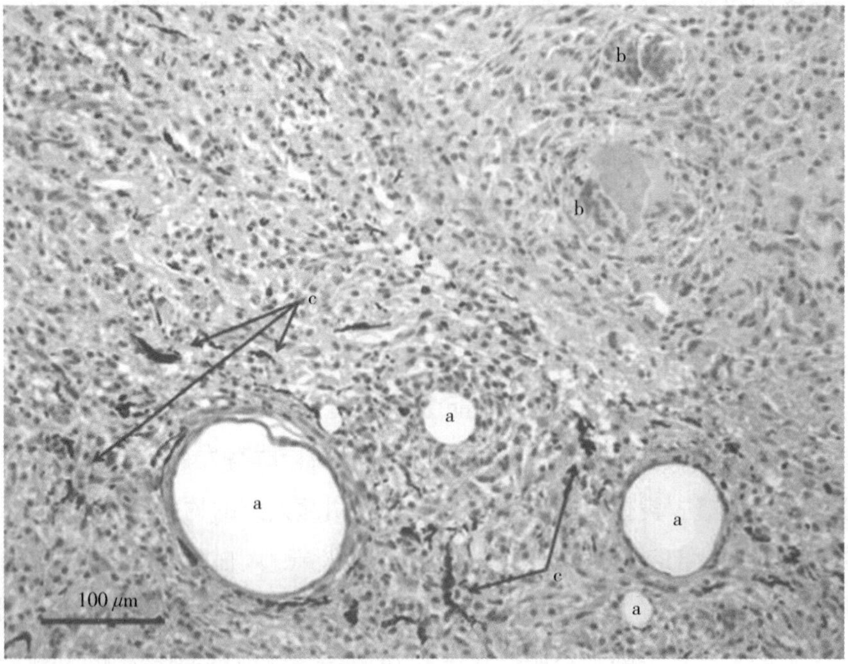

图 13-3　幽门盲囊和肠壁间肉芽肿炎症的组织 H&E 染色

a. 异常油滴　b. 多核巨大细胞　c. 肉芽肿基质中的黑色素沉积

（引自 Trygve T. Poppe）

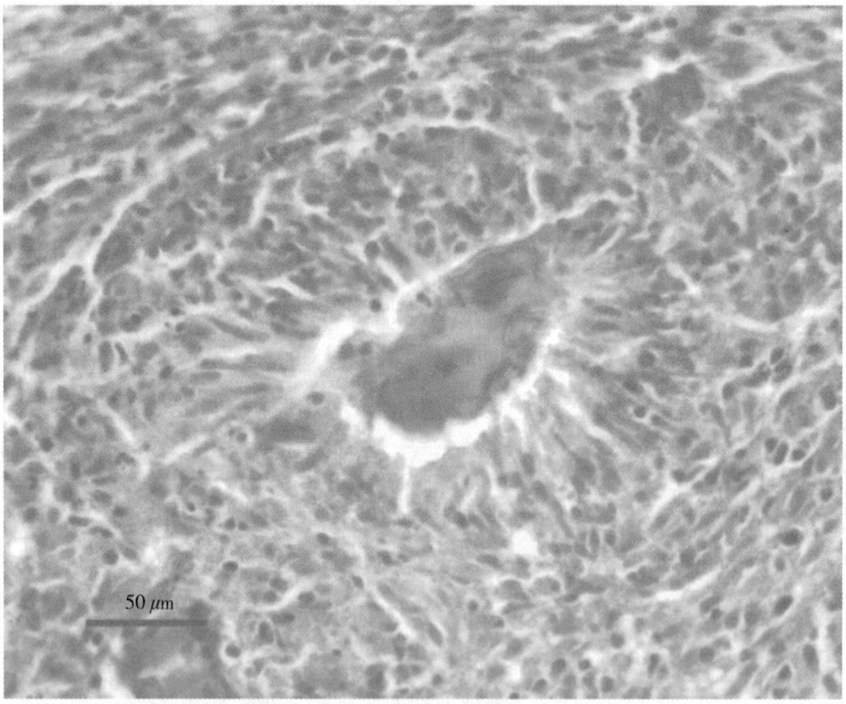

图 13-4　注射部位肉芽肿组织的斯普伦多雷-何博礼现象（马休猩红蓝染色）

（引自 Trygve T. Poppe）

第六节　其他器官的病变

病变还发生在肌肉、眼色素膜和肾脏等组织器官。肉眼可见，鲑腹腔的病变、肌肉坏死并伴有鱼肉中色素沉积（Koppang 等，2005）。组织学显示，鱼肉中出现异常油滴以及异常黑色斑点。眼部炎症与大多数内脏器官（包括脊柱在内）普遍的炎症反应一样（Haugarvoll 等，2010），也是疫苗接种后的全身性副作用（Koppang 等，2004）。在炎性病变中出现核扁平的球形细胞，是功能紊乱的 B 细胞，称为 Mott 细胞，先前已在接种疫苗的鱼体内发现了这种细胞，因此在常规诊断中又称之为“疫苗细胞”。

第七节　骨骼病变

有多种病因可引起养殖的鱼出现骨骼畸形（扁平椎）。鱼小、生长快、接种疫苗的时间、高温养殖和可利用的磷较低等均可造成骨骼病变。此外，鲑在秋季由河水转入海水比在春季转入海水产生的副作用多，包括骨骼病变（Fjelldal 等，2012）。脊柱的病变包括“驼背”（脊柱前部）、“短尾”（脊柱后部）或整个脊柱变形。水温和接种疫苗的时机是导致扁平椎的两个因素，实验证明，较小的未银化的鲑苗，过早接种疫苗（此时水温较高）会造成腹腔内迅速病变，到鲑上市时发现骨骼畸形的发生率较高，与较晚接种和在低温下接种的鲑相比，这些鱼长得也最小（Berg 等，2006）。另有研究也证实了疫苗接种会导致脊柱畸形（Aunsmo 等，2008；Haugarvoll 等，2010）。系列的比对研究证明，延长从疫苗接种到转入海水的时间（2～5 个月）会增加鲑骨骼病变发生的风险（Vågsholm 和 Djupvik，1998）。以上结果清楚地表明，可通过科学的疫苗接种计划以及熟知鱼类生物学特性来减少疫苗接种的负面影响。

第八节　自身免疫

在小鼠中腹腔注射类似疫苗油类佐剂的某些烃油可诱导产生自身系统免疫，大西洋鲑接种油类佐剂疫苗后，也会产生系统性自身免疫这样的副作用（Koppang 等，2008），接种过疫苗的鱼出现广泛的副作用可能与系统性自身免疫有关（Haugarvoll 等，2010）。系统性自身免疫有多个方面，例如，免疫复合物沉积在肾小球基底膜上，并产生各种自身抗体形成的肾小球性肾炎。此外，Mott 细胞的出现也很可能与系统性自身免疫相关。

第九节　非鲑鳟鱼类的病变

海鲈、海鲷、罗非鱼、斑点叉尾鮰、巨鲶和大西洋鳕等养殖鱼类也接种针对其养殖环境中严重病原的疫苗，因这些鱼类很少使用佐剂疫苗，因此罕有接种疫苗后出现副作用的报道。

研究表明，海鲈接种油类佐剂疫苗后 24～48h，体内的中性粒细胞和巨噬细胞被活化，其后，嗜酸性粒细胞、淋巴细胞、中性粒细胞和巨噬细胞的数量持续增加，可达免疫后 60d（Afonso 等，2005）。虽然海鲈在接种疫苗 11 个月后出现肉眼可见的病变，但从未达到鲑的病变程度。

虽然油类佐剂疫苗可能会对养殖的鳕产生不利的副作用，但其病变程度通常比鲑的缓和，从动物福利和消费者的角度看，在可接受的范围内。由于鳕的肝脏很大，为了减少肝损害，疫苗注射在正确部位非常关键。鳕接种疫苗后的典型病变特点是疫苗被快速包裹，内脏器官和体壁之间的粘连通常比鲑轻微，此外，鳕中很少发生黑化病变和肌肉黑化的现象（Maira 等，2007）。

参考文献

Afonso，A.，Gomes，S.，da Silva，J.，*et al.*，2005. Side effects in sea bass（*Dicentrarchus labrax* L.）due to intraperitoneal vaccination against vibriosis and pasteurellosis. *Fish Shellfish Immunol* 19：1-16.

Aunsmo，A.，Guttvik，A.，Midtlyng，P. J.，*et al.*，2008. Association of spinal deformity and vaccine-induced abdominal lesions in harvest-sized Atlantic salmon，*Salmo salar* L. *J Fish Dis* 31：514-524.

Berg，A.，Rødseth，O. M.，Tangerås，A. and Hansen，T. 2006. Time of vaccination influences development of adhesions，growth and spinal deformities in Atlantic salmon *Salmo salar*. *Dis Aquat Org* 69：239-248.

Bjørge，M. H.，Nordgreen，J.，Janczak，A. M.，*et al.*，2011. Behavioural changes following intraperitoneal vaccination in Atlantic salmon（*Salmo salar*）. *Appl Anim Behav Sci* 133：127-135.

Colquhoun，D. J.，Skjerve，E. and Poppe，T. T. 1998. *Pseudomonas fluorescens*，infectious pancreatic necrosis virus and environmental stress as potential factors in the development of vaccine related adhesions in Atlantic salmon，*Salmo salar* L. *J Fish Dis* 21：355-364.

Fjelldal，P. G.，Hansen，T.，Breck，O.，*et al.*，2012. Vertebral deformities in farmed Atlantic salmon（*Salmo salar* L.）：etiology and pathology. *J Appl Ichthyol* 28：433-440.

Haugarvoll，E.，Bjerkås，I.，Szabo，N. N.，*et al.*，2010. Manifestations of systemic autoimmunity in vaccinated salmon. *Vaccine* 28：4961-4969.

Horne，M. T.，Roberts，R. J.，Tatner，M. and Ward，P. 1984. The effects of the use of potassium alum adjuvant in vaccines against vibriosis in rainbow trout，*Salmo gairdneri* Richardson. *J Fish Dis* 7：91-99.

Ingilæ，M.，Arnesen，J. A.，Lund，V. and Eggset，G. 2000. Vaccination of Atlantic halibut *Hippoglossus hippoglossus* L.，and spotted wolffish *Anarhichas minor* L.，against atypical *Aeromonas salmonicida*. *Aquaculture* 183：31-44.

Koppang，E. O.，Haugarvoll，E.，Hordvik，I.，*et al.*，2004. Granulomatous uveitis associated with vaccination in the Atlantic salmon. *Vet Pathol* 41：122-130.

Koppang，E. O.，Haugarvoll，E.，Hordvik，I.，*et al.*，2005. Vaccine-associated granulomatous inflammation and melanin accumulation in Atlantic salmon，*Salmo salar* L.，white muscle. *J Fish Dis* 28：13-22.

Koppang，E. O.，Bjerkås，I.，Haugarvoll，E.，*et al.*，2008. Vaccination-induced systemic autoimmunity in farmed Atlantic salmon. *J Immunol* 181：4807-4814.

Koskela，J.，Rahkonen，R.，Pasternack，M. and Knuutinen，H. 2004. Effect of immunization with two commercial vaccines on feed intake，growth and lysozyme activity in European whitefish（*Coregonus lavaretus* L.）. *Aquaculture* 234：41-50.

Lillehaug，A. 1989. A survey on different procedures used for vaccinating salmonids against vibriosis in Norwegian fish farming. *Aquaculture* 83：217-226.

Lillehaug，A. 1991. Vaccination of Atlantic salmon（*Salmo salar* L.）against cold-water vibriosis—duration of protection and effect on growth rate. *Aquaculture* 92：99-107.

Lillehaug，A.，Lunder，T. and Poppe，T. T. 1992. Field testing of adjuvanted furunculosis vaccines in Atlantic salmon，*Salmo salar* L. *J Fish Dis* 15：485-496.

Maira，C.，Lystad，Y.，Schrøder，M. B.，*et al.*，2007. Bieffektvurderinger hos torsk etter stikkvaksinering—

bruk av en vurderingsskala tilpasset torsk. *Nor Fiskeoppdr* 32: 52-55.

Midtlyng, P. J. and Lillehaug, A. 1998. Growth of Atlantic salmon *Salmo salar* after intraperitoneal administration of vaccines containing adjuvants. *Dis Aquat Org* 32: 91-97.

Midtlyng, P. J., Reitan, L. J. and Speilberg, L. 1996. Experimental studies on the efficacy and side-effects of intraperitoneal vaccination of Atlantic salmon (*Salmo salar* L.) against furunculosis. *Fish Shellfish Immunol* 6: 335-350.

Mulvey, B., Landolt, M. L. and Busch, R. A. 1995. Effects of potassium aluminium sulphate (alum) used in an *Aeromonas salmonicida* bacterin on Atlantic salmon, *Salmo salar* L. *J Fish Dis* 18: 495-506.

Mutoloki, S., Brudeseth, B., Reite, O. B. and Evensen, Ø. 2006a. The contribution of *Aeromonas salmonicida* extracellular products to the induction of inflammation in Atlantic salmon (*Salmo salar* L.) following vaccination with oil-based vaccines. *Fish Shellfish Immunol* 20: 1-11.

Mutoloki, S., Reite, O. B., Brudeseth, B., *et al.*, 2006b. A comparative immunopathological study of injectionsite reactions in salmonids following intraperitoneal injection with oil-adjuvanted vaccines. *Vaccine* 24: 578-588.

Olsen, A. B., Birkbeck, T. H., Nilsen, H. K., *et al.*, 2006. Vaccine-associated systemic *Rhodococcus erythropolis* infection in farmed Atlantic salmon *Salmo salar*. *Dis Aquat Org* 72: 9-17.

Poppe, T. T. and Breck, O. 1997. Pathology of Atlantic salmon *Salmo salar* intraperitoneally immunized with oil-adjuvanted vaccine. A case report. *Dis Aquat Org* 29: 219-226.

Pylkkö, P., Lyytikäinen, T., Ritola, O. and Pelkonen, S. 2000. Vaccination influences growth of Arctic charr. *Dis Aquat Org* 43: 77-80.

Rønsholdt, B. and McLean, E. 1999. The effect of vaccination and vaccine components upon short-time growth and feed conversion efficiency in rainbow trout. *Aquaculture* 174: 213-221.

Skrudland, A., Taksdal, T. and Kvellestad, A. 2002. Uhell og skade ved vaksinering av laks. *Nor Vet Tidsskr* 114: 494-495.

Sørum, U. and Damsgård, B. 2004. Effects of anaesthetisation and vaccination on feed intake and growth in Atlantic salmon (*Salmo salar* L.). *Aquaculture* 232: 333-341.

Vågsholm, I. and Djupvik, H. O. 1998. Risk factors for spinal deformities in Atlantic salmon, *Salmo salar* L. *J Fish Dis* 21: 47-53.

第十四章　鱼类疫苗学的发展趋势

Øystein Evensen

本章概要　SUMMARY

大多数用于注射免疫的鱼类疫苗都是灭活的全病毒/全细菌疫苗，通常配以矿物油或植物油等佐剂使用。浸泡疫苗通常不含佐剂，并且大多数是细菌疫苗，但无论用于注射或浸泡，活疫苗均较少使用。未来的疫苗将在现有配方的基础上更加精细化以减少机体局部和全身的副作用。疫苗制剂和免疫调节剂将有助于直接提升免疫应答，特别是针对胞内病原的免疫反应，同时，DNA 疫苗也可能成为未来疫苗产品的重要组成。经由黏膜表面的免疫接种将会大大改善和提升，其重点是增强病原入侵部位或病原初始复制点的局部免疫反应，以诱导产生免疫保护作用。未来十年，研制胞内细菌疫苗和病毒疫苗将是鱼类疫苗研究重点，而 DNA 疫苗技术将会在上述疫苗的研发中发挥作用。同时，新型改进的疫苗制剂、口服加强免疫，甚至活疫苗制剂都将发挥作用。

第一节　分子技术

分子生物学的发展促进了病毒学的研究，同时也加强了预防病毒感染的疫苗研发。应用分子生物学技术将病毒基因去除（敲除）、失活和插入，对开展病毒的毒力机制以及对免疫具有价值的病毒组分研究极为重要。当病原难以培养，且对免疫具有重要作用基因已明确的情况下，采用新技术研制疫苗相比传统疫苗有一定的优势。

第二节　重组疫苗

分子生物学的新技术本应在重组疫苗的研发方面大放异彩，然而，事实并非如此。目前，用于鱼类的重组亚单位疫苗很少，不仅如此，恒温动物甚至人类的重组亚单位疫苗也非常少。

这有几个原因，如前所述重组亚单位疫苗的免疫原性较差。DNA 疫苗技术是疫苗学领域新的热点，但也存在一定问题，不仅是诱导免疫反应的疫苗剂量问题，病原变异导致的疫苗无效等问题也阻碍了 DNA 疫苗的长远发展。同时，DNA 疫苗的安全问题也引起关注。值得注意的是用于人类的治疗性 DNA 疫苗比预防性 DNA 疫苗预期效果好。目前有两种获批的 DNA 疫苗，即大西洋鲑传染性造血器官坏死病疫苗和狗黑素瘤疫苗（Bergman 和 Wolchok，2008），另外还有一种用于预防秃鹰西尼罗病毒感染的 DNA 疫苗（Chang 等，2007）。第一个获批的马西尼罗病毒 DNA 疫苗已经停止销售。

一、亚单位疫苗

大肠杆菌（*Escherichia coli*）菌株被用作感受态细胞转入带有编码保护性抗原基因序列，经过发酵后，提纯获得保护性抗原。鱼类疫苗学中的一个经典例子是，应用大肠杆菌研发的大西洋鲑传染性胰腺坏死（IPN）亚单位疫苗于 1995 年首次获得批准。大肠杆菌使用方便，尤其是被广泛应用于分子生物学研究中，因此，许多研究实验室都掌握了正确表达转基因所需要的技术和工具。

重组疫苗也可以应用酵母菌（*Saccharomyces cerevisiae*）生产，例如智利的传染性鲑贫血症（ISA）亚单位疫苗。蚕、菜粉蝶幼虫、植物细胞和昆虫细胞等也可作为载体生产重组疫苗。目前，水产养殖中还没有基于这些基因重组方法的商业疫苗，但已开展多项实验研究，如菜粉蝶幼虫（*Trichoplusia ni*）生产 IPN 疫苗（Shivappa 等，2005），植物表达诺达病毒衣壳蛋白，大肠杆菌表达病毒性出血性败血症病毒 G 蛋白（Lorenzen 等，1993）。病毒性出血性败血症病毒也可在昆虫细胞中增殖（Lorenzen 和 Olesen，1995）。但重组表达的病毒性出血性败血症病毒 G 蛋白诱导免疫反应的效力尚未测定，重组蛋白诱导的免疫反应不稳定，产生的免疫保护水平较低（Lorenzen 等，1993）。

重组亚单位疫苗比传统灭活疫苗有优势。用载体生物生产保护性抗原，再从载体生物中获得抗原，经过下游多个步骤纯化以制备重组疫苗，若致病微生物难以培养，那么重组亚单位疫苗就更有其特殊的优势，因此重组亚单位疫苗适用于某些病毒类病原微生物。一般认为，重组亚单位疫苗安全性好，但其免疫原性比灭活的全细胞/病毒疫苗低（Webster 等，1977），因此，需要使用佐剂提高重组亚单位疫苗的免疫原性。

养殖鱼类可用的亚单位疫苗较少，只有 IPN 和 ISA 疫苗两个例子。对于高等脊椎动物，重组亚单位疫苗需要市场许可，因而数量有限。一个例子是 B 型肝炎病毒疫苗；还有一种是猫传染性白血病疫苗，是在金丝雀天花病毒中表达的传染性白血病毒（FeLV）蛋白 P45，已被许可上市。重组亚单位疫苗尽管在鱼类疫苗学是不可预知的，但有些难以在体外培养的病毒，如鱼呼肠孤病毒（引起大西洋鲑心脏和骨骼肌炎症）和鱼心肌炎病毒（引起大西洋鲑心肌综合征），重组亚单位疫苗可能是这些病毒疫苗的优先选择方法。

二、转基因疫苗

数十年来，基于体外无数次传代培养研制的减毒活疫苗，被用于免疫接种高等脊椎动物。传代培养会使病原体致病基因发生突变，而失去致病力，但机制尚不清楚。通过应用分子生物学技术移除/敲除病原微生物特定的基因或部分基因，使其毒力衰减，失去致病性。

转基因疫苗作为活疫苗免疫接种动物，在动物体内也可复制，但与野生型相比其复制水平较低（Vaughan 等，1993；Marsden 等，1996，1998）。因此，转基因疫苗能激活体液免疫和细胞免疫应答（Marsden 等，1998），也可用于诱导黏膜免疫反应。转基因疫苗的主要优点是能引起广泛的免疫反应，且通常可诱导中、高水平的免疫保护（Vivas 等，2004）。

细菌性减毒活疫苗可通过合理设计，改变或敲除致病菌中编码特定酶的基因而获得。这些酶可能是合成某些氨基酸所必须的，如，缺乏 *aroA* 基因的菌株不能产生芳香族氨基酸（Vaughan 等，1993）。减毒也可通过敲除或改变编码毒力基因实现，这对控制疾病发生极为重要。

遗传稳定性是评价减毒活疫苗安全性的关键。一般认为，应用传统（传代）方法减毒比分子生物学方法减毒研制的疫苗，毒力恢复或增加的风险更高，原因是传代减毒通常出现基因组的点突变，而分子生物学技术可以“敲除”整个基因。

鱼类减毒活疫苗的例子是敲除 *aroA* 基因（Marsden 等，1996）而降低毒力的杀鲑气单胞菌（*Aeromonas salmonicida*）疫苗。实验证明，该杀鲑气单胞菌突变株对预防疖疮病可产生短期的良好的保护作用（Marsden 等，1998）。由于现有的含佐剂的灭活杀鲑气单胞菌疫苗既安全，保护效果又较好，而且考虑到减毒疫苗免疫保护的持续性问题，减毒疫苗一直未进入实际应用（图 14-1）。

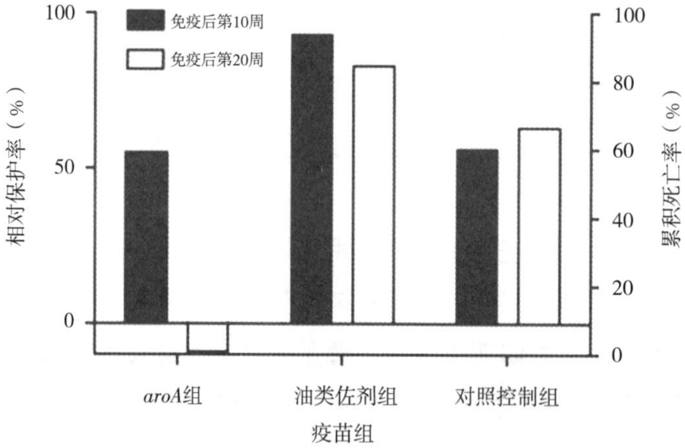

图 14-1　大西洋鲑免疫 *aroA* 缺失株疫苗和油类佐剂疫苗后第 10 和 20 周攻毒后的相对保护率（RPS）结果
免疫后 10 周 2 种疫苗 RPS 都很高，但在免疫后 20 周 *aroA* 免疫组出现负 RPS，而油类疫苗组 RPS 仍超过 80%
（引自 Evensen 2009，未公开发表）

反向遗传学是研发病毒性减毒疫苗较好的选择，但该方法并不适用于所有病毒。弹状病毒属的传染性造血器官坏死病毒（IHNV）和病毒性出血性败血症病毒（VHS），是通过敲除 *NV* 基因而使其失去致病性（Brémont，2005）。这样的 *NV* 基因缺失株分别免疫鱼类预防 IHN 和 VHS 病，在虹鳟和牙鲆中获得很高的免疫保护率。此外，还可以采用初次-加强免疫的策略提高疫苗的保护效果。还有研究显示，通过反向遗传学研制的鲑胰腺病病毒（SPDV）无毒突变株疫苗，对鳟昏睡病具有较好的免疫保护作用（Mérour 等，2013）。

三、载体疫苗

第三种类型的重组疫苗是将致病微生物的一个或多个致病因子或保护性抗原的基因转入无毒活微生物内。这种重组疫苗的载体可以是病毒或细菌。牛痘病毒最初用来根除人类的天花，后来被用作疫苗载体，因其基因组很大，所以可以接受外来基因，且特性好，无致病性。

载体疫苗可以刺激机体产生多种免疫反应，重组载体疫苗可以刺激体液免疫和细胞免疫应答，通常灭活疫苗是不能诱导这些免疫反应的。由非致病性的肠道细菌组成的口服疫苗可以定植于肠道上皮细胞中，同时产生各种微生物的重要抗原，从而能够刺激产生黏膜免疫保护。有些非保护性抗原促进了疾病的发生（如猫传染性腹膜炎和猫白血病病毒），载体疫苗可以排除免疫抑制抗原和非保护性抗原。

在有鳍鱼类中，载体疫苗的研究较少。鱼体接种载体疫苗后，也会产生针对载体或载体抗原的免疫反应。已经存在的抗载体病毒的抗体可以中和或抑制该载体，这会减弱机体对载体携带的外源抗原的免疫反应。载体疫苗是一种很好的疫苗类型，也适用于有鳍鱼类疫苗的研制。有研究表明，鲑胰腺病病毒结构蛋白复制子疫苗能产生对 ISA 的免疫保护作用（Wolf 等，2013）。

植物也可以作为载体使用，这意味着疫苗可以添加在饲料中，这对动物疫苗的应用极为有利。口服疫苗既可以减少注射器及其他设备的使用，也可以减少疫苗接种操作人员的培训需求。但是，口服疫苗产生的保护效果仍较差，这是植物载体疫苗研发的重要限制因素。

四、DNA 疫苗

DNA 疫苗已有单独一章介绍。很难预测 DNA 疫苗在水产养殖中推广使用的程度。已有研究表明，DNA 疫苗具有较好的预防许多病毒感染的效果（Biering 和 Salonius，2014），因此，像鲑鳟鱼类胰腺病的 DNA 疫苗一样，很可能会有更多的 DNA 疫苗推向市场。所以未来 DNA 疫苗将有可能成为水产养殖疫苗产品的重要组成部分。

DNA 疫苗的缺点主要是有关基因的整合，这是一个安全性问题，不仅关乎接种疫苗的动物的安全性，对食品安全也意义重大。产生的问题还有，接种疫苗的动物体内会产生抗 DNA 的抗体，这种抗体会产生自身免疫性疾病，但目前的研究表明，几乎不存在相关的担忧。

第三节　标记疫苗

在严格的监管制度下，传统的疾病控制措施一般采取扑杀已感染的群体。然而如果应用血清学方法，就区分不开接种过疫苗的或感染的动物（Vannier 等，2007），但接种重组疫苗的就可以与感染的动物区分开，这种疫苗被称为标记疫苗或 DIVA（区分感染和免疫动物）疫苗。标记疫苗通常是转基因疫苗（基因敲除疫苗），或者疫苗缺失某一抗原，而感染动物对其又产生免疫反应。上述四类重组疫苗均可作为标记疫苗使用，但目前为止，标记疫苗还未用于水产养殖。

第四节 黏膜疫苗

黏膜是许多病原体在机体全身分布的通道，因此，激发黏膜免疫反应，可达到在早期阻止感染的目的，是一种很好的策略。哺乳动物中 IgA 是黏膜免疫反应产生的抗体。在虹鳟中，IgT 被鉴定为黏膜抗体（Zhang 等，2010）。局部（如肠道）的疫苗接种如何诱导其他部位（如鳃）产生免疫保护效果还有待探究。一般来说，通过一次黏膜免疫很难获得良好的免疫保护效果。各种类型的疫苗测试发现，浸泡疫苗比口服疫苗能更好地激活黏膜免疫反应。通常灭活疫苗经口服免疫引起的机体免疫反应较差（Rombout 和 Viswanath，2014）。

目前，由于使用含佐剂的注射疫苗产生副作用且越来越引起重视，另外还存在注射或者经黏膜表面接种需要加强免疫等问题，未来对口服疫苗或黏膜疫苗的需求将会越来越多。

第五节 寄生虫疫苗

寄生虫对所有生物产品的生产均可造成巨大损失，而疫苗对预防寄生虫感染效果甚微（Vercruysse 等，2007）。部分的原因是寄生虫药物治疗的效果好，还有一部分的原因是很难获得较好的抗寄生虫感染的保护效果。

寄生虫疫苗研发主要有两种方法：一是使用活的、减毒的寄生虫；二是亚单位疫苗，其为特异性抗原，由重组技术将寄生虫的内源性蛋白质或酶克隆而来，也称为隐性抗原疫苗。活的、减毒的寄生虫疫苗不适用于养殖鱼类。众所周知的隐性抗原疫苗例子是预防牛虱（微小牛蜱 *Boophilus microplus*）病疫苗。疫苗是以 Bm86 蛋白作为抗原（Willadsen 等，1992），该蛋白是寄生虫肠道细胞膜的组成部分，宿主不会因为接触该蛋白而感染牛虱病。通过重组技术研制出 Bm86 抗原蛋白的亚单位疫苗免疫牛，当微小牛蜱吸牛血时，牛血液中循环的抗体将与牛蜱的肠道内上皮相关细胞的蛋白结合，从而破坏牛蜱的肠道功能。这种疫苗减少了经济损失，同时也降低了由寄生虫传播其他疾病的风险。

类似的方法已被用于研制预防鲑鳟鱼类海虱［鲑疮痂鱼虱（*Lepeophtheirus salmonis*）和智利鱼虱（*Caligus rogercrossey*）］病疫苗。初步研究表明，与未接种疫苗的鲑相比，接种疫苗的鱼体表附着阶段的海虱（附着幼体）数量减少了 40%。到目前，以上的研究还没有大规模应用于养殖生产中。此外，寄生虫疫苗还面临很多挑战：首先，鉴定和生产合适的抗原有难度。其次，附着阶段的海虱摄吸鱼血的程度仍有争议，如果海虱没有摄吸血液，那么循环系统中的抗体就应输送到皮肤黏液或表面才能起作用。最近发现，鱼类也有黏膜免疫球蛋白（IgT），表明黏膜诱导产生的免疫反应将有利于抗寄生虫的免疫应答，这会是黏膜疫苗研制和应用的发展前景（Zhang 等，2010）。

第六节 繁殖控制疫苗

养殖的鲑可能会从养殖区域中逃离到自然水域（或称逃亡鱼），这会对溯河洄游的当地大西洋鲑种群构成威胁，为此探讨通过接种疫苗控制逃亡鱼的性成熟。通过人工合成促性腺激素释放激素（GnRH）制作疫苗，免疫接种后刺激鱼体产生抗 GnRH 的抗体，实现控制性成熟的目的（Delves 和 Roitt，2005）。由于抗原的分子量较低，一般需要连接载体蛋白，同时使用佐剂以提高抗 GnRH 的抗体水平。另外，也可用下丘脑和垂体的黄体生成素（LH）为抗原制作疫苗。这种内分泌调控可能影响生殖，但也有可能影响其他内分泌功能。

第七节　疫苗配方改良

新型佐剂和载体分子的相关领域研究已经取得了长足的进展，所面临的挑战主要是增强黏膜和皮肤的免疫。难度在于如何才能保护抗原在胃肠道前部不被降解，确保抗原在淋巴组织附近释放，这可通过以生物可降解聚合物制成的微胶囊包被抗原来完成。近十年来，鱼类疫苗领域最重要的进步是注射型油类疫苗中的佐剂使用量减少。目前，油类疫苗的通常注射剂量是 50μL，而最初油类疫苗的注射剂量是现在的 4 倍多，因此，近年来佐剂的注射量减少得极为显著。油类佐剂注射剂量的减少有助于改善大西洋鲑免疫的副作用，尤其在挪威和智利。对比一下，目前鱼类疫苗的注射剂量为 1.0～1.25μL/g（体重为 40～50g 鱼），而人类的注射剂量仅为 0.1μL/g（10kg 体重的小孩注射 1mL）。

第八节　免疫调节

细胞因子是免疫系统中的淋巴细胞、巨噬细胞和其他细胞等多种细胞分泌的蛋白质，有的细胞因子参与调节免疫反应的强度和特性或范围。细胞因子也会使免疫反应发生偏离，协助免疫效应细胞发挥作用。辅助性 T 细胞（Th）通过分泌各种细胞因子诱导产生不同类型的免疫应答，因而在免疫系统中起重要作用，主要有 Th1、Th2 和 Th17 细胞（Mutoloki 等，2014）。利用分子技术体外生产细胞因子可作为免疫调节物或刺激物，然而，将细胞因子作为疫苗免疫调节剂，必须解决几个重要的问题。首先，细胞因子是在精细的信号转导网络中发挥作用，过度表达对宿主有害，反而会促使疾病发生而不会预防疾病；此外，细胞因子不稳定，且半衰期短，添加细胞因子很难实现长效作用。

在过去十年间，开展了有关病原的各种细胞受体的研究，主要有 Toll 样受体，以及各种双链核糖核酸（dsRNA）化合物的细胞受体和三磷酸核糖核酸（RNA）传感器的胞质受体。这些受体的配体，包括 dsRNA（如 poly I:C）、CpG 以及其他参与这些受体/传感器系统的合成配体，它们可作为疫苗免疫调节剂使用。这些受体/感受器分子在低等脊椎动物中是非常保守的，已经在细胞水平和个体水平开展了免疫刺激的相关研究（Jørgensen，2014）。推测，很可能在细胞因子用作免疫佐剂之前，Toll 样受体配体在疫苗中已作为免疫调节剂使用。这两种免疫调节剂，特别是针对灭活疫苗，都有可能诱导和改变抗原的免疫应答。

第九节　细胞因子和组织损伤相关分子模式（DAMPs）佐剂

近些年，已在多种鱼类中鉴定出许多种细胞因子基因，尽管哺乳动物中对细胞因子的佐剂效应进行了深入的研究，但很少有关注细胞因子作为鱼类疫苗佐剂的研究。主要原因可能是大多数鱼类细胞因子的免疫调节作用机制还未知晓，在了解这些细胞因子在免疫过程中的调控作用之前，只能是反复试验。不管怎样，目前已经开展了某些鱼类的细胞因子作为疫苗佐剂的探索性研究。

有研究表明，虹鳟腹腔注射白介素（IL-1β）-衍生肽后 2d 再感染 VHSV，死亡率降低（Peddie 等，2003）。腹腔注射 IL-1β-衍生肽后 1～3d，可以诱导白细胞迁移进入腹腔，但并未进一步探讨该多肽用作佐剂的可能性。鲤中研究了 IL-1β 的佐剂效果，在添加或未添加 IL-1β 重组 C-端多肽的条件下，鲤腹腔注射灭活的嗜水气单胞菌（*A. hydrophila*）。免疫后 3 周，与仅免疫灭活疫苗的鲤相比，注射添加 IL-1β 重组多肽组鱼体内凝集抗体效价明显升高（Yin 和 Kwang，2000）。组织损伤相关分子模式 *Hsp*70（热休克蛋白 70）基因插入编码海豚链球菌（*Streptococcus inia*）抗原的 DNA 疫苗中，该疫苗在攻毒实验中显示出了良好的免疫保护效果（Hu 等，2012）。

在哺乳动物中，白细胞介素-8（IL-8）是 CXC 型趋化因子（α 趋化因子），在病原感染或 IL-1β、肿瘤坏死因子 α（TNF-α）等细胞因子的刺激下，由巨噬细胞、单核细胞、上皮细胞、中性粒细胞和成纤维细胞等多种细胞产生。趋化因子不仅能招募更多的细胞进入炎症部位，而且还能调节招募细胞的免疫功能，因此在哺乳动物中被广泛用作病毒疫苗的试验佐剂。在虹鳟及其他鱼类中，IL-8 已经被鉴定出来，并证实了其趋化作用（Harun 等，2008）。将编码 VHSV 糖蛋白基因的质粒疫苗和编码 IL-8 的质粒共同注射入虹鳟体内，研究 IL-8 的佐剂效应（Jimenez 等，2006；Sanchez 等，2007），结果表明，编码 IL-8 质粒（$pIL\text{-}8^{+}$）与 VHSV 疫苗同时免疫，虹鳟脾脏中的 IL-1β 表达显著增加，并且接种部位细胞浸润明显增强；此外，单独注射 $pIL\text{-}8^{+}$，虹鳟脾脏中 TNF-α、IL-11 和 TGF-β 的表达量显著上升（Jimenez 等，2006）。进一步研究 VHSV DNA 疫苗和/或 $pIL8^{+}$ 免疫后虹鳟 CC 趋化因子的表达情况，发现 IL-8 被用作佐剂时，CK5A、CK6、CK7 和 CK5B 等趋化因子基因的表达也随之变化（Sanchez 等，2007）。以上结果表明，IL-8 能够调节机体的早期免疫应答反应，并可能成为鱼类病毒疫苗的有效佐剂。

第十节 结　语

在本章中没有涉及疫苗研发和依据宿主基因组测序评估宿主对疫苗接种的反应。几种模式鱼类已完成了全基因组测序，鲑的基因组也即将公布，特别是对免疫或自然感染后保护性免疫反应的精准辨别，这将是鱼类疫苗学的新时代。随着高通量测序等新工具的出现，加上转录组和蛋白质组水平的强大分析能力，很有可能在未来十年，在探索鱼类对疫苗接种的免疫应答机理方面的认知将向前迈出一大步。

参考文献

Bergman，P. J. and Wolchok，J. D. 2008. Of Mice and Men（and Dogs）：development of a xenogeneic DNA vaccine for canine oral malignant melanoma. *Cancer Ther* 6：817-826.

Biering，E. and Salonius，K. 2014. DNA vaccines，in *Fish Vaccination*（eds R. Gudding，A. Lillehaug and

Ø. Evensen). Oxford, John Wiley & Sons, Ltd., 47-55.

Brémont, M. 2005. Reverse genetics on fish rhabdoviruses: tools to study the pathogenesis of fish rhabdoviruses. *Curr Top Microbiol Immunol* 292: 119-141.

Chang, G. J., Davis, B. S., Stringfield, C. and Lutz, C. 2007. Prospective immunization of the endangered California condors (*Gymnogyps californianus*) protects this species from lethal West Nile virus infection. *Vaccine* 25: 2325-2330.

Delves, P. J. and Roitt, I. M. 2005. Vaccines for the control of reproduction—status in mammals, and aspects of comparative interest. *Dev Biol* (*Basel*) 121: 265-273.

Harun, N. O., Zou, J., Zhang, Y. A., *et al.*, 2008. The biological effects of rainbow trout (*Oncorhynchus mykiss*) recombinant interleukin-8. *Dev Comp Immunol* 32: 673-681.

Hu, Y. H., Dang, W., Zhang, M. and Sun, L. 2012. Japanese flounder (*Paralichthys olivaceus*) Hsp70: adjuvant effect and its dependence on the intrinsic ATPase activity. *Fish Shellfish Immunol* 33: 829-834.

Jimenez, N., Coll, J., Salguero, B. F. J. and Tafalla, C. 2006. Co-injection of interleukin 8 with the glycoprotein gene from viral haemorrhagic septicemia virus (VHSV) modulates the cytokine response in rainbow trout (*Oncorhynchus mykiss*). *Vaccine* 24: 5615-5626.

Jørgensen, J. 2014. The innate immune response in fish, in *Fish Vaccination* (eds R. Gudding, A. Lillehaug and Ø. Evensen). Oxford, John Wiley & Sons, Ltd., 85-103

Lorenzen, N. and Olesen, N. J. 1995. Multiplication of VHS virus in insect cells. *Vet Res* 26: 428-432.

Lorenzen, N., Olesen, N. J., Vestergård-Jørgensen, P. E., *et al.*, 1993. Molecular cloning and expression in *Escherichia coli* of the glycoprotein gene of VHS virus, and immunization of rainbow trout with the recombinant protein. *J Gen Virol* 74: 623-630.

Marsden MJ, Vaughan LM, Fitzpatrick RM, *et al.*, 1998. Potency testing of a live, genetically attenuated vaccine for salmonids. *Vaccine* 16: 1087-1094.

Marsden, M. J., Vaughan, L. M., Foster, T. J. and Secombes, C. J. 1996. A live (*ΔaroA*) *Aeromonas salmonicida* vaccine for furunculosis preferentially stimulates T-cell responses relative to B-cell responses in rainbow trout (*Oncorhynchus mykiss*). *Infect Immun* 64: 3863-3869.

Mérour, E., Lamoureux, A., Bernard, J., *et al.*, 2013. A fully attenuated recombinant salmonid alphavirus becomes pathogenic through a single amino acid change in the E2 glycoprotein. *J Virol* 87: 6027-6030.

Mutoloki, S., Jørgensen, J. and Evensen, Ø. 2014. The adaptive immune response in fish, in *Fish Vaccination* (eds R. Gudding, A. Lillehaug and Ø. Evensen). Oxford, John Wiley & Sons, Ltd., 85-103.

Peddie, S., McLauchlan, P. E., Ellis, A. E. and Secombes, C. J. 2003. Effect of intraperitoneally administered IL-1 beta-derived peptides on resistance to viral haemorrhagic septicaemia in rainbow trout *Oncorhynchus mykiss*. *Dis Aquat Org* 56: 195-200.

Rombout, J. H. W. M. and Kiron, V. 2014. Mucosal vaccination of fish, in *Fish Vaccination* (eds R. Gudding, A. Lillehaug and Ø. Evensen). Oxford, John Wiley & Sons, Ltd., 56-67.

Sanchez, E., Coll, J. and Tafalla, C. 2007. Expression of inducible CC chemokines in rainbow trout (*Oncorhynchus mykiss*) in response to a viral haemorrhagic septicemia virus (VHSV) DNA vaccine and interleukin 8. *Dev Comp Immunol* 31: 916-926.

Shivappa, R. B., McAllister, P. E., Edwards, G. H., *et al.*, 2005. Development of a subunit vaccine for infectious pancreatic necrosis virus using a baculovirus insect/larvae system. *Dev Biol* (*Basel*) 121: 165-174.

Vannier, P., Capua, I., Le Potier, M. F., *et al.*, 2007. Marker vaccines and the impact of their use on diagnosis and prophylactic measures. *Rev Sci Tech* 26：351-372.

Vaughan，L. M.，Smith，P. R. and Foster，T. J. 1993. An aromatic-dependent mutant of the fish pathogen *Aeromonas salmonicida* is attenuated in fish and is effective as a live vaccine against the salmonid disease furunculosis. *Infect Immun* 61：2172-2181.

Vercruysse，J.，Schetters，T. P. M.，Knox，D. P.，*et al.*，2007. Control of parasitic disease using vaccines：an answer to drug resistance? *Rev Sci Tech* 26：105-115.

Vivas，J.，Riaño，J.，Carracedo，B.，*et al.*，2004. The auxotrophic *aroA* mutant of *Aeromonas hydrophila* as a live attenuated vaccine against *A. salmonicida* infections in rainbow trout（*Oncorhynchus mykiss*）. *Fish Shellfish Immunol* 16：193-206.

Webster，R. G.，Glezen，P.，Hannoun，C. and Laver，W. G. 1977. Potentiation of the immune response to influenza virus subunit vaccines. *J Immunol* 119：2073-2077.

Willadsen，P.，Kemp，D. H.，Cobon，G. S. and Wright，I. G. 1992. Successful vaccination against *Boophilus microplus* and *Babesia bovis* using recombinant antigens. *Mem Inst Oswaldo Cruz* 87（Suppl 3）：289-294.

Wolf，A.，Hodneland，K.，Frost，P.，*et al.*，2013. A hemagglutinin-esterase-expressing salmonid alphavirus replicon protects Atlantic salmon（*Salmo salar*）against infectious salmon anemia（ISA）. *Vaccine* 31：661-669.

Yin，Z. and Kwang，J. 2000. Carp interleukin-1β in the role of an immuno-adjuvant. *Fish Shellfish Immunol* 10：375-378.

Zhang，Y. A.，Salinas，I.，Li，J.，*et al.*，2010. IgT，a primitive immunoglobulin class specialized in mucosal immunity. *Nat Immunol* 11：827-835.

第十五章　弧菌病的疫苗接种

Duncan J. Colquhoun and Atle Lillehaug

本章概要　SUMMARY

经典弧菌病的疫苗接种是一个成功的案例，它激发了人们研发鱼类其他疾病的疫苗。弧菌病的暴发可能发生在世界各地生活在海水中的许多鱼种上。弧菌病（vibriosis）除了鳗弧菌（*Vibrio anguillarum*）外，还包括病海鱼弧菌（*V. ordalii*）、灿烂弧菌（*V. splendidus*）、创伤弧菌（*V. vulnificus*）、杀鲑弧菌（*V. salmonicida*）和粘摩替亚菌（*Moritella viscosa*）的感染。一般来说，弧菌病疫苗是灭活的全菌菌苗，脂多糖（LPS）是全菌疫苗的主要抗原成分。这些疫苗通过浸泡/浸浴或腹腔注射的方式进行免疫，两种接种方式均可产生显著的保护水平。在各种鉴定的弧菌病血清型中，目前只有血清型 O1、O2α 和 O2β 有商业疫苗。也有杀鲑弧菌和粘摩替亚菌的商业疫苗，其他血清型鳗弧菌和其他弧菌的疫苗多为自体疫苗。

第一节　弧菌病

在现代水产养殖中，弧菌病的预防接种是疾病预防发展中最全面的成功代表之一。这一成功与宿主的广泛分布以及相关病原体的广泛地域分布有关，但最特别的是它与开发的疫苗产品的有效性有关。弧菌病疫苗研究开展了鱼类疫苗学领域最早的一些疫苗实验。

与疫苗学有关的“弧菌病”一词，一般是指由鳗弧菌（*Vibrio anguillarum*）和病海鱼弧菌（*V. ordalii*）引起的感染。然而，由其他弧菌以及已经被归为其他属的密切相关细菌引起的鱼类疾病也包括在这一范畴中，包括由杀鲑弧菌（*V. salmonicida*）引起的冷水弧菌病、由粘摩替亚菌（*Moritella viscosa*）引起的冬季溃疡以及由灿烂弧菌（*V. splendidus*）、创伤弧菌（*V. vulnificus*）等引起的感染。

第二节 发生流行与危害

弧菌病［鳗弧菌（*V. anguillarum*）和病海鱼弧菌（*V. ordalii*）］是一种世界范围内影响多种鱼类的疾病，感染对象包括野生和养殖的鱼类。弧菌属海洋微生物，且疾病的暴发只发生在海水和半咸水里，通常是在温带地区的夏季。患该疾病的鱼体表现通常是出血性败血症，并且可能出现皮肤溃疡和脓肿。在野生鱼类中，浅水种类如绿鳕（Egidius，1987）和胡瓜鱼（Yignisis 等，2007）中报道过疾病暴发。弧菌病会造成海水养殖的毁灭性损失。海洋鱼类和大西洋鲑以及太平洋鲑都有感染风险，虹鳟尤其敏感（Lillehaug 等，2003）。在一些水产养殖种类中已报道有弧菌病的暴发，包括亚洲香鱼（*Plecoglossus altivelis*）和黄鲕（Sakata 等，1989），温水性种类如欧洲鲈（*Dicentrarchus labrax*）、海鲷（*Sparus aurata*）和大菱鲆（*Scopththalmus maximus*）（Larsen 等，1994）以及冷水性种类如大西洋鳕（*Gadus morhua*）和大比目鱼（Lillehaug 等，2003）。

在易感鱼类和地理分布上，冷水弧菌病［杀鲑弧菌（*V. salmonicida*）］比经典弧菌病的传播范围要小得多。这种疾病几乎只感染大西洋鲑，尽管偶尔的情况下，会感染人工养殖的虹鳟和大西洋鳕。在挪威，杀鲑弧菌已经成为一种危害极其严重的鱼类病原体，并且在 20 世纪 80 年代，冷水弧菌病是最常见的疾病，因此抗生素被广泛使用。在杀鲑弧菌流行的高峰期之前（1977—1984 年），杀鲑弧菌造成的相关损失估计已经超过挪威水产养殖业的疾病相关费用的 80%（Poppe 等，1985）。自从 1987 年引进有效的疫苗，挪威每年的发病情况已经从 1987 年的 200 多起疫情急剧下降到 20 世纪 90 年代末的几乎为零。除了挪威，英国的苏格兰（Bruno 等，1985）、冰岛、丹麦的法罗群岛（Dalsgaard 等，1988），以及美国和加拿大（O’Halloran 和 Henry，1993）都报道了冷水弧菌病。

冬季溃疡病是一种与粘摩替亚菌（*Moritella viscosa*）有关的疾病。与冷水弧菌病一样，这种疾病主要分布在北大西洋，而且在冬季最常观察到。粘摩替亚菌是挪威和冰岛鱼类养殖中危害严重的病原，主要感染海水养殖的大西洋鲑（Lunder 等，1995），以及少量的海产虹鳟（Lillehaug 等，2003）。鱼类皮肤的血液循环障碍与机械受伤可能导致诱发了粘摩替亚菌的感染。尽管粘摩替亚菌的疫苗已广泛接种，但在本书写作期间，这种疾病仍然是挪威水产养殖中造成的经济损失最严重的细菌性疾病。患病鱼体长时间出现溃疡，如果水温升高，鱼体可能会恢复正常。除了直接死亡，在很大程度上，经济损失与最终鱼片产品的降级有关，因为这些鱼片有溃疡/瘢痕。

在鱼类致病性弧菌中，只有创伤弧菌（*V. vulnificus*）才会在人类身上引起严重的疾病（Austin，2010）。然而，人畜共患方面强调了在水生诊断细菌学中科学能力的重要性。创伤弧菌有三种鉴定的生物型，其中生物 1 型和 3 型通常会导致人类的疾病（Colodner 等，2004），而生物 2 型（血清型 A 和 E）几乎是（但不完全）仅与鱼类疾病有关（Fouz 等，2010）。创伤弧菌最常在人工养殖的鳗中导致疾病（Amaro 等，1992；Dalsgaard 等，1999），其他海洋鱼类也会受到影响（Li 等，2006）。这种细菌似乎在暖温带的海洋环境中和温带地区的海洋环境中无处不在，而且在远东地区和西欧，鳗的疾病问题也有相当多的报道。

弧菌其余重要的病原体还包括以下种类：哈维氏弧菌（*V. harveyi*），其与海洋有鳍鱼类和甲壳类动物死亡率的相关性越来越高（Robertson 等，1998）；溶藻弧菌（*V. alginolyticus*）（Austin 等，1993），灿烂弧菌（*V. splendidus*）（Thomson 等，2005）以及蛤弧菌（*V. tapetis*）（Bergh 和 Samuelsen，2007）。应该承认，和鳗弧菌一样，这些物种也是水生微生物群中无处不在的成员，但只有少数菌株可能是致病的。

第三节 病原学

弧菌属是革兰氏阴性、弯曲或直的棒状细菌，具有发酵作用，通常运动活泼，但不全是运动型的，有附着的极生鞭毛（Farmer 等，2005）。弧菌属可能在淡水中发现（Bockemuhl 等，1986；Ivanova 等，2001），但其通常生活于海水和咸水环境。虽然已在淡水环境中发现各种弧菌属引起的疾病（Giorgetti 和 Ceschia，1982；Fouz 和 Amaro，2003；Kitao 等，2006），但最常见的还是在海水和咸水环境中。一些弧菌是水生生物的条件致病菌，其导致的水生动物全身感染被统称为弧菌病。因此，严格地说，弧菌病的所有病原体都应该属于弧菌属（*Vibrio*）、弧菌科（Vibrionaceae）、弧菌目（Vibrionales）、丙型变形菌纲（Gammaproteobacteria）。然而，细菌命名是一个动态的领域，导致鱼类系统性疾病的细菌种类，以前被归为弧菌属的成员，现在已经被转移到弧菌科的其他属中。例如粘弧菌（*Vibrio viscosus*），现在叫粘摩替亚菌（*Moritella viscosa*）（Benediktsdottir 等，2000）；杀鲑弧菌（*Vibrio salmonicida*）学名现在叫 *Allivibrio salmonicida*（Urbanczyk 等，2007），后者的命名变化没有被科学界普遍接受，是由于与另一种重要的鱼类病原体——杀鲑气单胞菌（*Aeromonas salmonicida*），有混淆的可能性，因为两个物种的学名缩写都是 *A. salmonicida*。因此，本章节的“弧菌病”一词包括目前隶属于弧菌属的一些细菌种类引起的疾病和一种隶属于 Moritellaceae 科的细菌引起的疾病（表 15-1）。

表 15-1 鱼类弧菌病有关的主要细菌

细菌种类	主要宿主种类	相关文献
鳗弧菌（*Vibrio anguillarum*）	种类多样*	Austin 等（1995） Grisez 和 Ollevier（1995） Kitao 等（1984） Larsen 等（1988） Larsen 等（1994） Pedersen 和 Larsen（1995） Pedersen 等（1999） Silva-Rubio 等（2008） Song 等（1988） Sørensen 和 Larsen（1986） Toranzo 和 Barja（1990）
病海鱼弧菌（*Vibrio ordalii*）	太平洋鲑（*Oncorhynchus* spp） 香鱼（*Plectoglossus altivelis*） 大西洋鲑（*Salmo salar*） 大西洋鳕（*Gadus morhua*）	Colquhoun 等（2004） Muroga 等（1986） Schiewe 和 Crosa（1981） Schiewe 等（1981）

（续）

细菌种类	主要宿主种类	相关文献
哈维氏弧菌（*Vibrio harveyi*）	多种海水鱼类和虾类	Austin 和 Zhang（2006）
灿烂弧菌（*Vibrio splendidus*）	娇扁隆头鱼（*Symphodus melops*）	Jensen 等（2003）
蛤弧菌（*Vibrio tapetis*）	娇扁隆头鱼（*Symphodus melops*）	Jensen 等（2003）
创伤弧菌（*Vibrio vulnificus*）	欧洲鳗鲡（*Anguilla anguilla*）	Amaro 和 Biosca（1996）
溶藻弧菌（*Vibrio alginolyticus*）	种类多样（暖水性）	Balebona 等（1998） Colorni 等（1981） Kusada 等（1986） Lee（1995）
杀鲑弧菌［*Vibrio*（*Aliivibrio*）*salmonicida*］	大西洋鲑（*Salmo salar*） 虹鳟（*Oncorhynchus mykiss*） 大西洋鳕（*Gadus morhua*）	Lillehaug（1990） Lillehaug 等（1990） Schrøder 等（1992）
粘摩替亚菌［*Moritella viscosa*（*V. viscosus*）］	大西洋鲑（*Salmo salar*） 虹鳟（*Oncorhynchus mykiss*）	Lunder 等（1995） Grove 等（2010）

注：* 血清型间宿主特异性的程度已被鉴定。

虽然“弧菌病”这个词可用于描述由众多的或多或少与弧菌相关的种类引起的疾病（表 15-1），但对许多人来说，“经典弧菌病”这个词是专指由鳗弧菌和病海鱼弧菌引起的感染。鳗弧菌以前的“灰色”分类学位置最近得到了解决，将这个细菌从利斯顿氏菌属（*Listonella*）回归到弧菌属（*Vibrio*）（Thompson 等，2011）。鳗弧菌是世界性分布的，它是欧洲、亚洲、美洲和大洋洲（Munday 等，1992；Austin 和 Austin，2007；Silva-Rubio 等，2008）很多不同鱼类（Austin 和 Austin，2007）的弧菌病病原体。最初为鳗弧菌建立的几个血清型方案被统一于一个定义 23 种血清型的单一系统（Pedersen 等，1999）。此外，通常发现的与已知血清型不相应的菌株主要来自环境。到目前为止，大部分的鱼类病原鳗弧菌分离株属于血清型 O1 和 O2（包括各种亚型），尽管血清型 O3 也与多种鱼类的疾病相关，包括黑鲈、香鱼、鳗鲡（Tiainen 等，1997）以及智利的鲑（Silva-Rubio 等，2008）。鳗弧菌某一血清型内也存在一定程度的宿主-种特异性的现象（Colquhoun，未发表结果）。在水产养殖引进的新鱼类中可能发现新的血清/基因型（Mikkelsen 等，2011），可以预期，已知致病弧菌属的多样性会随着水产养殖的扩大和多样化而增加。

在其作为一个独立物种的描述之前（Schiewe 等，1981），病海鱼弧菌最初被描述为鳗弧菌生物 2 型（Schiewe 和 Crosa，1981），实际上，病海鱼弧菌和鳗弧菌是高度相关的，虽然病海鱼弧菌拥有一个相对小一些的基因组（Naka 等，2011）。较小的基因组表明，病海鱼弧菌可能正在经历基因组退化，这与从环境到一种专有细胞内生活方式的进化过程是一致的（Naka 等，2011）。而病海鱼弧菌已确认也是弧菌病的一种病原，其报道的分布范围比鳗弧菌更受限制。病海鱼弧菌已经被报道为北美、南美以及日本和澳大利亚的鱼类病原体（Schiewe 和 Crosa，1981；Wards 等，1991；Colquhoun 等，2004），最近挪威也有报道（http：//www. vetinst. no/eng/Publications/Fish-Health-Report/The-Farmed-Fish-HealthReport-2005）。

第四节　发病机理

与所有传染性疾病一样，弧菌病是病原体、宿主和环境间相互作用的结果。水温无疑是与弧菌病暴发有关的主要因素之一，在温带区域的夏季，大多数野生的和养殖的鱼类的弧菌病发生与相对较高的水温有关。毋庸置疑，高水温与细菌毒性和活性的增加有关，也与感染鱼种群的免疫力降低以及鱼体应激升高有关。某些情况下，鱼类分级或免疫鱼过程中的鱼类处理，也可能会打破平衡倾向于病原体，导致疾病的暴发。高水温导致细菌毒性增加的规律要排除由杀鲑弧菌（*V. salmonicida*）和粘摩替亚菌（*M. viscosa*）导致的冷水弧菌病和冬季溃疡病。这些细菌与低温有关，高于 10℃ 时很少引起疾病（Colquhoun 等，2002；Björnsdóttir 等，2011）。杀鲑弧菌在低温下，其铁螯合作用大大增加，温度高于 10℃，很少或没有含铁细胞的产生（Colquhoun 和 Sørum，2001）。毫无疑问，这种关系的原因是以宿主/病原体平衡为基础，低温降低了宿主的免疫活性。

通常，弧菌病被描述为扩散性的全身感染，其肉眼可见，非特异性的外部体征可能包括坏死性病变以及鱼体上不同程度的皮肤出血，有或者没有眼球突出症。在鱼体内部，经常可能观察到广泛的炎症、出血和腹水。由于弧菌病通常表现为和其他细菌种类不同的特征，因此许多鱼类致病性的弧菌很有可能共享许多基本的毒理机制，这些与更具体的特征一起可能会产生强毒的表型。种属特异性的发病机制无疑是多种因素之间复杂的相互作用，这些因素中很少是毒性绝对需要的。胞外毒素，如蛋白酶（Denkin 和 Nelson，2004；Varina 等，2008）和溶血素，无疑在弧菌病感染中起着重要作用，如肌肉注射细菌胞外产物导致病灶的形成，这与在自然感染期间观察到的病理变化一致。

尽管鳗弧菌（*V. anguillarum*）有许多潜在的毒力因子，但是这些毒力因子的特性尚不清楚（Naka 等，2011）。有人对这种细菌的毒力因子进行了综述（Frans 等，2011），其中包括转运鞭毛系统（Milton 等，1996；Ormonde 等，2000）。Karlsen 等（2008）证明，杀鲑弧菌的鞭毛在海水中很活跃，但细菌在相当于宿主体内生理水平的盐度下，其鞭毛却是不活跃的，这表明鞭毛在细菌定植中起作用。鳗弧菌和溶藻弧菌（*V. alginolyticus*）对鱼类的黏液具有高度趋化作用，特别是在鱼类的肠内黏液中（O'Toole 等，1996；Bordas 等，1998），并且 Knudsen 等（1999）证明鳗弧菌和杀鲑弧菌能与鲑黏液进行血清型特异性结合。Denkin 和 Nelson（1999）研究表明，鲑黏液诱导鳗弧菌蛋白酶的产生。人们已经在各种弧菌种类中鉴定出和毒力有关的质粒（Crosa 等，1977）以及染色体结合的铁螯合系统（Lemos 等，1988）。杀鲑弧菌中也鉴定出温度依赖性的含铁细胞介导的铁螯合（Colquhoun 和 Sørum，2001）。对于许多种类的细菌来说，包括弧菌，形成生物膜的能力是很常见的，并且生物膜可能在细菌定植和持久性中发挥作用（Marco-Noales 等，2001）。鳗弧菌（Croxatto 等，2002；Wang 等，2002）、病海鱼弧菌（*V. ordalii*）（Naka 等，2011）以及溶藻弧菌（Chen 等，2012）中已经鉴定出生物膜相关基因。研究表明，创伤弧菌（*Vibrio vulnificus*）在感染的鳗鲡皮肤表面建立了生物膜，并且有关研究还描述了虾病相关的各种虾体表面的哈维氏弧菌（*Vibrio harveyi*）生物膜（Karunasagar 等，1996）。Tunsjø 等（2009）认为，温度相关的附着可能是粘摩替亚菌（*Moritella viscosa*）

的一种毒力因子，而这位作者也提出鞭毛在细菌定植中的作用，并且鞭毛参与假定的组织坏死因子和假定的蛋白酶导致的组织损伤（Tunsjø 等，2011）。

由此可见，鱼类致病性弧菌能够迅速侵入组织。鱼类的鳃被认定为粘摩替亚菌的入侵门户（Løvoll 等，2009），而杀鲑弧菌在浸浴攻毒仅 2h 后在鱼类血液中被识别（Bjelland 等，2012）。鳗弧菌是在皮肤（Kanno 等，1990）和后肠（Olsson 等，1996）中定植，随后入侵到组织中。

第五节 疫 苗

大多数情况下，弧菌病疫苗都是灭活的全菌疫苗，用于浸泡或腹腔注射。许多疫苗都可以按照这两种方式进行免疫接种。近来，虽然油类佐剂注射弧菌疫苗已经成功用于海洋鱼类，这类佐剂配方疫苗比非佐剂配方的疫苗的保护水平更高、更持久，但大多数可注射的弧菌疫苗是作为多组分疫苗的一部分进行注射，这些多组分疫苗不含佐剂或含有油类佐剂。目前在可用的商品化疫苗市场中，最多组分的许可疫苗似乎是用于挪威的大西洋鲑疫苗，含有 7 个抗原成分，其中有 5 个细菌性组分，包括鳗弧菌（*V. anguillarum*）血清型 O1 和 O2a、杀鲑弧菌（*V. salmonicida*）、粘摩替亚菌（*Moritella viscosa*）和杀鲑气单胞菌（*Aeromonas salmonicida*）。油类佐剂疫苗提高了保护水平，这可能与储存效应增加和油滴中抗原缓慢释放相结合以及油类本身普遍具有的免疫刺激有关。

虽然商业疫苗配方很保密，但大多数商品可用的弧菌疫苗是基于灭活的（无论是甲醛灭活还是热杀灭）肉汤培养的全细菌细胞。毫无疑问，制造商根据养分有效性和培养温度改进培养条件，优化相关抗原的生产。

脂多糖（LPS）绝对是鱼类致病弧菌疫苗中最重要的抗原成分之一，对于任何一种弧菌疫苗而言，一个重要特征是选择 LPS 相关的抗原（即相关的细菌种类融合以及疫苗配方内的血清型）。口服来自病海鱼弧菌（*V. ordalii*）纯化的 LPS 100 pg 或浸泡于 500 ng/ mL LPS（Velji 等，1990）能使银大麻哈鱼免受感染。毫无疑问，弧菌有一定程度的交叉保护，这种交叉保护可能是相当广泛的，例如，杀鲑弧菌疫苗似乎能对杀鲑气单胞菌提供一定程度的保护（Hoel 等，1998）。因此，期望通过对细菌及其血清型进行深入研究，最大限度地认知疫苗的免疫交叉效果。

目前，世界范围内销售的大多数商业弧菌疫苗都是针对由鳗弧菌或病海鱼弧菌引起的疾病。尽管有很多鳗弧菌血清型被鉴定出来，但只有血清型 O1、O2a（O2α）和 O2b（O2β）似乎被认为是危害严重的病原体，需要以商业规模生产这些疫苗。世界范围内养殖的大西洋鲑案例中，接种疫苗的策略似乎是高度可变的，一些国家（如挪威）为所有的海产大西洋鲑接种含有血清型 O1 和 O2a 的疫苗，而大多数其他国家的大西洋鲑似乎接种水平较低（Hastein 等，2005）。鲑科鱼类的常规疫苗接种并没有广泛出现在苏格兰、法罗群岛或爱尔兰、塔斯马尼亚岛或新西兰，而在北美的东部和西部有一部分的疫苗接种比例。目前，智利和美国西北部养殖的大西洋鲑中接种了一定比例的病海鱼弧菌疫苗。目前用于鲑科鱼类疫苗的两株病海鱼弧菌为智利株和北美株。虽然抗原不同，但这些菌株似乎都有一定程度的抗原交叉反应（Silva-Rubio 等，2010）。海产虹鳟针对弧菌病的接种疫苗种类在不同国家有所不同，但当接种疫苗时，虹鳟最常接种免疫的是含有鳗弧菌血清型 O1（世界范围内）和 O2a 的疫苗。

海水鱼似乎易受鳗弧菌血清型的影响，这种血清型也会导致鲑科鱼类的疾病（O1 和 O2a），尽管这些物种对血清型的敏感性相差很大。地中海地区和日本的海水鱼最常接种免疫鳗弧菌血清型 O1 疫苗。海水鱼类似乎也明显受到非致病的或低毒力的血清型影响（如血清型 O2b）。

挪威似乎是唯一一个在大范围内免疫接种海水养殖鲑的国家，鲑接种针对由杀鲑弧菌引起的冷水弧菌病和粘摩替亚菌引起的冬季溃疡的疫苗。尽管这些细菌在其他国家存在，但只有冰岛对这些细菌进行了一定程度的疫苗接种。尽管从来没有使用现代分子生物学、流行病学方法分析杀鲑弧菌，但研究者们描述了抗原（Schrøder 等，1992）和质粒的变化（Sørum 等，1990）。使用表型和遗传方法相结合的方法，人们在粘摩替亚菌（*M. viscosa*）中发现了一种以前未被认可的多样性（Grove 等，2010）：有一个菌株对大西洋鲑有亲和力。基因多样性是否代表粘摩替亚菌的抗原多样性，以及这种多样性是否可能与目前疫苗相对较差的保护有关，目前还不清楚。

用于细分市场的弧菌疫苗至今仍对国际商品疫苗公司缺乏吸引力，比如新物种的初始培养，以及鱼类的小规模养殖。弧菌疫苗通常在大学和其他研究实验室生产，也有专门从事自体疫苗生产的公司。西班牙巴伦西亚大学研发的针对创伤弧菌（*V. vulnificus*）血清型 A 和 E（Esteve-Gassent 等，2004b）的浸泡疫苗被证明具有很强的保护作用，但该浸泡疫苗从未进入商业生产，这是由于商品化鳗鲡养殖的市场比较有限。

虽然许多针对哈维氏弧菌（*V. harveyi*）和溶藻弧菌（*V. alginolyticus*）的疫苗已经实验性生产，并且相当多的科技出版物已经专门探讨哈维氏弧菌疫苗的可行性，但是仍然没有哈维氏弧菌商品化的疫苗产品面世。

第六节　疫苗接种程序

目前可用（和已用）的弧菌疫苗要么是浸泡的，要么是腹腔注射的。尽管腹腔注射疫苗通常被认为是给鱼体提供了更好更持久的保护，但这个过程花费相对高，而且需要接种疫苗的鱼超过最小尺寸，具体取决于鱼的种类，但一般体重大于 20g。因此，该接种途径的选择是基于鱼须的类型和大小/发育阶段，以及接种鱼的价值。

由于弧菌是与鲑海水生长阶段有关，因此，在海水中生长的高价值的鲑有机会在无弧菌的淡水环境中生长到一定尺寸，这时腹腔注射免疫是一种可行的选择。这一程序也保证鲑在疫苗免疫和海上转移之间有足够的时间产生最佳的保护性免疫反应。在这种情况下，淡水养殖阶段的鲑一次接种疫苗通常足以获得直到 18 个月收获或更长时间的保护。

然而，海水鱼类在一个包含许多可能的致病性弧菌属的环境中孵化，一旦鱼类免疫成熟，就需要疫苗进行保护。这意味着最初的浸泡接种通常是在早期阶段（体重大约 1 g）进行，接着是一次或多次的加强浸泡疫苗接种或腹腔注射免疫。当供水量不大时，使用养殖水处理系统，如臭氧化或紫外线辐射水体，可以显著减少早期生命阶段的鱼类暴露于病原微生物中（Liltved 等，1995）。因此，疾病暴发的风险将降低，从而使鱼体在接种疫苗前可以形成免疫力，并且转移到海水网箱的小鱼在面对一个更加“敌对”的环境时，可以提前得到保护。

第七节 疫苗效果

由于疫苗的多样性，以及以前各种形式的弧菌病疫苗被证明是有效的，近年来几乎没有发表有关现代弧菌疫苗获得保护的文献。根据商业疫苗提供的产品信息，浸渍/浸泡疫苗的相对保护率（RPS）似乎可以超过40%，而商业化注射鳗弧菌疫苗的RPS大于75%，报道的杀鲑弧菌疫苗的RPS大于90%，冬季溃疡的病原体粘摩替亚菌的保护水平明显较低(60%～73%)。

人们普遍认为，弧菌病疫苗对大多数的细菌物种和相关鱼类都有好处，虽然已经报道了鱼体的保护程度和保护期是多变的。浸泡免疫通常会得到一个有限的令人满意的保护，并且免疫预防必须至少重复一次，在鱼体收获前，这种重复取决于疫苗、鱼的种类和养殖周期的长短。最大程度的保护和最长持续时间的保护无疑来源于腹腔注射（IP）疫苗，特别是和油类佐剂相关的疫苗。然而，这种疫苗接种方法并不总是可用的，因为用于保护鱼的尺寸太小不适合进行IP疫苗接种或者疫苗成本高于产品价值。

保护的时间长短是多变的，取决于多种因素，包括接种方式、佐剂类型、水温等。制造商似乎不愿在产品信息表上提供“持续保护时间”的信息。然而，通常是在鱼体尺寸太小时（体重大约1g)，最初浸泡免疫大约1个月后要重复一次浸泡免疫。加强免疫可能取决于鱼的种类，但当鱼体重为5g时建议加强免疫。鱼类合适尺寸的重复浸泡免疫或单一的腹腔注射可能是鱼类到达商品尺寸前获得保护的相关替代方法。鳗鲡收获之前，可能需要通过反复接种疫苗来获得对创伤弧菌的保护（Esteve-Gassent等，2004a)。对于注射接种的疫苗来说，特别是油类佐剂疫苗，在溯河产卵的鱼中，单一的IP疫苗接种应该能为鱼体提供免疫保护直到收获。

由鳗弧菌和病海鱼弧菌引起的弧菌病的疫苗接种通常是成功的，在接种疫苗的鱼群中很少出现疾病暴发。在转移至海水之前，能在淡水中接种疫苗的鱼类，其弧菌病的危害是最小的，因此可以在感染之前，使鱼体形成一种特异性的保护性反应，这主要是对鲑科鱼类而言，尤其是大西洋鲑和海水养殖的虹鳟。在海水鱼中，疫苗通常是非常有效的，特别是使用油类佐剂的腹腔注射疫苗，疫苗使用之后这种疾病便不会在短期内发生。对于海水鱼类来说，在可能已存在病原菌株的环境（如未处理的海水）中饲养，接种程序本身可能对鱼体有应激，因此接种疫苗的鱼形成免疫反应之前就引起了弧菌病暴发。

目前，海水养殖的大西洋鲑针对粘摩替亚菌的疫苗在实验室实验中的确获得了显著的保护(RPS≥60%)，并且无疑有助于降低海上养殖大西洋鲑的损失。然而，这种疾病仍造成损失，而且这种细菌经常从接种疫苗的群体中分离到。由于挪威具有比较大的市场，因此疫苗公司要继续研究改进粘摩替亚菌疫苗的配方。

第八节 副作用

疫苗接种的副作用在很大程度上与油类佐剂疫苗的使用有关（Midtlyng和Lillehaug，1998)。然而，即使是浸泡接种也会在一定程度上造成鱼体生长发育迟缓（Lillehaug，1991)，这可能是由于鲑接近银化的时期，鱼体很脆弱，在这段时间内接种免疫，会对鱼体造成应激。

第九节 管理条例

关于弧菌疫苗以及鱼类其他的弧菌感染没有特定的管理条例。水产养殖中引起疾病的弧菌在全球范围均有分布，疫苗是一项基本的防控技术，且大多数疫苗产品是由灭活菌苗构成。

参考文献

Amaro，C. and Biosca，E. 1996. *Vibrio vulnificus* biotype 2，pathogenic for eels，is also an opportunist pathogen of humans. *Appl Environ Microbiol* 62：1454-1457.

Amaro，C.，Biosca，E. G.，Esteve，C.，*et al.*，1992. Comparative study of phenotypic and virulence properties in *Vibrio vulnificus* biotype-1 and biotype-2 obtained from a European eel farm experiencing mortalities. *Dis Aquatic Org* 13：29-35.

Austin，B. 2010. Vibrios as causal agents of zoonoses. *Vet Microbiol* 140：310-317.

Austin，B. and Austin，D. A. 2007. *Bacterial Fish Pathogens. Disease of Farmed and Wild Fish*. 4th edn. Chichester，Springer Praxis.

Austin，B. and Zhang，Z. H. 2006. *Vibrio harveyi*：a significant pathogen of marine vertebrates and invertebrates. *Lett Appl Microbiol* 43：119-124.

Austin，B.，Stobie，M.，Robertson，P. A. W.，*et al.*，1993. *Vibrio alginolyticus* —the cause of gill disease leading to progressive low-level mortalities among juvenile turbot，*Scophthalmus maximus* L in a Scottish aquarium. *J Fish Dis* 16：277-280.

Austin，B.，Alsina，M.，Austin，D. A.，*et al.*，1995. Identification and typing of *Vibrio anguillarum*：a comparison of different methods. *Syst Appl Microbiol* 18：285-302.

Balebona，M. C.，Zorilla，I.，Morinigio，M. A. and Borrego，J. J. 1998. Survey of bacterial pathogens affecting farmed gilthead seabream（*Sparus aurata* L.）in southwestern Spain from 1990 to 1996. *Aquaculture* 166：19-35.

Benediktsdóttir，E.，Verdonck，L.，Sproer，C.，*et al.*，2000. Characterization of *Vibrio viscosus* and *Vibrio wodanis* isolated at different geographical locations：a proposal for reclassification of *Vibrio viscosus* as *Moritella viscosa* comb. nov. *Int J Syst Evol Microbiol* 50：479-488.

Bergh，O. and Samuelsen，O. B. 2007. Susceptibility of corkwing wrasse *Symphodus melops*，goldsinny wrasse *Ctenolabrus rupestis*，and Atlantic salmon *Salmo salar* smolt，to experimental challenge with *Vibrio tapetis* and *Vibrio splendidus* isolated from corkwing wrasse. *Aquacult Int* 15：11-18.

Bjelland，A. M.，Johansen，R.，Brudal，E.，*et al.*，2012. *Vibrio salmonicida* pathogenesis analyzed by experimental challenge of Atlantic salmon（*Salmo salar*）. *Microb Pathog* 52：77-84.

Björnsdóttir，B.，Guðmundsdóttir，T. and Guðmundsdóttir，B. K. 2011. Virulence properties of *Moritella viscosa* extracellular products. *J Fish Dis* 34：333-343.

Bockemuhl，J.，Roch，K.，Wohlers，B.，*et al.*，1986. Seasonal distribution of facultatively enteropathogenic vibrios（*Vibrio cholerae*，*Vibrio mimicus*，*Vibrio parahaemolyticus*）in the freshwater of the Elbe River at Hamburg. *J Appl Bacteriol* 60：435-442.

Bordas，M. A.，Balebona，M. C.，Rodriguez-Maroto，J. M.，*et al.*，1998. Chemotaxis of pathogenic *Vibrio* strains towards mucus surfaces of gilt-head sea bream (*Sparus aurata* L.) . *Appl Environ Microbiol* 64: 1573-1575.

Bruno, D., Hastings, T. S., Ellis, A. E. and Wootten, R. 1985. Outbreak of a cold-water vibriosis in Atlantic salmon in Scotland. *Bull Eur Assoc Fish Patol* 5: 62-63.

Chen, C., Wang, Q. B., Liu, Z. H., *et al.*, 2012. Characterization of role of the toxR gene in the physiology and pathogenicity of *Vibrio alginolyticus*. *Antonie Van Leeuwenhoek* 101: 281-288.

Colodner, R., Raz, R., Meir, I., *et al.*, 2004. Identification of the emerging pathogen *Vibrio vulnificus* biotype 3 by commercially available phenotypic methods. *J Clin Microbiol* 42: 4137-4140.

Colorni, A., Paperna, I. and Gordin, H. 1981. Bacterial infections in gilthead seabream *Sparus aurata* cultured in Eilat. *Aquaculture* 23: 257-267.

Colquhoun, D. J. and Sørum, H. 2001. Temperature dependent siderophore production in *Vibrio salmonicida*. *Microb Pathog* 31: 213-219.

Colquhoun, D. J., Alvheim, K., Dommarsnes, K., *et al.*, 2002. Relevance of incubation temperature for *Vibrio salmonicida* vaccine production. *J Appl Microbiol* 92: 1087-1096.

Colquhoun, D. J., Aase, I. L., Wallace, C., *et al.*, 2004. First description of *Vibrio ordalii* from Chile. *Bull Eur Assoc Fish Patol* 24: 185-188.

Crosa, J. H., Schiewe, M. H. and Falkow, S. 1977. Evidence for plasmid contribution to the virulence of fish pathogen *Vibrio anguillarum*. *Infect Immun* 18: 509-513

Croxatto, A., Chalker, V. J., Lauritz, J., *et al.*, 2002. VanT, a homologue of *Vibrio harveyi* LuxR, regulates serine, metalloprotease, pigment, and biofilm production in *Vibrio anguillarum*. *J Bacteriol* 184: 1617-1629.

Dalsgaard, I., Hoi, L., Siebeling, R. J. and Dalsgaard, A. 1999. Indole-positive *Vibrio vulnificus* isolated from disease outbreaks on a Danish eel farm. *Dis Aquat Org* 35: 187-194.

Dalsgaard, I., Jürgens, O. and Mortensen, A. 1988. *Vibrio salmonicida* isolated from farmed Atlantic salmon in the Faroe Islands. *Bull Eur Assoc Fish Patol* 8: 53-54.

Denkin, S. M. and Nelson, D. R. 1999. Induction of protease activity in *Vibrio anguillarum* by gastrointestinal mucus. *Appl Environ Microbiol* 65: 3555-3560.

Denkin, S. M. and Nelson, D. R. 2004. Regulation of *Vibrio anguillarum* empA metalloprotease expression and its role in virulence. *Appl Environ Microbiol* 70: 4193-4204.

Egidius, E. 1987. Vibriosis: Pathogenicity and pathology. A review. *Aquaculture* 67: 15-28.

Esteve-Gassent, M. D., Fouz, B. and Amaro, C. 2004a. Efficacy of a bivalent vaccine against eel diseases caused by *Vibrio vulnificus* after its administration by four different routes. *Fish Shellfish Immunol* 16: 93-105.

Esteve-Gassent, M. D., Fouz, B., Barrera, R. and Amaro, C. 2004b. Efficacy of oral reimmunisation after immersion vaccination against *Vibrio vulnificus* in farmed European eels. *Aquaculture* 231: 9-22.

Farmer, J. J., Janda, J. M., Brenner, F. W., *et al.*, 2005. Genus *Vibrio* Pacini 1854, 411AL, in *Bergey's Manual of Systematic Bacteriology*, Vol. 2 (ed. G. M. Garrity) . The *Proteobacteria*, Part B, The *Gammaproteobacteria*. New York, Springer, 494-546.

Fouz, B. and Amaro, C. 2003. Isolation of a new serovar of *Vibrio vulnificus* pathogenic for eels cultured in freshwater farms. *Aquaculture* 217: 677-682.

Fouz, B., Llorens, A., Valiente, E. and Amaro, C. 2010. A comparative epizootiologic study of the two fish-pathogenic serovars of *Vibrio vulnificus* biotype 2. *J Fish Dis* 33: 383-390.

Frans, I., Michiels, C. W., Bossier, P., *et al.*, 2011. *Vibrio anguillarum* as a fish pathogen: virulence factors, diagnosis and prevention. *J Fish Dis* 34: 643-661.

Giorgetti, G. and Ceschia, G. 1982. Vibriosis in rainbow trout, *Salmo gairdneri* Richardson, in freshwater in northeastern Italy. *J Fish Dis* 5: 125-130.

Grisez, L. and Ollevier, F. 1995. Comparative serology of the marine fish pathogen *Vibrio anguillarum*. *Appl Environ Microbiol* 61: 4367-4373.

Grove, S., Wiik-Nielsen, C. R., Lunder, T., *et al.*, 2010. Previously unrecognised division within *Moritella viscosa* isolated from fish farmed in the North Atlantic. *Dis Aquat Org* 93: 51-61.

Håstein, T., Gudding, R. and Evensen, Ø. 2005. Bacterial vaccines for fish—an update of the current situation worldwide. *Dev Biol* (*Basel*) 121: 55-74.

Hoel, K., Reitan, L. J. and Lillehaug, A. 1998. Immunological cross reactions between *Aeromonas salmonicida* and *Vibrio salmonicida* in Atlantic salmon (*Salmo salar* L.) and rabbit. *Fish Shellfish Immunol* 8: 171-182.

Ivanova, E. P., Zhukova, N. V., Gorshkova, N. M. and Chaikina, E. L. 2001. Characterization of *Aeromonas* and *Vibrio* species isolated from a drinking water reservoir. *J Appl Microbiol* 90: 919-927.

Jensen, S., Samuelsen, O. B., Andersen, K., *et al.*, 2003. Characterization of strains of *Vibrio splendidus* and *V. tapetis* from corkwing wrasse *Symphodus melops* suffering vibriosis. *Dis Aquat Org* 53: 25-31.

Kanno, T., Nakai, T. and Muroga, K. 1990. Scanning electron-microscopy on the skin surface of ayu *Plecoglossus altivelis* infected with *Vibrio anguillarum*. *Dis Aquat Org* 8: 73-75.

Karlsen, C., Paulsen, S. M., Tunsjø, H. S., *et al.*, 2008. Motility and flagellin gene expression in the fish pathogen *Vibrio salmonicida*: effects of salinity and temperature. *Microb Pathog* 45: 258-264.

Karunasagar, I., Otta, S. K. and Karunasagar, I. 1996. Biofilm formation by *Vibrio harveyi* on surfaces. *Aquaculture* 140: 241-245.

Kitao, T., Aoki, T. and Muroga, K. 1984. Three new serotypes of *Vibrio anguillarum*. *Bull Jap Soc Sci Fish* 50: 1955.

Kitao, T., Aoki, T., Fukudome, M., *et al.*, 2006. Serotyping of *Vibrio anguillarum* isolated from diseased freshwater fish in Japan. *J Fish Dis* 6: 175-181.

Knudsen, G., Sørum, H., Press, CMcL. and Olafsen, J. A. 1999. *In situ* adherence of *Vibrio* spp. to cryosections of Atlantic salmon, *Salmo salar* L., tissue. *J Fish Dis* 22: 409-418.

Kusada, R., Yokoyama, J. and Kawai, K. 1986. Bacteriological study on cause of mass mortalities in cultured black seabream fry. *Bull Jap Soc Sci Fish* 52: 1745-1751.

Larsen, J. L., Rasmussen, H. B. and Dalsgaard, I. 1988. Study of *Vibrio anguillarum* strains from different sources with emphasis on ecological and pathobiological properties. *Appl Environ Microbiol* 54: 2264-2267.

Larsen, J. L., Pedersen, K. and Dalsgaard, I. 1994. *Vibrio anguillarum* serovars associated with vibriosis in fish. *J Fish Dis* 17: 259-267.

Lee, K. K. 1995. Pathogenesis studies on *Vibrio alginolyticus* in the grouper, *Epinephalus malabaricus*, Bloch et Schneider. *Microb Pathog* 19: 39-48.

Lemos, M. L., Salinas, P., Toranzo, A. E., *et al.*, 1988. Chromosome-mediated iron uptake system in pathogenic strains of *Vibrio anguillarum*. *J Bacteriol* 170: 1920-1925.

Li, G. F., Zhao, D. H., Huang, L., *et al.*, 2006. Identification and phylogenetic analysis of *Vibrio vulnificus* isolated from diseased *Trachinotus ovatus* in cage mariculture. *Aquaculture* 261: 17-25.

Lillehaug, A. 1990. A field trial of vaccination against cold-water vibriosis in Atlantic salmon (*Salmo salar* L.). *Aquaculture* 84: 1-12.

Lillehaug, A. 1991. Vaccination of Atlantic salmon (*Salmo salar* L) against cold-water vibriosis—duration of protection and effect on growth rate. *Aquaculture* 92：99-107.

Lillehaug，A.，Sørum，R. H. and Ramstad，A. 1990. Cross-protection after immunization of Atlantic salmon，*Salmo salar* L.，against different strains of *Vibrio salmonicida*. *J Fish Dis* 13：519-523.

Lillehaug，A.，Lunestad，B. T. and Grave，K. 2003. Epidemiology of bacterial diseases in Norwegian aquaculture-a description based on antibiotic prescription data for the ten-year period 1991 to 2000. *Dis Aquat Org* 53：115-125.

Liltved，H.，Hektoen，H. and Efraimsen，H. 1995. Inactivation of bacterial and viral fish pathogens by ozonation or Uv irradiation in water of different salinity. *Aquacult Eng* 14：107-122.

Løvoll，M.，Wiik-Nielsen，C. R.，Tunsjø，H. S.，*et al.*，2009. Atlantic salmon bath challenged with *Moritella viscosa*—pathogen invasion and host response. *Fish Shellfish Immunol* 26：877-884.

Lunder，T.，Evensen，Ø.，Holstad，G. and Håstein，T. 1995. Winter ulcer in the Atlantic salmon *Salmo salar*. Pathological and bacteriological investigations and transmission experiments. *Dis Aquat Org* 23：39-49.

Marco-Noales，E.，Milan，M.，Fouz，B.，*et al.*，2001. Transmission to eels，portals of entry，and putative reservoirs of *Vibrio vulnificus* serovar E（biotype 2）. *Appl Environ Microbiol* 67：4717-4725.

Markestad，A. and Grave，K. 1997. Reduction of antibacterial drug use in Norwegian fish farming due to vaccination. *Dev Biol Stand* 90：365-369.

Midtlyng，P. J. and Lillehaug，A. 1998. Growth of Atlantic salmon *Salmo salar* after intraperitoneal administration of vaccines containing adjuvants. *Dis Aquat Org* 32：91-97.

Mikkelsen，H.，Lund，V.，Larsen，R. and Seppola，M. 2011. Vibriosis vaccines based on various sero-subgroups of *Vibrio anguillarum* O2 induce specific protection in Atlantic cod（*Gadus morhua* L.）juveniles. *Fish Shellfish Immunol* 30：330-339.

Milton，D. L.，O'Toole，R.，Horstedt，P. and Wolf-Watz，H. 1996. Flagellin A is essential for the virulence of *Vibrio anguillarum*. *J Bacteriol* 178：1310-1319.

Munday，B. L.，Carson，J.，Whittington，R. and Alexander，J. 1992. Serological responses and immunity produced in salmonids by vaccination with Australian strains of *Vibrio anguillarum*. *Immunol Cell Biol* 70：391-395.

Muroga，K.，Jo，Y. and Masumura，K. 1986. *Vibrio ordalii* isolated from diseased ayu（*Plecoglossus altivelis*）and rockfish（*Sebastes schlegeli*）. *Fish Pathol* 21：239-243.

Naka，H.，Dias，G. M.，Thompson，C. C.，*et al.*，2011. Complete genome sequence of the marine fish pathogen *Vibrio anguillarum* harboring the pJM1 virulence plasmid and genomic comparison with other virulent strains of *V. anguillarum* and *V. ordalii*. *Infect Immun* 79：2889-2900.

O'Halloran，J. and Henry，R. 1993. *Vibrio salmonicida*（Hitra disease）in New Brunswick. *Bull Aquacult Assoc Can* 93：96-98.

Olsson，J. C.，Jobörn，A.，Westerdahl，A.，*et al.*，1996. Is the turbot，*Scophthalmus maximus*（L.）intestine a portal of entry for the fish pathogen *Vibrio anguillarum*? *J Fish Dis* 19：225-234.

Ormonde，P.，Horstedt，P.，O'Toole，R. and Milton，D. L. 2000. Role of motility in adherence to and invasion of a fish cell line by *Vibrio anguillarum*. *J Bacteriol* 182：2326-2328.

O'Toole，R.，Milton，D. L. and Wolf-Watz，H. 1996. Chemotactic motility is required for invasion of the host by the fish pathogen *Vibrio anguillarum*. *Mol Microbiol* 19：625-637.

Pedersen，K. and Larsen，J. L. 1995. Evidence for the existence of distinct populations of *Vibrio anguillarum* serogroup O1 based on plasmid contents and ribotypes. *Appl Environ Microbiol* 61：2292-2296.

Pedersen，K.，Grisez，L.，van Houdt，R.，*et al.*，1999. Extended serotyping scheme for *Vibrio anguillarum* with the definition and characterization of seven provisional O-serogroups. *Curr Microbiol* 38：183-189.

Poppe, T. T., Håstein, T. and Salte, R. 1985. "Hitra disease" (Haemorrhagic syndrome) in Norwegian salmon farming: present status, in *Fish and Shellfish Pathology* (ed. A. E. Ellis). London, Academic Press, 223-229.

Robertson, P. A. W., Calderon, J., Carrera, L., *et al.*, 1998. Experimental *Vibrio harveyi* infections in *Penaeus vannamei* larvae. *Dis Aquat Org* 32: 151-155.

Sakata, T., Matsuura, M. and Shimokawa, Y. 1989. Characteristics of *Vibrio damsela* isolated from diseased yellowtail *Seriola-quinqueradiata*. *Nippon Suisan Gakkaishi* 55: 135-141.

Schiewe, M. H. and Crosa, J. H. 1981. Molecular characterisation of *Vibrio anguillarum* biotype 2. *Can J Microbiol* 27: 1011-1018.

Schiewe, M. H., Trust, T. J. and Crosa, J. H. 1981. *Vibrio ordalii* sp. nov.: A causative agent of vibriosis in fish. *Curr Microbiol* 6: 343-348.

Schrøder, M. B., Espelid, S. and Jørgensen, T. Ø. 1992. Two serotypes of *Vibrio salmonicida* isolated from diseased cod (*Gadus morhua* L.): virulence, immunological studies and vaccination experiments. *Fish Shellfish Immunol* 2: 211-221.

Silva-Rubio, A. A., Avendaño-Herrera, R., Jaureguiberry, B. B., *et al.*, 2008. First description of serotype O3 in *Vibrio anguillarum* strains isolated from salmonids in Chile. *J Fish Dis* 31: 235-239.

Silva-Rubio, A., Acevedo, C., Magarinos, B., *et al.*, 2010. Antigenic and molecular characterization of *Vibrio ordalii* strains isolated from Atlantic salmon *Salmo salar* in Chile. *Dis Aquat Org* 79: 27-35.

Song, Y. L., Chen, S. N. and Kou, G. H. 1988. Serotyping of *Vibrio anguillarum* strains isolated from fish in Taiwan. *Fish Pathol* 23: 185-189.

Sørensen, U. B. S. and Larsen, J. L. 1986. Serotyping of *Vibrio anguillarum*. *Appl Environ Microbiol* 1986; 51: 593-597.

Sørum, H., Hvaal, A. B., Heum, M., *et al.*, 1990. Plasmid profiling of *Vibrio salmonicida* for epidemiologic studies of cold-water vibriosis in Atlantic salmon (*Salmo salar*) and cod (*Gadus morhua*). *Appl Environ Microbiol* 56: 1033-1037.

Thompson, F. L., Thompson, C. C., Dias, G. M., *et al.*, 2011. The genus *Listonella* MacDonell and Colwell 1986 is a later heterotypic synonym of the genus *Vibrio* Pacini 1854 (Approved Lists 1980) —a taxonomic opinion. *Int J Syst Evol Microbiol* 61: 3023-3027.

Thomson, R., Macpherson, H. L., Riaza, A. and Birkbeck, T. H. 2005. *Vibrio splendidus* biotype 1 as a cause of mortalities in hatchery-reared larval turbot, *Scophthalmus maximus* (L.). *J Appl Microbiol* 99: 243-250.

Tiainen, T., Pedersen, K. and Larsen, J. L. 1997. *Vibrio anguillarum* serogroup O3 and *V. anguillarum*-like serogroup O3 cross-reactive species—comparison and characterization. *J Appl Microbiol* 82: 211-218.

Toranzo, A. E. and Barja, J. 1990. A review of the taxonomy and seroepizootiology of *Vibrio anguillarum*, with special reference to aquaculture in the northwest of Spain. *Dis Aquat Org* 9: 73-82.

Tunsjø, H. S., Paulsen, S. M., Berg, K., *et al.*, 2009. The winter ulcer bacterium *Moritella viscosa* demonstrates adhesion and cytotoxicity in a fish cell model. *Microb Pathog* 47: 134-142.

Tunsjø, H. S., Wiik-Nielsen, C. R., Grove, S., *et al.*, 2011. Putative virulence genes in *Moritella viscosa*: activity during *in vitro* inoculation and *in vivo* infection. *Microb Pathog* 50: 286-292.

Urbanczyk, H., Ast, J. C., Higgins, M. J., *et al.*, 2007. Reclassification of *Vibrio fischeri*, *Vibrio logei*, *Vibrio salmonicida* and *Vibrio wodanis* as *Aliivibrio fischeri* gen. nov., comb. nov., *Aliivibrio logei* comb. nov., *Aliivibrio salmonicida* comb. nov. and *Aliivibrio wodanis* comb. nov. *Int J Syst Evol Microbiol* 57: 2823-2829.

Varina, M., Denkin, S. M., Staroscik, A. M. and Nelson, D. R. 2008. Identification and characterization of epp, the secreted processing protease for the *Vibrio anguillarum* EmpA metalloprotease. *J Bacteriol* 190: 6589-6597.

Velji, M. I., Albright, L. J. and Evelyn, T. P. T. 1990. Protective immunity in juvenile coho salmon *Oncorhynchus kisutch* following immunization with *Vibrio ordalii* lipopolysaccharide or from exposure to live *V. ordalii* cells. *Dis Aquat Org* 9: 25-29.

Wang, S. Y., Lauritz, J., Jass, J. and Milton, D. L. 2002. A ToxR homolog from *Vibrio anguillarum* serotype O1 regulates its own production, bile resistance, and biofilm formation. *J Bacteriol* 184: 1630-1639.

Wards, B. J., Patel, H. H., Anderson, C. D. and Delisle, G. W. 1991. Characterization by restriction endonuclease analysis and plasmid profiling of *Vibrio ordalii* strains from salmon (*Oncorhynchus tshawytscha* and *Oncorhynchus nerka*) with vibriosis in New-Zealand. *New Zeal J Mar Fresh* 25: 345-350.

Yiagnisis, M., Vatsos, I. N., Kyriakou, C. and Alexis, M. 2007. First report of *Vibrio anguillarum* isolation from diseased big scale sand smelt, *Atherina boyeri* Risso 1810, in Limnos, Greece. *Bull Eur Assoc Fish Patol* 27: 61-69.

第十六章　疖疮病的疫苗接种

Paul J. Midtlyng

本章概要　SUMMARY

疖疮病是指由杀鲑气单胞菌杀鲑亚种（*Aeromonas salmonicida* subsp. *salmonicida*）引起的鲑科鱼类的致死性败血症，也指在多种海淡水鱼类上由非典型菌株（其他杀鲑气单胞菌亚种）引起的疾病，造成严重的皮肤溃疡和损伤。典型的疖疮病菌株（杀鲑亚种 subsp. *salmonicida*）在全球具有明显一致的基因型和表型，而非典型疖疮菌株的基因型和表型具有更丰富的多样性。典型和非典型疖疮菌株都具有多种胞内和胞外致病性毒力因子，对其已有几十年的深入研究。一般而言，单一毒力因子都不会导致宿主发生感染和死亡。对典型和非典型疖疮菌株保护性抗原强弱的研究结果显示出不一致性，甚至相互矛盾，但整体累积的证据表明，菌株 A 层（外膜蛋白 A）可能是最重要的抗原。典型疖疮病的油类佐剂注射疫苗已获得许可，在最大的鲑科鱼类养殖业上得以应用，该疫苗经常添加其他病原菌和病毒抗原联合使用。然而到目前为止，用于浸泡或口服免疫的疖疮病疫苗和活疫苗并未取得成功。尽管典型疖疮病疫苗能针对非典型疖疮病提供一定的交叉保护，但还是需要对不同宿主、不同地域菌株的非典型疖疮病，有针对性地研制油类佐剂疖疮疫苗。

第一节　引　言

对鱼类疾病来说，疖疮病广义上是指由杀鲑气单胞菌（*A. salmonicida*）在海水和淡水有鳍鱼类上引起多种临床症状的疾病。然而，大多数感染的鱼并没有表现出病名所示的症状（疖疮、骨骼肌出血性坏死）。典型的疖疮病主要发生在鲑科鱼类，由杀鲑气单胞菌杀鲑亚种（*A. salmonicida* subsp. *salmonicida*）引起；非典型疖疮病是由其他杀鲑气单胞菌亚种（*achromogenes* 亚种、*masouicida* 亚种、*smithia* 亚种、*pectinolytica* 亚种）在很多海水和淡水的鲑科以及非鲑科鱼类中引起的疾病。因此，区别介绍典型疖疮病和非典型疖疮病不仅是因为要对疖疮病的疫苗接种进行系统的回顾，更是因为典型和非典型疖疮病的预防措施存在很大差异。

最近，Cipriano 和 Austin（2011）对典型疖疮病和由非典型疖疮病菌株以及运动型气单胞菌引起的鱼类疾病进行了详细的描述。本章主要介绍典型疖疮病及其疫苗防控，最后简要介绍非典型杀鲑气单胞菌疫苗的研发进展。

第二节 发生流行与危害

对鲑科鱼类疖疮病最初的描述是在淡水养殖褐鳟发生的一种严重疾病（Emmerich 和 Weibel，1894），从褐鳟体内分离出一株似霍乱弧菌（*Vibrio cholera*）的细菌，命名为杀鲑细菌（*Bacterium salmonicida*）。该研究发表后，该病及相关病原菌的分离陆续在北美、英格兰和苏格兰被报道，病原菌造成河流中的鱼类流行病发生。几十年来，已有文献均认为疖疮病是一种仅在淡水或半咸水中发生的疾病。然而，海水养殖鲑疖疮病的首次暴发（Novotny，1978）以及苏格兰鲑海水养殖场出现的新的常发疖疮病（Munro 和 Waddell，1984）均表明，随着鲑海水养殖系统的发展，疖疮病的发病模式已发生变化。1985 年，挪威某海水鲑养殖场的一批隐性感染疖疮病的鲑鱼苗，导致了该病的暴发流行，证实了上述观点（Egidius，1987）。

典型的疖疮病可能是研究最深入的鱼类疾病，也是为数不多的有自己专著资料的动物疾病之一（Bernoth 等，1997），这些资料为进一步研究该病提供了参考。目前除了澳大利亚和智利，引起典型疖疮病的致病菌在其他所有野生和养殖鲑科鱼类中均有报道。然而，该病临床发生和发病率主要取决于增殖或养殖鲑种类的常规管理模式，特别是疫苗接种（见下文）。

典型疖疮病的临床表现是引起败血症，导致高死亡率，存活下来的鱼有慢性疾病症状，鱼体几乎没有形成自然免疫力。该病流行季节是春季至深秋和初冬。在使用有效疫苗以前，估计欧洲的鲑工厂化养殖场每年因疖疮病导致的损失累计超过 20%（Ellis，1997b），这给水产养殖业带来毁灭性的后果，不仅会导致直接的经济损失，还会因较长的停药期使鱼不能及时收获上市而造成经济损失，同时长期用药使病原菌产生耐药性，导致该病的药物治疗无效。

第三节 病 原 学

典型疖疮病的病原是杀鲑气单胞菌杀鲑亚种（*A. salmonicida* subsp. *salmonicida*），属革兰氏阴性菌，非运动型，兼性厌氧，短杆状菌，可生长于多种非选择性琼脂培养基（20～22℃），有自主凝集现象（Austin 和 Austin，2007）。杀鲑亚种（subsp. *salmonicida*）在大豆胰蛋白酶或酵母胰蛋白酶琼脂（TSA/TYA）培养基上进行初次培养或再次培养时，具有布朗扩散产生色素的特征，这常被用于该菌确认，但是也有具高致病性但无色素的杀鲑亚种分离株的报道（Koppang 等，2000）。与非典型菌株不一样，不同临床暴发模式、分离时间、发病鱼的种类或地理来源的杀鲑亚种表型均高度一致，所有的杀鲑亚种都像是一个菌株克隆（Cipriano 和 Austin，2011）。在很早以前就已经报道临床无症状但隐性感染的鱼的运动或迁移是疖疮病地区性传播的主要途径，因此菌株间高度一致性与其传播途径有关（Horne，1928）。致病菌是感染典型疖疮病的必要条件，但临床上疾病的发生和致病往往是由各种环境因素、养殖管理或其他胁迫因素导致鱼体免疫功能下降而造成的（Bernoth，1997）。杀鲑气单胞菌杀鲑亚种是专性寄生菌，在集约化养殖条件下，如果未接种疫苗，隐性感染的鲑很难在整个养殖周期中存活下来而不出现临床症状。与之相比，野生鲑种群疖疮病暴发的频率要低得多，还与水流、温度和种群密度等当地的条件密切相关。此外，人工感染野生幼鲑，其最小致死剂量（小于 50cfu）的实验结果也与上述情况一致（Nordmo 和 Ramstad，1997）。

疖疮病的致病菌不仅从感染的个体传播，还会从死鱼中大量释放。该病很容易通过鱼体接触传播或经水传播。此外，杀鲑气单胞菌悬浮在水中，易与脂滴结合并聚集在水面上，而鲑经常在水面游动造成传播（Enger，1997）。疖疮病垂直传播（从亲代到后代）的风险非常低，如果防控措施到位，用碘消毒鱼卵即可有效预防垂直传播。

第四节　致病机理与毒力

典型疖疮病急性的临床表现和病理特征为由于细菌性败血症而死亡，炎症反应极少，鱼体表面无特异性变化：皮肤暗淡、食欲不振、嗜睡，有时皮肤少量出血。病鱼内部表现为充血和血管疏松，浆膜表面出血，脾脏和肾脏肿大。出现亚急性或慢性症状的鱼，其骨骼肌系统出现多发性出血性坏死，可能发展或合并为较大的肌肉疖疮或皮下疖疮。疖疮病的组织病理学特征是鱼类的心脏、脾脏、肾脏发生局部坏死，且坏死部分包裹着散布在血管周围的菌落，这对造血组织、肾脏、肝脏和心肌有急性毒性作用（McCarthy 和 Roberts，1980）。所有这些症状可以解释其发病机理，即疖疮病病原菌产生的一些细胞外毒素的综合作用导致鱼体发生严重的凝血障碍（Salte 等，1993）。

杀鲑气单胞菌的致病性和毒力因子已有大量研究和综述报道（Ellis，1991，1997a）。相关论文中提及的以及应用实验分子遗传技术研究的毒力因子列于表 16-1。

细菌的表面成分通常被认为是附着和入侵的重要因子。分子生物学研究表明，杀鲑气单胞菌表面成分的相关基因缺失后，细菌无法形成极性鞭毛或侧生鞭毛（Reith 等，2008）。然而，还有研究已经证实，杀鲑气单胞菌的细胞表面有较短鞭毛或菌毛，有助于细菌黏附肠黏液细胞，但并未发现其在入侵宿主细胞是必需的（Boyd 等，2008；Danacay 等，2010）。

杀鲑气单胞菌外膜蛋白的毒力作用已被广泛关注，其中研究最多的是晶体表面蛋白阵列（S层，之前被命名为 A 层），由 A 蛋白分子组成并覆盖于细菌细胞外层膜上（Trust 等，1996；Noonan 和 Trust，1997）。虽然有些 S 层缺失的菌株在实验环境下具有致病性，但几乎所有的致病菌株都存在 S 层，并且已经证明 S 层可抵御巨噬细胞对该菌的杀伤作用。同时也研究了杀鲑气单胞菌的铁调节外膜蛋白的毒力作用，但其作为保护性抗原的作用尚未被证实（见下文）。

表 16-1　杀鲑气单胞菌（*A. salmonicida*）致病因子和毒力因子及其作为保护性抗原的研究

（自 Ellis 1997a，1997c；Tomás，2012；Beaz-Hidalgo 和 Figueras 2013）

致病因子和毒力因子	参考文献
细胞表面成分	
菌毛	Masada 等（2002）；Boyd 等（2008）；Danacay 等（2010）

（续）

致病因子和毒力因子	参考文献
外膜蛋白（OMPs）	Hirs 和 Ellis（1994）；Bricknell 等（1999）；O'Dowd 等（1999）；Ebanks 等（2005）
S层（A层）	Ellis 等（1988）；Garduno 等（2000）；Lund 等（2003a）
孔蛋白	Lutwyche 等（1995）；Merino 等（2005）
O-抗原（脂多糖 LPS）	Wang 等（2006）
荚膜多糖	Bricknell 等（1997）；Wang 等（2004）；Magnadóttir（2010）
胞外（分泌）毒素	
GCAT	Lee 和 Ellis（1990，1991）；Vipond 等（1998）
细胞溶素（分泌 LPS）	Ellis 等（1988）
丝氨酸蛋白酶	Ellis 等（1988）；Ellis（1991）；Vipond 等（1998）
金属蛋白酶	Hussain 等（2000）；Prost（2001）；Burr 等（2005）
胞内结构	
毒素分泌系统	Burr 等（2005）；Danacay 等（2006）；Fast 等（2009）
铁获取系统	Ebanks 等（2004，2005）；Najimi 等（2009）
群体感应系统	Bruhn 等（2005）；Schwenteit 等（2011）

Ellis（1997c）总结了典型疖疮病细菌胞外产物（ECPs）的毒力研究，典型疖疮病病原菌的胞外毒素（复合分泌脂多糖的 GCAT，还有蛋白酶）主要作用是破坏组织，导致宿主死亡。这些毒素由胞内分泌系统分泌释放，因此，高效的胞内分泌系统也被认为是重要的毒力因子（Danacay 等，2006）。杀鲑气单胞菌在体内或某些培养基中生长会产生多糖荚膜（Garduno 和 Kay，1995），推测其作用是使病原菌逃避宿主免疫系统的识别，并与外膜蛋白 A 协同抵抗宿主的非特异性防御（Magnadottir，2010）。

典型疖疮病病原入侵宿主的靶器官/组织的途径仍有待进一步研究，但经一系列复杂的实验，Jutfelt 等（2006，2008）证明，致病菌可在鲑肠的上皮细胞移动，细菌产生的胞外产物和分泌的 LPS 会影响该过程。

第五节　抗　原

Ellis（1997c）总结了杀鲑气单胞菌大部分的致病因子和毒力因子在疖疮病疫苗研发的优点，结论是“抗原的特性及其诱导的抗疖疮病的免疫反应参数不清楚”。后续工作仍存在不同观点，在某种程度上有些结果是矛盾的，例如关于铁调节外膜蛋白 IROMP 在毒力和疫苗保护中的重要性问题（Hirst 和 Ellis，1994；Vipond 等，1998；Bricknell 等，1999）。

从疖疮病疫苗研究方面看，Ellis（1997c）得出的“杀鲑气单胞菌 A 蛋白诱导的免疫反应没有保护作用”这一结论尚待商榷。Lund 等（2003a）和 Romstad 等（2013）的研究结果提出了相反的结论，S 层缺失的减毒杀鲑气单胞菌不能产生免疫保护作用。此外，A 层缺失的菌株再次附着鱼体后，会在体外（Garduno 等，2000）和体内（Arnesen 等，2010）出现毒力恢复。目前看来，用于典型疖疮病疫苗的抗原应该来自可分泌上述列出的所有致病因子和毒力因子的菌株。杀鲑气单胞菌基因组序列及其注释信息已经公布（Reith 等，2008；Charette 等，2012），未来使用反向遗传学技术，将有助于阐释该菌的致病机制和毒力机制，进而促进对杀鲑气单胞菌感染和临床疾病的了解。这些方法也将有助于更准确地鉴定和筛选杀鲑气单胞菌的抗原，以研制保护效果较好的疫苗。

第六节　疫苗的研发

首个报道疖疮病疫苗接种实验是氯仿灭活的疖疮病病原菌培养物添加到饲料投喂克氏鳟（*Salmo clarkii*），发现鱼体内产生抗体并且实验攻毒后存活率提高（Duff 等，1942）。如 Midtlyng（1997）和 Evelyn（1997）所述，历时约 20 年才成功研制出矿物油佐剂的杀鲑气单胞菌（*A. salmonicida*）疫苗（Krantz 等，1964）。由一加拿大创业公司（Paterson 等，1985）生产的第一个鱼类疖疮病疫苗历经波折，在很长一段时间内，人们普遍认为疖疮病疫苗并不成熟，也没有重要的商业价值。随着杀鲑气单胞菌毒力和疖疮病免疫学研究取得突破进展，该疫苗的价值才被认可（Horne 和 Ellis，1988）。

自典型疖疮病危害挪威鲑养殖业的发展起，便开始了该病有效疫苗的独立研发，最早开始研究的是位于奥斯陆的挪威兽医研究所。研究显示，鲑腹腔注射铝佐剂疫苗，实验室条件下（Erdal 和 Reitan，1992）和野外生长（Lillehaug 等，1992）的鲑均获得明显的保护作用，并且保护水平良好，但在接种后的第二年夏天，野外生长的鲑保护水平显著下降。自 1991 年起，对更多的候选疖疮病疫苗和疫苗配方进行了大量的研究，其中单价和三联（疖疮病、弧菌病和冷水弧菌病）的油类佐剂配方疫苗，在免疫接种后的 3 个月和 6 个月，以及大规模的野外研究中均获得了较好的保护水平和更持久的保护作用（Midtlyng，1996；Midtlyng 等，1996b）。基于 1991 年的研究成果，挪威鲑养殖业随即开始强制执行疫苗大规模接种，这一举措产生了良好群体免疫效应，显著降低了抗菌药物的使用（Markestad 和 Grave，1997）。在苏格兰，疖疮病疫苗接种也使得鲑幼鱼从转至海水养殖到收获期的存活率大幅上升，在 20 世纪 80 年代后期鲑的存活率约为 65%；但 1994 年和次年，鲑转入海水中的存活率已超过 90%（Munro 和 Gauld，1997）。基于在行业中取得的明显成效，油类佐剂疫苗控制疖疮病的免疫策略逐渐被认可（Smith 和 Hiney，2000）。

虽然小规模实验已经证明，典型疖疮病的口服疫苗和浸泡疫苗会产生一定程度的保护，但浸泡攻毒和野外实验发现，其不能使鱼获得足够长的免疫保护时间，在整个鲑科鱼类工厂化养殖条件下不能够持续起作用（Midtlyng，1996）。因此，疖疮病浸泡或口服疫苗未获得商业许可。关于活疫苗预防典型疖疮病的研究已有论文发表，但其实验结果未能使该活疫苗获得批准（Vaughan 等，1993；Thornton 等，1994；Marsden 等，1998；Grove 等，2003）。

在全球范围内，注射油类佐剂疫苗已经完全成为疖疮病疫苗供应的主导产品，它还可与一种或多种弧菌抗原混合使用，如果用于大西洋鲑还可以含有其他更多的细菌或病毒组分。由于必须使用油类佐剂来制备疖疮病疫苗，因此融合其他抗原的多联、油类佐剂配方的疫苗主导了全球范围的鲑科鱼类商业化疫苗的应用和发展。然而其他抗原并非一定需要如此强效的佐剂才有效，已有一种含铝的单价疖疮病疫苗在加拿大获得许可。

第七节 疫苗接种程序

所有商业化疖疮病疫苗均采用腹腔注射，最好选择 1 龄鲑或幼鲑，经重度镇静或麻醉鱼体后进行免疫接种。

研究显示，使用注射疫苗，鲑需要 300～500℃・d（度日）甚至更久才能形成抗疖疮病的免疫力（Midtlyng 等，1996b；Eggset 等，1997），而使用油类佐剂配方疫苗获得的免疫保护可超过1 500℃・d（Midtlyng 等，1996b）。在进行大规模疫苗接种的鲑中几乎不发生疖疮病，这表明油类佐剂疫苗诱导的对典型疖疮病的免疫力可以持续整个生产周期。

不同疫苗接种程序的比较研究发现，只有腹腔注射含佐剂抗原的大西洋鲑有明显的抗体反应，并且头肾白细胞被激活（Midtlyng 等，1996a）。鱼体接种疫苗后，免疫效果持续至少 16 周，而非注射的疫苗免疫组效果很低。该研究结果使疫苗研发聚焦于注射型疖疮病疫苗，随之中止了浸泡型和口服型疫苗的研发。然而，有许多研究使用非油类疖疮病疫苗抗原和制剂，通过口服或其他方式加强免疫，也获得了较好的效果（Durbin 等，1999，Ackermann 等，2000；Bebak-Williams 和 Bullock，2002；Irie 等，2005）。

当今大西洋鲑养殖中，注射疖疮病多联疫苗的时间通常是在秋天幼鲑转移前的 6～8 周和春天幼鲑转移至海水前的 4～5 个月，会随着水温不同而有差异。在预期的繁殖群体中，在不减弱疫苗保护效果的情况下，乳化型油类疫苗可接种于鱼体的结缔组织，这样就可避免注射部位造成的副作用影响鱼类性腺发育（Treasurer 和 Cox，2008）。鲑转移到浮筏网箱养殖后也可接种疖疮病疫苗，但需要专门的设备和熟练掌握网箱操控、麻醉和复苏程序，以免造成意外损失。

第八节 效　　果

与 20 世纪 80 年代末的观点不同（Ellis 等，1988），研究发现抗杀鲑气单胞菌的抗体水平和疫苗的免疫保护相关（Midtlyng 等，1996a，1996b），这一发现后来也逐步得到证实（Hoel 等，1998；Bricknell 等，1999；Villumsen 等，2012；Romstad 等，2012，2013）。其实，在很早以前就已通过高免血清证实抗疖疮病抗体在免疫保护中的作用（Spence 等，1965）。

Midtlyng 等（1996）研究表明，三联的油类疖疮病疫苗（杀鲑气单胞菌杀鲑亚种 *A. salmonicida* subsp. *salmonicida*；鳗弧菌 *V. anguillarum* O1α；杀鲑弧菌 *V. salmonicida*）接种大西洋鲑，相对保护率达 87%（相对于注射生理盐水对照组的存活率）。免疫保护水平逐步升高，在鱼体免疫后 6～26 周显著增强。Hoel 等（1997）研究发现，某种弧菌及其免疫机制提高了多联疫苗中鲑抗疖疮病的免疫力。多因子实验设计和分析发现，杀鲑弧菌作用最大，抗杀鲑弧菌的抗体可与杀鲑气单胞菌整个菌体及其 LPS 结合，但不与杀鲑气单胞菌的外膜蛋白 A 反应。进一步的研究发现，这种现象可能是由抗体与菌体表面抗原的交叉反应以及佐剂对胸腺依赖型抗原的免疫效应产生的（Hoel 等，1998）。

Midtlyng 等（1996a）证明，使用疖疮病油类佐剂疫苗可以非特异性地持续激活大西洋鲑的头肾白细胞。这个功能很重要，因为疖疮病病原菌能够在未免疫鱼的巨噬细胞内存活并杀死巨噬细胞（Olivier 等，1992），但被杀鲑气单胞菌抗原和油类佐剂激活的巨噬细胞可以杀死致病的杀鲑气单胞菌（Chung 和 Secombes，1987）。

综上所述，由乳化抗原诱导免疫细胞的激活和高水平的血清抗体是有效预防疖疮病的先决条件。免疫应答机制的一个重要方面就是所谓的佐剂缓释效应，指的是由油类佐剂包裹的抗原能在注射部位停留较长时间，这种抗原的持久性与局部炎症反应的发生是一致的。研究发现，抗原的刺激可以持续到免疫后 6～9 个月（Mutoloki 等，2004）。

第九节　副 作 用

疖疮病油类佐剂疫苗诱导有效的免疫保护作用，也会产生免疫接种后的不良反应，特别是在注射部位形成炎症和肉芽肿病变。有关这些副作用的研究已有很多报道，这里只列举部分研究：涉及病理学变化（Midtlyng，1996；Poppe 和 Breck，1997）、不同的接种剂量所产生的副作用（Midtlyng 和 Lillehaug，1998；Rønsholt 和 McLean，1999；Aunsmo 等，2008a）、接种方法的改进（Aunsmo 等，2008b）以及详细描述组织病变过程（Mutoloki 等，2006）等。疖疮病疫苗除了使某些鱼体产生炎症反应外，会在后期产品加工时留下可见的瘢痕或色素，还会造成鱼体生长减缓，但实验证明这种影响有差异，与地域、温度和养殖场等因素有关（Midtlyng Lillehaug，1998；Aunsmo 等，2008）。到目前为止，尚未发现疖疮病疫苗中起保护作用的抗原会导致这些副作用。

第十节　非典型疖疮病的疫苗接种

非典型杀鲑气单胞菌菌株感染可导致多种临床病症，在各大洲均有报道，包括未发生典型疖疮病的智利和澳大利亚，也有非典型性疖疮病的报道。非典型疖疮病感染的鱼类范围非常广，常以其主要的临床症状命名，如传染性皮炎（鲑）、金鱼溃疡病、鲤赤皮病等。尽管大多数鱼类的主要症状是溃疡和皮肤损伤，但也常发生急性感染性败血症、器官的严重病变和高死亡率，这些症状和典型疖疮病流行病学特征非常相似。有关非典型疖疮病和其致病菌可参读 Wiklund 和 Dalsgaard（1998），Guðmundsdottir（1998）以及 Guðmundsdottir 和 Bjornsdottir（2007）等撰写的综述。

采用典型疖疮病的疫苗技术控制非典型疖疮病较难，因此需要不断开展非典型疖疮病菌株毒力因子开发疫苗的研究，目前大部分研究来自冰岛和挪威的研究团队。据 Lund 和 Mikkelsen（2004），以及 Gunnlaugsdottir（1997，1998）报道，像典型疖疮病一样，在非典型疖疮病的研究中，很多都关注 S 层（A 蛋白）和分泌蛋白酶的致病性和毒力作用。总的来说，非典型疖疮病的致病和毒力因子具有明显的多样性，在大多数菌株中，导致急性死亡和肌肉组织坏死的因子似乎没有典型疖疮病明显。在斑点狼鱼幼鱼中，每尾鱼接种少于 100 cfu 的某种非典型疖疮病致病菌株会导致死亡，但浸浴感染非典型菌株不会造成死亡（Guðmundsdóttir 和 Magnadóttir，1997；Lund 等，2008b）。在大菱鲆中，每尾注射 10^7～10^9 cfu 才能导致较高的死亡率（Santos 等，2005）。鉴于此，再加上典型的溃疡临床症状，可见典型和非典型疖疮病菌株在侵染性和败血症的流行方面存在差异。

表 16-2 列出了已发表的淡水和海水鱼类非典型疖疮病疫苗相关的研究文章，约有 20 篇，全面详细的介绍可参读 Guðmundsdottir 和 Bjornsdottir（2007）的报道。

总结这些研究成果发现，虽然已报道典型疖疮病疫苗可以诱导鲑产生对非典型疖疮病一定程度的交叉保护，但对其他鱼类的结果并不理想（Lund 等，2002；Björnsdóttir 等，2005，Lund 等，2008b），非典型疖疮病疫苗的研发不能使用典型疖疮病病原菌的变异菌株（Guðmundsdóttir 和 Guðmundsdóttir，1997）。此外，在某些非鲑科海水鱼类上分离的非典型疖疮病菌株，免疫接种和攻毒后发现具有一定程度的交叉保护。挪威的研究人员基于以上研究结果认为，非典型疖疮病疫苗应该是物种特异性的，疫苗中应该含有几种不同菌株的抗原，才能应对水产养殖生产中出现的多样病原（Lund 等，2002，2008b）。这一结论对非典型疖疮病疫苗研发在经济可行性方面提出了严峻的挑战。

表 16-2　已发表的淡水和海水鱼非典型疖疮病免疫接种的研究

鲑科和白鲑属	Guðmundsdóttir 和 Guðmundsdóttir（1997）；Guðmundsdóttir 等（1997）；Guðmundsdóttir 和 Magnadóttir（1997）；Lönnström 等（2001）；Pylkkö 等（2002）
鲤科	Evenberg 等（1988）；Sinyakov 等（2001）；Maurice 等（2004）；Robertson 等（2005）
大西洋鳙鲽	Ingilæ 等（2000）；Guðmundsdóttir 等（2003）；Lund 等（2008b）
斑点狼鱼	Ingilæ 等（2000）；Lund 等（2002，2003b）；Grøntvedt 等（2003）；Grøntvedt 和 Espelid（2004）
大西洋鳕	Mikkelsen 等（2004）；Lund 等（2006，2008a）；Arnesen 等（2010）
大菱鲆	Björnsdóttir 等（2005）；Santos 等（2005）

第十一节　法律问题和管理条例

鱼类疖疮病没有被列在《OIE 水生动物卫生法典》中，欧盟 91/67 指令下关于欧洲经济区的鲜活水产品的贸易和转让的要求也没有相关规定。然而挪威和瑞典等国家将典型疖疮病列在国家级疾病名录上，保留了相关法律文件，供临床疾病暴发和发生动物流行病事件时使用。挪威法律规定，育苗场的鲑必须注射疖疮病疫苗；在智利和加拿大，典型和非典型疖疮病均被列于鱼类疾病法规范围。

欧洲市场上关于疖疮病油类佐剂疫苗的效力测试，须按照《欧洲药典》描述的方法进行（Anonymous，2008），除非是有记录在案且获准的其他测试方法。然而这些测试是基于腹腔注射攻毒，并且一直被认为无法测定出次级配方的效果（Midtlyng，2005），而基于测定抗体反应检测疫苗效果的研究获得了较好的结果（Romstad 等，2013），这极有可能替代现有检测方法。致死临界点的检测也有类似的情况。欧洲市场的疖疮病疫苗，包括不含油类佐剂配方的疫苗，必须遵循兽医疫苗（EMA/CVMP/IWP/206555/2010）和鱼类疫苗（EMA/CVMP/IWP/314550/2010）的指南。

疖疮病疫苗通常被许可主要用于养殖鲑科鱼（大西洋鲑和虹鳟），但按照处方“级联”原则，在其他鱼类中合法使用这些产品也是可行的。在欧洲，这一原则也适用于需要紧急使用疖疮病疫苗的情况，尽管某些国家尚未获得此类疫苗的常规营销授权。同样，在智利紧急使用的疖疮病疫苗，可由农畜业部（SAG）授权使用。在根本没有可用的许可疫苗情况下，多数国家允许生产和使用被批准的由发病农场、地区或生产环节中分离到的致病菌株制备的自体疫苗，对非典型疖疮病进行免疫接种。

参考文献

Ackermann, P. A., Iwama, G. K. and Thornton, J. C. 2000. Physiological and immunological effects of adjuvanted *Aeromonas salmonicida* vaccines on juvenile rainbow trout. *J Aquat Anim Health* 12: 157-164.

Anonymous. 2008. Furunculosis vaccine (inactivated, oil-adjuvanted, injectable) for salmonids. *European Pharmacopoeia Monographs No*. 1521.

Arnesen, K. R., Mikkelsen, H., Schrøder, M. B. and Lund, V. 2010. Impact of reattaching various *Aeromonas salmonicida* A-layer proteins on vaccine efficacy in Atlantic cod (*Gadus morhua*). *Vaccine* 8: 4703-4708.

Aunsmo, A., Guttvik, A., Midtlyng, P. J., *et al*., 2008a. Association of spinal deformity and vaccine induced abdominal lesions in harvest sized Atlantic salmon, *Salmo salar*. *J Fish Dis* 31: 515-524.

Aunsmo, A., Larssen, R. B., Valle, P. S., *et al*., 2008b. Improved field trial methodology for quantifying vaccination side-effects in farmed Atlantic salmon (*Salmo salar* L.). *Aquaculture* 284: 19-24.

Austin, B. and Austin, D. A. 2007. *Bacterial Fish Pathogens*. *Disease of Farmed and Wild Fish*. 4th edn. Chichester, Springer Praxis.

Beaz-Hidalgo, R. and Figueras, M. J. 2013. *Aeromonas* spp. whole genomes and virulence factors implicated in fish disease. *J Fish Dis* 36: 371-388.

Bebak-Williams, J. and Bullock, G. L. 2002. Vaccination against furunculosis in Arctic char: efficacy of a commercial vaccine. *J Aquat Anim Health* 14: 294-297.

Bernoth, E.-M. 1997. Furunculosis, the history of the disease and of disease research, in *Furunculosis-Multidiciplinary Fish Disease Research* (eds E.-M. Bernoth, A. E. Ellis, P. J. Midtlyng, *et al*.,). London, Academic Press, 1-20.

Bernoth, E.-M., Ellis, A. E., Midtlyng, P. J., *et al*., (eds) 1997. *Furunculosis-Multidiciplinary Fish Disease Research*. London, Academic Press.

Björnsdóttir, B., Guðmundsdóttir, S., Bambir, S. H. and Guðmundsdóttir, B. K. 2005. Experimental infection of turbot, *Scophthalmus maximus* (L.) by *Aeromonas salmonicida* subsp. *achromogenes* and evaluation of cross protection induced by a furunculosis vaccine. *J Fish Dis* 28: 181-188.

Boyd, J. M., Danacay, A., Knickle, L. C., *et al*., 2008. Contribution of type IV pili to the virulence of *Aeromonas salmonicida* subsp. *salmonicida* in Atlantic salmon (*Salmo salar* L.). *Infect Immun* 76: 1445-1455.

Bricknell, I. R., Bowden, T. J., Lomax, J. and Ellis, A. E. 1997. Antibody response and protection in Atlantic salmon (*Salmo salar*) immunized with an extracellular polysaccharide of *Aeromonas salmonicida*. *Fish Shellfish Immunol* 7: 1-16.

Bricknell, I. R., King, J. A., Bowden, T. J. and Ellis, A. E. 1999. Duration of protective antibodies, and the correlation with protection in Atlantic salmon (*Salmo salar* L), following vaccination with an *Aeromonas salmonicida* vaccine containing iron-regulated outer membrane protein and secretory polysaccharide. *Fish Shellfish Immunol* 9: 139-151.

Bruhn, J. B., Dalsgaard, I., Nielsen, K. F., *et al*., 2005. Quorum sensing signal molecules (acylated homoserine lactones) in Gram-negative fish pathogenic bacteria. *Dis Aquat Org* 65：43-52.

Burr，S. E.，Pugovkin，D.，Wahli，T.，*et al.*，2005. Attenuated virulence of an *Aeromonas salmonicida* subsp. *salmonicida* type III secretion mutant in a rainbow trout model. *Microbiology* 151：2111-2118.

Charette，S. J.，Brochu，F.，Boyle，B.，*et al.*，2012. Draft genome sequence of the virulent strain 01-B526 of the fish pathogen *Aeromonas salmonicida*. *J Bacteriol* 194：722-723.

Chung，S. and Secombes，C. J. 1987. Activation of rainbow trout macrophages. *J Fish Biol* 31A：51-56.

Cipriano，R. C. and Austin，B. 2011. Furunculosis and other aeromonad diseases，in *Fish Diseases and Disorders. Vol.* 3：*Viral，Bacterial and Fungal Infections*，2nd edn（eds P. T. K. Woo and D. B. Bruno）. Wallingford，CABI，424-483.

Danacay，A.，Knickle，L. C.，Solanky，K. S.，*et al.*，2006. Contribution of the type Ⅲ secretion system（TTSS）to virulence of *Aeromonas salmonicida* subsp. *salmonicida*. *Microbiology* 152：1847-1856.

Danacay，A.，Boyd，J. M.，Fast，M. D.，*et al.*，2010. *Aeromonas salmonicida* type I pilus system contributes to host colonization but not invasion. *Dis Aquat Org* 88：199-206.

Duff，D. C. B. 1942. The oral immunization of trout against *Bacterium salmonicida*. *J Immunol* 44：87-94.

Durbin，M.，McIntosch，D.，Smith，P. D.，*et al.*，1999. Immunization against furunculosis in rainbow trout with iron-regulated outer membrane protein vaccines：relative efficacy of immersion，oral，and injection delivery. *J Aquat Anim Health* 11：68-75.

Ebanks，R. O.，Dacanay，A.，Goguen，M.，*et al.*，2004. Differential proteomic analysis of *Aeromonas salmonicida* outer membrane proteins in response to low iron and *in vivo* growth conditions. *Proteomics* 4：1074-1085.

Ebanks，R. O.，Goguen，M.，McKinnon，S.，*et al.*，2005. Identification of the major outer membrane proteins of *Aeromonas salmonicida*. *Dis Aquat Org* 68：29-38.

Eggset，G.，Mikkelsen，H. and Killie，J. E. A. 1997. Immunocompetence and duration of immunity against *Vibrio salmonicida* and *Aeromonas salmonicida* after vaccination of Atlantic salmon（*Salmo salar* L.）at low and high temperatures. *Fish Shellfish Immunol* 7：247-260.

Egidius，E. 1987. *Import of furunculosis to Norway with Atlantic salmon smolts from Scotland*. International Council for the Exploration of the Sea（ICES），Mariculture Committee Report no. C. M. 1987/F：8.

Ellis，A. E. 1991. An appraisal of the extracellular toxins of *Aeromonas salmonicida* subsp. *salmonicida*. *J Fish Dis* 14：265-277.

Ellis，A. E. 1997a. Furunculosis protective antigens and mechanisms，in *Furunculosis-Multidiciplinary Fish Disease Research*（eds E. -M. Bernoth，A. E. Ellis，P. J. Midtlyng，*et al.*，）. London，Academic Press，382-404.

Ellis，A. E. 1997b. Immunization with bacterial antigens：Furunculosis. *Dev Biol Stand* 90：107-116.

Ellis，A. E. 1997c. The extracellular toxins of *Aeromonas salmonicida* subsp. *salmonicida*，in *Furunculosis-Multidiciplinary Fish Disease Research*（eds E. -M. Bernoth，A. E. Ellis，P. J. Midtlyng，*et al.*，）. London，Academic Press，248-268.

Ellis，A. E.，Burrows，A. S. and Stapleton，K. J. 1988. Lack of relationship between virulence of *Aeromonas salmonicida* and the putative virulence factors，A-layer，extracellular proteases，and extracellular hemolysins. *J Fish Dis* 11：309-323.

Emmerich，R. and Weibel，E. 1894. Über eine durch Bakterien erzeugte Seuche unter den Forellen. *Arch Hyg* 21：1-21.

Enger，Ø. E. 1997. Survival and inactivation of *Aeromonas salmonicida* outside the host-a most superficial way of life，in *Furunculosis-Multidiciplinary Fish Disease Research*（eds E. -M. Bernoth，A. E. Ellis，P. J. Midtlyng，*et al.*,). London, Academic Press, 159-177.

Erdal, J. I. and Reitan, L. J. 1992. Immune response and protective immunity after vaccination of Atlantic salmon (*Salmo salar* L.) against furunculosis. *Fish Shellfish Immunol* 2: 99-108.

Evelyn, T. P. T. 1997. A historical review of fish vaccinology. *Dev Biol Stand* 90: 3-12.

Evenberg, D., de Graaff, P., Lugtenberg, B. and van Muiswinkel, W. B. 1988. Vaccine-induced protective immunity against *Aeromonas salmonicida* tested in experimental carp erythrodermatitis. *J Fish Dis* 1: 337-350.

Fast, M. D., Tse, B., Boyd, J. M. and Johnson, S. C. 2009. Mutation in the *Aeromonas salmonicida* subsp. *salmonicida* type Ⅲ secretion system affect Atlantic salmon leucocyte activation and downstream immune responses. *Fish Shellfish Immunol* 27: 721-728.

Garduno, R. A. and Kay, W. W. 1995. Capsulated cells of *Aeromonas salmonicida* grown *in vitro* have different functional properties than capsulated cells grown *in vivo*. *Can J Microbiol* 41: 941-945.

Garduno, R. A., Moore, A. R., Olivier, G., *et al.*, 2000. Host cell invasion and intracellular residence by *Aeromonas salmonicida*: role of the S-layer. *Can J Microbiol* 46: 660-668.

Grøntvedt, R. N. and Espelid, S. 2004. Vaccination and immune responses against atypical *Aeromonas salmonicida* in spotted wolffish (*Anarhichas minor* Olafsen) juveniles. *Fish Shellfish Immunol* 16: 271-285.

Grøntvedt, R. N., Lund, V. and Espelid, S. 2003. Atypical furunculosis in spotted wolffish (*Anarhichas minor* O.) juveniles: bath vaccination and challenge. *Aquaculture* 232: 69-80.

Grove, S., Høie, S. and Evensen, Ø. 2003. Distribution and retention of antigens of *Aeromonas salmonicida* in Atlantic salmon (*Salmo salar L.*) vaccinated with a Δ-*aroA* mutant or formalin-inactivated bacteria in oil-adjuvant. *Fish Shellfish Immunol* 15: 349-358.

Guðmundsdóttir, B. K. 1998. Infections by atypical strains of the bacterium *Aeromonas salmonicida*: a review. *Búvisindi*; *Icel Agric Sci* 12: 61-72.

Guðmundsdóttir, B. K. and Björnsdóttir, B. 2007. Vaccination against atypical furunculosis and winter ulcer disease of fish. *Vaccine* 25: 5512-5523.

Guðmundsdóttir, B. K. and Guðmundsdóttir, S. 1997. Evaluation of cross protection by vaccines against atypical and typical furunculosis in Atlantic salmon (*Salmo salar L.*). *J Fish Dis* 20: 343-350.

Guðmundsdóttir, B. K. and Magnadóttir, B. 1997. Protection of Atlantic salmon (*Salmo salar* L.) against an experimental infection of *Aeromonas salmonicida* subsp. *achromogenes*. *Fish Shellfish Immunol* 7: 55-69.

Guðmundsdóttir, B. K., Jonsdóttir, H., Steinthorsdóttir, V., *et al.*, 1997. Survival and humoral antibody response of in Atlantic salmon, *Salmo salar L.*, vaccinated against *Aeromonas salmonicida* subsp. *achromogenes*. *J Fish Dis* 20: 351-360.

Guðmundsdóttir, S., Lange, S., Magnadóttir, B. and Guðmundsdóttir, B. K. 2003. Protection against atypical furunculosis in Atlantic halibut *Hippoglossus hippoglossus* (L.): comparison of a commercial furunculosis vaccine and an autogenous vaccine. *J Fish Dis* 26: 331-338.

Gunnlaugsdóttir, B. and Guðmundsdóttir, B. K. 1997. Pathogenicity of atypical *Aeromonas salmonicida* in Atlantic salmon compared with protease production. *J Appl Bacteriol* 83: 542-551.

Hirst, I. D. and Ellis, A. E. 1994. Iron-regulated outer membrane proteins of *Aeromonas salmonicida* are important protective antigens in Atlantic salmon against furunculosis. *Fish Shellfish Immunol* 4: 29-45.

Hoel, K., Salonius, K. and Lillehaug, A. 1997. *Vibrio* antigens of polyvalent vaccines enhance the humoral immune response to *Aeromonas salmonicida* antigens in Atlantic salmon (*Salmo salar* L.). *Fish Shellfish Immunol* 7: 71-80.

Hoel, K, Reitan, L. J. and Lillehaug, A. 1998. Immunological cross reactions between *Aeromonas salmonicida* and *Vibrio salmonicida* in Atlantic salmon (*Salmo salar* L.) and rabbit. *Fish Shellfish Immunol* 8: 171-182.

Horne, J. H. 1928. Furunculosis in trout and the importance of carriers in the spread of the disease. *J Hyg* 28: 67-78.

Horne, M. T. and Ellis, A. E. 1988. Commercial production and licensing of vaccines for fish, in *Fish Vaccination* (ed. A. E. Ellis). London, Academic Press, 47-54.

Hussain, I., Mackie, C., Cox, D., *et al.*, 2000. Suppression of the humoral immune response of Atlantic salmon, *Salmo salar* L. by the 64 kDa serine protease of *Aeromonas salmonicida*. *Fish Shellfish Immunol* 10: 359-373.

Ingilæ, M., Arnesen, J. A., Lund, V. and Eggset, G. 2000. Vaccination of Atlantic halibut (*Hippoglossus hippoglossus* L.) and spotted wolffish (*Anarhichas minor* O.) against atypical *Aeromonas salmonicida*. *Aquaculture* 183: 31-44.

Irie, T., Watarai, S., Iwasaki, T. and Kodama, H. 2005. Protection against experimental *Aeromonas salmonicida* infection in carp by oral immunisation with bacterial antigen entrapped liposomes. *Fish Shellfish Immunol* 18: 235-242.

Jutfelt, F., Olsen, R. E., Glette, J., *et al.*, 2006. Translocation of viable *Aeromonas salmonicida* across the intestine of rainbow trout, *Oncorhynchus mykiss* (Walbaum). *J Fish Dis* 29: 255-262.

Jutfelt, F., Sundh, H., Glette, J., *et al.*, 2008. The involvement of *Aeromonas salmonicida* virulence factors in bacterial translocation across the rainbow trout, *Oncorhynchus mykiss* (Walbaum) intestine. *J Fish Dis* 31: 141-151.

Koppang, E. O., Fjølstad, M., Melgård, B., *et al.*, 2000. Non-pigment-producing isolates of *Aeromonas salmonicida* subspecies *salmonicida*: isolation, identification, transmission and pathogenicity in Atlantic salmon, *Salmo salar* L. *J Fish Dis* 23: 39-48.

Krantz, G. E., Reddecliff, J. M. and Heist, C. E. 1964. Immune response of trout to *Aeromonas salmonicida*. Part I. Development of agglutinating antibodies and protective immunity. *Prog Fish Cult* 26: 3-10.

Lee, K. K. and Ellis, A. E. 1990. Glycerophospholipid: cholesterol acyltransferase complexed with lipopolysaccharide (LPS) is a major lethal exotoxin and cytolysin of *Aeromonas salmonicida*: LPS stabilizes and enhances toxicity of the enzyme. *J Bacteriol* 172: 5382-5393.

Lee, K. K. and Ellis, A. E. 1991. Role of the lethal extracellular cytolysin of *Aeromonas salmonicida* in the pathology of furunculosis. *J Fish Dis* 14: 453-460.

Lillehaug, A., Lunder, T. and Poppe, T. T. 1992. Field testing of adjuvanted furunculosis vaccines in Atlantic salmon, *Salmo salar L*. *J Fish Dis* 15: 485-496.

Lönnström, L.-G., Rahkonen, R., Lundén, T., *et al.*, 2001. Protection, immune response and side-effects in European whitefish (*Coregonus lavaretus* L.) vaccinated against vibriosis and furunculosis. *Aquaculture* 200: 271-284.

Lund, V. and Mikkelsen, H. 2004. Genetic diversity among A-proteins of atypical strains of *Aeromonas salmonici*da. *Dis Aquat Org* 61: 257-262.

Lund, V., Arnesen, J. A. and Eggset, G. 2002. Vaccine development for atypical furunculosis in spotted wolffish *Anarhichas minor* O.: comparison of efficacy of vaccines containing different strains of atypical *Aeromonas salmonicida*. *Aquaculture* 204: 33-44.

Lund, V., Arnesen, J. A., Coucheron, D., *et al.*, 2003a. The *Aeromonas salmonicida* A-layer protein is an important protective antigen in oil-adjuvanted vaccines. *Fish Shellfish Immunol* 15：367-372.

Lund，V.，Espelid，S. and Mikkelsen，H. 2003b. Vaccine efficacy in spotted wolffish *Anarhichas minor*：relationship to molecular variation in A-layer protein of atypical *Aeromonas salmonicida*. *Dis Aquat Org* 56：31-42.

Lund，V.，Børdal，S.，Kjellsen，O.，*et al.*，2006. Comparison of antibody responses in Atlantic cod（*Gadus morhua* L.）to *Aeromonas salmonicida* and *Vibrio anguillarum*. *Dev Comp Immunol* 30：1145-1155.

Lund，V.，Arnesen，J. A.，Mikkelsen，H.，*et al.*，2008a. Atypical furunculosis vaccines for Atlantic cod（*Gadus morhua*）；vaccine efficacy and antibody responses. *Vaccine* 26：679-679.

Lund，V.，Mikkelsen，H. and Schrøder，M. B. 2008b. Comparison of atypical furunculosis vaccines in spotted wolffish（*Anarhichas minor* O.）and Atlantic halibut（*Hippoglossus hippoglossus* L.）*Vaccine* 26：2833-2840.

Lutwyche，P.，Exner，M. M.，Hancock，R. E. W. and Trust，T. J. 1995. A conserved *Aeromonas salmonicida* porin provides protective immunity to rainbow trout. *Infect Immun* 63：3137-3142.

Magnadóttir，B. 2010. Immunological control of fish diseases. *Mar Biotechnol* 12：361-379.

Markestad，A. and Grave，K. 1997. Reduction of antibacterial drug use in Norwegian fish farming due to vaccination. *Dev Biol Stand* 90：365-369.

Marsden，M.，Vaughan，L. M.，Fitzpatrick，R. M.，*et al.*，1998. Potency testing of a live，genetically attenuated vaccine for salmonids. *Vaccine* 16：1087-1094.

Masada，C. L.，LaPatra，S. E.，Morton，A. W. and Strom，M. S. 2002. An *Aeromonas salmonicida* type VI pilin is required for virulence in rainbow trout *Oncorhynchus mykiss*. *Dis Aquat Org* 51：13-25.

Maurice，S.，Nussinovitch，A.，Jaffe，N.，*et al.*，2004. Oral immunization of *Carassius auratus* with modified recombinant A-layer proteins entrapped in alginate beads. *Vaccine* 23：450-459.

McCarthy，D. H. and Roberts，R. J. 1980. Furunculosis of fish—the present state of our knowledge，in *Advances in Aquatic Microbiology*（eds M. R. Droop and H. W. Jannasch）. London，Academic Press，293-341.

Merino，S.，Vilches，S.，Canals，R.，*et al.*，2005. A C1q-binding 40kDa porin from *Aeromonas salmonicida*：cloning，sequencing，role in serum susceptibility and fish immunoprotection. *Microb Pathog* 38：227-237.

Midtlyng，P. J. 1996. A field study on intraperitoneal vaccination of Atlantic salmon（*Salmo salar L.*）against furunculosis. *Fish Shellfish Immunol* 6：553-565.

Midtlyng，P. J. 1997. Vaccination against furunculosis，in *Furunculosis-Multidiciplinary Fish Disease Research*（eds E. M. Bernoth，A. E. Ellis，P. J. Midtlyng，G. Olivier and P. R. Smith）. London，Academic Press，382-404.

Midtlyng，P. J. 2005. Critical assessment of regulatory standards and tests for fish vaccines. *Dev Biol*（*Basel*）121：219-226.

Midtlyng，P. J. and Lillehaug，A. 1998. Growth of Atlantic salmon *Salmo salar* after intraperitoneal administration of vaccines containing adjuvants. *Dis Aquat Org* 32：91-97.

Midtlyng，P. J.，Reitan，L. J.，Lillehaug，A. and Ramstad，A. 1996a. Protection，immune responses and side-effects in Atlantic salmon（*Salmo salar* L.）vaccinated against furunculosis by different procedures. *Fish Shellfish Immunol* 6：599-613.

Midtlyng，P. J.，Reitan，L. J. and Speilberg，L. 1996b. Experimental studies on the efficacy and side-effects of intraperitoneal vaccination of Atlantic salmon（*Salmo salar* L.）against furunculosis. *Fish Shellfish Immunol* 6：335-350.

Mikkelsen，H.，Schrøder，M. B. and Lund，V. 2004. Vibriosis and atypical furunculosis vaccines；efficacy and side-effects in Atlantic cod（*Gadus morhua* L）. *Aquaculture* 242：81-91.

Munro，A. L. S. and Gauld，J. A. 1997. *Scottish fish farms annual production survey* 1996. SOAFD Marine

Laboratory Aberdeen Report 1997.

Munro, A. L. S. and Waddell, I. F. 1984. *Furunculosis: experience of its control in the sea water cage culture of Atlantic salmon in Scotland*. International Council of the Exploration of the Sea, Mariculture Committee Report 1984; CM 1984 /F: 32.

Mutoloki, S., Alexandersen, S. and Evensen, Ø. 2004. Sequential study of antigen persistence and concomitant inflammatory reactions relative to side-effects and growth of Atlantic salmon (*Salmo salar* L.) following intraperitoneal injection with oil-adjuvanted vaccines. *Fish Shellfish Immunol* 16: 633-644.

Mutoloki, S., Brudeseth, B., Reite, O. B. and Evensen, Ø. 2006. The contribution of *Aeromonas salmonicida* extracellular products to the induction of inflammation in Atlantic salmon (*Salmo salar* L.) following vaccination with oil-based vaccines. *Fish Shellfish Immunol* 20: 1-11.

Najimi, M., Lemos, M. L. and Osorio, C. R. 2009. Identification of iron regulated genes in the fish pathogen *Aeromonas salmonicida* subsp. *salmonicida*. Genetic diversity and evidence of conserved iron uptake systems. *Vet Microbiol* 133: 377-382.

Noonan, B. and Trust, T. J. 1997. The synthesis, secretion and role in virulence of the paracrystalline surface protein layers of *Aeromonas salmonicida* and *A. hydrophila*. *FEMS Microbiol Lett* 154: 1-7.

Nordmo, R. and Ramstad, A. 1997. Comparison of different challenge methods to evaluate the efficacy of furunculosis vaccines in Atlantic salmon, *Salmo salar* L. *J Fish Dis* 20: 119-126.

Novotny, A. 1978. Vibriosis and furunculosis in marine cultured salmon in Puget Sound, Washington. *Mar Fish Rev* 40: 52-55.

O'Dowd, A. M., Bricknell, I. R., Secombes, C. J. and Ellis, A. E. 1999. The primary and secondary antibody responses to IROMP antigens in Atlantic salmon (*Salmo salar* L.) immunized with A+ and A− *Aeromonas salmonicida* bacterins. *Fish Shellfish Immunol* 9: 125-138.

Olivier, G., Moore, A. and Fildes, J. 1992. Toxicity of *Aeromonas salmonicida* cells to Atlantic salmon, *Salmo salar*, peritoneal macrophages. *Dev Comp Immunol* 16: 49-61.

Paterson, W. D., Lall, S. P., Airdrie, D., *et al.*, 1985. Prevention of disease in salmonids by vaccination and dietary modification. *Fish Pathol* 20: 427-434.

Poppe, T. T. and Breck, O. 1997. Pathology of Atlantic salmon *Salmo salar* intraperitoneally immunized with oil-adjuvanted vaccine. A case report. *Dis Aquat Org* 29: 219-226.

Prost, M. 2001. Extracellular proteases ECP of *Aeromonas salmonicida* and their significance in the pathogenesis and immunoprophylaxis of fish furunculosis. *Med Weter* 57: 79-82.

Pylkkö, P., Lyytikainen, T., Ritola, O., *et al.*, 2002. Temperature effect on the immune defense functions of Arctic char *Salvelinus alpinus*. *Dis Aquat Org* 52: 47-55.

Reith, M. E., Singh, R. K., Curtis, B., *et al.*, 2008. The genome of *Aeromonas salmonicida* subsp. *Salmonicida* A449: insights into the evolution of a fish pathogen. *BMC Genomics* 9: 427.

Robertson, P. A. W., Austin, D. A. and Austin, B. 2005. Prevention of ulcer disease in goldfish by means of vaccination. *J Aquat Anim Health* 17: 203-209.

Romstad, A. B., Reitan, L. J., Midtlyng, P., *et al.*, 2012. Development of an antibody ELISA for potency testing of furunculosis (*Aeromonas salmonicida* subsp. *salmonicida*) vaccines in Atlantic salmon (*Salmo salar* L). *Biologicals* 40: 67-71.

Romstad, A. B., Reitan, L. J., Midtlyng, P., *et al.*, 2013. Antibody responses correlate with antigen dose and *in vivo* protection for oil-adjuvanted, experimental furunculosis (*Aeromonas salmonicida* subsp. *salmonicida*) vaccines in Atlantic salmon (*Salmo salar* L.) and can be used for batch potency testing of vaccines. *Vaccine* 31: 791-796.

Rønsholt, B. and McLean, E. 1999. The effect of vaccination and vaccine components upon short-term growth and feed conversion efficiency in rainbow trout. *Aquaculture* 174: 213-221.

Salte, R., Norberg, K., Ødegård, O. R., *et al.*, 1993. Exotoxin-induced consumptive coagulopathy in Atlantic salmon *Salmo salar* L.; inhibitory effects of exogenous antithrombin and α-2 macroglobulin on *Aeromonas salmonicida* serine protease. *J Fish Dis* 16: 561-568.

Santos, Y., Garcia-Marquez, S., Pereira, P. G., *et al.*, 2005. Efficacy of furunculosis vaccines in turbot, *Scophthalmus maximus* (L.): evaluation of immersion, oral and injection delivery. *J Fish Dis* 28: 165-172.

Schwenteit, J., Gram, L., Nielsen, K. F., *et al.*, 2011. Quorum sensing in *Aeromonas salmonicida* subsp. *achromogenes* and the effect of the auto inducer synthease AsaI on bacterial virulence. *Vet Microbiol* 147: 389-397.

Sinyakov, M., Dror, M., Margel, M. and Avtalion, R. R. 2001. Immunogenicity of the *Aeromonas salmonicida* A-protein in goldfish (*Carassius auratus* L.). *Isr J Aquacult-Bamidgeh* 53: 110-114.

Smith, P. and Hiney, M. 2000. Oil-adjuvanted furunculosis vaccines in commercial fish farms: a preliminary epizootiological investigation. *Aquaculture* 190: 1-9.

Spence, K. D., Pilcher, K. S. and Fryer, J. L. 1965. Active and passive immunization of certain salmonids fishes against *Aeromonas salmonicida*. *Can J Microbiol* 11: 397-405.

Thornton, J. C., Garduno, R. A. and Kay, W. W. 1994. The development of live vaccines for furunculosis lacking the A-layer and O-antigen of *Aeromonas salmonicida*. *J Fish Dis* 17: 195-204.

Tomás, J. M. 2012. The main *Aeromonas* pathogenic factors. *ISRN Microbiology*, article ID 256261.

Treasurer, J. and Cox, C. 2008. Intraperitoneal and dorsal median sinus vaccination effect on growth, immune response, and reproductive potential in farmed Atlantic salmon *Salmo salar*. *Aquaculture* 275: 51-57.

Trust, T. J., Noonan, B., Chu, S. J., *et al.*, 1996. A molecular approach to understanding the pathogenesis of *Aeromonas salmonicida*; relevance to vaccine development. *Fish Shellfish Immunol* 6: 269-276.

Vaughan, L. M., Smith, P. R. and Foster, T. J. 1993. An aromatic-dependent mutant of the fish pathogen *Aeromonas salmonicida* is attenuated in fish and is effective as a live vaccine against the salmonid disease furunculosis. *Infect Immun* 61: 2172-2181.

Villumsen, K. R., Dalsgaard, I., Holten-Andersen, L. and Raida, M. K. 2012. Potential role of specific antibodies as important vaccine induced protective mechanism against *Aeromonas salmonicida* in rainbow trout. *PloS One* 7: e46733.

Vipond, R., Bricknell, I. R., Durant, E., *et al.*, 1998. Defined deletion mutants demonstrate that the major secreted toxins are not essential for the virulence of *Aeromonas salmonicida*. *Infect Immun* 66: 1990-1998.

Wang, S., Laroque, S., Vinogradov, E., *et al.*, 2004. Structural studies of the capsular polysaccharide and lipopolysaccharide O-antigen of *Aeromonas salmonicida* strain 80204-1 produced under *in vitro* and *in vivo* growth conditions. *Eur J Biochem* 271: 4507-4516.

Wang, S., Li, J., Vinogradov, E. and Altman, E. 2006. Structural studies of the core region of *Aeromonas salmonicida* subsp. *salmonicida* lipopolysaccharide. *Carbohydr Res* 341: 109-117.

Wiklund, T. and Dalsgaard, I. 1998. Occurrence and significance of atypical *Aeromonas salmonicida* in non-salmonid and salmonid species: a review. *Dis Aquat Org* 32: 49-69.

第十七章　发光杆菌病的疫苗接种

Jesús L. Romalde

本章概要　SUMMARY

发光杆菌病是一种细菌性败血症，以前又被称为巴斯德菌病或“假结核病”。因在疾病慢性形成过程中，患病鱼的脏器，特别是在脾脏和肾脏中出现由细菌积聚形成的白色结节而得名。嗜盐性细菌，美人鱼发光杆菌杀鱼亚种（*Photobacterium damselae* subsp. *Piscicida*）(原称为杀鱼巴斯德菌（*Pasteurella piscicida*）是该病的致病菌。该病在欧洲、亚洲和美洲等均有报道，且可感染多种经济鱼类，如白鲈（*Morone americana*）、条纹鲈（*M. saxatilis*）、杂交条纹鲈（*M. saxatilis*× *M. chrysops*）、黄尾鰤（*Seriola quinqueradiata*）、海鲷（*Sparus aurata*）、海鲈（*Dicentrarchus labrax*）和塞内加尔鳎（*Solea senegalensis*）。已研制出该病的多种疫苗，在欧洲的某些国家，十余年来，通过使用商品化富含该致病菌胞外产物（ECP）的菌苗免疫海鲷和海鲈，已经成功地控制了该病，目前，日本已研发出一些疫苗保护黄尾鰤抵御该病。

第一节　发生流行与危害

1963 年，在美国切萨皮克湾上游的白鲈和条纹鲈中发现了美人鱼发光杆菌杀鱼亚种(*Photobacterium damselae* subsp. *piscicida*)。Sniezsko 等（1964）从发病鲈的血液和脏器中分离出一株独特的细菌，基于其形态和生化特性，将其归为巴斯德菌属。然而，该细菌的分类地位一直受到争议，直到最近才被明确。尽管该菌曾被归至不同的属（Romalde，2002），但基于其 rRNA 和 DNA-DNA 杂交研究，确认引起鱼类巴斯德菌病的病原为美人鱼发光杆菌杀鱼亚种(Gauthier 等，1995)。

虽然该菌最初分离自患病的白鲈和条纹鲈的野生种群，但从 1969 年起，造成了日本养殖黄尾鰤（*Seriola quinqueradiata*）幼鱼的巨大损失。1970 年，该菌被认为是造成美国德克萨斯州加尔维斯顿湾的野生鲱（*Brevoortia tyranus*）和鲻（*Mugil cephalus*）严重损失的主要原因（Lewis 等，

1970)。几年后，巴斯德菌病成为日本黄尾鰤、香鱼（*Plecoglosus altivelis*）和黑鲷（*Mylio macrocephalus*）养殖中的严重问题（Kusuda 和 Miura，1972；Muroga 等，1977；Ohnishi 等，1982）。随后，该病传播到其他鱼类中，在红鲷（*Acanthopagrus schlegeli*）（Yasunaga 等，1983）、红石斑鱼（*Epinephelus akaara*）（Ueki 等，1990）和杂交条纹鲈（*M. saxatilis* × *M. chrysops*）（Hawke 等，2003）等中的流行被相继报道。

自 1990 年，美人鱼发光杆菌杀鱼亚种造成了欧洲的地中海国家海水养殖的金头鲷（*Sparus aurata*）、海鲈（*Dicentrarchus labrax*）和塞内加尔鳎（*Solea senegalensis*）的经济损失（Ceschia 等，1990；Toranzo 等，1991；Balebona 等，1992；Zorrilla 等，1999；Baptista 等，1996）。后来该病也造成法国的鲻（*Mugil cephalus*）、竹筴鱼（*Trachurus trachurus*）和红鲂鮄（*Aspitrigla cuculus*）（P. Daniel，个人交流）以及中国台湾的军曹鱼（*Rachycentron canadum*）（Liu 等，2003）的死亡。此外，在意大利，表征健康的鲻中也检测到该病原（Serracca 等，2011）。

巴斯德菌病在较高的水温（大于 23℃）和盐度（20～30），以及水质较差的夏季易流行（Fryer 和 Rohovec，1984；Frerichs 和 Roberts，1989；Magariños 等，2001）。Janssen 和 Surgalla（1968）以及 Toranzo 等（1982）的实验发现，美人鱼发光杆菌杀鱼亚种在水中仅存活 4～5d，表明水体不是该细菌的主要储存库。然而，海水和底质中的生存试验表明，虽然该菌离开鱼体后在水体环境中不能长期处于可培养的状态，但它可以进入一种有活力但非可培养的休眠状态（Magariños 等，1994b）。这些结果均表明，尽管美人鱼发光杆菌杀鱼亚种非常不稳定，但水生环境构成了该病原的储存库和传播媒介。此外，研究还证明，处于活的非可培养状态的美人鱼发光杆菌杀鱼亚种对鱼类仍有潜在的致病性，当营养条件适宜时，其能快速复苏并增殖。

血清学技术已成功应用于检测、诊断以及鉴定该病原。该技术应用特异性的抗血清能够鉴定出美人鱼发光杆菌杀鱼亚种，另外，不论菌株的地域来源，该种菌在血清学上均是同源的（Magariños 等，1992a；Salati 等，1994），这也使该菌无法建立血清型分类。抗原的均一性有助于研发血清学诊断工具，其中有许多已商品化生产，有的甚至可以从无症状鱼体中快速检测出该病原（Romalde 等，1995a，1995b，1999），抗原的均一性也有助于研发有效的疫苗（Magariños 等，1994c，1999；Bakopoulos 等，2003；Romalde 等，2005；Gravningen 等，2008）。

第二节　病原学

美人鱼发光杆菌杀鱼亚种（*Photobacterium damselae* subsp. *piscicida*）的特征是无运动力、革兰氏阴性、杆状，依不同的培养条件，该菌可由球形变至长杆形。氧化酶和过氧化氢酶阳性、发酵型，对弧菌抑制剂 O/129 敏感。精氨酸水解酶、脂肪酶和磷脂酶呈阳性反应，对吲哚、硝酸盐还原、脲酶、明胶酶、淀粉酶和硫化氢的产生呈阴性反应。在碳水化合物的利用方面，该菌只能利用葡萄糖、甘露糖、半乳糖和果糖产酸。美人鱼发光杆菌杀鱼亚种的生长温度范围 15～32.5℃，最适生长温度为 22.5～30℃。美人鱼发光杆菌杀鱼亚种为同一表型（Magariños 等，1992a）。有研究报道认为，典型的美人鱼发光杆菌杀鱼亚种为无荚膜的球杆菌（Koike 等，1975），然而，Bonet 等（1994）研究发现，大多数美人鱼发光杆菌杀鱼亚种具有由99.6%的碳水化合物和0.4%的蛋白质组成的荚膜，这对细菌发病机制起着重要作用。

血清学上，美人鱼发光杆菌杀鱼亚种构成一个高度相似的类群（Kusuda 和 Fukuda，1980；Nomura 和 Aoki，1985；Magariños 等，1992a）。这种血清学同质性表现在所有菌株脂多糖（LPS）电泳结果相同，即含有高分子量的 O 抗原呈阶梯状条带。此外，在所有的分离菌株的总蛋白和外膜蛋白的电泳中观察到一致的条带模式。这些 LPS 和膜蛋白均与美人鱼发光杆菌杀鱼亚种分离株的免疫学特性相关。

应用核型和不同的 PCR 技术、AFLP（放大片段长度多态性）、RAPD（随机放大多态 DNA）、ERIC（肠杆菌重复基因序列的序列）和 REP（重复的外基因组序列）等 DNA 指纹图谱技术发现，美人鱼发光杆菌杀鱼亚种内主要存在两个克隆谱系，一个为欧洲分离株，另一个为日本和美国分离株（Magariños 等，1997，2000；Thyssen 等，1999；Mancuso 等，2007）。这些结果说明，该菌存在遗传异质性，可用于该菌的流行病学研究。

通过 SSH（抑制消减杂交）技术分析不同区域菌株的基因组证实了这种遗传的高度异质性，同时表明结合共轭因子（ICE）可将毒力基因转移至该病原中（Juíz-Río 等，2005；Osorio 等，2008）。

通过分析发现，不同美人鱼发光杆菌杀鱼亚种菌株所含的质粒不同，大部分欧洲分离株含有一个大约 50Mu 的质粒，日本分离株常见 37Mu 的质粒（Magariños 等，1992a）。

第三节 致病机理

美人鱼发光杆菌杀鱼亚种的临床症状已在日本养殖的黄尾鰤中有详细报道（Kubota 等，1970；Fukuda 和 Kusuda，1981），与加利西亚（西班牙西北部）以及地中海地区养殖海鲷的临床症状相似（Toranzo 等，1991）。巴斯德菌病的外表症状通常不明显，一般无体表病变。某些感染的鱼可能出现体色变黑或变浅，头和鳃部轻微出血。鱼体内病理变化与疾病的急性和慢性病变有关。急性发病时，器官严重的病理变化较少，慢性病变的特点是器官上会形成白色结节或肉芽肿。通过病理学观察，肉芽肿组织中含有细菌菌落和坏死的吞噬细胞。此外，脾脏、肾脏和肝脏中有大量巨噬细胞聚集，其中许多都是已吞噬大量菌体的坏死的巨噬细胞（Nelson 等，1989；Noya 等，1995a）。这些结果表明，美人鱼发光杆菌杀鱼亚种能在宿主巨噬细胞中存活并且不会被降解。此外，鳃中也观察到含有美人鱼发光杆菌杀鱼亚种菌体并变性的巨噬细胞，以及坏死的各种类型细胞，造成严重的呼吸功能损伤（Noya 等，1995a）。

美人鱼发光杆菌杀鱼亚种对多种海洋鱼类均有高致病性，无宿主特异性，（Magariños 等，1995）。有研究发现，海鲈和海鲷在不同生长阶段对该病的易感性差异与巨噬细胞和中性粒细胞有关，宿主的这些细胞仅能在成鱼阶段发挥高效吞噬并杀灭细菌的作用（Noya 等，1995b；Skarmeta 等，1995）。总之，对美人鱼发光杆菌杀鱼亚种不同分离株高致病性的研究表明，该病原一定具有重要的致病机制。

关于细菌表面特性方面，已报道美人鱼发光杆菌杀鱼亚种具有细胞膜适度疏水性，并且呈弱凝集性或无凝集性（Magariños 等，1996a）。侵入特性分析结果表明，该病原菌呈中度侵染性。有趣的是，研究发现该菌能够在宿主细胞中存活至少 6d，并且可在细胞间传播，能在鱼体单细胞层形成噬斑。此外，细胞松弛素 D 实验结果表明，宿主细胞的细胞骨架参与了细菌的入侵过程（Magariños 等，1996a）。

美人鱼发光杆菌杀鱼亚种分离株的多糖荚膜在病原致病力方面发挥重要作用（Magariños 等，1996b）。实验结果表明，由于荚膜的存在，阻抗了血清对病原菌的杀菌作用，而使得该菌具有较低的半致死剂量 LD_{50}。

病原菌对铁离子的利用能力对其在宿主体内的生长繁殖以及引起宿主感染至关重要。据报道（Magariños 等，1994a），美人鱼发光杆菌杀鱼亚种具有铁离子获取系统，由一个铁载体和至少 3 个高分子量的铁调控外膜蛋白组成。此外，美人鱼发光杆菌杀鱼亚种菌体可以直接将血红素和血红蛋白结合到铁调节膜受体上，将其作为菌体唯一的铁来源（Magariños 等，1994a；do Vale 等，2002）。目前已知，与致病力相关的铁载体合成，是由于在该菌携带质粒上插入了编码毒力岛的基因（Osorio 等，2008）。

鱼类病原菌分泌的胞外产物（ECPs）是重要的毒力因子，胞外产物能够促进病情进一步发展或抵抗宿主的防御机制，这对病原菌是有利的。实验证明，美人鱼发光杆菌杀鱼亚种分离株的 ECPs 可以使金头鲷、海鲈、大菱鲆和虹鳟等多种鱼类死亡。这些 ECPs 含有对小鼠致命的毒素，并具有很强的磷脂酶活性、细胞毒性和溶血性（Magariños 等，1992b）。

研究发现，由 AIP56 质粒编码的分子量为 56 ku 的美人鱼发光杆菌杀鱼亚种外毒素蛋白与该菌的毒力有关，可促进海鲈的巨噬细胞和中性粒细胞凋亡（Vale 等，2005）。因此，该病原是通过影响巨噬细胞和中性粒细胞，逃避宿主的吞噬作用，从而在宿主体内大量增殖。凋亡反应涉及激活半胱天冬酶途径和线粒体途径，表明存在内源和外源两种凋亡途径（Costa-Ramos 等，2011）。另外，Liu 等（2011）从美人鱼发光杆菌杀鱼亚种中纯化出一种分子量为 34.3 ku 的毒性金属蛋白酶，该酶在 pH 为 6.0～8.0 时活性最大，对军曹鱼的 LD_{50} 为 6.8 mg/g（鱼体重）。

第四节　疫　苗

在 20 世纪 80 年代以前，药物一直是治疗鱼类巴斯德菌病最有效的方法。但之后研究发现了该菌存在由质粒或染色体编码的对各种抗生素和其他药物的抗性基因（Magariños 等，1992b；Kim 等，1993）。此外，美人鱼发光杆菌杀鱼亚种在感染过程中，存在一段巨噬细胞内寄生的时期（Kusuda 和 Salati，1993），这也可以解释在疾病暴发时使用药物没有疗效的原因。

几十年来，已研制了不同种鱼类巴斯德菌病的疫苗（Romalde 和 Magariños，1997）。这些疫苗通常是常规培养基培养的菌体，经热或甲醛灭活的全菌疫苗。为提高这些疫苗的免疫保护率，使用浸泡免疫结合超声波接种方法（Navot 等，2005），以及注射疫苗时混合使用矿物油或非矿物油类佐剂（Le Breton，2009）。

为获得更好的保护效果，还有研究使用新型培养基（即葡萄糖和/或富盐培养基，或缺铁培养基），使细菌产生更多的 ECPs/LPS，或使用粗制的多糖荚膜等细菌自身成分等作为抗原（Magariños 等，1994c，1999；Bakopoulos 等，2003；Gravningen 等，2008）。

在一些国家，商品化的美人鱼发光杆菌杀鱼亚种疫苗可以从制药公司的动物保健部门购买，如 Pharmaq 公司、Hipra SA 公司、Merck 公司、Novartis AG 公司以及 Pfizer Inc 公司等。其他兽药公司，如 Acuitec SL，拥有授权研制自体疫苗的实验室。

处于实验阶段的疫苗有各种稳定的缺铁载体和 *aro*-A 缺失突变株（Kusuda 和 Hamaguchi，1988；Thune 等，2003），后者已在美国获得专利，用于杂交条纹鲈的免疫接种。

研究鉴定出美人鱼发光杆菌杀鱼亚种入侵过程中的相关蛋白，体外过表达该蛋白，与甲醛灭活的菌苗配伍使用（Barnes 等，2005），所研制的疫苗已被加拿大的 Aqua Health 公司商业化生产销售，名为 Photogen。该疫苗对海鲈和黄尾鰤有免疫保护作用，但具有剂量依赖性。还有疫苗的研究旨在利用美人鱼发光杆菌杀鱼亚种胞外产物中的保护性蛋白制备亚单位疫苗，目前尚未获得满意结果（Barnes 等，2005）。

有研究报道基于免疫蛋白质组学的研究结果，鉴定出美人鱼发光杆菌杀鱼亚种的相关蛋白，通过大肠杆菌原核表达制备亚单位候选疫苗，结果表明在军曹鱼中亚单位疫苗优于全菌疫苗的保护效果（Ho 等，2011）。

第五节　疫苗接种规程

疫苗接种日益成为水产养殖的重要部分，是公认的防控各种流行疾病的有效方法。“疫苗接种策略”是指实施免疫防控所确定的疫苗类型、接种方式、接种时间和再次免疫等具体内容。

疫苗的研发和商业化的重要因素包括疫苗接种的操作方法和程序，这些应纳入接种鱼类的日常生产管理中，与该疾病发生的典型生态学和流行病学相联系（即季节性发生、鱼体大小、发病宿主和疾病发生的地理范围）。

任何疫苗在实施接种策略之前，确定鱼体免疫系统在形态和功能上均已成熟是非常重要的。大菱鲆、鲽、海鲷、海鲈或鳎等严格意义上是完全海水中生活的鱼类，其孵化期短；孵化后，脆弱的未成熟幼体在海水环境中，免疫系统发育不完善。与上述鱼类不同，鲑科鱼类具有较大的卵黄，孵化期长。此外，还需要考虑，严格的海水鱼类与鲑科鱼类相比，其从孵化到开口饲喂之间的时间非常短。

有关海水鱼类疫苗接种最早发育期的资料很少，因为大多数海水鱼类在孵化时，其淋巴系统尚在发育，淋巴系统的功能在孵化后 70～100d 才成熟，因此非特异性免疫是幼鱼防御机制中最重要的组成部分。一般来说，如果鱼在较早生长阶段接种了疫苗，免疫保护期将会很短，需要在免疫后约一个月进行再次免疫。此外，如果鱼在成长期内发生疾病，必须在鱼体重达 30～50g 时进行第三次接种。本章已叙述过，免疫接种富含胞外产物的菌苗使海鲷和海鲈获得了对发光杆菌病的最佳免疫保护，接种程序如下：在仔鱼阶段（体重平均 0.05g）进行第一次浸泡免疫；在幼鱼阶段（体重 1～2 g）进行第二次接种；随后，在养殖场对生长期的鱼体进行口服或注射加强免疫（Romalde 等，2005；Le Breton，2009）。

有关使用免疫增强剂预防美人鱼发光杆菌杀鱼亚种的研究表明，葡聚糖添加于饲料中可预防或减少金头鲷巴斯德菌病的死亡率（Couso 等，2003）。另一方面，饲料成分改变或者在饲料中加入预防美人鱼发光杆菌杀鱼亚种的添加剂，对鱼体既有积极作用也有消极影响。例如，饲料中添加精氨酸显著提高了塞内加尔鳎的先天性免疫反应（Costas 等，2011），而用植物油代替鱼油导致了海鲷免疫应答失调，造成鱼体对细菌感染应答反应变弱（Montero 等，2010）。

发光杆菌病与高温有关。Magariños 等（2001）研究表明，无症状带菌的金头鲷亲鱼繁育的幼鱼也会携带病菌，当水温高于 18℃ 时，幼鱼会发生发光杆菌病；而当水温低于 18℃ 时，虽然鱼生长变慢，但有助于控制发光杆菌病（Magariños 等，2001）。

综上所述，疫苗接种、控制水温和饲料中使用葡聚糖等措施相结合有助于更好地预防美人鱼发光杆菌杀鱼亚种引起的发光杆菌病。

第六节 效 果

已有诸多研究对热或甲醛灭活疫苗预防发光杆菌病的免疫效果进行了评价（Romalde 和 Magariños，1997）。然而，这些被评估的疫苗配方均存在重复性差以及难以大规模生产的问题。Fukuda 和 Kusuda（1981）等研究了黄尾鰤的被动免疫接种，结果显示保护期非常短。

在海鲷和海鲈中，鱼体重为 0.5～2 g 时，免疫接种富含 ECP 的菌苗获得了最佳免疫保护，相对保护率（RPS）高达 75%（Magariños 等，1994c）。该疫苗接种金头鲷幼鱼（50 日龄）也有效，RPS 为 84%～90%（Magariños 等，1999）。该疫苗已获得授权并商品化，在欧洲国家成功应用。

鳎免疫接种该富含 ECP 的疫苗后，也能有效预防该病（Romalde 等，2005）。鳎和海鲷使用预防发光杆菌病和弧菌病的二联疫苗免疫后，同样获得了保护效果（Moriñigo 等，2002；Arijo 等，2005）；黄尾鰤和杜氏鰤（*Seriola dumerili*）接种发光杆菌病和乳球菌病的二联疫苗后，也出现了类似结果（Wada 等，2008）。然而研究还发现，使用美人鱼发光杆菌杀鱼亚种和海洋屈挠杆菌（*Tenacibaculum maritimum*）的二联疫苗时，两种抗原存在拮抗效应（Romalde 等，2005）。

第七节 副作用

正如本书其他章节所述，疫苗接种的副作用在很大程度上与使用油类佐剂疫苗或接种处理过程中的应激反应有关。就目前所知，仅有一项关于海鲈腹腔接种发光杆菌病疫苗的特异性副作用研究（Afonso 等，2005）。研究发现，腹腔注射油类佐剂疫苗的鱼体，其淋巴细胞稳定增长，中性粒细胞和巨噬细胞数量增加。然而，除了一些肉眼可见的肉芽肿外，鱼体病变属中度，由此推测海鲈腹腔注射该疫苗产生的副作用没有其他鱼类严重。

第八节 管理条例

目前常用的鱼类发光杆菌病疫苗均是基于常规技术研制的灭活疫苗，因此尚无该疫苗相关的专门管理条例。

参考文献

Afonso, A., Gomes, S., da Silva, J., *et al.*, 2005. Side effects in sea bass (*Dicentrarchus labrax* L.) due to intraperitoneal vaccination against vibriosis and pasteurellosis. *Fish Shellfish Immunol* 19: 1-16.

Arijo, S., Rico, R., Chabrillon, M., *et al.*, 2005. Effectiveness of a divalent vaccine for sole, *Solea senegalensis* (Kaup), against *Vibrio harveyi* and *Photobacterium damselae* subsp. *piscicida*. *J Fish Dis* 28: 33-38.

Bakopoulos, V., Volpatti, D., Gusmani, L., *et al.*, 2003. Vaccination trials of sea bass, *Dicentrarchus labrax* (L.), against *Photobacterium damsela* subsp. *piscicida*, using novel vaccine mixtures. *J Fish Dis* 26: 77-90.

Balebona, M. C., Moriñigo, M. A., Sedano, J., *et al.*, 1992. Isolation of *Pasteurella piscicida* from sea bass in southwestern Spain. *Bull Eur Assoc Fish Pathol* 12: 1-3.

Baptista, T., Romalde, J. L. and Toranzo, A. E. 1996. First epizootic of pasteurellosis in Portugal affecting cultured gilthead sea bream (*Sparus aurata*). *Bull Eur Assoc Fish Pathol* 16: 92-95.

Barnes, A. C., dos Santos, N. M. S. and Ellis, A. E. 2005. Update on bacterial vaccines: *Photobacterium damselae* subsp. *piscicida*. *Dev Biol* (*Basel*) 121: 75-84.

Bonet, R., Magariños, B., Romalde, J. L., *et al.*, 1994. Capsular polysacaccharide expressed by *Pasteurella piscicida* grown *in vitro*. *FEMS Microbiol Lett* 124: 285-290.

Ceschia, G., Giorgetti, G. and Bovo, G. 1990. Grave epizoozia da *Pasteurella piscicida* in specie eurihaline del Nord Adriatico. *Boll Soc Ital Patol Ittica* 4: 11-17.

Costa-Ramos, C., do Vale, A., Ludovico, P., *et al.*, 2011. The bacterial exotoxin AIP56 induces fish macrophage and neutrophil apoptosis using mechanisms of the extrinsic and intrinsic pathways. *Fish Shellfish Immunol* 30: 173-181.

Costas, B., Conceição, L. E. C., Dias, J., *et al.*, 2011. Dietary arginine and repeated handling increase disease resistance and modulate innate immune mechanisms of Senegalese sole (*Solea senegalensis* Kaup, 1858). *Fish Shellfish Immunol* 31: 838-847.

Couso, N., Castro, R., Magariños, B., *et al.*, 2003. Effect of oral administration of glucans on the resistance of gilthead sea bream to pasteurellosis. *Aquaculture* 219: 99-109.

do Vale, A., Magariños, B., Romalde, J. L., *et al.*, 2002. Binding of haemin by the fish pathogen *Photobacterium damselae* subsp. *piscicida*. *Dis Aquat Org* 48: 109-115.

do Vale, A., Silva, M. T., dos Santos, N. M., *et al.*, 2005. AIP56, a novel plasmid-encoded virulence factor of *Photobacterium damselae* subsp. *piscicida* with apoptogenic activity against sea bass macrophages and neutrophils. *Mol Microbiol* 58: 1025-1038.

Frerichs, G. N. and Roberts, R. J. 1989. The bacteriology of teleosts, in *Fish Pathology* (ed. R. J. Roberts). London, Bailliere Tindall, 289-319.

Fryer, J. L. and Rohovec, J. S. 1984. Principal bacterial diseases of cultured marine fish. *Helgoländer Meeresunters*

37：533-545.

Fukuda，Y. and Kusuda，R. 1981. Efficacy of vaccination for pseudotuberculosis in cultured yellowtail by various routes of administration. *Bull Jap Soc Sci Fish* 47：147-150.

Gauthier G，Lafay B，Ruimi R，*et al.*，1995. Small subunit rRNA sequences and whole DNA relatedness concur for the reassignment of *Pasteurella piscicida* （Sniezsko *et al.*） Janssen and Surgalla to the genus *Photobacterium* as *Photobacterium damselae* subsp. *piscicida* comb. nov. *Int J Syst Bacteriol* 45：139-144.

Gravningen，K.，Sakai，M.，Mishiba，T. and Fujimoto，T. 2008. The efficacy and safety of an oil-based vaccine against *Photobacterium damsela* subsp. *piscicida* in yellowtail （*Seriola quinqueradiata*）：a field study. *Fish Shellfish Immunol* 24：523-529.

Hawke，J. P.，Thune，R. L.，Cooper，R. K.，*et al.*，2003. Molecular and phenotypic characterization of strains of *Photobacterium damselae* subsp. *piscicida* from hybrid striped bass cultured in Louisiana，USA. *J Aquat Anim Health* 15：189-201.

Ho，L.-P.，Lin，J. H.-Y.，Liu，H.-C.，*et al.*，2011. Identification of antigens for the development of a subunit vaccine against *Photobacterium damselae* ssp. *piscicida*. *Fish Shellfish Immunol* 30：412-419.

Janssen，W. A. and Surgalla，M. J. 1968. Morphology，physiology，and serology of *Pasteurella* species pathogenic for white perch. *J Bacteriol* 96：1606-1610.

Juíz-Río，S.，Osorio，C. R.，de Lorenzo，V. and Lemos，M. L. 2005. Subtractive hybridization reveals a high genetic diversity in the fish pathogen *Photobacterium damselae* subsp. *piscicida*：evidence of a SXT-like element. *Microbiology* 151：2659-2669.

Kim，E.-H.，Yoshida，T. and Aoki，T. 1993. Detection of R plasmids encoded with resistance to florfenicol in *Pasteurella piscicida*. *Fish Pathol* 28：165-170.

Koike，Y.，Kuwahara，A. and Fujiwara，H. 1975. Characterization of *Pasteurella piscicida* isolated from white perch and cultivated yellowtail. *Jap J Microbiol* 19：241-247.

Kubota，S. S.，Kimura，M. and Egusa，S. 1970. Studies of a bacterial tuberculoidosis of the yellowtail. I. Symptomatology and histopathology. *Fish Pathol* 4：11-18.

Kusuda，R. and Fukuda，Y. 1980. Agglutinating antibody titers and serum protein changes of yellowtail after immunization with *Pasteurella piscicida* cells. *Bull Jap Soc Sci Fish* 46：103-108.

Kusuda，R. and Hamaguchi，M. 1988. The efficacy of attenuated live bacterin of *Pasteurella piscicida* against pseudotuberculosis in yellowtail. *Bull Eur Assoc Fish Pathol* 8：51-53.

Kusuda，R. and Salati，F. 1993. Major bacterial diseases affecting mariculture in Japan. *Annu Rev Fish Dis* 3：69-85.

Kusuda，R. and Miura，W. 1972. Characteristics of a *Pasteurella* sp. pathogenic for pond cultured ayu. *Fish Pathol* 7：51-57.

Le Breton，A. D. 2009. Vaccines in Mediterranean aquaculture：practice and needs. *Op Mediterr* 86：147-154.

Lewis，D. H.，Grumbles，L. C.，McConnell，S. and Flowers，A. I. 1970. *Pasteurella*-like bacteria from an epizootic in menhaden and mullet in Galveston Bay. *J Wild Dis* 6：160-162.

Liu，P. C.，Liu，J. Y. and Lee，K. K. 2003. Virulence of *Photobacterium damselae* subsp. *piscicida* in cultured cobia *Rachycentron canadum*. *J Basic Microbiol* 43：499-507.

Liu，P. C.，Chuang，W. H. and Lee，K. K. 2011. Purification of a toxic metalloprotease produced by the pathogenic *Photobacterium damselae* subsp. *piscicida* isolated from cobia （*Rachycentron canadum*）. *Z Naturforsch* 66：287-295.

Magariños，B.，Romalde，J. L.，Bandín，I.，*et al.*，1992a. Phenotypic，antigenic and molecular characterization of *Pasteurella piscicida* strains isolated from fish. *Appl Environ Microbiol* 58：3316-3322.

Magariños，B.，Santos，Y.，Romalde，J. L.，*et al.*，1992b. Pathogenic activities of live cells and extracellular products of the fish pathogen *Pasteurella piscicida*. *J Gen Microbiol* 138：2491-2498.

Magariños，B.，Romalde，J. L.，Lemos，M. L.，*et al.*，1994a. Iron uptake by *Pasteurella piscicida* and its role in pathogenicity for fish. *Appl Environ Microbiol* 60：2990-2998.

Magariños，B.，Romalde，J. L.，Barja，J. L. and Toranzo，A. E. 1994b. Evidence of a dormant but infective state of the fish pathogen *Pasteurella piscicida* in seawater and sediment. *Appl Environ Microbiol* 60：180-186.

Magariños，B.，Romalde，J. L.，Santos，Y.，*et al.*，1994c. Vaccination trials on gilhead sea bream（*Sparus aurata*）against *Pasteurella piscicida*. *Aquaculture* 120：201-208.

Magariños，B.，Toranzo，A. E. and Romalde，J. L. 1995. Different susceptibility of gilthead sea bream and turbot to *Pasteurella piscicida* infection by the water route. *Bull Eur Assoc Fish Pathol* 15：88-90.

Magariños，B.，Romalde，J. L.，Noya，M.，*et al.*，1996a. Adherence and invasive capacities of the fish pathogen *Pasteurella piscicida*. *FEMS Microbiol Lett* 138：29-34.

Magariños，B.，Bonet，R.，Romalde，J. L.，*et al.*，1996b. Influence of the capsular layer on the virulence of *Pasteurella piscicida* for fish. *Microb Pathogen* 21：289-297.

Magariños，B.，Osorio，C. R.，Toranzo，A. E. and Romalde，J. L. 1997. Applicability of ribotyping for intraspecific classification and epidemiological studies of *Photobacterium damselae* subsp. *piscicida*. *Syst Appl Microbiol* 20：634-639.

Magariños，B.，Romalde，J. L.，Barja，J. L.，*et al.*，1999. Protection of gilthead sea bream against pasteurellosis at the larval stages. *Bull Eur Assoc Fish Pathol* 19：159-161.

Magariños，B.，Toranzo，A. E.，Barja，J. L. and Romalde，J. L. 2000. Existence of two geographically-linked clonal lineages in the bacterial fish pathogen *Photobacterium damselae* subsp. *piscicida* evidenced by random amplified polymorphic DNA analysis. *Epidemiol Infect* 125：213-219.

Magariños，B.，Couso，N.，Noya，M.，*et al.*，2001. Effect of temperature on the development of pasteurellosis in carrier gilthead sea bream（*Sparus aurata*）. *Aquaculture* 195：17-21.

Mancuso，M.，Avendaño-Herrera，R.，Zaccone，R.，*et al.*，2007. Evaluation of different DNA-based fingerprinting methods for typing *Photobacterium damselae* subsp. *piscicida*. *Biol Res* 40：85-92.

Montero，D.，Mathlouthi，F.，Tort，L.，*et al.*，2010. Replacement of dietary fish oil by vegetable oils affects immunity and expression of pro-inflammatory cytokines genes in gilthead sea bream *Sparus aurata*. *Fish Shellfish Immunol* 29：1073-1081.

Moriñigo，M. A.，Romalde，J. L.，Chabrillon，M.，*et al.*，2002. Effectiveness of a divalent vaccine for gilthead sea bream（*Sparus aurata*）against *Vibrio alginolyticus* and *Photobacterium damselae* subsp. *piscicida*. *Bull Eur Assoc Fish Pathol* 22：298-303.

Muroga，K.，Sugiyama，T. and Ueki，N. 1977. Pasteurellosis in cultured black sea bream，*Mylio macrocephalus*. *J Fac Fish Anim Husb Hiroshima Univ* 16：17-21.

Navot，N.，Kimmel，E. and Avtalian，R. R. 2005. Immunisation of fish by bath immersion using ultrasound. *Dev Biol*（*Basel*）121：135-142.

Nelson，J. S.，Kawahara，E.，Kawai，K. and Kusuda，R. 1989. Macrophage infiltration in pseudotuberculosis of yellowtail，*Seriola quinqueradiata*. *Bull Mar Sci Fish Kochi Univ* 11：17-22.

Nomura，J. and Aoki，T. 1985. Morphological analysis of lipopolysaccharide from Gram negative fish pathogenic bacteria. *Fish Pathol* 20：193-197.

Noya，M.，Magariños，B.，Toranzo，A. E. and Lamas，J. 1995a. Sequential pathology of experimental pasteurellosis in gilthead sea bream, *Sparus aurata*. A light- and electron-microscopic study. *Dis Aquat Org* 21: 177-186.

Noya, M., Magariños, B. and Lamas, J. 1995b. Interactions between peritoneal exudate cells (PECs) of gilthead sea bream (*Sparus aurata*) and *Pasteurella piscicida*. *A morphological study*. *Aquaculture* 131: 11-21.

Ohnishi, K., Watanabe, K. and Jo, Y. 1982. *Pasteurella* infection in young black sea bream. *Fish Pathol* 16: 207-210.

Osorio, C. R., Marrero, J., Wozniak, R. A. F., *et al.*, 2008. Genomic and functional analysis of ICE*Pda*Spa1, a fish-pathogen-derived SXT-related integrating conjugative element that can mobilize a virulence plasmid. *J Bacteriol* 190: 3353-3361.

Romalde, J. L. 2002. *Photobacterium damselae* subsp. *pisicida*: an integrated view of a bacterial fish pathogen. *Int Microbiol* 5: 3-9.

Romalde, J. L. and Magariños, B. 1997. Immunization with bacterial antigens: pasteurellosis. *Dev Biol* (*Basel*) 90: 167-177.

Romalde, J. L., Magariños, B., Fouz, B., *et al.*, 1995a. Evaluation of BIONOR mono-kits for rapid detection of bacterial fish pathogens. *Dis Aquat Org* 21: 25-34.

Romalde, J. L., le Breton, A., Magariños, B. and Toranzo, A. E. 1995b. Use of BIONOR Aquarapid-Pp kit for the diagnosis of *Pasteurella piscicida* infections. *Bull Eur Ass Fish Pathol* 15: 64-66.

Romalde, J. L., Magariños, B., Lores, F., *et al.*, 1999. Assessment of a magnetic bead-EIA based kit for the rapid diagnosis of fish pasteurellosis. *J Microbiol Methods* 38: 147-154.

Romalde, J. L., Ravelo, C., López-Romalde, S., *et al.*, 2005. Vaccination strategies to prevent emerging diseases for Spanish aquaculture. *Dev Biol* (*Basel*) 121: 85-95.

Salati, F., Giorgetti, G. and Kusuda, R. 1994. Comparison among strains of *Pasteurella piscicida* from Japan, Italy and USA. *Riv Itali Acquac* 29: 133-139.

Serracca, L., Ercolini, C., Rossini, I., *et al.*, 2011. Occurrence of both subspecies of *Photobacterium damselae* in mullets collected in the river Magra (Italy). *Can J Microbiol* 57: 437-440.

Skarmeta, A. M., Bandín, I., Santos, Y. and Toranzo, A. E. 1995. *In vitro* killing of *Pasteurella piscicida* by fish macrophages. *Dis Aquat Org* 23: 51-57.

Sniezsko, S. F., Bullock, G. L., Hollis, E. and Boone, J. G. 1964. *Pasteurella* sp. from an epizootic of white perch (*Roccus americanus*) in Chesapeake Bay tidewaters areas. *J Bacteriol* 88: 1814-1815.

Thune, R. L., Fernández, D. H., Hawke, J. P. and Miller, R. 2003. Construction of a safe, stable, efficacious vaccine against*Photobacterium damselae* ssp. *piscicida*. *Dis Aquat Org* 57: 51-58.

Thyssen, A., Goris, J., Pedersen, K., *et al.*, 1999. Phenotypic and genotypic characterization of *Photobacterium damselae* subsp. *piscicida*, in *Proceedings of the 9th International Conference of the EAFP*: *Diseases of Fish and Shellfish*, Rhodes, Greece.

Toranzo, A. E., Barja, J. L. and Hetrick, F. M. 1982. Survival of *Vibrio anguillarum* and *Pasteurella piscicida* in estuarine and fresh waters. *Bull Eur Ass Fish Pathol* 3: 43-45.

Toranzo, A. E., Barreiro, S., Casal, J. F., *et al.*, 1991. Pasteurellosis in cultured gilthead sea bream (*Sparus aurata*): first report in Spain. *Aquaculture* 99: 1-15.

Ueki, N., Kayano, Y. and Muroga, K. 1990. *Pasteurella piscicida* in juvenile red grouper. *Fish Pathol* 25: 43-44.

Wada, Y., Oyamatsu, T., Ng, J., *et al.*, 2008. Bivalent vaccine protects yellowtails, amberjack against pasteurellosis. *Glob Aquac Advocate* 11 (May-June): 51-52.

Yasunaga, N., Katai, K. and Tsukahara, J. 1983. *Pasteurella piscicida* from an epizootic of cultured red sea bream. *Fish Pathol* 18: 107-110.

Zorrilla, I., Balebona, M. C., Moriñigo, M. A., *et al.*, 1999. Isolation and characterization of the causative agent of pasteurellosis, *Photobacterium damselae* ssp. *piscicida*, from sole, *Solea senegalensis* (Kaup). *J Fish Dis* 22: 167-172.

第十八章　鮰肠道败血症的疫苗接种

Phillip H. Klesius and Julia W. Pridgeon

本章概要　SUMMARY

鮰爱德华菌（*Edwardsiella ictaluri*）是一种重要的水产养殖致病菌，可引起鮰科和非鮰科鱼肠道败血症（ESC）。该病的减毒活疫苗已经获批商品化生产，用于对鱼体实施大规模浸浴免疫接种。商品化的减毒活疫苗是利用利福平抗性策略生产，另一种减毒活疫苗的研发是基于新霉素抗性策略。在美国，该病的灭活疫苗还未上市。本章介绍该病的发生流行、危害、病原学、致病机制、疫苗、疫苗接种程序、疫苗效果和疗效。

第一节　危　　害

鮰爱德华菌（*Edwardsiella ictaluri*）是鮰肠道败血症（ESC）的致病菌，是造成养殖斑点叉尾鮰经济损失最严重的疾病之一（Hawke，1979；Hawke 等，1981）。2002 年，Wagner 等报道，肠道败血症和柱状杆菌病是美国养殖鮰最主要的疾病。据报道，中国养殖的黄颡鱼（*Pelteobagrus fulvidraco*）也能感染鮰爱德华菌（Ye 等，2009；Liu 等，2010），每年的累积死亡率达 50%。越南也有人工养殖巴沙鱼（*Pangasius hypophthalmus*）感染鮰爱德华菌的报道（Ferguson 等，2001；Crumlish 等，2002；Dung 等，2008）。在印度尼西亚，鮰爱德华菌也是养殖巴沙鱼的致病菌（Yuasa 等，2003），并且东南亚鮰养殖业中的 ESC 暴发也越来越严重。在美国，每年 ESC 造成的经济损失为5 000万～8 000万美元（Russo 等，2009）。疾病的暴发造成的鱼类产量减少、疾病治疗费用和死亡率增加等均是导致经济损失的原因。

第二节　发生流行

鮰爱德华菌比迟缓爱德华菌（*E. tarda*）的宿主范围小。最初，该菌被认为是鮰科鱼类特有的病原菌（Plumb，2000）。随后发现非鮰科鱼类也易感染该病，因此鮰爱德华菌的宿主范围比最初记载的广（Waltman 等，1985）。除斑点叉尾鮰和越南巴沙鱼及黄颡鱼外，鮰爱德华菌还可感染云斑鮰（*Ameriurus nebulosus*）、白鲶（*Ictalurus catus*）、玫瑰无须鲃（*Puntius conchonius*）（Hawke 等，1979，1981；Humphrey 等，1986）、绿刀鱼（*Eigemannia virens*）和孟加拉斑马鱼（*Danio devario*）（Kent 和 Lyons，1982；Waltman 等，1985）。蟾胡鲶（*Clarias batrachua*）、蝌蚪石鮰（*Noturus gyrinus*）、虹鳟（*Oncorhynchus mykiss*）和高体波鱼（*Rosbara hetermorpha*）也出现肠道败血症或类似症状（Kasornchandra 等，1987；Reid 和 Boyle，1989；Klesius 等，2003；Keskin 等，2004）。从蝌蚪石鮰上分离的鮰爱德华菌腹腔注射感染斑点叉尾鮰，死亡率达 60%（Klesius 等，2003）。斑马鱼（*D. rerio*）已经成为研究该病的模式鱼类（Petrie-Hanson 等，2007），斑马鱼感染鮰爱德华菌后的死亡率（13.3%）显著低于斑点叉尾鮰（100%）。此外，在实验条件下该菌还可感染大鳞大麻哈鱼（*Oncorhynclus tshawytsha*）、罗非鱼（*Oreochromis aureus*）和欧洲鲶（*Silurus gianis*）（Plumb 和 Sanchez，1983；Plumb 和 Huge，1987；Baxa 等，1990），这些鱼中并无鮰爱德华菌自然感染的报道。之前的研究认为鮰爱德华菌血清型和生化型均是同类的（Plumb 和 Klesius，1988；Plumb 和 Vinitnantharat，1993；Waltman 等，1986；Bertolini 等，1990）。

第三节　病原学

典型的鮰肠道败血症通常是急性发病，患病鱼出现游动异常、鳃苍白、眼球突出、腹部肿大和死亡快等症状。慢性的肠道败血症通常病程较慢，并在头部形成“头颅穿孔”病灶，该病灶可能停留在局部，或发展为败血症进而死亡。患急性肠道败血症存活下来的鱼可能会发展为慢性肠道败血症，在这个进程中，鮰爱德华菌可从鱼体的大脑、头肾和脑病变处恢复感染。肠道败血症可发生在鮰的所有生活阶段，包括仔鱼、幼鱼乃至商品鱼阶段。在美国，肠道败血症通常发生在春季和秋季的养殖鮰中，此季节该病的死亡率达到高峰（Plumb 和 Sanchez，1983）。鮰爱德华菌的最适生长温度是 25～30℃，Klesius（1992a）推测，水温从 11～17℃ 到 20～30℃ 的急剧变化可能加速鮰爱德华菌的水平传播（Bly 和 Clem，1991），导致肠道败血症更易暴发。

有关肠道败血症的病原学所知甚少。Klesius（1992b）研究表明，健康的斑点叉尾鮰感染鮰爱德华菌后，经 Romet©-30（磺胺二甲嘧啶与二甲氧苄氨嘧啶的复方药，译者注）治疗后，仍是鮰爱德华菌的携带者，可能是易感肠道败血症鱼类的病原储库。因此，无症状携带鮰爱德华菌的鱼可能会传播该疾病到其他养殖鮰中，特别是在集约化养殖的鮰的不同生长阶段中。另一方面，Klesius（1994）观察到感染鱼的同类相食，或是爱德华菌从死鱼上游离出来并再次感染其他鱼，两者皆是鮰爱德华菌的水平传播方式，并导致疾病的发生。鮰爱德华菌其他的传播途径是经水和含有感染病原的斑点叉尾鮰养殖池塘的底泥传播（Plumb 和 Quinlan，1986），也可能通过被病菌污染的网具和设施以及池塘里感染死亡的鱼体传播鮰爱德华菌。Newton 等（1989）实验室条件下研究发现，与感染鮰爱德华菌的斑点叉尾鮰同居，可能是其他鱼类感染的途径。

环境、宿主和微生物是影响鮰养殖中肠道败血症发病频率的主要因素。水质差、残饵、排泄物、病鱼残骸、高密度养殖、营养不良、投喂不当等都可能导致肠道败血症暴发。渔业养殖区域杂乱、卫生消毒不当以及抗生素类药物的过度使用等都会有引入病原的风险，耐药微生物转移至陆生动物和人类的风险增加是一个严峻的公共卫生问题。全球范围内，集约化养殖导致了鮰产业肠道败血症暴发风险升高，在中国，集约化养殖也是导致黄颡鱼死亡率上升和产业经济损失的重要原因（Deng 等，2008），肠道败血症发病的时间与柱状黄杆菌病发生时间大致相同。鮰爱德华菌和多子小瓜虫（*Ichthyophthirus multifillis*）也都是鮰的主要病原，一项研究表明，小瓜虫寄生的斑点叉尾鮰感染鮰爱德华菌后，其死亡率显著高于未感染小瓜虫的斑点叉尾鮰，小瓜虫的感染使细菌更易入侵斑点叉尾鮰并增加鱼的死亡率（Xu 等，2012），这一事例说明两种病原混合感染会增强斑点叉尾鮰病原的致病力。

第四节　致病机理

鮰爱德华菌可以通过几种途径入侵鱼体并引起肠道败血症，而最有可能的两条途径是通过鮰的肠道和鼻孔（Shotts 等，1986）。鮰爱德华菌进入鼻孔并转移至嗅觉神经，然后进入脑膜、头骨以及皮肤（Morrison 和 Plumb，1994）。鮰爱德华菌侵染嗅觉神经并导致神经系统病变，可能是由于细菌的凝集素附着于鼻黏膜上特定的糖基，如 d-甘露糖、N-乙酰神经氨酸等，造成感染（Wolfe 等，1994）。此外将斑点叉尾鮰浸泡于^{35}S-标记的鮰爱德华菌中感染 1h 后研究发现，鮰爱德华菌会在鱼鳃中累积并通过鳃进入鱼体（Nusbaum 和 Morrison，1996），因此推测，鮰爱德华菌通过鮰鳃的巨噬细胞进入血液循环，迅速地通过血液循环转移到前肾，被前肾巨噬细胞类群内化。另一研究表明，鮰爱德华菌最初通过斑点叉尾鮰擦伤的皮肤入侵，随后发生系统性感染（Menanteau-Ledouble 等，2011）。浸浴感染后，鮰爱德华菌数量增加，这表明鮰爱德华菌可能经皮肤入侵肌肉层，因此鮰皮肤擦伤会更易患肠道败血症。Santander 等（2011）研究发现，鮰浸泡感染后，鮰爱德华菌的环腺苷 3′5′-单磷酸受体蛋白（Crp）突变株能定植于鮰的鳃、脾和肾中。

利用 Illumina RNA 测序技术研究鮰爱德华菌感染后，鮰肠道的转录应答差异表明，在感染后 3h、24h 和 3d，发现了1 633个差异表达的基因，鉴定出多个参与应答的基因，研究显示肌动蛋白-细胞骨架的聚合和连接调控作用在鮰爱德华菌入侵鱼体以及炎症反应中具有重要作用（Li 等，2012）。

第五节　毒力因子

目前已发现鮰爱德华菌的多种毒力因子。其中脂多糖（LPS）是重要的毒力因子（Weete 等，1988；Newton 和 Triche，1993a；Klesius 和 Shoemaker，1999；Lawrence 等，2001，2003；Arias 等，2003；Thune 等，2007），Saeed 和 Plumb（1986）研究发现，鮰爱德华菌的脂多糖虽然能够刺激鮰产生较高的抗体滴度，但攻毒后不能使鮰获得免疫保护。鞭毛也被认为是鮰爱德华菌的毒力因子（Newton 和 Triche，1993b；Panangala 等，2009a），鮰爱德华菌基因组显示可以编码至少 4 种鞭毛蛋白，与沙雷氏菌（*Serratia marcescens*）的鞭毛蛋白非常相似。另一种毒力因子为外膜蛋白（Newton 等，1990；Klesius 和 Horst，1991；Vinitnantharat 等，1993；Bader 等，2004）。从鮰爱德华菌的胞外产物或菌体外部抗原中，鉴定出分子量分别为 35ku 和 58.5ku 的两种外膜蛋白（Klesius 和 Horst，1991；Klesius，1993）。细胞外膜蛋白被鉴定为主要的免疫原，能够产生强烈的血清抗体反应。Klesius 和 Sealey（1996）研究发现，鮰爱德华菌的胞外产物能够促进斑点叉尾鮰巨噬细胞的趋化作用和趋化动力学效应。

溶血素和软骨素酶被认为是鮰爱德华菌可能的毒力因子。然而，研究发现，野生型鮰爱德华菌与溶血素基因缺失的鮰爱德华菌突变体在溶血活力上无显著差别（Williams 和 Lawrence，2005）。Copper 等（1996）的研究认为，软骨素硫酸酯酶在鮰爱德华菌毒力中发挥作用，制备的软骨素硫酸酯酶基因缺失的鮰爱德华菌突变株并不能引起鮰发病。此前，Waltman 等（1986）发现鮰爱德华菌能降解硫酸软骨素，这可能是形成“头颅穿孔”病变特征的原因。

第六节 疫苗和免疫力

用于鱼类的疫苗大多数都是灭活疫苗或减毒疫苗。在美国，有 10 种鱼类疫苗已获得批号，包括 5 种灭活细菌疫苗、1 种灭活病毒疫苗、1 个减毒活病毒疫苗和 3 个减毒活细菌疫苗。表 18-1 列出了当前美国获得批号的鱼类疫苗。大多数疫苗都用于鲑科鱼类，两种疫苗可用于鮰。目前，没有任何一种获得批号的疫苗可用于罗非鱼、澳洲肺鱼、黄鲈、海水鱼类和其他美国养殖鱼类。与家禽和畜牧行业相比，鱼类养殖生产者可使用的疫苗较少。因此，必须认识到研发鱼类疫苗以及提高现有细菌和病毒病疫苗的效力是非常必要的。

表 18-1 截至 2013 年 2 月美国农业部许可的鱼类疫苗

产品名字	产品号	许可发布日期
诺华动物健康公司		
杀鲑气单胞菌菌苗	2035.02	2005 年 6 月 1 日
节细菌疫苗，活体培养物	1K11.00	2005 年 6 月 1 日
杀鲑气单胞菌-鳗弧菌-奥达利弧菌-杀鲑弧菌-菌苗	2138.02	2005 年 6 月 1 日
柱状黄杆菌菌苗	2974.00	2005 年 6 月 1 日
鲑贫血症病毒疫苗，灭活疫苗	4A45.20	2005 年 6 月 1 日
耶尔森氏菌菌苗	2638.00	2005 年 6 月 1 日
鳗弧菌-奥达利弧菌-菌苗	2858.03	2005 年 6 月 1 日
鲤疱疹病毒Ⅲ型疫苗，基因工程活疫苗	1443.20	2005 年 6 月 1 日
默克动物健康公司		
鮰爱德华菌疫苗，减毒活疫苗	1531.00	2009 年 10 月 26 日
柱状黄杆菌疫苗，减毒活疫苗	17F1.00	2009 年 10 月 26 日

表 18-2 列出了鮰爱德华菌疫苗的优缺点。已研发出一些鮰爱德华菌疫苗并且进行了评估研究（Plumb，1988；Vinitnantharat 和 Plumb，1992，1993；Plumb 等，1994；Nusbaum 和 Morrison，1996；Thune 等，1992，1997a，1997b），证明这些疫苗是安全的。然而，由于灭活鮰爱

德华菌疫苗针对肠道败血症的保护效果不足，野外试验结果不稳定，目前在美国鮰养殖业，灭活鮰爱德华菌疫苗未被许可使用。研究发现，鮰爱德华菌疫苗可以刺激鱼体产生特异性抗体，但不产生对该病的免疫保护（Klesius 和 Sealy，1995；Thune 等，1997a）。

表 18-2　鮰爱德华菌疫苗的优缺点

优点	缺点
安全	注射免疫
全抗原组分	免疫保护效果持续时间短
蛋白变性甲醛灭活	
对野生株有交叉保护	需要加强免疫
生产成本低	抗体介导的免疫
发酵生产	
保质期长	对胞内病原保护力低或无保护
冻干	

表 18-3　鮰爱德华菌减毒活疫苗的优缺点

优点	缺点
失去毒力	在欧盟国家存在安全和管理条例问题
改变了脂多糖	
全抗原成分	
细胞免疫	
能在鱼体内复制	
在宿主体内可复制 7～10d	
具有检测标记分子	
免疫保护持续时间长	
可免疫孵化后 7～10d 的幼鱼	
对野生株具有交叉保护	
易于大量生产	
生产成本低	
发酵生产	

表 18-3 列出了鮰爱德华菌减毒疫苗的优缺点。尽管有安全性风险，但由于其有诸多优点，减毒活菌疫苗的研发和使用正变得越来越具有吸引力。与灭活疫苗相比，普遍认为减毒活疫苗存在的风险更高，主要原因是其可能从减毒状态恢复到致病状态。然而，这种隐患目前还没有报道，AQUAVAC-ESC® 的减毒疫苗已在美国鮰养殖业中使用 10 年以上，并未出现关于菌体毒力恢复的情况。通过对使用减毒鮰爱德华菌疫苗的风险/效益评估的分析发现，其带来的预防效益远超过了不用疫苗的情况。Favre 和 Viret（2006）构建了欧洲法规框架对人类口服减毒疫苗进行生物安全评估，这也为兽医实践中评估减毒疫苗提供了有价值的参考。抗生素耐药性等特异性遗传标记基因整合入减毒疫苗株，会有助于通过适当的细菌学方法，区分减毒疫苗菌株与野生型亲本菌株，这对野外监测减毒疫苗的生物安全性具有重要意义。

早期的研究结果为鮰爱德华菌减毒活疫苗的设计提供了指导（Klesius 等，2001）。首次使用辐射和异源诱变技术制备鮰爱德华菌突变株，但是γ射线诱导产生鮰爱德华菌突变体并接种于鸡胚中的研究未获成功(Klesius,个人交流)。Norqvist 等(1989) 采用利福平 (200μg/mL) 和链霉素 (100μg/mL) 诱变制备出抗生素突变的减毒活鳗弧菌 (*Vibrio anguillarum*) 株(Van1000)，该突变菌株能够诱导虹鳟产生对同源或异源鳗弧菌的免疫保护，仅需浸浴免疫接种 Van1000 一次，虹鳟产生的免疫保护作用能维持至少 17 周。并且免疫后第 12 周，该疫苗能诱导虹鳟产生对杀鲑气单胞菌 (*Aeromonas salmonicida*) 的免疫保护，证明 Van1000 菌株是安全的减毒疫苗。

Schurig 等 (1991) 研制了牛布鲁氏菌 (*Brucella abortus*) 减毒活疫苗，对牛的布氏菌病有效。该疫苗是通过增加利福平浓度后多次传代制备获得，利福平能抑制细菌 DNA 依赖性 RNA 聚合酶 (Wehrli 等，1968)。

Klesius 和 Shoemaker (1999) 证明，不同鮰爱德华菌菌株制备的活疫苗产生的对肠道败血症的免疫保护力存在微小差异。用鮰爱德华菌活疫苗株 EILO 免疫鮰，再以鮰分离出的鮰爱德华菌株攻毒，结果显示产生了显著的免疫保护效果，这一结果对研制有效的肠道败血症疫苗具有重要意义。研究发现，18 株鮰爱德华菌的蛋白表达谱具有较高同质性 (70%相似或更高)，脂肪酸甲酯谱相似度达 95% (Panangala 等，2006，2009b)。研究还表明，疫苗能够在二次攻毒时保护鮰免受感染 (Shoemaker 和 Klesius，1997)，表明该保护性免疫可能是由细胞介导产生的。

Klesius 和 Shoemaker (1999) 从患病胡子鲶中分离得到鮰爱德华菌，采用利福平诱变技术制备出非斑点叉尾鮰来源的鮰爱德华菌 EILO 减毒菌株。该株减毒活疫苗通过浸浴方式可进行大规模接种且对鱼体应激极小 (Shoemaker 等，1999，2009；Klesius 等，2001；Klesius 和 Shoemaker，2000)。该减毒活疫苗对鮰也有效，因此，美国农业部授权许可 Intervet/Schering-Plough Animal Health 公司生产名为 AQUAVAC-ESC© 的减毒活疫苗 (Klesius 和 Shoemaker，2000)，根据 AQUAVAC-ESC©说明书，该产品可用于 7～10 日龄或更大的鮰浸泡免疫 (Klesius 和 Shoemaker，1999；Wise 等，2000；Wise，2006)，对 10～48 日龄的鮰 RPS 为 57%～94%，该减毒疫苗的毒力缺失是由于改变了该菌脂多糖成分 (Klesius 和 Shoemaker，1999；Arias 等，2003)。

该疫苗浸泡免疫鮰发眼卵效果较好 (Klesius 等，2000，2004；Shoemaker 等，2002，2007)，在免疫接种后 60d 攻毒，相对保护率达 57%；但在孵化后的第 7 天加强免疫，其相对保护率下降至 27.3%，这一结果表明加强免疫的额外抗原量减弱了减毒活疫苗产生的免疫保护作用。有研究报道，减毒疫苗可诱导机体的免疫抑制 (Salmbandamurthy 和 Jacobs，2005)，可以肯定的是，减毒活疫苗在鱼卵和鱼苗中产生的持续免疫保护作用说明其成功诱导了发眼卵和仔稚鱼的免疫反应，发眼卵的免疫接种无异常死亡。

鮰发眼卵浸泡免疫柱状黄杆菌 (*F. columnare*) 减毒疫苗单苗，或免疫柱状黄杆菌和鮰爱德华菌减毒二联疫苗 (Shoemaker 等，2007)，接种后 137d，用柱状黄杆菌攻毒，柱状黄杆菌减毒单苗的相对保护率为 76.8%；接种后 109d，用柱状黄杆菌攻毒，二联减毒疫苗的相对保护率为 56.7%；接种后 116d，用鮰爱德华菌攻毒，二联减毒疫苗的相对保护率为 66.7%，这些结果表明，减毒疫苗能使鱼苗或仔稚鱼产生极为有效的免疫保护。

鮰爱德华菌减毒活疫苗可以激活鱼体的免疫反应，与自然感染类似，因为大多数抗原都可在体内复制。疫苗可以同时激活鱼体的先天性免疫系统和获得性免疫系统，在激活体液免疫和细胞免疫的同时，减毒活疫苗可诱导局部免疫反应和全身免疫反应（Carrias 等，2008），通常减毒活疫苗比灭活疫苗能更有效地激活细胞免疫。Shoemaker 等（1997）研究发现鮰免疫接种低剂量的鮰爱德华菌后其巨噬细胞杀灭鮰爱德华菌的效率更高。还有研究表明，鮰爱德华菌能在鮰的巨噬细胞中存活和复制，并导致巨噬细胞裂解（Russo 等，2009），与未接种疫苗的鮰巨噬细胞相比，受免鱼的血清对鮰爱德华菌的调理作用增加了巨噬细胞的杀菌能力，接种过疫苗的鮰，其巨噬细胞比未接种疫苗鮰巨噬细胞的吞噬和杀菌力更强。

为了探究鮰爱德华菌减毒活疫苗的免疫保护作用的机制，Pridgeon 等（2010）用聚合酶链式反应（PCR）和消减 cDNA 杂交技术，从鮰爱德华菌的减毒活菌株与毒性亲本菌株中鉴定出 41 个差异表达序列标签（ESTs）。结果显示，在减毒活菌株中 33 个 ESTs 比亲本菌株上调表达高至少 3 倍；22%的上调表达基因与产生对肠道败血症的保护性免疫有关。另一研究发现，减毒活疫苗免疫的鮰相比对照组有 167 个已知的功能基因上调表达，40 个基因下调表达；微阵列分析表明，鮰接种后 48h 溶菌酶 C 基因上调表达最显著，而肌微管素相关蛋白 1a 和细胞色素 *P450 2J27* 基因下调表达。因此认为，减毒活疫苗接种后，这些差异表达的基因在鱼体免疫应答中发挥重要的作用（Pridgeon 等，2012a）。AQUA-ESC© 疫苗浸泡接种鮰后，全基因表达谱分析显示，有 52 个独特的基因上调表达；微阵列分析发现，鮰载脂蛋白 *A*-1 基因在接种 48h 后上调表达最显著（Pridgeon 等，2012b）。

减毒活疫苗能激活机体较强的长期记忆 T 细胞（Salerno-Goncalves 和 Sztein，2006），刺激机体释放细胞因子，并产生细胞毒性 T 细胞（Secombes，2008）。该反应是初始 CD4 T 细胞和抗原递呈细胞相互作用产生的，会诱导细胞因子生成；随后辅助 T 细胞亚群激活和分化，使机体释放不同类型的细胞因子并最终产生对病原的特异性免疫。

一般来说，减毒活细菌疫苗对许多野生菌株均可产生保护作用（Klesius 和 Shoemaker，1999），这是减毒活疫苗优于灭活细菌疫苗的重要方面，因为灭活疫苗通常对不同菌株的免疫交叉保护效果较弱。灭活疫苗能够刺激产生特异性抗体反应（Seder 和 Hill，2000），但研究发现，甲醛灭活的鮰爱德华菌疫苗保护效果欠佳（Nusbaum 和 Morrison，1996），可能是因为甲醛的灭活方式改变了鮰爱德华菌的表面抗原（Bader 等，1997）或灭活的鮰爱德华菌无法入侵宿主细胞产生免疫刺激（Nusbaum 和 Morrison，1996）。此外，灭活疫苗产生的免疫保护期较短，而减毒活疫苗则较长，这是由于减毒的细菌可在鱼体组织内复制（Norqvist 等，1989）。减毒菌株在宿主中存在的时间越长，宿主获得的免疫保护越强（Salmbandarmurthy 和 Jacobs，2005），这可能与活疫苗诱导的免疫反应强度和/或活疫苗所表达抗原的功能有关。有研究报道，鮰浸泡免疫减毒活鮰爱德华菌后，免疫保护持续超过 4 个月（Klesius 和 Shoemaker，1999；Shoemaker 等，2009）。由于减毒活细菌疫苗能更有效地诱导细胞免疫，因此对如鮰爱德华菌这类革兰氏阴性的胞内病原会产生更有效的免疫保护。

鮰爱德华菌的减毒突变体已通过营养缺陷体、转座子插入和化学/药物诱变方法制备获得（表 18-4）。营养缺陷体是通过插入含有耐药性基因的 DNA 片段使 *aroA* 基因失活的菌株，应用等位基因交换将自杀载体替换 *aroA* 基因，筛选出 *aroA* 突变体，因其需要芳香代谢物而无法在鱼体内存活（Thune 等，1999）。研究表明，*aroA* 减毒株的毒力比野生型菌株弱 5 log_{10}；*aroA* 减毒疫苗浸泡免疫鮰 48～72h 后，鱼体内未检测到活的突变菌体；*aroA* 减毒疫苗对肠道败血症的保护效果较弱（RPS 为 54.1%～63.8%）（Thune 等，1999）。有研究制备出鮰爱德华菌的 *purA* 突变株并评估其在鮰体内的减毒效果、持久性和有效性，结果显示，该减毒株比野生型菌株毒力减弱了 5 log_{10}，鮰浸泡接种 48h 后，体内可检测到 *purA* 突变体。野生型鮰爱德华菌攻毒后，RPS 达到 63.3%（Lawrence 等，1997）。

表 18-4　鮰爱德华菌的减毒活疫苗

疫苗	鱼类	减毒方法	接种方式	鱼体年龄或大小	相对保护率（免疫后周数）
鮰爱德华菌[1]	鮰	利福平抗性	浸泡	7～10 日龄	60～100（4）
鮰爱德华菌[2]	鮰	新生霉素抗性	注射	10g	93～100（4～6）
			浸泡		90（4～6）
鮰爱德华菌[3]	鮰	*aroA*-缺失	浸泡	8 个月	54.1～63.8（4）
鮰爱德华菌[4]	鮰	*purA*-缺失	浸泡	5g	67（3）
鮰爱德华菌[5]	鮰	LPS-缺失	注射	6 个月	94（4）
			浸泡		0（4）

注：*来源：1 Klesius 和 Shoemaker，1999；2 Pridgeon 等，2011；3 Thune 等，1999；4 Lawrence 等，1997；5 Lawrence 等，2001。

有研究应用转座子诱变产生脂多糖 O 侧链（O LPS）缺失的鮰爱德华菌突变体，该突变株毒性大幅减弱（Lawrence 等，2001）。鮰浸泡免疫 14d，可在鱼体内检测到该突变体（Lawrence 和 Barnes，2005），腹腔注射可产生免疫保护（RPS 为 90%），但浸泡免疫 RPS 为 0%。减毒活疫苗 RE-33 浸泡免疫鮰，RPS 达到 100%（Klesius 和 Shoemaker，1999），而脂多糖 O 抗原缺失突变疫苗的 RPS 却是 0%（Lawrence 和 Barnes，2005）。

研究发现，一种新的减毒活疫苗浸泡和腹腔注射均获得较好效果（Pridgeon 和 Klesius，2011）。这种疫苗是通过新生霉素抗性技术从有致病力的鮰爱德华菌株中筛选出的减毒活疫苗，新生霉素能与 DNA 促旋酶结合，阻断腺苷三磷酸酶（ATPase）的活力（Sugino 等，1978）。鮰浸泡接种后 15d、30d 和 60d，RPS 均大于 90%；鮰腹腔注射免疫后 15d、30d 和 60d，RPS 分别是 100%、93% 和 100%。研究表明，该减毒疫苗可以安全、有效地预防鮰肠道败血症（Pridgeon 和 Klesius，2011）。

Santander 等（2011）研究发现，敲除鮰爱德华菌的 *crp* 基因研发抗生素敏感的浸泡型减毒活疫苗是一种有效的疫苗研发策略。结果表明，鮰爱德华菌 crp-10 突变体菌株毒力弱，能在鱼体淋巴组织中定植，并能诱导斑马鱼和鮰产生对鮰爱德华菌感染的免疫保护。鮰浸浴和口服免疫后，对肠道败血症的保护率分别达到 92% 和 100%。浸泡免疫鮰，免疫保护作用至少持续 6 周。

第七节　管理条例（美国）

在美国，疫苗研发和生物安全评估是由美国农业部（USDA）动植物检查局兽医生物制剂中心监管。疫苗的许可需要对疫苗研发中的所有标准进行全面的评估，包括对水生动物和环境的生物安全，以及对疫苗纯度和效果的评估。在美国，关于对宿主动物的免疫原性/功效、生物安全、毒力恢复、脱落/传播、免疫干扰以及其他方面的研究方案可以在网址 http：//www.aphis.usda.gov/animal _ health/vet _ biologics 上获得，这些要求适用于传统改良的活疫苗产品，以及重组 DNA 技术研制的疫苗。

第八节　疫苗接种应用

疫苗不仅应降低鱼类的死亡率和发病率，而且还应该增加额外的经济效益，例如促进鱼体生长以及降低饲料系数。在经济方面，在野外试验中，接种过 AQUAVAC-ESC® 疫苗的鮰每公顷利润比未接种的鮰多 3 000～4 000 美元（Wise 和 Terhune，2001）。也有研究发现，AQUAVAC-ESC® 疫苗能够提高鮰鱼苗的存活率，可延长在育苗池的时间，推后放入幼鱼培育池，从而增加利润（Carrias 等，2008）。显然，养殖者使用 AQUAVAC-ESC® 疫苗或类似的减毒疫苗明显增加了经济效益。

对某种病原是否进行免疫接种取决于养殖者是否愿意冒疾病暴发的风险（Thorarinsson 和 Powell，2006），疫苗接种是一种防止疾病暴发影响养殖者经济效益的保险策略。养殖时间越长，发生重大经济损失的风险也越大。在成鱼阶段暴发高死亡率和发病率的疾病产生的损失远大于鱼苗阶段。如果成鱼阶段或上市阶段的鱼死亡 50%，就可能造成重大的损失，甚至导致养殖场破产。接种减毒疫苗可以减少疾病风险并产生利润，这些额外利润就能够抵消接种疫苗的成本。如果在某个养殖场频繁暴发某一种疾病，那么拒绝接种疫苗可能是最大的风险。定期进行科学的疫苗接种将会降低疾病暴发风险，长远看会因此获得更高的利润，而通常使用药物治疗不会达到如此好的效果。

有关减毒活疫苗使用情况的调查研究表明，2009 年，9.7% 鱼苗接种了鮰爱德华菌减毒活疫苗。生产企业养殖接种疫苗鮰的面积增大，平均为 510.6 英亩*（acres）。在接种过肠道败血症疫苗鱼的生产商中，41.9% 的认为接种疫苗的鱼存活率较高。对养殖者关于疫苗预防肠道败血症预期的调查显示，52.4% 的养殖者期望 100% 的鱼能得到保护（Bebak 和 Wagner，2012）。

参考文献

Arias，C. R.，Shoemaker，C. A.，Evans，J. J. andKlesius，P. H. 2003. A comparative study of *Edwardsiella ictaluri* parent（EILO）and *E. ictulari* rifampicin-mutant（RE-33）isolates using lipopolysaccharide，outer membrane proteins，fatty acids，Biolog，API 20E and genomic analysis. *J Fish Dis* 26：415-421.

Bader，J. A.，Vinitnantharat，S. and Klesius，P. H. 1997. Comparison of whole cell antigens of pressure-and formalin-killed *Flexibacter columnare* from channel catfish（*Ictalurus punctutus*）. *Am J Vet Res* 58：985-988.

Bader，J. A.，Shoemaker，C. A. and Klesius，P. H. 2004. Immune response induced by N-lauroylsarcosine extracted outer membrane proteins of an isolate *Edwardsiella ictaluri* in channel catfish. *Fish Shellfish Immunol* 16：

* 1 英亩≈4 047 m²。——编者注

281-428.

Baxa，D. W.，Groff，J. M.，Wishkovsky，A. and Hedrick，R. P. 1990. Susceptibility of nonictalurid fishes to experimental infection with *Edwardsiella ictaluri*. *Dis Aquat Org* 8：113-117.

Bebak J，Wagner B. Use of vaccination against enteric septicemia of catfish and columnaris disease by the US catfish industry. *J Aquat Anim Health* 2012；24：30-36.

Bertolini，J. M.，Capriano，R. C.，Pyle，S. W. and McLaughlin，J. J. A. 1990. Serological investigation of the fish pathogen *Edwardsiella ictaluri*，cause of septicemia of catfish. *J Wildl Dis* 26：246-252.

Bly，J. E. and Clem，L. W. 1991. Temperature-mediated processes in teleost immunity：*in vitro* immunosuppression induced by *in vivo* low temperature in channel catfish. *Vet Immunol Immunopathol* 28：365-377.

Carrias，A. A.，Terhune，J. S.，Sayles，C. A. and Chappel，C. A. 2008. Effects of an extended hatchery phase and vaccination against enteric septicemia of catfish on the production of channel catfish，*Ictalurus punctatus* fingerlings. *J World Aquat Soc* 39：259-266.

Copper，R. K.，Shotts，E. B. and Nolan，L. K. 1996. Use of a minitransposon to study chondroitinase activity association with *Edwardsiella ictaluri*. *J Aquat Anim Health* 8：319-324.

Crumlish，M.，Dung，T. T.，Turnbull，J. F.，*et al.*，2002. Identification of *Edwardsiella ictaluri* from diseased freshwater catfish，*Pangasius hypophthalmus*（Sauvage），cultured in the Mekong Delta，Vietnam. *J. Fish Dis* 25：733-736.

Deng，X. Y.，Luo，W.，Tan，S. H.，*et al.*，2008. Isolation and identification of bacteriosis pathogen-*Edwardsiella tarda* from yellow catfish（*Pelteobagrus fulvidraco*）with red head disease. *Oceanol Limnol Sinica* 39：511-515.

Dung，T. T.，Haesebrouck，F.，Tuan，N. A.，*et al.*，2008. Antimicrobial susceptibility pattern of *Edwardsiella ictaluri* isolates from natural outbreaks of bacillary necrosis of *Pangasianodon hypophthalmus* in Vietnam. *Microbiol Drug Resist* 14：311-316.

Favre，D. and Viret，J.-F. 2006. Biosafety evaluation of recombinant live oral bacterial vaccines in the context of European regulation. *Vaccine* 24：3856-3864.

Ferguson，H. W.，Turnbull，J. F.，Shinn，A. K.，*et al.*，2001. Bacillary necrosis in farmed *Pangasius hypophthalmus*（Sauvage）from the Mekong Delta，Vietnam. *J Fish Dis* 24：509-513.

Hawke，J. 1979. A bacterium associated with disease of pond cultured channel catfish，*Ictalurus punctatus*. *J Fish Res Board Canada* 36：1508-1512.

Hawke，J.，McWhorter，A.，Steignwalt，A. and Brenners，D. 1981. *Edwardsiella ictulari* sp nov.，the cause agent of enteric septicemia of catfish. *Int J Sys Bacteriol* 31：396-400.

Humphrey，J. D.，Lancaster，C.，Gudkovs，N. and McDonald，W. 1986. Exotic bacterial pathogens *Edwardsiella* tarda and *Edwardsiellia ictaluri* from imported ornamental fish *Betta splendens* and *Puntius conchonius*，respectively：isolation and quarantine significance. *Aust Vet J* 63：368-371.

Kasornchandra，J.，Roger，W. A. and Plumb，J. A. 1987. *Edwardsiella ictaluri* from walking catfish，*Clarias batrahus L.* in Thailand. *J Fish Dis* 10：137-138.

Kent，M. and Lyons，J. 1982. *Edwardsiella ictaluri* in the green knife fish，*Eigemannia virescens*. *Fish Health News* 2：2.

Keskin，O.，Secer，S.，Izgur，M.，*et al.*，2004. *Edwardsiella ictaluri* infection in rainbow trout（*Oncorhynchus mykiss*）. *Turkish J Vet Anim Sci* 28：649-653.

Klesius，P. H. 1992a. Carrier state of channel catfish infected with *Edwardsiella ictaluri*. *J Aquat Anim Health* 4：

227-230.

Klesius, P. 1992b. Immune system of channel catfish: an overture of immunity to *Edwardsiella ictaluri*. *Annu Rev Fish Dis* 325-338.

Klesius, P. H. 1993. Rapid enzyme-linked immunosorbent test for detecting antibodies to *Edwardsiella ictaluri* in channel catfish, *Ictalurus punctatus*, using exoantigen. *J Aquat Anim Health* 36: 359-369.

Klesius, P. H. 1994. Transmission of *Edwardsiella ictaluri* from infected, dead to noninfected channel catfish. *J Aquat Anim Health* 6: 180-182.

Klesius, P. H. and Horst, M. H. 1991. Characterization of a major outer-membrane antigen of *Edwardsiella ictaluri*. *J Aquat Anim Health* 3: 181-187.

Klesius, P. H. and Sealey, W. M. 1995. Characteristics of serum antibody in enteric septicemia of catfish. *J Aquat Anim Health* 7: 205-210.

Klesius, P. H. and Sealey, W. M. 1996. Chemotactic and chemokinetic responses of channel catfish macrophages to exoantigen from *Edwardsiella ictaluri*. *J Aquat Anim Health* 8: 314-318.

Klesius, P. H. and Shoemaker, C. A. 1999. Development and use of modified live *Edwardsiella ictaluri* vaccine against enteric septicemia of catfish. *Adv Vet Med* 41: 523-537.

Klesius, P. H. and Shoemaker, C. A. 2000. Modified live *Edwardsiella ictaluri* against enteric septicemia in channel catfish. US Patent No. 6, 019, 981.

Klesius, P. H., Shoemaker, C. A. and Evans, J. J. 2000. *In ovo* methods for utilizing live *Edwardsiella ictaluri* against enteric septicemia in channel catfish. US Patent No. 6, 153, 202.

Klesius, P. H., Shoemaker, C. A., Evans, J. J. and Lim, C. 2001. Vaccines: prevention of diseases in aquatic animals, in *Nutrition and Fish Health* (eds C. Lim and C. Webster). Binghamton, New York, Haworth Press, 317-335.

Klesius, P., Lovy, J., Evans, J., *et al.*, 2003. Isolation of *Edwardsiella ictaluri* from tadpole madtom in a southwestern New Jersey River. *J Aquat Anim Health* 15: 295-301.

Klesius, P. H., Evans, J. J. and Shoemaker, C. A. 2004. Warmwater fish vaccinology in catfish production. *Anim Health Res Rev* 5: 305-311.

Lawrence, M. L. and Barnes, M. M. 2005. Tissue persistence and vaccine efficacy of an O side chain mutant strain of *Edwardsiella ictaluri*. *J Aquat Anim Health* 17: 228-232.

Lawrence, M. L., Copper, R. K. and Thune, R. L. 1997. Attenuation, persistence and vaccine potential of an *Edwardsiella ictaluri* purA mutant. *Infect Immun* 65: 4642-4751.

Lawrence, M. L., Barnes, M. M. and Williams, M. L. 2001. Phenotype and virulence of a transponson-derived lipopolysaccharide O side-chain mutant strain of *Edwardsiella ictulari*. *J Aquat Anim Health* 13: 291-299.

Lawrence, M. L., Barnes, M. M., Azadi, P. and Reeks, B. Y. 2003. The *Edwardsiella ictaluri* O polysaccharide biosynthesis gene cluster and the role of O polysaccharide in resistance to normal catfish serum and catfish neutrophils. *Microbiology* 149: 1402-1421.

Li, C., Zhang, Y., Wang, R., *et al.*, 2012. RNA-seq analysis of mucosal immune response reveals signatures of intestinal barrier disruption and pathogen entry following *Edwardsiella ictaluri* infection in channel catfish, *Ictalurus punctatus*. *Fish Shellfish Immunol* 32: 816-827.

Liu, J. Y., Li, A. H., Zhou, D. R., *et al.*, 2010. Isolation and characterization of *Edwardsiella ictaluri* strains as pathogens from diseased yellow catfish *Pelteobagrus fulvidraceo* (Richardson) cultured in China. *Aquac Res* 41: 1835-1844.

Menanteau-Ledouble, S., Karsi, A. and Lawrence, M. L. 2011. Importance of skin adhesion as a primary site adhesion for *Edwardsiella ictaluri* and impact on invasion and systematic infection in channel catfish *Ictalurus punctatus*. *Vet Microbiol* 148: 425-430.

Morrison, E. E. and Plumb, J. A. 1994. Olfactory organ of channel catfish as a site of experimental *Edwardsiella ictaluri* infection. *J Aquat Anim Health* 6: 101-109.

Newton, J. C. and Triche, P. L. 1993a. Electrophoretic and immunochemical characterization of lipopolysaccharide of *Edwardsiella ictaluri* from channel catfish. *J Aquat Anim Health* 5: 246-253.

Newton, J. C. and Triche, P. L. 1993b. Isolation and characterization of flagella from *Edwardsiella ictaluri*. *J Aquat Anim Health* 5: 16-22.

Newton, J. C., Wolfe, L. G., Grizzle, J. M. and Plumb, J. A. 1989. Pathology of experimental enteric septicemia in channel catfish, *Ictalurus punctatus* (Rafinesque), following immersion exposure to *Edwardsiella ictaluri*. *J Fish Dis* 12: 335-347.

Newton, J. C., Blevins, W. T., Wilt, G. R. and Wolfe, L. G. 1990. Outer membrane protein profiles of *Edwardsiella ictulari* from fish. *Am J Vet Res* 51: 264-273.

Norqvist, A., Hagstrom, A. and Wolf, H. 1989. Protection of rainbow trout against vibriosis and furunculosis by the use of attenuated strains of *Vibrio anguillarum*. *Appl Environ Microbiol* 55: 1400-1405.

Nusbaum, K. E. and Morrison, E. E. 1996. Entry of 35S-labeled *Edwardsiella ictaluri* into channel catfish. *J Aquat Anim Health* 8: 146-149.

Panangala, V. S., Shoemaker, C. A., McNulty, S. T., *et al.*, 2006. Intra- and interspecific phenotypic characteristics of fish-pathogenic *Edwardsiella ictaluri* and *E. tarda*. *Aquac Res* 37: 49-60.

Panangala, V. S., Russo, R., Van Santen, V. L., *et al.*, 2009a. Organization and sequence of four flagellin-encoding genes of *Edwardsiella ictaluri*. *Aquac Res* 40: 1135-1147.

Panangala, V. S., Shoemaker, C. A., Klesius, P. H., *et al.*, 2009b. Cross-protection elicited in channel catfish (*Ictalurus punctatus* Rafinesque) immunized with low dose of virulent *Edwardsiella ictaluri* strains. *Aquac Res* 40: 915-926.

Petrie-Hanson, L., Romano, C. L., Mackey, R. B., *et al.*, 2007. Evaluation of zebrafish *Danio rerio* as a model for enteric septicemia of catfish (ESC). *J Aquat Anim Health* 19: 151-158.

Plumb, J. A. 1988. Vaccination against *Edwardsiella ictaluri*, in *Fish Vaccination* (ed. A. E. Ellis). London, Academic Press, 152-161.

Plumb, J. A. 2000. Catfish bacterial diseases, in *Health Maintenance and Principal Microbial Diseases of Cultured Fishes* (ed. J. A. Plumb). Ames, Iowa State University Press, 187-194.

Plumb, J. A. and Sanchez, D. 1983. Susceptibility of five species of fish to *Edwardsiella ictaluri*. *J Fish Dis* 6: 261-266.

Plumb, J. A. and Quinlan, E. E. 1986. Survival of *Edwardsiella ictaluri* in pond water and bottom mud. *Prog Fish Cult* 48: 212-214.

Plumb, J. A. and Huge, V. 1987. Susceptibility of European catfish (*Silurus gianis*) to *Edwardsiella ictaluri*. *J Appl Ichthyol* 3: 45-48.

Plumb, J. A. and Klesius, P. H. 1988. An assessment of the antigenic homogeneity of *Edwardsiella ictaluri* using monoclonal antibody. *J Fish Dis* 11: 499-510.

Plumb, J. A. and Vinitnantharat, S. 1993. Vaccination of channel catfish, *Ictalurus punctatus* (Rafinesque) by immersion and oral booster against *Edwardsiella ictaluri*. *J Fish Dis* 16: 65-71.

Plumb, J. A., Vinitnantharat, S. and Paterson, W. D. 1994. Optimum concentration of *Edwardsiella ictaluri* vaccine in feed for oral vaccination of channel catfish. *J Aquat Anim Health* 6: 118-121.

Pridgeon, Y. and Klesius, P. 2011. Development of a novobiocin-resistant *Edwardsiella ictaluri* as a novel vaccine in channel catfish (*Ictalurus punctatus*). *Vaccine* 29: 5631-5637.

Pridgeon, Y., Russo, R., Shoemaker, C. and Klesius, P. 2010. Identification of *in vitro* upregulated genes in a modified live vaccine strain of *Edwardsiella ictaluri* compared to a virulent parent strain. *Comp Immunol Microbiol Infect Dis* 33 (6): e31-40.

Pridgeon, Y., Yeh, H. H., Shoemaker, C., *et al.*, 2012a. Global gene expression in channel catfish after vaccination with an attenuated *Edwardsiella ictaluri*. *Fish Shellfish Immunol* 32: 524-533.

Pridgeon, Y., Yeh, H. H., Shoemaker, C. and Klesius, P. 2012b. Global transcription analysis of vaccinated channel catfish following challenge with virulent *Edwardsiella ictaluri*. *Vet Immunol Immunopathol* 146: 53-61.

Reid, W. S. and Boyle, J. A. 1989. Plasmid homologies in *Edwardsiella ictaluri*. *Appl Environ Microbiol* 55: 3253-3255.

Russo, R., Shoemaker, C. A., Panangala, V. S. and Klesius, P. H. 2009. *In vitro* and *in vivo* interaction of macrophages from vaccination and non-vaccination channel catfish (*Ictalurus punctatus*) to *Edwardsiella ictaluri*. *Fish Shellfish Immunol* 26: 543-552.

Saeed, M. O. and Plumb, J. A. 1986. Immune response of channel catfish to lipopolysaccharides and whole cell *Edwardsiella ictaluri* vaccine. *Dis Aquat Org* 2: 21-26.

Salerno-Goncalves, R. and Sztein, M. B. 2006. Cell-mediated immunity and challenges for vaccine development. *Trends Microbiol* 14: 536-542.

Salmbandamurthy, V. K. and Jacobs, W. R. 2005. Live attenuated mutants of *Mycobacterium tuberculosis* as candidate vaccines against tuberculosis. *Microbiol Infect* 5: 955-961.

Santander, J., Mitra, A. and Curtiss, R. 2011. Phenotype, virulence and immunogenicity of *Edwardsiella ictaluri* cyclic adenosine 3′5′ monophosphate receptor protein (Crp) mutants in catfish host. *Fish Shellfish Immunol* 31: 1142-1153.

Schurig, G. G., Roop II, R. M., Bagchi, T., *et al.*, 1991. Biological properties of RB51; a stable rough strain of *Brucella abortus*. *Vet Microbiol* 28: 171-188.

Secombes, C. 2008. Will advances in fish immunology change vaccination strategies? *Fish Shellfish Immunol* 25: 409-416.

Seder, R. A. and Hill, A. V. S. 2000. Vaccines against intracellular infections requiring cellular immunity. *Nature* 406: 793-797.

Shoemaker, C. A. and Klesius, P. H. 1997. Protective immunity against enteric septicemia in channel catfish, *Ictalurus punctatus* (Rafineque), following controlled exposure to *Edwardsiella ictaluri*. *J Fish Dis* 20: 101-108.

Shoemaker, C. A., Klesius, P. H. and Plumb, J. A. 1997. Killing of *Edwardsiella ictaluri* by macrophages from channel catfish immune and susceptible to enteric septicemia of catfish. *Vet Immunol Immunopathol* 58: 181-190.

Shoemaker, C. A., Klesius, P. H. and Bricker, J. M. 1999. Efficacy of a modified live *Edwardsiella ictaluri* vaccine in channel catfish as young as seven days post hatch. *Aquaculture* 176: 189-193.

Shoemaker, C. A., Klesius, P. H. and Evans, J. J. 2002. *In ovo* method for utilizing live *Edwardsiella ictaluri* vaccine against enteric septicemia in channel catfish. *Aquaculture* 203: 221-222.

Shoemaker, C. A., Klesius, P. H. and Evans, J. J. 2007. Immunization of eyed channel catfish, *Ictalurus punctatus*, eggs with monovalent *Flavobacterium columnare* vaccine and bivalent *F. columnare* and *Edwardsiella ictaluri* vaccine. *Vaccine* 25：1126-1131.

Shoemaker, C. A., Klesius, P. H., Evans, J. J. and Arias, C. R. 2009. Use of modified live vaccines in aquaculture. *J World Aquac* 40：573-585.

Shotts, E. B., Blazer, V. S. and Waltman, W. D. 1986. Pathogenesis of experimental *Edwardsiella ictaluri* infections in channel catfish (*Ictalurus punctatus*). *Can J Fish Aquat Sci* 43：36-42.

Sugino, A., Higgins, N. P., Brown, P. O., *et al.*, 1978. Energy coupling in DNA gyrase and the mechanism of action of novobiocin. *Proc Natl Acad Sci USA* 75：4838-4842.

Thorarinsson, R. and Powell, D. B. 2006. Effects of disease risk, vaccine efficacy, and market price on the economics of fish vaccination. *Aquaculture* 256：42-49.

Thune, R. L., Hawke, J. P. and Johnson, M. C. 1992. Studies on vaccination of channel catfish *Ictalurus punctatus*, against *Edwardsiella ictaluri*. *J Appl Aquac* 3：11-24.

Thune, R. L., Hawke, J. P., Fernandez, D. H., *et al.*, 1997a. Immunization with bacterial antigens：edwardsiellosis. *Dev Biol Stand* 90：125-134.

Thune, R. L., Collins, L. A. and Pena, M. A. 1997b. A comparison of immersion, immersion/oral combination and injection vaccination of channel catfish *Ictalurus punctatus* against *E. ictaluri*. *J World Aquac Soc* 28：193-201.

Thune, R. L., Fernandez, D. H. and Battista, R. 1999. An aroA mutant of *Edwardsiella ictaluri* is safe and efficacious as a live attenuated vaccine. *J Aquat Anim Health* 11：358-372.

Thune, R. L., Fernandez, D. L., Benoit, J. L., *et al.*, 2007. Signature-tagged mutagenesis of *Edwardsiella ictaluri* identifies virulence-related genes, including a *Salmonella* pathogenicity island 2 class of type Ⅲ secretion systems. *Appl Environ Microbiol* 73：7934-7946.

Vinitnantharat, S. and Plumb, J. A. 1992. Kinetics of the immune response of channel catfish to *Edwardsiella ictaluri*. *J Aquat Anim Health* 4：207-214.

Vinitnantharat, S. and Plumb, J. A. 1993. Protection of channel catfish *Ictalurus punctatus* following natural exposure to *Edwardsiella ictaluri* and effects of feeding antigen on antibody titer. *Dis Aquat Org* 15：31-34.

Vinitnantharat, S., Plumb, J. A. and Brown, A. E. 1993. Isolation and purification of an outer membrane of *Edwardsiella ictaluri* and its antigenicity to channel catfish (*Ictalurus punctatus*). *Fish Shellfish Immunol* 3：401-409.

Wagner, B. A., Wise, D. J., Khoo, L. H. and Terhune, J. S. 2002. The epidemiology of bacterial diseases in food-size channel catfish. *J Aquat Anim Health* 14：263-272.

Waltman, W. D., Shotts, E. B. and Blazer, V. S. 1985. Recovery of *Edwardsiella ictaluri* from danio (*Danio devario*). *Aquaculture* 46：63-66.

Waltman, W. D., Shotts, E. B. and Hsu, T. C. 1986. Biochemical characteristics of *Edwardsiella ictaluri*. *Appl Environ Microbiol* 51：101-104.

Weete, J. D., Blevins, W. T., Chitrakorn, S., *et al.*, 1988. Chemical characterization of lipopolysaccharide from *Edwardsiella ictaluri* from channel catfish. *Can J Microbiol* 34：1224-1229.

Wehrli, W., Knusel, P., Schmid, K. and Staehelin, M. 1968. Interaction of rifamycin with bacterial RNA polymerase. *Proc Natl Acad Sci USA* 61：667-673.

Williams, M. L. and Lawrence, M. L. 2005. Identification and characterization of a two-component hemolysin from *Edwardsiella ictaluri*. *Vet Microbiol* 108：281-289.

Wise, D. J. 2006. Vaccination shown to improve production efficiencies. *Catfish J*：6-7.

Wise, D. J. and Terhune, J. 2001. The relationship between vaccine dose and efficacy in channel catfish *Ictalurus punctutus* vaccinated with live attenuated strain of *Edwardsiella ictaluri* (RE-33). *J World Aquac Soc* 32: 177-183.

Wise, D. J., Klesius, P. H., Shoemaker, C. A. and Wolters, W. R. 2000. Vaccination of mixed and full-sib families of channel catfish *Ictalurus punctatus* against enteric septicemia of catfish with a live attenuated *Edwardsiella ictaluri* isolate (RE-33). *J World Aquac Soc* 31: 206-212.

Wolfe, K., Plumb, J. A. and Morrison, E. E. 1994. Lectin binding characteristics of the olfactory mucosa of channel catfish: potential factors in attachment of *Edwardsiella ictulari*. *J Aquat Anim Health* 10: 348-360.

Xu, D. H., Shoemaker, C. A., Martins, M. L., *et al.*, 2012. Enhanced susceptibility of channel catfish to the bacterium *Edwardsiella ictaluri* after parasitism by *Ichthyophthirius multifiliis*. *Vet Microbiol* 158: 216-219.

Ye, S., Qiao, H. L. G. and Zhangshi, L. 2009. First case of *Edwardsiella ictaluri* infection in China farmed yellow catfish *Pelteobagrus fulvidraco*. *Aquaculture* 292: 6-10.

Yuasa, K., Kholidin, E. B. and Panigoro Nand Hatai, K. 2003. First isolation of *Edwardsiella ictaluri* from cultured striped catfish *Pangasius hypophthalmus* in Indonesia. *Fish Pathol* 38: 181-183.

第十九章　耶尔森菌病的疫苗接种

Andrew Bridle and Barbara Nowak

本章概要　SUMMARY

耶尔森菌病是由鲁克氏耶尔森菌（*Yersinia ruckeri*）引起的疾病，该菌主要感染孵卵阶段的鲑科鱼类。该病的商品化疫苗已有40多年的历史，这些疫苗均为灭活菌苗，主要通过浸泡接种，且需要加强免疫以达到对鱼体的长期保护。自20世纪80年代末以来，主要由鲁克氏耶尔森菌生物2型的变异株造成鱼类感染，现已成为导致美国和欧洲耶尔森菌病的主要株型。由于新变异株的出现，使得用于疫苗生产的菌株也随之发生了相应的改变。

第一节　耶尔森菌病

鱼类耶尔森菌病是由肠杆菌科中的鲁克氏耶尔森菌（*Yersinia ruckeri*）感染所致，可引起虹鳟肠炎红嘴症状，因而该病又称肠炎红嘴病。20世纪50年代后期，肠炎红嘴病在美国爱达荷州哈德曼峡谷的虹鳟中首次发现（Rucker，1966）。鲁克氏耶尔森菌最可能在上世纪70年代末或80年代早期由美国传入欧洲（Horne和Barnes，1999），目前，该病在世界范围所有主要鲑科鱼类养殖地区流行。由于鲁克氏耶尔森菌流行的地域日益扩大，并可感染虹鳟之外的其他鱼类，因此该病现多称为“耶尔森菌病”，从而逐渐代替了“肠炎红嘴病”。肠炎红嘴病疫苗是20世纪70年代美国首批获得许可的鱼类商业疫苗之一。

第二节　发生流行与危害

鲁克氏耶尔森菌广泛存在于淡水水体和鱼体内。据报道，已在欧洲、北美、南美、南非、亚洲以及澳大利亚和新西兰等地的鱼类样品中分离出鲁克氏耶尔森菌，主要的鱼类包括鲑科鱼类、鳗、金鱼、鲽、鲟、尼罗罗非鱼和大菱鲆。该病是鲑科鱼类养殖国家面临的十分严重的问题，如果不及时治疗，鲑的死亡率可达30%～79%（Horne和Barnes，1999）。耶尔森菌病主要感染育苗场的小规格虹鳟（北半球）或小规格大西洋鲑（澳大利亚），这些小规格鱼转移至流水网箱或池塘，如果有病原存在且水质、水温难以控制，则容易发病。耶尔森菌病也会影响2龄银化鲑等大鱼的产量。

在多数国家该菌感染虹鳟，而在澳大利亚该菌感染养殖的大西洋鲑，存在很大差异，Hagerman菌株（引起肠炎红嘴病，认为是澳大利亚的外来菌株）已经成为北半球主要流行的鲁克氏耶尔森菌。已发现一株新的菌株，疫苗对其无效，非运动型、吐温-80阴性，是生物2型的鲁克氏耶尔森分离菌株（Austin等，2003）。这表明生物2型的表型与欧洲大陆和英国的菌株不同（Wheeler等，2009）。

第三节　病原学

鲁克氏耶尔森菌属于肠杆菌科，革兰氏阴性、无芽孢、无荚膜、杆状、兼性厌氧、可运动。虽然与黑死病的病原——鼠疫耶尔森菌（*Y. pestis*）同属，但该菌只感染鱼类。根据16S rRNA系统进化分析，鲁克氏耶尔森菌与其他耶尔森菌的差异，使其在该属内的分类地位受到了质疑。在此之前，就对鲁克氏耶尔森菌的分类存在争议，因为鲁克氏耶尔森菌的表型与肠杆菌科其他菌属相似性高于耶尔森菌属。随后，经过分析大量的鲁克氏耶尔森菌分离株，认定其确属耶尔森菌属。

最近，利用高通量全基因组测序技术比较分析了耶尔森菌属的基因组，发现鲁克氏耶尔森菌是在耶尔森菌属内进化最远的一个种（Chen等，2010）。研究表明，鲁克氏耶尔森菌的全基因组大小为3.58Mbp到3.89Mbp，显著小于耶尔森菌属内其它成员的4.6Mbp到4.8Mbp。利用耶尔森菌属中所有成员的核心蛋白进行系统发育分析，耶尔森菌属分成三大进化分支：第一分支仅有鲁克氏耶尔森菌，是位置最远的分支；第二分支包括鼠疫耶尔森菌和假结核耶尔森菌；第三分支则包含小肠结肠炎耶尔森菌。

鲁克氏耶尔森菌的生化特性与肠杆菌科其他成员高度一致，发酵葡萄糖、氧化酶阴性，硝酸盐还原型；采用试剂盒API 20E系统等常规的生化测试，即可鉴定肠杆菌科的鲁克氏耶尔森菌（Frerichs，1993）。最初的研究表明，从疾病暴发时分离的鲁克氏耶尔森菌属同一菌株（Ross等，1966；Busch，1973）。早期分子特性分析采用15种酶位点的多种同工酶电泳技术对鲁克氏耶尔森菌分离株进行鉴定，结果发现，47株分离株中有42株为同类型，表明该菌遗传多样性非常低且自然株型很少（Schill等，1984）。最近，采用血清型分型、抗生素耐药性分型、生物分型、脉冲场凝胶电泳（PFGE）、外膜蛋白（OMP）分析等高分辨率表型分析技术和分子生物技术研究了鲁克氏耶尔森菌的基因多样性（Wheeler等，2009；Bastardo等，2011；Tinsley等，2011a）。其中，血清型分型虽然还有需要改进之处，但其仍是细菌种内分类最常用的方法。使用热稳定O抗原的特异性抗血清将鲁克氏耶尔森菌分离株分为5种血清型（O1、O2、O5、O6、O7）（Davies，1990）。血清型O1可分为O1a和O1b两个亚型；而血清型O2可分为3个亚型，即O2a、O2b和O2c（Romalde等，1993）。

鲁克氏耶尔森菌非运动型变异株的出现是重大发现，该突变株发现于20世纪80年代，其缺乏分泌磷脂酶的能力，属O1血清型（Davies和Frerichs，1989），最初该变异菌株被称为

EX5，鉴定为生物 2 型。该菌导致接种了当时流行 O1 血清型（生物 1 型）鲁克氏耶尔森菌疫苗的虹鳟发病（Austin 等，2003）。如今，生物 1 型 Hagerman 株，O1a 血清型的疫苗在美国和欧洲成功应用 30 多年，生物 2 型鲁克氏耶尔森菌变异菌株成为耶尔森菌病的主要致病菌株。

第四节 致病机理

耶尔森菌病给全球鲑科鱼类养殖造成了巨大的损失（Tobback 等，2007），遗憾的是，仍未对该病及其致病机理完全了解清楚（Tobback 等，2007）。鲁克氏耶尔森菌的毒力菌株能够引起鲑科鱼类全身性败血症并且导致死亡（Tobback 等，2009a）。病原菌感染后，患鱼鳃中很快出现大量病菌并在器官中迅速传播，这表明鳃可能是鲁克氏耶尔森菌入侵鱼体的一个重要途径，而且鲁克氏耶尔森菌的致病力与免疫逃避有关（Tobback 等，2009a）。该菌是一种兼性胞内感染菌，可在巨噬细胞中存活，这进一步证明上述观点（Ryckaert 等，2010）。虹鳟浸泡感染鲁克氏耶尔森菌后 24～72h，肠道内即可检测到大量菌体；感染 5d 后，出现全身性感染（Mendez 和 Guijarro，2013）。

鲁克氏耶尔森菌能产生溶血素等胞外产物（ECPs），ECPs 在致病机理中发挥重要作用（Romalde 和 Toranzo，1993）。研究发现，溶血素/细胞溶素 YhlA 能溶解红细胞和培养的鱼类细胞，有助于从宿主获取铁元素，因此在该菌致病性中发挥重要作用（Fernandez 等，2007a）。体外研究其毒力因子的特性发现，鲁克氏耶尔森菌致病菌株对鱼鳃和肠黏液有较高的黏附能力，并且具有抗血清活性（Tobback 等，2009b）。

研究发现，鲁克氏耶尔森菌的某些基因在耶尔森菌病的致病机理中发挥重要作用，通铁载体的铁获取、蛋白质水解活性、溶血活力以及抵御免疫反应等均与该菌的致病力有关（Fernandez 等，2007b）。BarA-UvrY 双组件系统（TCS）可能是通过调节对上皮细胞的侵染性，以及免疫细胞诱导的氧化应激敏感性，参与鲁克氏耶尔森菌感染虹鳟（Dahiya 和 Stevenson，2010）。应用体内表达技术（IVET）在该菌中鉴定出相邻的两个基因，它们是参与摄取和降解 L-半胱氨酸且与致病力有关的新系统的标志（Mendez 等，2011）。

感染鲁克氏耶尔森菌后，虹鳟脾脏中白介素 1（IL-1）家族、IL-2（Diaz-Rosales 等，2009）、IL-17 家族（Wang 等，2009）、IL-6、IL-8、IL-10 以及 IL-11 等促炎和抗炎症细胞因子基因显著上调表达（Raida 和 Buchmann，2008）。然而，虹鳟浸泡感染鲁克氏耶尔森菌后，血液中的某些白介素基因（包括 IL-1β、IL-8 和 IL-10）出现上调表达，基因上调与血液中的细菌载量呈正相关，但却未产生免疫保护作用（Raida 等，2011b）。

已经证明 Cathelicidins（抗菌肽）在大西洋鲑的耶尔森菌病致病机理中发挥着重要作用（Bridle 等，2011，2012）。Cathelicidins 是含组织蛋白酶区域的抗菌肽，大西洋鲑有 *asCATH1* 和 *asCATH2* 两个抗菌肽基因（Chang 等，2006）。大西洋鲑感染耶尔森菌后，其抗菌肽基因显著上调表达。例如，在感染 72h 和 96h 时，大西洋鲑脾脏中，*asCATH1* 基因表达量约为对照组的 30 倍；*asCATH2* 基因的表达从感染后 72h 的 48 倍增加到 96h 的 98 倍。相对于鳃，脾脏中 *asCATH1* 基因和 *asCATH2* 基因均提前上调表达且表达量增加显著（Bridle 等，2011）。然而，不同的鲁克氏耶尔森菌菌株毒力不尽相同，发病也有差异（Tobback 等，2009b）。因此，在耶尔森菌病中观察到的 *Cathelicidins* 基因应答反应程度可能只特异性地与塔斯马尼亚的鲁克氏耶尔森菌 O1b 血清型相关（Bridle 等，2011）。

第五节　疫　苗

1976 年，福尔马林灭活的鲁克氏耶尔森菌疫苗获得了生产许可，这是首个预防耶尔森菌病的商业疫苗，也是第一个商业化的鱼类疫苗。该疫苗是以灭活的全菌制备而成，主要用于孵化阶段鱼的浸泡接种。耶尔森菌病疫苗通常推荐浸泡免疫接种，有的耶尔森菌疫苗加强免疫可采用无油佐剂注射接种或口服接种。

鲁克氏耶尔森菌生物 2 型菌株的出现，造成接种生物 1 型菌疫苗的鲑科鱼类发病，促使二价的商品化耶尔森菌病疫苗研发，二价疫苗含有“Hagerman”O1 生物 1 型菌株和生物 2 型菌株（EX5 样菌株）。从虹鳟分离出的鲁克氏耶尔森菌的生物 1 型分离株和生物 2 型分离株的抗原性和交叉保护研究表明，O 抗原是鲁克氏耶尔森菌主要的免疫原性分子，且单价（生物 1 型）疫苗对生物 2 型分离株的免疫保护十分有限（Tinsley 等，2011b）。此外，虹鳟接种二价疫苗获得的针对生物 1 型分离株的免疫保护效果与单价疫苗相同（Deshmukh 等，2012）。

自从 Klontz（1963）研究发现虹鳟口服鲁克氏耶尔森菌疫苗相对保护率达到 87%后，针对鲁克氏耶尔森菌疫苗效果的研究不断出现。脂多糖的 O 抗原是已知的免疫原，从鲁克氏耶尔森菌、温和气单胞菌（*Aeromonas sobria*）和枯草芽孢杆菌（*Bacillus subtilus*）等益生菌中分离出 O 抗原与其他亚细胞成分共同腹腔注射虹鳟，能够对该病提供高水平的免疫保护（Abbass 等，2010）。此外，Fernandez 等（2003）研究了免疫 Yrp1 蛋白酶的类毒素获得预防该病的保护效果，结果发现，鲑科鱼类肌肉注射该类毒素，能够有效预防耶尔森菌病。

较有争议的疫苗研制方法是诱导 *aroA* 基因突变导致功能失调的减毒活疫苗。基因 *aroA* 编码 5-烯醇丙酮基莽草酸-3-磷酸合成酶，该酶参与鲁克氏耶尔森菌芳香氨基酸的生物合成。尽管 Temprano 等（2005）发现，与目前使用的细菌疫苗相比，这种减毒活疫苗能够提供更好的保护，但是转基因生物的环境安全问题已经或很有可能会继续限制减毒活疫苗的商业应用。

尽管预防耶尔森菌病的全菌疫苗已投入使用并研究了 30 多年，但其引起的免疫保护效果机制仍然知之甚少。普遍认为疫苗接种中产生特异性抗体非常重要，然而抗体介导预防鲁克氏耶尔森菌病的保护机制尚未明晰。抗体介导的免疫保护研究非常少，之前只有 Olesen 等（1991）用虹鳟抗血清成功地进行了被动免疫预防耶尔森菌病。之后 Lee 等（2000）的研究也强调了抗体介导免疫保护的重要性，该研究结果显示，鳟腹腔注射鸡胚中产生的抗耶尔森菌病的 IgY 抗体，免疫接种 4h 后攻毒，能保护鳟免受鲁克氏耶尔森菌感染。最近一项研究发现，鳟浸泡免疫耶尔森菌病疫苗后，其血浆的特异性 IgM 抗体滴度和杀菌活性显著增强，对疫苗免疫 3～14d 后的鳟攻毒，鱼血液中几乎没有鲁克氏耶尔森菌（Raida 等，2011a）。也有报道虹鳟腹腔注射接种耶尔森菌病的商业疫苗后，血清 IgM 滴度增加（Raida 和 Buchmann，2007）。疫苗诱导宿主体液或细胞因子发挥的免疫作用，能够解释为什么鱼体通过反复被动免疫不能获得免疫保护。研究发现，鲁克氏耶尔森菌能在巨噬细胞内生存（Ryckaert 等，2010），这可能也有助于该菌逃避抗体介导的宿主免疫反应。

第六节 疫苗接种流程

Klontz（1963）首次报道，通过口服全菌疫苗成功预防了虹鳟耶尔森菌病。Ross 和 Klontz（1965）以及 Anderson 和 Ross（1972）以浸泡免疫作为疫苗接种方式开展了进一步的研究。Anderson 和 Nelson（1974）的研究表明，在疫苗成分几乎没有改变的情况下，腹腔注射耶尔森菌病菌苗可以获得更好的保护。由于耶尔森菌病仍然是孵化阶段鲑科鱼类的主要疾病，浸泡免疫对鱼苗和幼鱼更易操作，因此其仍然是首选的疫苗接种方式。浸泡免疫过程通常是将体重不超过 5g 的鱼在疫苗中浸泡 30～60s。然而，在育苗场若要获得足够长的持续保护期就需要多次加强免疫，通常将疫苗掺入饲料中或者包裹在饲料中进行口服免疫更切实可行。当然，在相对保护率和持续保护期方面，腹腔注射免疫提供的保护更加有效（Anderson 和 Nelson，1974），但这种方式不合适小规格鱼（<20g）。

将疫苗稀释倍数增加（1∶100）并浸浴至少 1h，可作为浸泡免疫的替代方式，这是因为有些水产养殖者认为加长浸浴时间能减少鱼体应激，同时可以避免由于应激反应而降低疫苗功效，直到 Costa 等（2011）发现浸浴接种获得的持续免疫保护较差，塔斯马尼亚的鲑科鱼类养殖企业才将其惯例的浸浴接种转为前述标准的浸泡接种。

耶尔森菌病是鲑科鱼类孵化阶段的最常见的问题，该病可能自发育早期就开始影响鱼类生长，因此，一些疫苗生产商建议，最好在鲑科鱼类达到适合免疫接种体重 5g 前就进行浸泡免疫。为了解决这个问题，根据鱼体大小和年龄在 1 年内进行加强免疫，对于小规格鱼，较适合通过浸泡或口服加强免疫；对于大规格鱼，在水温不太高的情况下，可进行腹腔注射加强免疫。通常在鱼苗体重 1～3 g 时进行疫苗接种，5g 时再次免疫。在鲑养殖周期的其他时间段，采取再次加强免疫保证高水平免疫保护。最近，在欧洲和澳大利亚，转移至海水养殖的大西洋鲑暴发耶尔森菌病，这使得水产养殖者开始重新考虑腹腔注射免疫，或者将腹腔注射免疫并入现有的注射型疫苗和治疗方案，并评估它们的成本与效益。除了能获得较好的保护水平外，注射耶尔森菌病疫苗的另一个好处是不需要佐剂，这可避免油类佐剂造成注射部位周围组织损伤的问题。

第七节 疫苗效果

浸泡免疫鲁克氏耶尔森菌全菌灭活疫苗通常可获得 70%～100%的相对保护率，这也取决于鱼的种类、大小、攻毒剂量和接种持续时间或加强免疫。最早 Klontz（1963）的实验中，福尔马林灭活的鲁克氏耶尔森菌全菌疫苗，用于口服免疫虹鳟，获得了 87%的相对保护率。不久后，有研究发现，虹鳟浸泡于同样方法制备的全菌疫苗中也可获得对耶尔森菌病较好的免疫保护水平（Ross 和 Klontz，1965；Anderson 和 Ross，1972）。这些单价菌苗可提供较高的预防生物 1 型菌株的保护水平，3～10g 的鲑浸泡免疫 20～60s，获得的免疫保护持续时间通常为 120～300d，随后开始下降（Johnson 等，1982）。

在单价疫苗接种的虹鳟中暴发了耶尔森菌病，发现了生物 2 型菌株，因此需要研发针对生物 2 型菌株的新疫苗。很幸运，生物 2 型菌株疫苗和生物 2 型与生物 1 型菌株疫苗的联合使用均能提供对该病有效的免疫保护。有关评估这些疫苗预防生物 1 型和 2 型菌株交叉保护的研究结果显示，单价和二价疫苗均使用 1∶10 稀释的疫苗浸泡免疫虹鳟 30s，当虹鳟免疫接种商品化二价疫苗 28d 后，生物 2 型菌株腹腔注射攻毒，相对保护率为 87%～100%。免疫接种单价疫苗以相同的生物 2 型菌株攻毒后，相对保护率降低至 72%～100%，而单价疫苗和二价疫苗对生物 1 型菌株均提供了良好的保护，相对保护率分别为 81%和 92%（Tinsley 等，2011b）。Deshmukh 等（2012）研究了相同的商业单价疫苗和二价疫苗在虹鳟中对这两种类型菌株的交叉保护，但该研究更侧重疫苗长期保护的评估，结果显示，单价疫苗接种 4 个月和 6 个月后，腹腔注射生物 2 型菌株攻毒，相对保护率较低，分别为 13.5%和 29.5%；而二价疫苗接种 4 个月和 6 个月后攻毒，相对保护率分别为 42.5%和 52%。这些研究证明了二价疫苗的保护效果，同时强调需要采取加强免疫确保持续的有效保护。

大西洋鲑浸泡免疫耶尔森菌病疫苗后，鳃中有 17 个基因出现差异表达，这些生物学特征被认为是疫苗诱导免疫保护的指示因子（Bridle 等，2012），这些基因包括参与解毒与修复的基因（醇脱氢酶 3 和硫氧还原白结合蛋白基因）以及其他参与免疫应答的基因（两个 Ig 重链转录基因以及肝抗菌肽、胸腺肽类激素、E3 泛素蛋白连接酶、含蛋白 1 的 SAM 和 SH3 结构域、髓磷脂、淋巴细胞蛋白、LIM、肌动蛋白结合蛋白和 TRIM 基因）。接种疫苗的大西洋鲑在攻毒后的 72h，鳃中的免疫球蛋白重链基因和硒蛋白基因显著上调表达，这说明疫苗引起了宿主的免疫保护性反应（Bridle 等，2012）。

第八节　副作用

目前还没有耶尔森菌病疫苗副作用的相关报道，这可能是由于该疫苗没有使用佐剂（Hastein 等，2005），因为通常疫苗接种的副作用与油类佐剂有关（Midtlyng 和 Lillehaug，1998；Koppang 等，2005）。尽管耶尔森菌病疫苗没有副作用，但在免疫接种的鱼中携带细菌且无症状感染的比例很高，至少在大西洋鲑中是如此（Bridle，未发表），这表明虽然疫苗可以预防耶尔森菌病并减少死亡率，但该病的感染仍在种群中持续存在。

第九节　管理条例

鱼类疫苗的许可是依据《欧洲药典》及其指导文件中所述的法律和管理条例（Galligani 等，2005）。管理架构具有国家特定性，虽然会尽量协调一致，但不同国家和地区的许可和要求仍有所不同。在欧盟，只要与《欧洲药典》有关，疫苗试验和流程就必须符合已公布的标准（Midtlyng，2005）。有许多种耶尔森菌病疫苗在一些国家和地区被许可使用。在澳大利亚，耶尔森菌病疫苗没有注册就可获准允许在兽医的指导下使用。澳大利亚限制进口其他国家和地区生产的疫苗，例如，澳大利亚以外储存或处理过病原的设施中生产的疫苗，或者对澳大利亚有生物安全风险的疾病来源的疫苗，澳大利亚禁止进口。

参考文献

Abbass, A., Sharifuzzaman, S. and Austin, B. 2010. Cellular components of probiotics control *Yersinia ruckeri* infection in rainbow trout, *Oncorhynchus mykiss* (Walbaum). *J Fish Dis* 33: 31-37.

Anderson, D. P. and Ross, A. J. 1972. Comparative study of Hagerman redmouth disease oral bacterins. *Prog Fish Cult* 34: 226-228.

Anderson, D. P. and Nelson, J. R. 1974. Comparison of protection in rainbow trout (*Salmo gairdneri*) inoculated with and fed Hagerman redmouth bacterins. *J Fish Res Board Can* 31: 214-216.

Austin, D. A., Robertson, P. A. W. and Austin, B. 2003. Recovery of a new biogroup of *Yersinia ruckeri* from diseased rainbow trout (*Oncorhynchus mykiss*, Walbaum). *System Appl Microbiol* 26: 127-131.

Bastardo, A., Bohle, H., Ravelo, C., *et al.*, 2011. Serological and molecular heterogeneity among *Yersinia ruckeri* strains isolated from farmed Atlantic salmon *Salmo salar* in Chile. *Dis Aquat Org* 93: 207-214.

Bridle, A., Nosworthy, E., Polinski, M. and Nowak, B. F. 2011. Evidence of an antimicrobialimmunomodulatory role of Atlantic salmon cathelicidins during infection with *Yersinia ruckeri*. *PLoS One* 6: e23417.

Bridle, A. R., Koop, B. F. and Nowak, B. F. 2012. Identification of surrogates of protection against yersiniosis in immersion vaccinated Atlantic salmon. *PLoS One* 7: e40841.

Busch, R. A. 1973. *The serological surveillance of salmonid populations for presumptive evidence of specific disease association*. PhD thesis, University of Idaho.

Chang, C. I., Zhang, Y. A., Zou, J., *et al.*, 2006. Two cathelicidin genes are present in both rainbow trout (*Oncorhynchus mykiss*) and Atlantic salmon (*Salmo salar*). *Antimicrob Agents Chemother* 50: 185-195.

Chen, P. E., Cook, C., Stewart, A. C., *et al.*, 2010. Genomic characterization of the *Yersinia* genus. *Genome Biol* 11: R1.

Costa, A. A., Leef, M. J., Bridle, A. R., *et al.*, 2011. Effect of vaccination against yersiniosis on the relative percent survival, bactericidal and lysozyme response of Atlantic salmon, *Salmo salar*. *Aquaculture* 315: 201-206.

Dahiya, I. and Stevenson, R. M. W. 2010. The UvrY response regulator of the BarA-UvrY two-component system contributes to *Yersinia ruckeri* infection of rainbow trout (*Oncorhynchus mykiss*). *Arch Microbiol* 192: 541-547.

Davies, R. 1990. O-serotyping of *Yersinia ruckeri* with special emphasis on European isolates. *Vet Microbiol* 22: 299-307.

Davies, R. and Frerichs, G. N. 1989. Morphological and biochemical differences among isolates of *Yersinia ruckeri* obtained from wide geographical areas. *J Fish Dis* 12: 357-365.

Deshmukh, S., Raida, M. K., Dalsgaard, I., *et al.*, 2012. Comparative protection of two different commercial vaccines against *Yersinia ruckeri* serotype O1 and biotype 2 in rainbow trout (*Oncorhynchus mykiss*). *Vet Immunol Immunopathol* 145: 379-385.

Diaz-Rosales, P., Bird, S., Wang, T. H., *et al.*, 2009. Rainbow trout interleukin-2: cloning, expression and bioactivity analysis. *Fish Shellfish Immunol* 27: 414-422.

Fernandez, L., Lopez, J., Secades, P., *et al.*, 2003. *In vitro* and *in vivo* studies of the Yrp1 protease from *Yersinia ruckeri* and its role in protective immunity against enteric red mouth disease of salmonids. *Appl Environ Microbiol* 69: 7328-7335.

Fernandez, L., Prieto, M. and Guijarro, J. A. 2007a. The iron- and temperature-regulated haemolysin *yhia* is a virulence factor of *Yersinia ruckeri*. *Microbiology* 153: 483-489.

Fernandez, L., Mendez, J. and Guijarro, J. A. 2007b. Molecular virulence mechanisms of the fish pathogen *Yersinia ruckeri*. *Vet Microbiol* 125: 1-10.

Frerichs, G. N. 1993. Isolation and identification of fish bacterial pathogens, in *Bacterial Diseases of Fish* (eds V. Inglis, R. J. Roberts and N. R. Bromage). Oxford, Blackwell Scientific Publications, 270-272.

Galligani, G., David-Andersen, I. and Fossum, B. 2005. Regulatory constraints as seen from the pharmaceutical industry. *Dev Biol* (*Basel*) 121: 235-241.

Håstein, T., Gudding, R. and Evensen, Ø. 2005. Bacterial vaccines for fish-an update of the current situation worldwide. *Dev Biol* (*Basel*) 121: 55-74.

Horne, M. T. and Barnes, A. C. 1999. Enteric redmouth disease (*Yersinia ruckeri*), in *Fish Diseases and Disorders*, *Viral*, *Bacterial and Fungal Infections*, 2nd edn (eds P. T. K. Woo and D. W. Bruno). Wallingford, CABI, 455-477.

Johnson, K., Flynn, J. and Amend, D. 1982. Duration of immunity in salmonids vaccinated by direct immersion with *Yersinia ruckeri* and *Vibrio anguillarum* bacterins. *J Fish Dis* 5: 207-213.

Klontz, G. W. 1963. Oral immunization of rainbow trout against redmouth, in *Proceedings of the Northwest Fish Cultural Conference*, Olympia, Washington 5-6 December, p. 121.

Koppang, E. O., Haugarvoll, E., Hordvik, I., *et al.*, 2005. Vaccine-associated granulomatous inflammation and melanin accumulation in Atlantic salmon, *Salmo salar* L., white muscle. *J Fish Dis* 28: 13-22.

Lee, S. B., Mine, Y. and Stevenson, R. M. W. 2000. Effects of hen egg yolk immunoglobulin in passive protection of rainbow trout against *Yersinia ruckeri*. *J Agric Food Chem* 48: 110-115.

Mendez, J. and Guijarro, J. A. 2013. *In vivo* monitoring of *Yersinia ruckeri* in fish tissues: progression and virulence gene expression. *Environ Microbiol Reports* 5: 179-185.

Mendez, J., Reimundo, P., Perez-Pascual, D., *et al.*, 2011. A novel *cdsAB* operon is involved in the uptake of L-Cysteine and participates in the pathogenesis of *Yersinia ruckeri*. *J Bacteriol* 193: 944-951.

Midtlyng, P. J. 2005. Critical assessment of regulatory standards and tests for fish vaccines. *Dev Biol* (*Basel*) 121: 219-226.

Midtlyng, P. J. and Lillehaug, A. 1998. Growth of Atlantic salmon *Salmo salar* after intraperitoneal administration of vaccines containing adjuvants. *Dis Aquat Org* 32: 91-97.

Olesen, N. J. 1991. Detection of the antibody-response in rainbow-trout following immersion vaccination with *Yersinia ruckeri* bacterins by Elisa and passive-immunization. *J Appl Ichthyol* 7: 36-43.

Raida, M. K. and Buchmann, K. 2007. Temperature-dependent expression of immune-relevant genesin rainbow trout following *Yersinia ruckeri* vaccination. *Dis Aquat Org* 77: 41-52.

Raida, M. K. and Buchmann, K. 2008. Development of adaptive immunity in rainbow trout, *Oncorhynchus mykiss* (Walbaum) surviving an infection with *Yersinia ruckeri*. *Fish Shellfish Immunol* 25: 533-541.

Raida, M. K., Nylen, J., Holten-Andersen, L. and Buchmann, K. 2011a. Association between plasma antibody response and protection in rainbow trout *Oncorhynchus mykiss* immersion vaccinated against *Yersinia ruckeri*. *PLoS One* 6: e18832.

Raida, M. K., Holten-Andersen, L. and Buchmann, K. 2011b. Association between *Yersinia ruckeri* infection, cytokine expression and survival in rainbow trout (*Oncorhynchus mykiss*). *Fish Shellfish Immunol* 30: 1257-1264.

Romalde, J. L. and Toranzo, A. E. 1993. Pathological activities of *Yersinia-ruckeri*, the enteric redmouth (ERM) bacterium. *FEMS Microbiol Lett* 112: 291-300.

Romalde, J. L., Magariños, B., Barja, J. L. and Toranzo, A. E. 1993. Antigenic and molecular characterization of *Yersinia ruckeri*. Proposal for a new intraspecies classification. *System Appl Microbiol* 16: 411-419.

Ross, A. J. and Klontz, G. W. 1965. Oral immunization of rainbow trout (*Salmo gairdneri*) against an etiologic agent of redmouth disease. *J Fish Res Board Can* 22: 763-770.

Ross, A. J., Rucker, R. R. and Ewing, W. H. 1966. Description of a bacterium associated with redmouth disease of rainbow trout (*Salmo gairdneri*). *Can J Microbiol* 12: 763-770.

Rucker, R. 1966. Redmouth disease of rainbow trout (*Salmo gairdneri*). *Bull Off Int Epiz* 65: 825-830.

Ryckaert, J., Bossier, P., D'Herde, K., *et al.*, 2010. Persistence of *Yersinia ruckeri* in trout macrophages. *Fish Shellfish Immunol* 29: 648-655.

Schill, W. B., Phelps, S. R. and Pyle, S. W. 1984. Multilocus electrophoretic assessment of the genetic structure and diversity of *Yersinia ruckeri*. *Appl Environ Microbiol* 48: 975-979.

Temprano, A., Riano, J., Yugueros, J., *et al.*, 2005. Potential use of a *Yersinia ruckeri* O1 auxotrophic aroA mutant as a live attenuated vaccine. *J Fish Dis* 28: 419-427.

Tinsley, J. W., Austin, D. A., Lyndon, A. and Austin, B. 2011a. Novel non-motile phenotypes of *Yersinia ruckeri* suggest expansion of the current clonal complex theory. *J Fish Dis* 34: 311-317.

Tinsley, J. W., Lyndon, A. R. and Austin, B. 2011b. Antigenic and cross-protection studies of biotype 1 and biotype 2 isolates of *Yersinia ruckeri* in rainbow trout, *Oncorhynchus mykiss* (Walbaum). *J Appl Microbiol* 11: 8-16.

Tobback, E. A., Decostere, A., Hermans, K., *et al.*, 2007. *Yersinia ruckeri* infections in salmonid fish. *J Fish Dis* 30: 257-268.

Tobback, E., Decostere, A., Hermans, K., *et al.*, 2009a. Route of entry and tissue distribution of *Yersinia ruckeri* in experimentally infected rainbow trout *Oncorhynchus mykiss*. *Dis Aquat Org* 84: 219-228.

Tobback, E., Decostere, A., Hermans, K., *et al.*, 2009b. *In vitro* markers for virulence in *Yersinia ruckeri*. *J Fish Dis* 33: 197-209.

Wang, T., Martin, S. A. and Secombes, C. J. 2009. Two interleukin-17C-like genes exist in rainbow trout *Oncorhynchus mykiss* that are differentially expressed and modulated. *Dev Comp Immunol* 34: 491-500.

Wheeler, R. W., Davies, R. L., Dalsgaard, I., *et al.*, 2009. *Yersinia ruckeri* biotype 2 isolates from mainland Europe and the UK likely represent different clonal groups. *Dis Aquat Org* 84: 25-33.

第二十章　链球菌病和乳球菌病的疫苗接种

Julia W. Pridgeon and Phillip H. Klesius

本章概要　SUMMARY

链球菌病和乳球菌病是两种超急性全身性疾病，在水产养殖发达的国家，各种环境下的养殖鱼类和野生鱼类中均会发生。链球菌病和乳球菌病的致病菌是革兰氏阳性菌，分属链球菌属（*Streptococcus*）和乳球菌属（*Lactococcus*），包括无乳链球菌（*S. agalactiae*）、海豚链球菌（*S. iniae*）和格氏乳球菌（*L. garvieae*）。本章介绍该病的发生流行、危害、病原学、致病机制、疫苗、疫苗接种流程、疫苗效果和副作用等方面的内容，并重点介绍疫苗研发。

第一节　发生流行

链球菌病和乳球菌病是超急性全身性疾病，各种水生环境（淡水、河口和海洋）中的养殖鱼类和野生鱼类均可发生该病。多种鱼类均会感染链球菌病和乳球菌病，主要有罗非鱼、鲶、鲤、石斑鱼、鲻、鲳、鳟、鲑、鲆、海鲷、红鲷、尖吻鲈、河鲀、杂交条纹鲈、黄尾鰤、鲟、杜氏鰤和岩礁鱼类等。因此，在水产养殖业发达的国家中，链球菌病和乳球菌病是危害非常严重的细菌性疾病。

第二节　危　害

链球菌和乳球菌不仅是鱼类的主要致病菌，也会造成人类和其他脊椎动物的感染，例如猪、牛等。无乳链球菌（B 型链球菌）可感染婴幼儿、孕妇、老年人和免疫力低下的成年人并致病（Heath，2011）。无乳链球菌也会导致奶牛患乳腺炎（Watts 等，1984）。在水产养殖中，无乳链球菌会感染各种水生环境（淡水、河口和海洋）中的养殖鱼类和野生鱼类，链球菌病的暴发会导致养殖场鱼类大规模死亡，例如，2001 年 8 月和 9 月，该病造成科威特湾超过2 500 t 野生鲻死亡（Glibert 等，2002）；2009 年，链球菌病造成中国一些养殖场罗非鱼死亡率达 50%～70%（Lu，2010）。此外，2007 年至 2011 年，无乳链球菌引起澳大利亚昆士兰的大型昆士兰石斑鱼（*Epinephelus lanceolatus*）暴发疾病，这种石斑鱼可以长到 4m 并且体重可达 400 kg（Bowater 等，2012）。1997 年和 1998 年在中国台湾，虹鳟和灰鲻暴发乳球菌病，造成大量死亡（Chen 等，2002）。2001 年 5 月至 6 月间在土耳其，虹鳟（体重 100～150g）感染格氏乳球菌暴发了乳球菌病，死亡率达 80%（Diler 等，2002）。2003 年 8 月，中国河北省某养殖场 3～4 月龄的养殖牙鲆感染格氏乳球菌引起乳球菌病，死亡率达 25%～40%（Fang 等，2006）。2008 年到 2009 年在伊朗，从 7 个省的患病虹鳟中分离出 108 株革兰氏阳性菌，经鉴定，49 株（45%）为海豚链球菌，37 株（35%）为格氏乳球菌，22 株（19%）为链球菌属菌（Haghighi Karsidani 等，2010），表明伊朗虹鳟养殖场链球菌病和乳球菌病均比较严重。据估计，美国的水产养殖业仅海豚链球菌感染造成的年经济损失就达1 000万美元，而全球约为 1 亿美元（Shoemaker 等，2001）。

第三节 病原学

链球菌病和乳球菌病分别是由链球菌属和乳球菌属的革兰氏阳性菌感染引起，链球菌和乳球菌都属于链球菌科（细菌界、厚壁菌门、乳杆菌目）。据报道，作为鱼类致病菌的链球菌和乳球菌有多种，包括无乳链球菌（*S. agalactiae*）(Glibert 等，2002)、海豚链球菌（*S. iniae*）(Al Harbi，2011)、停乳链球菌（*S. dysgalactiae*）（Pourgholam 等，2011）、海豹链球菌（*S. phocae*）（Gonzalez-Contreras 等，2011）、副乳链球菌（*S. parauberis*）（Park 等，2009）、鮰链球菌（*S. ictaluri*）（Shewmaker 等，2007）、米氏链球菌（*S. milleri*）（Austin 和 Robertson，1993）、格氏乳球菌（*L. garvieae*）（Fang 等，2006）和乳酸乳球菌（*L. lactis*）（Gürpinar 和 Savasan，2010）。

1943 年，首次在挪威正常鱼体的菌群中发现了链球菌（Thjøtta Sømme，1943）。研究表明，α 型溶血性链球菌（Foo 等，1985）、β 型溶血性链球菌（Kusuda 等，1976；Minami 等，1979）以及 γ 型溶血性链球菌或非溶血性链球菌（Rasheed 和 Plumb，1984）都会感染鱼类。患病鱼体中分离出的无乳链球菌血清型属 Ⅰa 型、Ⅰb 型或者 Ⅱ 型（Bowater 等，2012；Guo 等，2012）。据报道，生物Ⅰ型和生物Ⅱ型海豚链球菌均可感染鱼类（Sheehan 等，2009）。

乳球菌病的病原是格氏乳球菌。1991 年，日本的患病黄尾鰤和鳗（Kusuda 等，1991）中分离出杀鰤肠球菌（*Enterococcus seriolicida*，为格氏乳球菌的异名；Eldar 等，1996）；2002 年，中国台湾的灰鲻（*Mugil cephalus*）暴发疾病，其病原为格氏乳球菌（Chen 等，2002）；2003 年，出现红海的野生隆头鱼格氏乳球菌感染的报道（Colorni 等，2003）；2004 年，葡萄牙的虹鳟（Pereira 等，2004）和日本的许氏平鲉中（Kobayashi 等，2004）也出现格氏乳球菌感染；2006 年报道，格氏乳球菌是导致 2003 年中国养殖牙鲆疾病暴发的病原（Fang 等，2006）；2007 年，希腊的养殖虹鳟中报道了多起由乳球菌导致的疾病暴发（Savvidis 等，2007）；2010 年，伊朗养殖的虹鳟暴发了由格氏乳球菌引起的疾病（Haghighi Karsidani 等，2010）；2011 年，西班牙的养殖虹鳟出现格氏乳球菌感染（Aguado-Urda 等，2011）。目前已知链球菌和乳球菌均能感染人类和其他动物（Wang 等，2007；MacFarquhar 等，2010），但是致病菌的来源仍不清楚。

第四节 致病机理

关于链球菌病和乳球菌病致病机理已有一些研究报道。链球菌外毒素感染养殖鰤（*Seriola* spp.）的研究结果表明，仅外毒素提取物可致鱼死亡；接种链球菌的弱毒力菌株或毒力菌株可以起到很好的保护作用，鰤不再发病（Kimura 和 Kusuda，1979，1982）。有研究报道，海豚链球菌产生的溶血素使血琼脂平板出现溶血（Fuller 等，2002）；转座子突变研究表明，海豚链球菌溶血素 S 的表达造成鱼体局部组织坏死（Fuller 等，2002）。应用斑马鱼模型对1 128个标记转座子突变体进行大规模筛选，鉴定出几种与海豚链球菌感染相关的荚膜合成基因（Miller 和 Neely，2005）。此外，以杂交条纹鲈为感染模型，通过基因敲除实验证明，荚膜基因 *cpsD*（B型链球菌荚膜聚合和输出必需基因）是海豚链球菌的一种毒力因子（Locke 等，2007a）。在牙鲆上也得到相同的结果，即海豚链球菌的荚膜多糖是感染牙鲆的毒力因子（Shutou 等，2007）。在杂交条纹鲈中，随机转座子诱变和高通量筛选研究显示，磷酸葡萄糖变位酶是海豚链球菌的毒力因子，该基因敲除的突变菌株致病性明显减弱，并在鱼体血液中很快被清除（Buchanan 等，2005）。在杂交条纹鲈上的研究表明，链球菌溶血素 S 是一种分泌型成孔细胞毒素，是海豚链球菌的毒力因子（Locke 等，2007b），此外，M-样蛋白在海豚链球菌的毒力方面发挥重要作用（Locke 等，2008）。全基因组焦磷酸测序和转座子诱变技术证明，多糖去乙酰化酶具有协助黏附和入侵鱼类宿主、抵抗溶菌酶活性、抗血清杀伤等功能，在海豚链球菌的致病过程中发挥作用（Milani 等，2010）。尽管已经报道了一些毒力因子，链球菌病和乳球菌病的致病机理仍有待阐明。

第五节 疫 苗

目前，已至少有一个商品疫苗用于预防无乳链球菌感染。2011 年，默克动物保健有限公司（Merck Animal Health Inc.）获得巴西监管部门的批准，开始销售名为 AQUAVAC® Strep Sa 的疫苗，这是一种灭活的油类佐剂疫苗，用于罗非鱼和其他易感鱼类预防无乳链球菌感染。然而，该疫苗只用于注射免疫体重超过 15g 的鱼，这显然非常耗费人力。然而，养殖户从育苗场拿到的仅为孵化后 1～2 周的鱼苗，体重 3～5g 或更小，不适合注射疫苗。因此，迫切需要一种针对小鱼的安全有效的无乳链球菌疫苗。有研究报道，福尔马林灭活的无乳链球菌与浓缩的胞外产物（大于 3 ku）混合免疫体重为 30g 的罗非鱼，接种 30d 后与未接种疫苗的罗非鱼相比，获得显著的免疫保护，相对保护率为 80%，但是对体重 5g 的罗非鱼却无免疫保护（Evans 等，2004）；该疫苗接种后的 47d、90d 和 180d，相对保护率分别降至 60%、55% 和 46%（Pasnik 等，2005）。有研究发现，与对照组相比，福尔马林灭活的无乳链球菌能提高受免罗非鱼产生抗体的效价，然而也有报道表明，无论是腹腔注射还是腹腔注射联合口服进行加强免疫（每月 1 次和 2 次），都未能使罗非鱼对无乳链球菌的感染产生保护作用（Manjaiang 等，2007）。将 85 株无乳链球菌应用 10 种脉冲场凝胶电泳（PFGE）基因分型（A-J）技术研发灭活全菌疫苗，发现无乳链球菌疫苗的免疫保护作用不仅与菌株的血清型有关，也与它们的

PFGE 基因分型有关（Chen 等，2012）。为了针对无乳链球菌感染提供广泛的免疫保护作用，有研究使用司帕沙星抗性实验研发出包含 30 株无乳链球菌分离株的多价减毒疫苗（Pridgeon 和 Klesius，2013），该多价疫苗免疫鱼体后，以这 30 株无乳链球菌亲本致病株攻毒后，可为体重 3～5g 和 15～20g 的尼罗罗非鱼提供显著的免疫保护（$P<0.001$）（Pridgeon 和 Klesius，2013）。

除了无乳链球菌，海豚链球菌也是水产养殖中重要的鱼类病原。目前，亚洲有一种疫苗可用于预防罗非鱼感染海豚链球菌（Sommerset 等，2005），在世界其他地区还没有可用的海豚链球菌疫苗。海豚链球菌的菌苗是研究最多的疫苗（Eldar 等，1997；Bercovier 等，1997；Romalde 等，1999）。在以色列，福尔马林灭活的海豚链球菌疫苗曾成功地用于虹鳟，然而有报道发现，这些菌苗不能提供针对海豚链球菌变异株（不同血清型）的感染保护（Bachrach 等，2001；Eyngor 等，2008）。福尔马林灭活的海豚链球菌与浓缩的胞外产物混合的疫苗可以给尼罗罗非鱼提供对海豚链球菌感染的部分免疫保护（Klesius 等，1999，2000，2006；Shoemaker 等，2006）。在实验室条件下，表达海豚链球菌 Sia10 抗原的 DNA 疫苗可以保护大菱鲆，分别用高剂量和低剂量的海豚链球菌攻毒，相对保护率分别为 73.9% 和 92.3%（对照组的累积死亡率分别为 92% 和 52%）（Sun 等，2010）。除了菌苗疫苗和 DNA 疫苗，磷酸葡萄糖变位酶和 M 样蛋白缺陷的海豚链球菌减毒活疫苗，可为鱼体提供针对同源海豚链球菌感染的免疫保护（Buchanan 等，2005；Locke 等，2008），目前尚不清楚这些减毒活疫苗是否能提供针对异源海豚链球菌的保护。有研究发现，一种高效减毒的耐新生霉素的链球菌疫苗（名为 ISNO）可以保护罗非鱼免受毒性海豚链球菌株的感染，保护期至少持续 6 个月（Pridgeon 和 Klesius，2011），接种 2 个月后，ISNO 能够在为罗非鱼提供针对强致病力的异源性海豚链球菌菌株（F3CB、102F1K、405F1K、IF6、ARS60）的保护，相对保护率分别为 78%、90%、100%、100% 和 100%（Pridgeon 和 Klesius，2011）。ISNO 疫苗的有效接种剂量范围比较宽，每尾鱼的接种量分别为 10^2 cfu、10^3 cfu、10^4 cfu、10^5 cfu、10^6 cfu、10^7 cfu 时，对应的相对保护率分别为 81%、94%、100%、100%、100%、100%（Pridgeon 和 Klesius，2011）。除了无乳链球菌和海豚链球菌外，停乳链球菌和海豹链球菌也是水产养殖中重要的新病原，目前还没有商业疫苗可用于预防这两种链球菌的感染。

在意大利、法国和英国，格氏乳球菌商品疫苗已应用于虹鳟养殖（Sommerset 等，2005）；在日本，也有一种格氏乳球菌商品疫苗应用于杜氏鰤/黄尾鰤（Sommerset 等，2005）。分别在实验室和野外条件下的研究发现，福尔马林灭活的格氏乳球菌能够使虹鳟分别获得持续免疫保护 5 个月和 3 个月，相对保护率均为 70%～80%（Bercovier 等，1997）。还有研究发现，黄尾鰤腹腔注射 KG－表型（有荚膜型）或 KG＋表型（无荚膜型）的福尔马林灭活格氏乳球菌时，可获得长期免疫保护（Ooyama 等，1999）。此外，加入不同佐剂与不加佐剂的格氏乳球菌菌苗的免疫效果比较研究发现，免疫第 4 周后，不加佐剂和加入非矿物油类佐剂（Aquamun）的疫苗均具有良好的免疫保护效果，相对保护率都高于 80%；然而，免疫 3 个月后，不加佐剂的疫苗相对保护率仅为 40%，而加 Aquamun 佐剂的疫苗在接种后 3 个月、6 个月和 8 个月的相对保护率分别高达 92%、90% 和 83%（Ravelo 等，2006）。与此相类似的研究也发现，与未加佐剂的疫苗相比，以弗氏不完全佐剂与福尔马林灭活的疫苗混合免疫虹鳟，鱼体获得的保护期更长（Kubilay 等，2008）。除了福尔马林灭活的格氏乳球菌疫苗外，一种格氏乳球菌减毒活疫苗处于实验室研究阶段，该候选疫苗菌株表面缺乏一种毒力相关荚膜，可为五条鰤提供持久的免疫保护作用（Ooyama 等，2006）。

第六节 疫苗接种和疫苗效果

与其他疫苗相同，链球菌病和乳球菌病疫苗常用的免疫接种方式为注射、浸泡和口服三种。据报道，链球菌病疫苗腹腔注射与口服相比，可为黄尾鰤提供更高水平的免疫保护（Iida 等，1982）；福尔马林灭活的β-溶血型链球菌疫苗通过腹腔注射和浸泡免疫，以致病性的链球菌感染攻毒，其相对保护率不低于 70%（Sakai 等，1987）。除了腹腔注射外，肌肉注射也用于鱼类疫苗接种，然而，在某些情况下，肌肉注射的效率明显低于腹腔注射。例如，罗非鱼腹腔注射福尔马林灭活的海豚链球菌疫苗（ARS-10），以同源分离株 ARS-10 攻毒，相对保护率为 45.6%，而肌肉注射免疫其保护率仅为 17.7%；同样地，罗非鱼腹腔注射 ARS-10 疫苗后，以异源分离株 ARS-60 攻毒，获得的相对保护率为 93.7%，而肌肉注射免疫保护率为 59.5%（Klesius 等，2000）。

同一疫苗进行腹腔注射和浸泡免疫获得的保护作用比较研究表明，腹腔注射通常比浸泡免疫的效果更好。例如，福尔马林灭活的无乳链球菌（ARS-KU-MU-11B）疫苗腹腔注射体重为 30g 的罗非鱼，其相对保护率为 80%，而浸泡免疫获得的相对保护率仅达到腹腔注射的一半（Evans 等，2004）。除了注射免疫和浸泡免疫，口服疫苗也可用于鱼类的免疫接种，比较福尔马林灭活链球菌疫苗以腹腔注射、口服和高渗浸泡三种接种方式免疫黄尾鰤的效果研究表明，腹腔注射方式效果最好，其次是高渗浸泡和口服（Iida 等，1982）。福尔马林灭活的海豚链球菌疫苗免疫虹鳟，腹腔注射获得 74%～100%的相对保护率，而浸泡免疫和口服免疫的相对保护率分别为 30%～45%和 9%～29%（Soltani 等，2007）。

评估包被和未包被的抗原口服免疫效果，探讨口服免疫作为虹鳟乳球菌病疫苗的替代免疫接种方式，结果表明，与其他包被方式比较，海藻酸包被的疫苗（50%的相对保护率）保护率最高；但是由于口服免疫的保护效果低于注射免疫，所以口服海藻酸包被的疫苗不宜作为初次免疫方式。格氏乳球菌水基疫苗初次免疫采用腹腔注射，以口服包被的疫苗进行加强免疫，虹鳟获得的相对保护率可增加至 87%（Kawai 和 Hatamoto，1999）。

对于活菌疫苗，浸泡和注射免疫效果同样有效。罗非鱼浸泡免疫海豚链球菌的减毒活疫苗 ISNO，致病海豚链球菌攻毒后，相对保护率为 88%，而腹腔注射相对保护率为 100%（Pridgeon 和 Klesius，2011）。

除了三种典型的疫苗接种方法外，还研究了应用皮肤多点穿刺技术进行浸泡免疫的新型接种方式，结果发现，β-溶血型链球菌疫苗免疫虹鳟幼鱼，该方法与腹腔注射同样具有较好的免疫保护效果。该方法采用多点穿刺器械在鱼体皮肤表面产生微小的创伤，然后将虹鳟浸泡在福尔马林灭活的海豚链球菌疫苗液中进行免疫接种，两周后进行攻毒，该方式与腹腔注射接种的虹鳟死亡率均为 40%，而未接种的对照组和浸泡组（无多点穿刺）虹鳟的死亡率均达到 80%（Nakanishi 等，2002）。

据报道，链球菌病疫苗对大菱鲆非常有效，腹腔注射免疫相对保护率可达 100%（Romalde 等，1999）；福尔马林灭活的海豚链球菌与其浓缩胞外产物（ECP）混合通过腹腔注射免疫罗非鱼（*Oreochromis niloticus*），保护率可达 91%（Klesius 等，1999）；罗非鱼腹腔注射福尔马林灭活的无乳链球菌与其浓缩胞外产物，对体重为 30g 和 5g 罗非鱼的相对保护率分别为 80%和 25%（Evans 等，2004）；福尔马林灭活的格氏乳球菌和 Aquamun 佐剂混合免疫虹鳟后 3 个月，相对保护率为 92%（Ravelo 等，2006）；福尔马林灭活的链球菌疫苗配以弗氏完全佐剂腹腔注射免疫海鲈，免疫后 10d、20d 和 30d，其相对保护率分别为 100%、92%和 74%（Wanman 等，2007）；一种表达 Sia10 抗原的 DNA 候选疫苗免疫大菱鲆后，使用高浓度和低浓度的海豚链球菌攻毒后，产生的相对保护率分别 73.9%和 92.3%（对照组中累积死亡率分别为 92%和 52%）；磷酸葡萄糖变位酶和 M 样蛋白缺失的海豚链球菌减毒活疫苗，可提供针对同源海豚链球菌感染的保护（Buchanan 等，2005；Locke 等，2008）；海豚链球菌减毒活疫苗（ISNO）可以预防海豚链球菌感染，对罗非鱼产生的免疫保护作用至少持续 6 个月（Pridgeon 和 Klesius，2011），此外，ISNO 接种 2 个月后，产生针对异源性海豚链球菌毒株（F3CB、102F1K、405F1K、IF6、ARS60）的保护，对罗非鱼的相对保护率分别为 78%、90%、100%、100%和 100%（Pridgeon 和 Klesius，2011）。

第七节 副作用

不含佐剂的链球菌病和乳球菌病灭活疫苗产生的免疫保护期通常较短，因此，通常会在水基疫苗中加入油类佐剂以延长疫苗的保护期。然而，有研究表明，佐剂疫苗会延缓鱼体的生长（Midtlyng 和 Lillehaug，1998）。而且，佐剂型疫苗会引起鱼体腹腔内部粘连和注射部位病变（Ronsholdt 和 McLean，1999）。因此，今后该病疫苗研制的重点不仅要考虑配方各参数对抗原的保持和缓释能力的影响，同时也要考虑减少对鱼体造成的应激影响。

第八节 管理条例

在美国，鱼类疫苗被归类为“生物制剂”，因此，受兽医生物学中心和美国农业部动植物检验服务中心（APHIS）的监管与许可。APHIS 监管兽医生物制剂（疫苗、菌苗、抗血清、诊断试剂盒以及其他生物来源的产品），以确保用于动物疾病的诊断、预防、治疗的兽医生物制剂是稳定、安全和有效的。APHIS 有关生物制剂的法规条例可以在以下网址上查阅：www.aphis.usda.gov/animal_health/vet_biologics/vb_regs_and_guidance.shtml。关于鱼类疫苗的联邦政府法典可在 www.aphis.usda.gov/animal_health/vet_biologics/vb_regs_and_guidance.shtml 上查阅。

参考文献

Aguado-Urda, M., Lopez-Campos, G. H., Gibello, A., *et al*., 2011. Genome sequence of *Lactococcus garvieae*

8831, isolated from rainbow trout lactococcosis outbreaks in Spain. *J Bacteriol* 193: 4263-4264.

Al Harbi, A. H. 2011. Molecular characterization of *Streptococcus iniae* isolated from hybrid tilapia [*Oreochromis niloticus* x *Oreochromis aureus*]. *Aquaculture* 312: 15-18.

Austin, B. and Robertson, P. A. 1993. Recovery of *Streptococcus milleri* from ulcerated koi carp (*Cyprinus carpio* L.) in the UK. *Bull Eur Assoc Fish Pathol* 13: 207-209.

Bachrach, G., Zlotkin, A., Hurvitz, A., *et al.*, 2001. Recovery of *Streptococcus iniae* from diseased fish previously vaccinated with a streptococcus vaccine. *Appl Environ Microbiol* 67: 3756-3758.

Bercovier, H., Ghittino, C. and Eldar, A. 1997. Immunization with bacterial antigens: infections with streptococci and related organisms. *Dev Biol Stand* 90: 153-160.

Bowater, R. O., Forbes-Faulkner, J., Anderson, I. G., *et al.*, 2012. Natural outbreak of *Streptococcus agalactiae* (GBS) infection in wild giant Queensland grouper, *Epinephelus lanceolatus* (Bloch), and other wild fish in northern Queensland, Australia. *J Fish Dis* 35: 173-186.

Buchanan, J. T., Stannard, J. A., Lauth, X., *et al.*, 2005. *Streptococcus iniae* phosphoglucomutase is a virulence factor and a target for vaccine development. *Infect Immun* 73: 6935-6944.

Chen, M., Wang, R., Li, L. P., *et al.*, 2012. Screening vaccine candidate strains against *Streptococcus agalactiae* of tilapia based on PFGE genotype. *Vaccine* 30: 6088-6092.

Chen, S., Liaw, L., Su, H., *et al.*, 2002. *Lactococcus garvieae*, a cause of disease in grey mullet, *Mugil cephalus* L., in Taiwan. *J Fish Dis* 25: 727-732.

Colorni, A., Ravelo, C., Romalde, J. L., *et al.*, 2003. *Lactococcus garvieae* in wild red sea wrasse *Coris aygula* (Labridae). *Dis Aquat Org* 56: 275-278.

Diler, O., Altun, S., Adiloglu, A. K., *et al.*, 2002. First occurrence of streptococcosis affecting farmed rainbow trout (*Oncorhynchus mykiss*) in Turkey. *Bull Eur Assoc Fish Pathol* 22: 21-26.

Eldar, A., Ghittino, C., Asanta, L., *et al.*, 1996. *Enterococcus seriolicida* is a junior synonym of *Lactococcus garvieae*, a causative agent of septicemia and meningoencephalitis in fish. *Curr Microbiol* 32: 85-88.

Eldar, A., Horovitcz, A. and Bercovier, H. 1997. Development and efficacy of a vaccine against *Streptococcus iniae* infection in farmed rainbow trout. *Vet Immunol Immunopathol* 56: 175-183.

Evans, J. J., Klesius, P. H. and Shoemaker, C. A. 2004. Efficacy of *Streptococcus agalactiae* (group B) vaccine in tilapia (*Oreochromis niloticus*) by intraperitoneal and bath immersion administration. *Vaccine* 22: 3769-3773.

Eyngor, M., Tekoah, Y., Shapira, R., *et al.*, 2008. Emergence of novel *Streptococcus iniae* exopolysaccharide-producing strains following vaccination with nonproducing strains. *Appl Environ Microbiol* 74: 6892-6897.

Fang, H., Chen, C. and Zhang, X. 2006. *Lactococcus garvieae* as a pathogen in flounder (*Paralichthys olivaceus* Temminck et Schlegel). *J Fish Sci China* 13: 403-409.

Foo, J. T., Ho, B. and Lam, T. J. 1985. Mass mortality in *Siganus canaliculatus* due to streptococcal infection. *Aquaculture* 49: 185-196.

Fuller, J. D., Camus, A. C., Duncan, C. L., *et al.*, 2002. Identification of a streptolysin S-associated gene cluster and its role in the pathogenesis of *Streptococcus iniae* disease. *Infect Immun* 70: 5730-5739.

Glibert, P. M., Landsberg. J. H., Evans, J. J., *et al.*, 2002. A fish kill of massive proportion in Kuwait Bay, Arabian Gulf, 2001: the roles of bacterial disease, harmful algae, and eutrophication. *Harmful Algae* 1: 215-231.

Gonzalez-Contreras, A., Magariños, B., Godoy, M., *et al.*, 2011. Surface properties of *Streptococcus phocae* strains isolated from diseased Atlantic salmon, *Salmo salar* L. *J Fish Dis* 34: 203-215.

Guo, Y., Zhang, D., Fan, H., *et al.*, 2012. Molecular epidemiology of *Streptococcus agalactiae* isolated from tilapia in Southern China. *J Fish China* 36: 399-406.

Gürpinar, S. and Savasan, S. 2010. Investigation of Gram-positive bacterial infections in culture fisheries in the Aegean Region. *Etlik Vet Mikrobiyol Derg* 21: 31-36.

Haghighi Karsidani, S., Soltani, M., Nikbakhat-Brojeni, G., *et al.*, 2010. Molecular epidemiology of zoonotic streptococcosis/lactococcosis in rainbow trout (*Oncorhynchus mykiss*) aquaculture in Iran. *Iran J Microbiol* 2: 198-209.

Heath, P. T. 2011. An update on vaccination against group B streptococcus. *Exp Rev Vaccines* 10: 685-694.

Iida, T., Wakabayashi, H. and Egusa, S. 1982. Vaccination for control of streptococcal disease in cultured yellowtail. *Fish Pathol* 16: 201-206.

Kawai, K. and Hatamoto, K. 1999. Encapsulation of oral delivery vaccine against *Lactococcus garvieae* infection in yellowtail *Seriola quinqueradiata*. *Bull Mar Sci Fish Kochi Univ* 19: 71-78.

Kimura, H. and Kusuda, R. 1979. Studies on the pathogenesis of streptococcal infection in cultured yellowtails *Seriola spp.*: effect of the cell free culture on experimental streptococcal infection. *J Fish Dis* 2: 501-510.

Kimura, H. and Kusuda, R. 1982. Studies on the pathogenesis of streptococcal infection in cultured yellowtails, *Seriola spp.*: effect of crude exotoxin fractions from cell-free culture on experimental streptococcal infection. *J Fish Dis* 5: 471-478.

Klesius, P. H., Shoemaker, C. A. and Evans, J. J. 1999. Efficacy of a killed *Streptococcus iniae* vaccine in tilapia (*Oreochromis niloticus*). *Bull Eur Assoc Fish Pathol* 19: 39-41.

Klesius, P. H., Shoemaker, C. A. and Evans, J. J. 2000. Efficacy of single and combined *Streptococcus iniae* isolate vaccine administered by intraperitoneal and intramuscular routes in tilapia (*Oreochromis niloticus*). *Aquaculture* 188: 237-246.

Klesius, P. H., Evans, J. J., Shoemaker, C. A. and Pasnik, D. J. 2006. A vaccination and challenge model using calcein marked fish. *Fish Shellfish Immunol* 20: 20-28.

Kobayashi, T., Ishitaka, Y., Imai, M. and Kawaguchi, Y. 2004. Pathological studies on *Lactococcus garvieae* infection of cultured rockfish *Sebastes schlegeli*. *Suisan Zoshoku* 52: 359-364.

Kubilay, A., Altun, S., Ulukoy, G., *et al.*, 2008. Immunization of rainbow trout (*Oncorhynchus mykiss*) against *Lactococcus garvieae* using vaccine mixtures. *Isr J Aquac* 60: 268-273.

Kusuda, R., Kawai, K., Toyoshima, T. and Komatsu, I. 1976. A new pathogenic bacterium belonging to the genus *Streptococcus* isolated from an epizootic disease of cultured yellowtails. *Nippon Suisan Gakkaishi* 42: 1345-1352.

Kusuda, R., Kawai, K., Salati, F., *et al.*, 1991. *Enterococcus seriolicida sp. nov.*, a fish pathogen. *Int J Syst Bacteriol* 41: 406-409.

Locke, J. B., Colvin, K. M., Datta, A. K., *et al.*, 2007a. *Streptococcus iniae* capsule impairs phagocytic clearance and contributes to virulence in fish. *J Bacteriol* 189: 1279-1287.

Locke, J. B., Colvin, K. M., Varki, N., *et al.*, 2007b. *Streptococcus iniae* beta-hemolysin streptolysin S is a virulence factor in fish infection. *Dis Aquat Org* 76: 17-26.

Locke, J. B., Aziz, R. K., Vicknair, M. R., *et al.*, 2008. *Streptococcus iniae* M-like protein contributes to virulence in fish and is a target for live attenuated vaccine development. *PloS One*: e2824.

Lu, M. 2010. Review of research on streptococcosis in tilapia. *S China Fish Sci* 6: 75-79.

MacFarquhar, J. K., Jones, T. F., Woron, A. M., *et al.*, 2010. Outbreak of late-onset group B *Streptococcus* in a neonatal intensive care unit. *Am J Infect Control* 38：283-288.

Manjaiang，P.，Areechon，N.，Srisapoome，P. and Mahasawas，S. 2007. Application of vaccine to prevent disease caused by *Streptococcus agalactiae* in Nile tilapia（*Oreochromis niloticus*）. *Proceedings of the* 45th *Kasetsart University Annual Conference*，Bangkok，174-182.

Midtlyng，P. J. and Lillehaug，A. 1998. Growth of Atlantic salmon *Salmo salar* after intraperitoneal administration of vaccines containing adjuvants. *Dis Aquat Org* 32：91-97.

Milani，C. J.，Aziz，R. K.，Locke，J. B.，*et al.*，2010. The novel polysaccharide deacetylase homologue Pdi contributes to virulence of the aquatic pathogen *Streptococcus iniae*. *Microbiology* 156：543-554.

Miller，J. D. and Neely，M. N. 2005. Large-scale screen highlights the importance of capsule for virulence in the zoonotic pathogen *Streptococcus iniae*. *Infect Immun* 73：921-934.

Minami，T.，Nakamura，M.，Ikeda，Y. and Ozaki，H. 1979. A beta-hemolytic *Streptococcus* isolated from cultured yellowtail. *Fish Pathol* 1：33-38.

Nakanishi，T.，Kiryu，I. and Ototake，M. 2002. Vaccine development of a new vaccine delivery method for fish：percutaneous administration by immersion with application of a multiple puncture instrument. *Vaccine* 20：3764-3769.

Ooyama，T.，Kera，A.，Okada，T.，*et al.*，1999. The protective immune response of yellowtail *Seriola quinqueradiata* to the bacterial fish pathogen *Lactococcus garvieae*. *Dis Aquat Org* 37：121-126.

Ooyama T，Shimahara Y，Nomoto R，*et al.*，2006. Application of attenuated *Lactococcus garvieae* strain lacking a virulence-associated capsule on its cell surface as a live vaccine in yellowtail *Seriola quinqueradiata* Temminck and Schlegel. *J Appl Ichthyol* 22：149-152.

Park，Y.，Nho，S.，Shin，G.，*et al.*，2009. Antibiotic susceptibility and resistance of *Streptococcus iniae* and *Streptococcus parauberis* isolated from olive flounder（*Paralichthys olivaceus*）. *Vet Microbiol* 136：76-81.

Pasnik，D. J.，Evans，J. J. and Klesius，P. H. 2005. Duration of protective antibodies and correlation with survival in Nile tilapia *Oreochromis niloticus* following *Streptococcus agalactiae* vaccination. *Dis Aquat Org* 66：129-134.

Pereira，F.，Ravelo，C.，Toranzo，A. E. and Romalde，J. L. 2004. *Lactococcus garvieae*，an emerging pathogen for the Portuguese trout culture. *Bull Eur Assoc Fish Pathol* 24：274-279.

Pourgholam，R.，Laluei，F.，Saeedi，A. A.，*et al.*，2011. Distribution and molecular identification of some causative agents of streptococcosis isolated from farmed rainbow trout（*Oncorhynchus mykiss*，Walbaum）in Iran. *Iran J Fish Sci* 10：109-122.

Pridgeon，J. W. and Klesius，P. H. 2011. Development and efficacy of a novobiocin-resistant *Streptococcus iniae* as a novel vaccine in Nile tilapia（*Oreochromis niloticus*）. *Vaccine* 29：5986-5993.

Pridgeon，J. W. and Klesius，P. H. 2013. Development of live attenuated *Streptococcus agalactiae* as potential vaccines by selecting for resistance to sparfloxacin. *Vaccine* 31：2705-2712.

Rasheed，V. and Plumb，J. A. 1984. Pathogenicity of a non-haemolytic group B *Streptococcus* sp. in gulf killifish（*Fundulus grandis* Baird and Girard）. *Aquaculture* 37：97-105.

Ravelo，C.，Magarinos，B.，Herrero，M. C.，*et al.*，2006. Use of adjuvanted vaccines to lengthen the protection against lactococcosis in rainbow trout（*Oncorhynchus mykiss*）. *Aquaculture* 251：153-158.

Romalde，J. L.，Magariños，B. and Toranzo，A. E. 1999. Prevention of streptococcosis in turbot by intraperitoneal vaccination：a review. *J Appl Ichthyol* 15：153-158.

Ronsholdt，B. and McLean，E. 1999. The effect of vaccination and vaccine components upon short-term growth and feed conversion efficiency in rainbow trout. *Aquaculture* 174：213-221.

Sakai, M., Kubota, R., Atsuta, S. and Kobayashi, M. 1987. Vaccination of rainbow trout *Salmo gairdneri* against beta-hemolytic streptococcal disease. *Nippon Suisan Gakkaishi* 53: 1373-1376.

Savvidis, G. K., Anatoliotis, C., Kanaki, Z. and Vafeas, G. 2007. Epizootic outbreaks of lactococcosis disease in rainbow trout, *Oncorhynchus mykiss* (Walbaum), culture in Greece. *Bull Eur Assoc Fish Pathol* 27: 223-228.

Sheehan, B., Labrie, L., Lee, Y., *et al.*, 2009. Streptococcal diseases in farmed tilapia. *Aqua Asia Paci* 5 (6): 26-29.

Shewmaker, P. L., Camus, A. C., Bailiff, T., *et al.*, 2007. *Streptococcus ictaluri* sp. nov., isolated from channel catfish *Ictalurus punctatus* broodstock. *Int J Syst Evol Microbiol* 57: 1603-1606.

Shoemaker, C. A., Klesius, P. H. and Evans, J. J. 2001. Prevalence of *Streptococcus iniae* in tilapia, hybrid striped bass, and channel catfish on commercial fish farms in the United States. *Am J Vet Res* 62: 174-177.

Shoemaker, C. A., Vandenberg, G. W., Désormeaux, A., *et al.*, 2006. Efficacy of a *Streptococcus iniae* modified bacterin delivered using OraljectTM technology in Nile tilapia (*Oreochromis niloticus*). *Aquaculture* 255: 151-156.

Shutou, K., Kanai, K. and Yoshikoshi, K. 2007. Virulence attenuation of capsular polysaccharide-deleted mutants of *Streptococcus iniae* in Japanese flounder *Paralichthys olivaceus*. *Fish Pathol* 42: 41-48.

Soltani, M., Alishahi, M., Mirzargar, S. and Nikbakht, G. 2007. Vaccination of rainbow trout against *Streptococcus iniae* infection: comparison of different routes of administration and different vaccines. *Iran J Fish Sci* 7: 129-140.

Sommerset, I., Krossøy, B., Biering, E. and Frost, P. 2005. Vaccines for fish in aquaculture. *Exp Rev Vaccines* 4: 89-101.

Sun, Y., Hu, Y., Liu, C. and Sun, L. 2010. Construction and analysis of an experimental *Streptococcus iniae* DNA vaccine. *Vaccine* 28: 3905-3912.

Thjøtta, T. and Sømme, O. M. 1943. *The Bacterial Flora of Normal Fish*. Oslo. (Det norske videnskapsakademii Oslo. I. Matematisk-naturvidenskapelig klasse. Skrifter 1942, no. 4).

Wang, C. Y., Shie, H. S., Chen, S. C., *et al.*, 2007. *Lactococcus garvieae* infections in humans: possible association with aquaculture outbreaks. *Int J Clin Pract* 61: 68-73.

Wanman, C. H., Tanmark, N. and Supamattaya, K. 2007. Production of killed vaccine from *Streptococcus* sp. and its application in sea bass (*Lates calcarifer*). *Songklanakarin J Sci Tech* 29: 1251-1261.

Watts, J. L., Nickerson, S. C. and Pankey, J. W. 1984. A case study of *Streptococcus* group G infection in a dairy herd. *Vet Microbiol* 9: 571-580.

第二十一章　鱼立克次氏体病的疫苗接种

Sergio H. Marshall and Jaime A. Tobar

本章概要　SUMMARY

鱼立克次氏体病（Piscirickettsiosis），又称为鲑鱼立克次氏体败血症（Salmonid rickettsial septicemia，SRS），由鲑鱼立克次氏体（*Piscirickettsia salmonis*）引起。该病自 1989 年发生以来，已严重影响智利鲑产业的可持续性发展。鲑鱼立克次氏体是一种兼性革兰氏阴性菌，但目前对与疫苗研发相关的该菌各发育阶段的毒力、致病机理等方面了解甚少。 疫苗接种是防治该病的关键，清楚了解病原的生物学和免疫学特征对研制有效疫苗至关重要。目前，对鲑鱼立克次体并未透彻了解，由于该病在智利鲑集约化养殖中非常严重，已研发了一些疫苗并商品化，但效果一般。

第一节　发生流行与危害

鲑鱼立克次氏体是鲑鱼立克次氏体败血症的病原，可感染鲑和其他鱼类。21 世纪初以前，该病仅感染银大麻哈鱼，此后，该病传播至所有养殖的鲑科鱼类以及其他商品鱼类中（Chen 等，1994；Athanassopoulou 等，2004；Arkush 等，2005；McCarthy 等，2005；Rojas 等，2009）。

尽管严格的疾病管理措施起到一定作用，但该病的问题依然存在。智利是该病感染最严重的地区，虽然已开展了一些传统疫苗和其他预防性措施的尝试，但该病仍持续流行，急需有效的新策略防止该病在地理和宿主范围上的扩大蔓延。

目前，该病的频繁暴发导致了鲑较高的死亡率，此外，患鱼肌肉组织严重损伤，影响出口商品质量（Marshall，未发表）。一方面，鲑鱼立克次氏体是一种功能多样、基因变异多和适应性强的微生物；另一方面，对该菌在环境中的生物学和生命周期特性尚未完全了解。目前，通过体外实验并优化条件后，了解了该病原的一些生物学特性，但与自然环境条件下的病原有很大差别。未见关于鲑鱼立克次体在高感染性的海洋环境中是如何存活的研究报道。有研究发现，鲑鱼立克次氏体能够在低密度海水中存活至少 40d，这表明了该病原的耐受特性，遗憾的是，尚未弄清该特性产生的机制（Olivares 和 Marshall，2010）。

鲑鱼立克次氏体最初在智利发现，随后在加拿大西部、挪威和爱尔兰的患病鲑中也鉴定出该病原（Brocklebank 等，1992；Evelyn，1992；Olsen 等，1993；Rodger 和 Drinan，1993；Palmer 等，1997；Almendras 和 Fuentealba，1997；Fryer 和 Hedrick，2003）。

伴随着各种集约化养殖模式的兴起，单一物种在高密度养殖条件下，传染性疾病病原很容易在宿主间传播，并引起疾病暴发而造成巨大经济损失。这就是智利鲑养殖的现状，该病已成为最大威胁。智利政府部门采取多种策略，通过改进养殖技术等方法减少该病所造成的损失，但所有这些措施的重心是提供不危害动物健康和福祉的最佳养殖环境。

第二节 病原学

鲑鱼立克次氏体最初被认为是一种严格的胞内寄生菌，直径 0.2～1.5μm，不运动，无荚膜，形态多样，通常为球形（Bravo 和 Campos，1989；Mauel 等，1999）。但这些特性不利于该菌在分子水平上的研究，因为该菌只在各种细胞系中生长（Peña 等，2010）。目前发现该菌能够在无细胞的培养基中生长，其被重新归为兼性胞内菌（Mauel 等，2008；Mikalsen 等，2008；Gómez 等，2009；Yáñez 等，2012）。

鱼立克次氏体属隶属丙型变形菌纲、硫发菌目、立克次氏体科，鱼立克次氏体科包含 6 种进化相关而表型不同的属（解烃菌属 *Cycloclasticus*、氢弧菌属 *Hydrogenovibrio*、硫化菌属 *Sulfurivirga*、硫碱微菌属 *Thioalkalimicrobium*、硫微螺菌属 *Thiomicrospira* 和鱼立克次氏体属 *Piscirickettsia*），而立克次体属包括一个种，即鲑鱼立克次氏体（Fryer 和 Lannan，2005）。野生型菌株 LF89（ATCC VR-1361）是从鱼体中分离出的第一个革兰氏阴性胞内细菌。该菌是 1989 年从智利南部感染死亡率高达 70%～90% 的银大麻哈鱼中分离到的（Bravo 和 Campos，1989）。最初，从表型上被归为甲型变形杆菌纲、立克次氏体目、立克次氏体科，与立克次氏体属（*Rickettsia*）、沃尔巴克氏体属（*Wolbachia*）、无形体属（*Anaplasma*）和埃利希氏体属（*Ehrlichia*）有关（Fryer 等，1990）。1992 年，依据该菌 16SrRNA 的分析结果将其重新归类为丙型变形杆菌、硫发菌目，组建了鲑鱼立克次氏体新属新种（*Piscirickettsia salmonis* gen. nov.，sp. nov.）（Fryer 等，1992）。

目前，关于鲑鱼立克次氏体分子特征的报道较少，其基因组序列也没有公布。大多数分子生物学研究主要集中在该菌的抗原性分子筛选，用于研发新疫苗。已有研究鉴定出一些与疫苗研发有关的伴侣分子，主要有 HSP60（ChaP. s/GroEL）、HSP70（DnaK）、HSP10（GroES）和 HSP16（Wilhelm 等，2005，2006；Marshall 等，2007）。另外，已鉴定出 OspA 抗原，研制出该抗原的重组疫苗具有一定的效果（Kuzyk 等，2001a）。

鲑鱼立克次氏体的检测是基于临床观察和病理结果，然后通过组织病理学、组织培养分离，结合免疫荧光或过氧化物酶联免疫染色以及斑点 DNA 杂交等结果确定（Noga，2010）。此外，对于临床检测，可用 PCR 技术对鲑鱼立克次氏体进行快速鉴定（Marshall 等，1998）。

第三节 致病机理

鲑鱼立克次氏体的重新分类鉴定，以及了解其在无细胞培养基中的培养特性，有利于更好地分析和理解疾病的致病机理。下面这些发现可能为鲑鱼立克次氏体病疫苗的抗原筛选提供思路：该病原能够通过复杂的细胞膜的“立克次氏体黏附复合体”（PAC）紧密黏附在靶标（包括卵）表面上（Larenas 等，2003）；该菌主要感染宿主的巨噬细胞，并通过巨噬细胞扩散（McCarthy 等，2005；Rojas 等，2009）；ChaPs 蛋白具有较强的免疫原性，作为疫苗抗原具有一定的保护潜力（Marshall 等，2007）；鲑鱼立克次氏体在体内和体外产生的小亚型都具有高传染性，是主要的感染体（Rojas 等，2008；Marshall 等，2012）；在体外能诱导巨噬细胞凋亡，这可能是逃避宿主免疫系统的策略（Rojas 等，2010）；部分基因组的分析表明存在外源插入序列，即鲑鱼立克次氏体基因组具有高度的可塑性（GenBank：AF184152；Marshall 等，2011），同时发现了编码毒素-抗毒素蛋白的两个操纵子，可严格调控宿主细胞活动（Gómez 等，2011，GenBank：JQ023629）；有研究证实了该病原可以逃逸鱼类巨噬细胞的吞噬体-溶酶体的融合，作为一种有效的在宿主体内的生存策略（Gómez 等，2013）；在不同的应激条件下该病原会产生生物膜（Marshall 等，2012）并表达Ⅳ型分泌系统的大多数基因，这均与其致病力直接相关（Gómez 等，2013），这些研究结果均会影响疫苗的研发。

鲑鱼立克次氏体不同分离株基因组测序的完成，将提供一些重要的生物学特性的相关信息（Marshall 等，2011），现在虽然已有几个基因组序列，但并未完成全部基因组的测序与注释。

第四节　疫苗和免疫接种

鲑鱼立克次氏体病疫苗的研发一直是鲑科鱼类研究的主题，主要聚焦于疫苗组分、攻毒设计和评估参数的研究（Bravo 和 Midtlyng，2007）。但是，目前还没有获得在养殖条件下实际可用疫苗的结果。除了未接种疫苗的对照组在内的经济风险外，鲑养殖场间的差异，如生物安全、天气、水环境、养殖种类、养殖密度，均使得疫苗的应用评估较为困难。

第五节　疫苗现状

第一个鲑鱼立克次氏体疫苗是福尔马林灭活的菌苗，用大鳞大麻哈鱼胚胎细胞系 CHSE-214 细胞培养获得的细胞培养物制备而成（Fryer，1992；Gaggero，1995）。通常，疫苗为乳化制剂，腹腔注射到银化前的苗种，目的是使鱼苗在入海养殖前，通过接种疫苗获得免疫保护。疫苗获得的保护效果与疫苗配方、制备方法、抗原剂量、攻毒剂量和接种后的攻毒时间等均相关，因此对疫苗保护效果的具体评估较为困难。不过，抗原剂量和接种后的攻毒时间是影响疫苗免疫结果的主要因素（Kuzyk 等，2001a；Birkbeck 等，2004b；Tobar 等，2011；Wilda 等，2012）。

首个鲑鱼立克次氏体的疫苗研究表明，甲醛灭活菌苗作为免疫原预防鲑鱼立克次氏体败血症是可行的，热灭活的菌苗可提高免疫保护水平。最初疫苗是在鱼类细胞系中制备抗原，但是产量非常有限，此外，考虑到佐剂的添加量至少为 20%（*v/v*），因此每尾鱼需要注射的菌苗浓度范围应在 10^3～10^4 的半数组织培养感染量（$TCID_{50}$）。使用以上的免疫方案，免疫接种体重为 18 g 的银大麻哈鱼，在免疫后 210℃ · d 进行攻毒，相对保护率较低，为－30% 到 15%

(Kuzyk 等，2001a)。改进鲑立克次氏体培养方法后，能够大量生产滴度更高的抗原（Birkbeck 等，2004a）用于疫苗效果实验，剂量设置比以前增加 4～5 个量级。体重为 100g 的大西洋鲑在免疫后2 000℃ · d 攻毒，甲醛灭活疫苗和热灭活疫苗产生的相对保护率分别提高至 70.7% 和 49.6%（Birkbeck 等，2004b）。这项研究揭示了该菌具有热不稳定性，存在免疫抑制蛋白，同时也表明高抗原浓度对免疫保护至关重要。

商业疫苗在可控条件下显示出了良好的效果。研究发现，体重为 30g 的大西洋鲑腹腔注射一种商品疫苗（Centrovet Ltda.，Santiago，Chile）后的存活率与接种疫苗后攻毒的时间有关，疫苗接种后 300℃ · d 或 600℃ · d 时进行攻毒，鱼的存活率为 45%～90%（Tobar 等，2011）。另一种商品疫苗（Recalcine SA，Santiago，Chile）免疫体重为 30g 的大西洋鲑，结果显示，鱼体接种后 370℃ · d 进行攻毒，对照组的死亡率达到 60% 时的相对保护率（RPS_{60}）为 86.4%（Wilda 等，2012）。

鲑鱼立克次氏体的重组蛋白抗原已用于制备疫苗并进行了评估，第一个报道有效的重组抗原是外膜蛋白 OspA，该蛋白是在大肠杆菌外膜定向表达文库中被发现的。感染鲑鱼立克次氏体病康复鱼的血清用于抗原鉴定（Kuzyk 等，2001a，2001b），已鉴定出一个对大肠杆菌表达有害的小分子量蛋白，利用已知的 T 细胞表位和融合分子进行抗原优化以消除毒性，该重组蛋白抗原通过腹腔注射免疫，攻毒后具有很高的保护作用（相对保护率从 50% 到 80% 不等），研究结果为发现和优化新的重组抗原奠定了基础。以康复鱼血清筛选抗原的方法，随后在嗜冷黄杆菌重组疫苗中也得到验证（Crump 等，2007）。这是因为鲑鱼立克次氏体与血清反应的蛋白是研制重组疫苗良好的候选抗原，也说明体液免疫反应产生高效免疫保护必不可少的部分（Kuzyk 等，2001a；Wilda 等，2012）。

已报道的其他重组疫苗的研究大多使用多种抗原混合配方，如热休克蛋白被认为是几种模式动物疫苗的有效佐剂（Wilhelm 等，2005）。由于抗原组成和鱼血清识别的蛋白种类不同，鲑科鱼类腹腔注射攻毒后的相对保护率各有差异，在 10% 到 85% 之间（Wilhelm 等，2006）。在抗原的鉴定研究中，发现了对细菌鞭毛组装有重要作用的 *fliC* 基因，虽然该蛋白未与鱼血清反应，但该基因可有助于解释鲑鱼立克次氏体在水中的传播。然而，还需要进行更多的研究证实鲑鱼立克次氏体是否在一定环境条件下具有运动性。

目前介绍的鲑鱼立克次氏体菌苗和重组疫苗均是注射型疫苗。通常，这些疫苗成本效益较好，减少了疫苗接种的个体不确定性，是当前使养殖鱼类获得免疫保护持续时间最长的的疫苗（Sommerset，2005）。然而，一些重要的问题仍然需要解决，如疫苗产生的副作用，在鲑中尤其严重，再如疫苗的免疫保护期等（Midtlyng 等，1996）。

目前，还未见鲑鱼立克次氏体减毒活疫苗的研究。然而，有研究报道一种非致病性的节杆菌（*Arthrobacter davidanieli*）能够诱导免疫反应，不仅可预防鲑鱼立克次氏体病，还可以预防由鲑肾杆菌引起的细菌性肾病。该疫苗在智利和加拿大获得许可（诺华动物保健公司 Novartis Animal Health），对细菌性肾病产生良好的保护（RPS 达到 80%），在养殖条件下使鲑鱼立克次氏体病的死亡率显著降低（Salonius 等，2005）。虽然该疫苗的作用机制尚未明释，推测该减毒活疫苗可能是一种免疫刺激剂而不是特异性的免疫诱导剂（Wietz 等，2012）。

鲑鱼立克次氏体病的口服疫苗接种在 20 世纪 80 年代早期就已有研究（Vandenberg，

2004；Rombout 等，2011）。为了阐明鱼类黏膜免疫，已经开展了大量的研究，然而鱼类可用的口服疫苗并不多（Plant 和 LaPatra，2011）。智利研发出的口服疫苗获得许可，并可在早期诱导产生保护，但与常规的油剂类菌苗相比，该口服疫苗的保护期较短。研究表明，口服疫苗免疫或者先腹腔注射免疫再口服加强免疫均可诱导大西洋鲑产生抗鲑鱼立克次氏体的特异性 IgM 抗体，且受免鱼的存活率达 85%～90%（Tobar 等，2011）。此外，实验证明通过腹腔注射进行初次免疫，口服疫苗作为加强免疫，可以延长免疫保护时间，且养殖者不需要增加人工操作就可以在海水养殖阶段对鲑进行加强免疫接种。口服免疫途径是无应激反应的疫苗接种方式，但存在如何提高免疫持续的问题。

第六节　发展前景

鲑鱼立克次氏体病疫苗的效果是水产养殖中最关注的话题之一。在已发表的论文中，因采用的攻毒方法不同，如疫苗接种后的攻毒时间、鱼体大小、水质条件和效果衡量单位的不同，较难建立统一的疫苗效果的评价标准（Smith 等，1997；Kuzyk 等，2001a；Birkbeck 等，2004b；Salonius 等，2005；Wilhelm 等，2006；Tobar 等，2011；Wilda 等，2012）。为了有效地研发疫苗，有必要设计符合通用的标准方法，该标准应该严格基于鱼的易感时期，对于诸如鲑等洄游鱼类，包括银化阶段和淡水阶段均应给予重点考虑。

在受免组或对照组中，细菌攻毒实验是复杂的，攻毒结果取决于所用菌株的种类、剂量和感染途径（Smith 等，1997）。此外，疫苗评估中的动物福利问题和“3Rs”原则也强调了其他客观保护指标的必要性（Midtlyng 等，2011）。虽然很难在鲑鱼立克次氏体这样的胞内病原上建立这些评估条件，但已有研究建立了一些有效的评估方法，将抗鲑鱼立克次氏体的特异性抗体作为疫苗保护效果的预测指标，使用乳化型疫苗进行免疫，鱼体产生的抗体效价与 RPS 值密切相关，呈疫苗剂量依赖型反应模式（Wilda 等，2012）。考虑到鲑鱼立克次氏体是胞内病原，首次发现体液免疫可能与免疫保护相关，这具有非常重要的意义，该结果与其他鲑鱼立克次氏体疫苗接种试验结果一致（Kuzyk 等，2001a；Wilhelm 等，2006；Tobar 等，2011），该方法也可应用于传染性鲑贫血等鲑的其他胞内病原疫苗的评估（Lauscher 等，2011）。然而，不能忽略细胞免疫在鲑鱼立克次氏体保护效果中的作用。

新的商业疫苗必须在实际应用中，具有持久和/或高程度的保护效果，对环境安全以及对鱼类无副作用。此外，疫苗必须无应激反应，口服疫苗接种在这方面已证明具有较好的应用前景。

要解决疫苗存在的问题必须考虑不同的免疫替代方法，例如口服疫苗、活疫苗或 DNA 疫苗。在野外养殖的条件下，注射方式不易操作，可以通过再次免疫延长免疫持续时间，又可以诱导产生更持久的免疫力，虽然目前还对其机理知之甚少（Lepa 等，2010）。

最初的研究显示，DNA 疫苗在免疫水平和实践方面均具有应用潜力，但尚未对其进行深入的评估研究。应用传统的真核生物启动子随机表达鲑鱼立克次氏体基因组中的基因，研究结果表明，该 DNA 疫苗可减少鲑的死亡率，同时也诱导产生了特异性 IgM，且抗体水平与疫苗剂量，以及加强免疫密切相关（Miquel 等，2003）。这些结果表明，通过研究可用的抗原来研制疫苗配方是可行的。鲑鱼立克次氏体的两种热休克蛋白 Hsp60 和 Hsp70 克隆至传统的哺乳动物表达载体中研制的疫苗，诱导鱼体产生了相应的抗体，RPS_{60} 值为 40.9%～54.5%，该研究表明，经典的哺乳动物启动子可在大西洋鲑中表达，并且 DNA 疫苗是有效的，可以通过适当的研究进行优化。

随着基因组测序、不断发展的生物信息学工具和抗原预测技术的发展，应用 DNA 疫苗预防鲑鱼立克次氏体病是一种可行的策略。但实现 DNA 疫苗的商业生产，必须深入研究 DNA 疫苗的安全性，以确保 DNA 疫苗载体对鱼体安全、有效可控并可跟踪。

参考文献

Almendras，F. E. and Fuentealba，I. C. 1997. Salmonid rickettsial septisemia caused by *Piscirickettsia salmonis*：a review. *Dis Aquat Org* 29：137-144.

Arkush，K. D.，McBride，A. M.，Mendonca，H. L.，*et al.*，2005. Genetic characterization and experimental pathogenesis of *Piscirickettsia salmonis* isolated from white seabass *Atractoscion nobilis*. *Dis Aquat Org* 63：139-149.

Athanassopoulou，F.，Groman，D.，Prapas，T. H. and Sabatakou，O. 2004. Pathological and epidemiological observations on rickettsiosis in cultured sea bass（*Dicentrarchus labrax* L.）from Greece. *J Appl Ichth* 20：525-529.

Birkbeck，T. H.，Griffen，A. A.，Reid，H. I.，*et al.*，2004a. Growth of *Piscirickettsia salmonis* to high titers in insect tissue culture cells. *Infect Immun* 72：3693-3694.

Birkbeck，T. H.，Rennie，S.，Hunter，D.，*et al.*，2004b. Infectivity of a Scottish isolate of *Piscirickettsia salmonis* for Atlantic salmon *Salmo salar* and immune response of salmon to this agent. *Dis Aquat Org* 60：97-103.

Bravo，S. and Campos，M. 1989. Síndrome del salmon Coho. *Chile Pesquero* 54：47-48.

Bravo，S. and Midtlyng，P. J. 2007. The use of fish vaccines in the Chilean salmon industry 1999-2003. *Aquaculture* 270：36-42.

Brocklebank，J. R.，Speare，D. J.，Armstrong，R. D. and Evelyn，T. 1992. British Columbia-septicemia suspected to be caused by a rickettsia-like agent in farmed Atlantic salmon. *Can Vet J* 33：407-408.

Chen，S. C.，Tung，M. C.，Chen，S. P.，*et al.*，1994. Systemic granulomas caused by a rickettsia-like organism in Nile tilapia，*Oreochronuis niloticus*（L.），from southern Taiwan. *J Fish Dis* 17：591-599.

Crump，E. M.，Burian，J.，Allen，P. D.，*et al.*，2007. Identification of a ribosomal L10-like protein from *Flavobacterium psychrophilum* as a recombinant vaccine candidate for rainbow trout fry syndrome. *J Mol Microbiol Biotechnol* 13：55-64.

Evelyn，T. P. T. 1992. Salmonid rickettsial septicemia，in *Diseases of Seawater Netpen-Reared Salmonid Fishes in the Pacific Northwest*（ed. M. L. Kent）. *Can Spec Pub Fish Aquat Sci* 116：18-19.

Fryer，J. L. and Hedrick，R. P. 2003. *Piscirickettsia salmonis*：a Gram-negative intracellular bacterial pathogen of fish. *J Fish Dis* 26：251-262.

Fryer，J. L. and Lannan，C. N. 2005. Family II. *Piscirickettsiaceae* fam. nov，in *Bergey's Manual of Systematic Bacteriology*，2nd edn，vol. 2（The *Proteobacteria*）part B（The *Gammaproteobacteria*）（eds D. J. Brenner，N. R. Krieg，J. T. Staley and G. M. Garrity）. New York，Springer，180.

Fryer，J. L.，Lannan，C. N.，Garcés，H. L.，*et al.*，1990. Isolation of Rickettsiales-like organism from diseased coho salmon (*Oncorhynchus kisutch*) in Chile. *Fish Pathol* 25: 107-114.

Fryer, J. L., Lannan, C. N., Giovannoni, S. J. and Wood, N. D. 1992. *Piscirickettsia salmonis* gen. nov. sp. nov. the causative agent of an epizootic disease in salmonid fishes. *Int J Sys Evol Micro* 42: 120-126.

Gaggero, A., Castro, H. and Sandino, A. M. 1995. First isolation of *Piscirickettsia salmonis* from coho salmon, *Oncorhynchus kisutch* (Walbaum), and rainbow trout, *Oncorhynchus mykiss* (Walbaum), during the freshwater stage of their life cycle. *J Fish Dis* 18: 277-280.

Gómez, F., Henríquez, V. and Marshall, S. H. 2009. Additional evidence of the facultative intracellular nature of the fish bacterial pathogen *Piscirickettsia salmonis*. *Arch Med Vet* 41: 261-267.

Gómez, F. A., Cárdenas, C., Henríquez, V. and Marshall, S. H. 2011. Characterization of a functional toxin/antitoxin module in the genome of the fish pathogen *Piscirickettsia salmonis*. *FEMS Microbiol Lett* 317: 83-92.

Gómez, F. A., Tobar, J. A., Henríquez, V., *et al.*, 2013. Evidence of the presence of a functional Dot/Icm Type IV-B secretion system in the fish bacterial pathogen *Piscirickettsia salmonis*. *PloS One* 8: 1-11.

Kuzyk, M. A., Burian, J., Machander, D., *et al.*, 2001a. An efficacious recombinant subunit vaccine against the salmonid rickettsial pathogen *Piscirickettsia salmonis*. *Vaccine* 19: 2337-2344.

Kuzyk, M. A., Burian, J., Thornton, J. C. and Kay, W. W. 2001b. OspA, a lipoprotein antigen of the obligate intracellular bacterial pathogen *Piscirickettsia salmonis*. *J Mol Microbiol Biotechnol* 3: 83-93.

Larenas, J. J., Bartholomew, J., Troncoso, O., *et al.*, 2003. Experimental vertical transmission of *Piscirickettsia salmonis* and *in vitro* study of attachment and mode of entrance into the fish ovum. *Dis Aquat Org* 56: 25-30.

Lauscher, A., Krossøy, B., Frost, P., *et al.*, 2011. Immune responses in Atlantic salmon (*Salmo salar*) following protective vaccination against infectious salmon anemia (ISA) and subsequent ISA virus infection. *Vaccine* 29: 6392-6401.

Lepa, A., Siwicki, A. K. and Terech-Majewska, E. 2010. Application of DNA vaccines in fish. *Pol J Vet Sci* 13: 213-215.

Marshall, S. H., Heath, S., Henríquez, V. and Orrego, C. 1998. Minimally invasive detection of *Piscirickettsia salmonis* in cultivated salmonids via the PCR. *Appl Environ Microbiol* 64: 3066-3069.

Marshall, S. H., Conejeros, P., Zahr, M., *et al.*, 2007. Immunological characterization of a bacterial protein isolated from salmonid fish naturally infected with *Piscirickettsia salmonis*. *Vaccine* 25: 2095-2102.

Marshall, S. H., Henríquez, V., Gómez, F. A. and Cárdenas, C. 2011. ISPsa2, the first mobile genetic element to be described and characterized in the bacterial facultative intracellular pathogen *Piscirickettsia salmonis*. *FEMS Microbiol Lett* 314: 18-24.

Marshall, S. H., Gómez, F. A., Ramírez, R., *et al.*, 2012. Biofilm generation by *Piscirickettsia salmonis* under growth stress conditions: a putative *in vivo* survival/persistence strategy in marine environments. *Res Microbiol* 163: 557-566.

Mauel, M. J., Giovannoni, S. J. and Fryer, J. L. 1999. Phylogenetic analysis of *Piscirikettsia salmonis* by 16S, internal transcribed spacer (ITS) and 23S ribosomal DNA sequencing. *Dis Aquat Org* 35: 115-123.

Mauel, M. J., Ware, C. and Smith, P. A. 2008. Culture of *Piscirickettsia salmonis* on enriched blood agar. *J Vet Diagn Invest* 20: 213-214.

McCarthy, U. M., Bron, J. E., Brown, L., *et al.*, 2005. Survival and replication of *Piscirickettsia salmonis* in rainbow trout head kidney macrophages. *Fish Shellfish Immunol* 25: 477-484.

Midtlyng, P. J., Reitan, L. J. and Speilberg, L. 1996. Experimental studies on the efficacy and side-effects of intraperitoneal vaccination of Atlantic salmon (*Salmo salar* L.) against furunculosis. *Fish Shellfish Immunol* 6:

335-350.

Midtlyng, P. J., Hendriksen, C., Balks, E., *et al.*, 2011. Three Rs approaches in the production and quality control of fish vaccines. *Biologicals* 39: 117-128.

Mikalsen, J., Skjærvik, O., Wiik-Nielsen, J., *et al.*, 2008. Agar culture of *Piscirickettsia salmonis*, a serious pathogen of farmed salmonid and marine fish. *FEMS Microbiol Lett* 278: 43-47.

Miquel, A., Müller, I., Ferrer, P., *et al.*, 2003. Immunoresponse of coho salmon immunized with a gene expression library from *Piscirickettsia salmonis*. *Biol Res* 36: 313-323.

Noga, E. J. 2010. *Fish Disease Diagnosis and Treatment* (2nd edn). Iowa, Wiley-Blackwell. Olivares, J. and Marshall, S. H. 2010. Determination of minimal concentration of *Piscirickettsia salmonis* in water columns to establish a fallowing period in salmon farms. *J Fish Dis* 33: 261-266.

Olsen, A. B., Evensen, Ø., Speilberg, L., *et al.*, 1993. Ny laksesykdom forårsaket av rickettsie. *Norsk Fiskeoppdrett* 12: 40-41.

Palmer, R., Ruttledge, M., Callanan, K. and Drinan, E. 1997. A piscirickettsiosis-like disease in farmed Atlantic salmon in Ireland-isolation of the agent. *Bull Eur Assoc Fish Pathol* 17: 68-72.

Peña, A. A., Bols, N. C. and Marshall, S. H. 2010. An evaluation of potential reference genes for stability of expression in two salmonid cell lines after infection with either *Piscirickettsia salmonis* or IPNV. *BMC Res Notes* 3: 101-110.

Plant, K. P. and LaPatra, S. E. 2011. Advances in fish vaccine delivery. *Dev Comp Immunol* 35: 1256-1262.

Rodger, H. D. and Drinan, E. M. 1993. Observation of a rickettsia-like organism in Atlantic salmon, *Salmo salar* L., in Ireland. *J Fish Dis* 16: 361-369.

Rojas, V., Olivares, J., Del Río, R. and Marshall, S. H. 2008. Characterization of a novel and genetically different small infective variant of *Piscirickettsia salmonis*. *Microb Pathog* 44: 370-378.

Rojas, V., Galanti, N., Bols, N. C. and Marshall, S. H. 2009. Productive infection of *Piscirickettsia salmonis* in macrophages and monocyte-like cells from rainbow trout, a possible survival strategy. *J Cell Biochem* 108: 631-637.

Rojas, V., Galanti, N., Bols, N. C., *et al.*, 2010. *Piscirickettsia salmonis* induces apoptosis in macrophages and monocyte-like cells from rainbow trout. *J Cell Biochem* 110: 468-476.

Rombout, J. H., Abelli, L., Picchietti, S., *et al.*, 2011. Teleost intestinal immunology. *Fish Shellfish Immunol* 31: 616-626.

Salonius, K., Siderakis, C., MacKinnon, A. M. and Griffiths, S. G. 2005. Use of *Arthrobacter davidanieli* as a live vaccine against *Renibacterium salmoninarum* and *Piscirickettsia salmonis* in salmonids. *Dev Biol (Basel)* 121: 189-197.

Smith, P. A., Contreras, J. R., Larenas, J. J., *et al.*, 1997. Immunization with bacterial antigens: piscirickettsiosis. *Dev Biol (Basel)* 90: 161-166.

Sommerset, I., Krossøy, B., Biering, E. and Frost, P. 2005. Vaccines for fish in aquaculture. *Exp Rev Vaccines* 4: 89-101.

Tobar, J. A., Jerez, S., Caruffo, M., *et al.*, 2011. Oral vaccination of Atlantic salmon (*Salmo salar*) against salmonid rickettsial septicaemia. *Vaccine* 29: 2336-2340.

Vandenberg, G. W. 2004. Oral vaccines for finfish: academic theory or commercial reality? *Anim Health Res Rev* 5: 301-304.

Wietz, M., Månsson, M., Bowman, J. S., *et al.*, 2012. Wide distribution of closely related, antibiotic-

producing Arthrobacter strains throughout the Arctic Ocean. *Appl Environ Microbiol* 78: 2039-2042.

Wilda, M., Lavoria, M. Á., Giráldez, A., *et al.*, 2012. Development and preliminary validation of an antibody filtration-assisted single-dilution chemiluminometric immunoassay for potency testing of *Piscirickettsia salmonis* vaccines. *Biologicals* 40: 415-420.

Wilhelm, V., Soza, C., Martínez, R., *et al.*, 2005. Production and immune response of recombinant Hsp60 and Hsp70 from the salmon pathogen *Piscirickettsia salmonis*. *Biol Res* 38: 69-82.

Wilhelm, V., Miquel, A., Burzio, L. O., *et al.*, 2006. A vaccine against the salmonid pathogen *Piscirickettsia salmonis* based on recombinant proteins. *Vaccine* 24: 5083-5091.

Yáñez, A. J., Valenzuela, K., Silva, H., *et al.*, 2012. Broth medium for the successful culture of the fish pathogen *Piscirickettsia salmonis*. *Dis Aquat Org* 97: 197-205.

第二十二章　细菌性肾病的疫苗接种

Diane G. Elliott, Gregory D. Wiens,
K. Larry Hammell and Linda D. Rhodes

本章概要　SUMMARY

鲑科鱼类的细菌性肾病（BKD）是由鲑肾杆菌（*Renibacterium salmoninarum*）引起的，虽然该病的研究持续了几十年，但其有效疫苗的研发依旧面临挑战。唯一获得商业许可的预防 BKD 的疫苗是注射型疫苗，其含有节细菌（*Arthrobacter davidanieli*）活菌体，是一种环境非致病性细菌，与鲑肾杆菌的系统发育较近。该活疫苗之所以有效果是因为节细菌表面具有与鲑肾杆菌的胞外多糖类似的碳水化合物。该疫苗腹腔注射大西洋鲑（*Salmo salar*）后具有显著的免疫保护作用，但是对大鳞大麻哈鱼（*Oncorhynchus tshawytscha*）效果甚微。因此，需要进一步研发针对多种鲑科鱼类的更有效的 BKD 疫苗。

第一节　引　言

细菌性肾病（BKD）由鲑肾杆菌（*Renibacterium salmoninarum*）引起，自 20 世纪 30 年代以来，一直是鲑科鱼类中严重的疾病（Belding 和 Merrill，1935；Smith，1964；Evelyn，1988），在 20 世纪 70 年代早期，首次报道鲑科鱼类可对该菌产生免疫反应（Evelyn，1971）。虽然有多年的研究，但只有一种该病的疫苗获得应用许可，其效果并不稳定（Rhodes 等，2004b；Alcorn 等，2005；Salonius 等，2005；Burnley 等，2010）。本章介绍 BKD 疫苗应用的现状，探讨该病疫苗的研发方向，包括鲑肾杆菌的特性和生物学特点，以及为研发更有效的疫苗开展的鲑科鱼类免疫系统的研究。

第二节　发生流行

在世界多地，包括北美、智利、欧洲和日本等，均报道在野生或养殖的鲑科鱼类中发现了鲑肾杆菌，并引发 BKD。已有多篇文献对该疾病和病原进行了综述介绍（Fryer 和 Sanders，1981；Austin 和 Austin，1987；Elliott 等，1989；Evelyn，1993；Evenden 等，1993；Fryer 和 Lannan，1993；Pascho 等，2002）。

所有的鲑科鱼类几乎都容易感染 BKD，但是不同物种的敏感性有差异。其中太平洋鲑，如红大麻哈鱼（*Oncorhynchus nerka*）、大麻哈鱼（*O. keta*）和大鳞大麻哈（*O. tshawytscha*）比大西洋鲑（*Salmo salar*）更易感；而淡水和洄游性鳟，如虹鳟（*Oncorhynchus mykiss*）、硬头鳟和褐鳟（*S. trutta*），以及红点鲑类，如强壮红点鲑（*Salvelinus confluentus*）和湖红点鲑（*S. namaycush*）不易感（Kawamura 等，1977；Sanders 等，1978；Sakai 等，1991；Starliper 等，1997；Meyers 等，2003；Jones 等，2007）。

仅在鲑科鱼类中观察到 BKD 临床症状（Fryer 和 Lannan，1993；Pascho 等，2002）。然而，通过培养、免疫学和分子生物学实验研究，鲑肾杆菌已在非鲑科鱼类和双壳软体动物中检测到（Sakai 和 Kobayashi，1992；Kent 等，1998；Starliper 和 Morrison，2000；Eissa 等，2006；Polinski 等，2010；Rhodes 等，2011）。非鲑科鱼类作为携带者可能增加了该菌水平传播的机会，这一点特别重要，因为这些鱼可能作为当地感染源并通过一些像鳗一样能在水陆迁移的鱼类在河流间传播病原（Chambers 等，2008）。

在养殖和野生的鱼类中均发现临床症状不明显的鲑肾杆菌感染（Lovely 等，1994；Elliott 等，1997；Suzuki 和 Sakai，2007），这严重影响到养殖鱼类的管理（Murray 等，2011）。通过多种灵敏的检测方法已经确认该病在鲑科鱼类中广泛流行（Elliott 等，1997；Bruno 等，2007；Sandell 和 Jacobson，2011）。BKD 在物种之间易感性的差异，可以运用到疫苗接种策略中，但是在鲑科鱼类养殖密集区域应该进行流行病学评估，以确保这些疫苗接种策略符合养殖生产管理规程（Murray 等，2012）。

第三节 危　害

鱼类集约化养殖方式常暴发 BKD，并且这种疾病传播随着鲑科鱼类养殖规模的扩大而扩散（Evenden 等，1993）。野生鱼类中也有 BKD 发生的报道（Smith，1964；Pippy，1969；MacLean 和 Yoder，1970；Mitchum 等，1979；Banner 等，1986；Holey 等，1998；Faisal 等，2010），在包括没有人为干预的自然产卵生长种群中也有发生（Evelyn 等，1973；Souter 等，1987）。

据报道，BKD 导致太平洋鲑和大西洋鲑产量的损失分别高达 80% 和 40%（Evenden 等，1993），但由于该病属慢性病，所以会影响死亡率的准确估算，这种情况在野生鱼类种群中更明显，因为病鱼更易被捕食，通常在野生鲑中不易发现 BKD 临床症状的病鱼（Mesa 等，1998）。已有研究表明，入海的银化鲑大大加剧了疾病扩散和暴发，造成 BKD 流行（Mesa 等，1999），并可能导致海水养殖鲑的感染死亡，并与从淡水设施系统转入海水养殖的 2 龄鲑死亡密切相关（Murray 等，2012）。此外，鲑肾杆菌感染造成鱼体的免疫抑制，加之二次感染，导致鲑的死亡。虽然 BKD 对水产养殖的影响很难量化，但此病会导致鲑生长缓慢、产量减少并提高疾病防控的成本（Burnley 等，2010；Munson 等，2010）。

细菌性肾病是最难控制的鱼类细菌性疾病之一（Elliott 等，1989）。目前该病的主要防控措施是化学药物治疗、加强管理和疫苗接种等。已研发出一些治疗 BKD 的药物，这些药物具有一定效果（Austin，1985；Brown 等，1990；Moffitt，1991，1992；Lee 和 Evelyn，1994；Moffitt 和 Kiryu，1999；Fairgrieve 等，2005），但无法完全消灭病原（Wolf 和 Dunbar，1959；Austin，1985；Moffitt，1992；Moffitt 和 Kiryu，1999），并且鲑肾杆菌可能产生抗药性（Bell 等，1988；Rhodes 等，2008）。改善养殖设施的卫生条件和加强养殖管理来增强生物安全性，可以降低 BKD 的感染概率（Maule 等，1996；Pascho 等，2002；Munson 等，2010；Murray 等，2012），并且实施诸如隔离或去除感染的亲鱼等措施，能够有效地降低该病在鲑子代的感染和流行（Pascho 等，1991；Guðmundsdóttir 等，2000；Meyers 等，2003；Munson 等，2010）。虽然 BKD 的疫苗接种还没有被广泛应用（Sommerset 等，2005；Bravo 和 Midtlyng，2007），但是疫苗接种将是预防该病有效的措施（Pascho 等，2002；Rhodes 等，2004b）。

第四节 病原学

鲑肾杆菌是典型的革兰氏阳性菌，通常成对出现，好氧、无芽孢、不运动，最适生长温度是 15～18℃（Sanders 和 Fryer，1980）。标准菌株在美国菌种保藏中心编号为 ATCC33209，其基因组由一个环状染色体组成，含有3 155 250个 DNA 碱基对，3 507个开放阅读框（ORFs；Wiens 等，2008）。该菌株缺乏整合噬菌体的区域或相关功能的质粒，但有 80 个插入序列分散在整个基因组中；通过移码突变、点突变、插入序列和假定缺失可以使大约 21%的预测 ORFs 失去活性。鲑肾杆菌基因组与节细菌（*Arthrobacter* sp.）FB24 菌株和节细菌（*Arthrobacter aurescens*）TC1 菌株的基因组有同线型扩展序列，但鲑肾杆菌的基因组要小得多，说明自共同祖先分支以后发生差异分化，导致鲑肾杆菌的基因组已显著减少。鲑肾杆菌基因组的代谢通路只存在两个核糖体基因位点和众多的假定基因，使其世代更替非常缓慢，接近 24h。

鲑肾杆菌分离菌株的生化特性和细胞壁组成很保守（Bruno 和 Munro，1986a；Fiedler 和 Draxl，1986），其表现在不能利用糖产酸（Sanders 和 Fryer，1980）以及不能合成菌体生长所需要的半胱氨酸（Ordal 和 Earp，1956；Daly 和 Stevenson，1985）。鲑肾杆菌 ATCC33209 的基因组缺乏合成丝氨酸、甘氨酸、半胱氨酸、天冬氨酸和蛋氨酸等重要功能基因（Wiens 等，2008）。基因组分析结果证实，该菌拥有功能性的核心中枢代谢途径，包括糖酵解、戊糖磷酸循环、三羧酸循环和丙酮酸循环。当对该菌导入相关基因时，其可能具备利用某些糖类和多元醇的能力，如葡萄糖、果糖、阿拉伯糖、葡萄糖酸盐、甘油和柠檬酸盐。鲑肾杆菌还能利用丙酮酸盐、乳酸盐、琥珀酸盐、苹果酸盐、甘油-3-磷酸、脯氨酸、丁酰 CoA 和脂肪酸等多种碳源物质获得能量。鲑肾杆菌还缺乏 *fabA* 和 *fabZ* 及其同系物，所以不能合成不饱和脂肪酸和饱和脂肪酸，表明该菌必须从鲑中获取这些化合物。

鲑肾杆菌产生过氧化氢酶（Ordal 和 Earp，1956），并且其基因组能编码超氧化物歧化酶、过氧化物酶和硫氧还蛋白过氧化物酶等抵抗氧自由基的酶类（Wiens 等，2008）。已有报道，该菌具有鲑肾杆菌蛋白水解活性（Ordal 和 Earp，1956；Smith，1964；Rockey 等，1991）、脱氧核糖核酸酶（DNAse）活性以及溶解鲑科鱼类红细胞的β-溶血活性（Bruno 和 Munro，1986a；Evenden 等，1993）。生物化学方法已鉴定出鲑肾杆菌中的两种特异性溶血素，其中一个已被克隆并命名为 *hly*（Evenden 等，1993；Grayson 等，1995，2001）。此外，从基因组中分析鉴定出 3 种候选溶血素，编码的 ORFs 为 RSal33209 _ 0811、RSal33209 _ 3195 和 RSal33209 _ 3047（Wiens 等，2008），但是目前这 3 种候选溶血素与鲑肾杆菌的毒力进化及其在鱼体内环境适应的关系尚不清楚。

鲑肾杆菌基因组包含一个由开放阅读框编码的 SrtD 同系物，这个同系物是革兰氏阳性菌产生的半胱氨酸转肽酶家族的成员之一，能够共价结合蛋白锚定到细胞膜肽聚糖表面。通过对鲑肾杆菌基因组的生物信息学分析，鉴定出 8 个含有假定转肽酶裂解序列的 ORFs。在体外实验中，用转肽酶抑制处理鲑肾杆菌，可降低细菌对纤维蛋白和鱼细胞的黏附，表明这些蛋白具有黏附作用（Sudheesh 等，2007）。

应用抗血清测定发现鲑肾杆菌不同分离株的抗原通常相似（Bullock 等，1974；Getchell 等，1985；Ficdlcr 和 Draxl，1986）。然而，研究挪威分离出的 8 株鲑肾杆菌，发现 5 株 *msa1* 和 *msa2* 基因发生了单个核苷酸突变，导致 p57（MSA）蛋白出现少量的抗原变异（Wiens 等，2002）。这些抗原变异菌株在基因序列上也存在变异特征，包括 *ETRA* 基因位点的一个串联重复、序列-4 16-23S rRNA 插入的 DNA 序列、在 *msa*1 5′区域中插入的更大的 *Xho* Ⅰ 片段以及 *msa3* 基因缺失。这些结果表明，鲑肾杆菌菌株间存在少量抗原和基因组变异，并且这种变异仅局限于一定的地理区域，这对于研发通用的疫苗是有利的。

第五节 致病机理

鲑肾杆菌主要感染皮肤、眼、眼眶后的组织以及大脑等（Hendricks 和 Leek，1975；Hoffmann 等，1984；Speare，1997；Ferguson，2006），典型的 BKD 是一个慢性的全身感染，鱼在 6～12 月龄前很少表现出临床症状（Evelyn，1993 年）。该疾病在组织学上表现为慢性肉芽肿炎症（Wolke，1975；Bruno，1986），当鱼体肾脏造血组织受到严重感染时，肉芽肿可在所有感染的组织中观察到。太平洋鲑感染后出现的肉芽肿是弥散性的且边缘不明显，而大西洋鲑中肉芽肿则被大量的上皮样巨噬细胞包裹（Evelyn，1993）。研究表明，鲑肾杆菌可在巨噬细胞中存活并可能增殖（Young 和 Chapman，1978；Bruno，1986；Bandín 等，1993；Flaño 等，1996；Dale 等，1997）。

鲑肾杆菌的水平和垂直传播方式致使其长期存在（Evelyn，1993），同时也使得控制管理策略更加复杂。水平传播的确切机制尚不清楚，但入侵途径主要有以下几方面：摄食被感染的鱼尸体（Wood 和 Wallis，1955）或有鲑肾杆菌的粪便（Balfry 等，1996），经胃肠道感染；通过眼或皮肤受伤部位感染（Hendricks 和 Leek，1975；Hoffmann 等，1984；Hayakawa 等，1989；Elliott 和 Pascho，2001）。该病在淡水（Mitchum 和 Sherman，1981；Bell 等，1984）和海水（Murray 等，1992；Evelyn，1993）中均可发生水平传播。种群方面，一个养殖公司或养殖场之间的来回转移、接触野生鱼的水域，或者接触过患病的虹鳟养殖场等，这种在时空上形成的短暂聚集也会发生水平传播。（Murray 等，2012）。鲑肾杆菌的垂直传播是由亲鱼的卵传给子代，至少是鱼卵携带病原（Evelyn 等，1984，1986）。通常在育苗场内，幼鱼间的水平传播会造成养殖和增殖群体的感染流行，并在其后的生长阶段进一步蔓延（Fenichel 等，2009）。

体内和体外试验已充分证明鲑肾杆菌的免疫抑制作用。卵子接触病原菌能增加鱼的易感性和死亡率，这表明该菌可诱导免疫耐受（Brown 等，1996）。鲑肾杆菌分泌大量的分子量为 57 ku 的可溶性胞外蛋白（MSA 或 p57 蛋白），该蛋白被认为是导致免疫抑制的主要介质（Turaga 等，1987；Fredriksen 等，1997），可以减少抗体的产生和细胞因子的干扰等。MSA 可以直接与免疫细胞相互作用（Campos-Perez 等，1997；Wiens 等，2002），清除 MSA 能促进机体产生抗体并降低易感性（Wood 和 Kaattari，1996；Piganelli 等，1999 b）。鲑肾杆菌毒力与 MSA 蛋白丰度（Senson 和 Stevenson，1999；O'Farrell 等，2000）或功能性 MSA 的基因拷贝数相关（Rhodes 等，2004a；Coady 等，2006），鉴于 MSA 具有毒力因子的主要特征，因此可以通过去除或中和鲑肾杆菌的 MSA 设计鲑肾杆菌疫苗。

鲑肾杆菌菌苗诱导的体液（抗体）反应与 BKD 的预防之间没有明显的相关性（Paterson 等，1981；Sakai 等，1989；Piganelli 等，1999b；Alcorn 等，2005）。抗原抗体复合物的形成可能产生有害的 BKD 病理效应（Sami 等，1992）。研究表明，针对鲑肾杆菌，细胞介导的免疫反应与体液免疫相比，对预防该病感染的作用更大（Hardie 等，1996；Kaattari 和 Piganelli，1997；Ellis，1999）。例如，体外研究发现，致敏的头肾白细胞（可能是辅助样 T 细胞）产生巨噬细胞活化相关的细胞因子，能诱导活化巨噬细胞并抑制鲑肾杆菌的生长，同时使巨噬细胞产生更强的呼吸暴发（Hardie 等，1996）。

鱼体对病原感染的反应取决于接触史、宿主遗传特征和细菌表型。虽然垂直传播可能诱导免疫耐受，但鲑肾杆菌同样可以诱导鱼体的记忆反应（Jansson 等，2003），这是疫苗接种很重要的特征。针对大鳞大麻哈鱼对鲑肾杆菌抗性的遗传变异研究表明，BKD 抗性增加具有选择性（Beacham 和 Evelyn，1992；Johnson 等，2003；Hard 等，2006），这种选择性可能已在五大湖的鲑中出现，因为该地区在过去的 40 年间一直流行此病（Purcell 等，2008）。抗性机制可能包括转铁蛋白紧密结合血清铁离子（Suzumoto 等，1977）和更强的先天性免疫反应（Ellis，1999）。由宿主反应引起的病理变化可能与细菌感染造成的直接损伤一样严重，Ⅲ型过敏反应中（Lumsden 等，2008）的抗原抗体复合物介导的肾小球性肾炎是 BKD 发病的标志（Sami 等，1992；O'Farrell 等，2000；Metzger 等，2010）。在患病存活下来的鱼中，表型有差异的鱼体，其免疫基因的表达模式没有显著差异（Metzger 等，2010）。与此结果相反的是，低 MSA 表达的鲑肾杆菌可诱导鲑干扰素诱导类基因的早期上调表达（Rhodes 等，2009），这表明有些免疫因子可作为疫苗研发的指标。

第六节　疫　苗

迄今为止，鲑肾杆菌疫苗研发工作大部分集中在全菌或裂解菌疫苗上，采用加热或甲醛灭活。早期研发的 BKD 疫苗通过急性攻毒结果显示，产生的免疫保护以及免疫反应效果不稳定（Paterson 等，1981；McCarthy 等，1984）。虽然鲑肾杆菌表面蛋白对诱导免疫起重要作用，但在灭活的菌体上保留的 MSA 蛋白，无论通过口服或腹腔注射均不能诱导太平洋鲑产生有效的免疫保护（Sakai 等，1993；Wood 和 Kaattari，1996；Piganelli 等，1999b；Alcorn 等，2005）。通过选择 MSA 低水平表达的菌株（Rhodes 等，2004b），利用加热的方法灭活菌体，可制备 MSA 含量低的菌苗制剂（Piganelli 等，1999a，1999b；美国专利 No. 5871751）。研究表明，MSA 含量低的菌苗对患病鱼具有较差的治疗效果（Rhodes 等，2004b）。

除了鲑肾杆菌灭活疫苗，还研究了低 MSA 含量和 MSA 正常含量的减毒活疫苗（Griffiths 等，1998；Daly 等，2001；Rhodes 等，2004b）。腹腔注射接种 MSA 正常含量的菌体比腹腔注射接种低 MSA 含量的菌株，虽然对大西洋鲑不能完全保护，但效果较好（Daly 等，2001）。

在尝试分离无毒力的鲑肾杆菌菌株作为候选疫苗过程中，发现节细菌（*Arthrobacter davidanieli*）作为疫苗可为大西洋鲑提供较为显著的免疫保护（Griffiths 等，1998；Salonius 等，2005；Burnley 等，2010）。实验室检测结果表明，这种疫苗比鲑肾杆菌减毒疫苗对大西洋鲑 BKD 的保护效果更强（Griffiths 等，1998）。然而与大西洋鲑的结果相反，节细菌活疫苗对太平洋鲑的保护效果有限（Rhodes 等，2004b），甚至没有效果（Alcorn 等，2005）。这种节细菌活疫苗 Renogen©[1] 已经在几个国家上市，而且是唯一获得商业许可的 BKD 疫苗。节细菌和鲑肾杆菌较近的系统发育关系可能是 Renogen©[1] 疫苗产生预防 BKD 的免疫原性的基础（Wiens 等，2008）。活疫苗 Renogen©[1] 能够产生免疫刺激作用，是由于节细菌表面的糖类与鲑肾杆菌胞外多糖相似（Griffiths 等，1998）。

细菌性肾病疫苗可配合使用的佐剂种类很多，从弗氏完全佐剂或弗氏不完全佐剂到 DNA 佐剂或者无佐剂（Pascho 等，1997；Griffiths 等，1998；Piganelli 等，1999b；Daly 等，2001；Rhodes 等，2004b；Alcorn 等，2005）。但目前尚未开展针对 BKD 疫苗佐剂的直接研究，所以佐剂能增强 BKD 疫苗效力的证据很少。

第七节　疫苗接种程序

预防 BKD 的疫苗大多数都是通过腹腔注射（或少有肌肉注射），口服疫苗的免疫保护作用也有报道。Piganelli 等（1999b）研究表明，银鲑口服经肠衣包被的缺失 MSA 鲑肾杆菌的热灭活微球疫苗，经鲑肾杆菌浸泡攻毒，其免疫保护效果比口服福尔马林灭活的全菌疫苗或腹腔注射 MSA 缺失疫苗的效果好。

获得商业化生产批文的 Renogen© 疫苗是节细菌菌体的冻干品，在使用前需用无菌生理盐水（0.9% NaCl）重悬和稀释。当进行 Renogen© 疫苗接种时，Renogen© 疫苗与其他抗原成分分别腹腔注射，因此，一般采用双筒免疫注射装置，即在单一位点注射，并且疫苗仅在注射时混合（图 22-1）（Burnley 等，2010）。若考虑与其他疫苗公司的疫苗抗原成分组合使用时，Renogen© 疫苗抗原成分分开注射的方式可能会限制其应用的方便性。鲑的疫苗接种都是在鱼苗入海水前进行，这时鱼的大小足以进行腹腔注射操作，保证鲑转移到海水前接受抗原刺激。节细菌疫苗的制造商建议，鲑接种疫苗后需 400℃·d 才能预防病原的感染。

疫苗在商业化养鱼场的野外测试主要是免疫方案的实施和攻毒，其中包括随机分组、设置对照组、效果评价以及大规模疫苗接种（要考虑网箱数量或者养殖地点的数量，Burnley 等，2010）。疫苗效果的评价常常受其他因素影响从而较难进行，比如海虱寄生以及养鱼设施的抗菌处理等对实验有影响等因素。

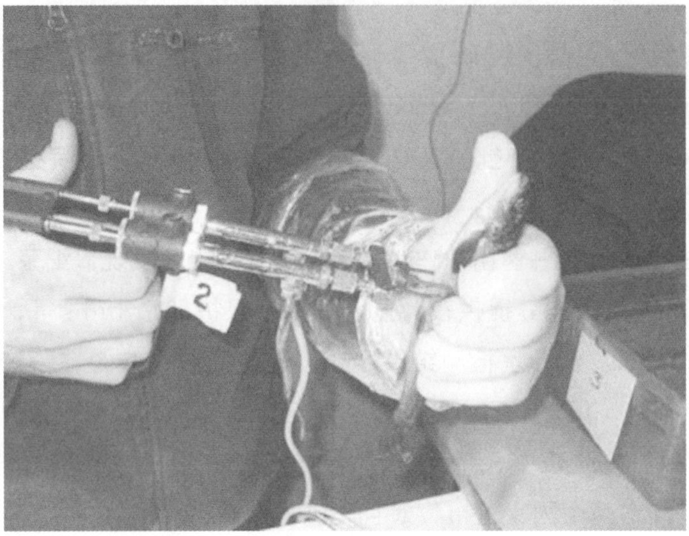

图 22-1 商品化细菌性肾病疫苗（包含节细菌的活菌）大规模接种的双筒接种装置，与甲醛灭活的其他病原疫苗联合使用
（来源：K. L. Hammell）

由于鲑肾杆菌感染可能发生在卵母细胞发育的早期（Bruno 和 Munro，1986b；Rhodes，未发表），因此尽可能早地对孵化后的鲑进行大规模疫苗接种，可以为其提供更好的免疫保护，防止该病的垂直和水平传播。尽管一些 DNA 疫苗已经显示了多年的保护效果（Kurath 等，2006），但其在鲑中观察到的免疫保护作用持续时间较短（Johnson 等，1982），这表明需要加强免疫，尤其是对于第一次接种的小鱼。尽管浸泡疫苗能刺激多器官的黏膜免疫应答，由于鲑肾杆菌的水平传播主要是通过胃肠黏膜（Balfry 等，1996）而不是鳃上皮细胞，因此口服疫苗具有可行性。

第八节 疫苗效果和副作用

尽管关于商品化的 Renogen© BKD 疫苗的疗效研究少有报道，但结果显示，疫苗效果因鱼种类不同有差异。实验室条件下，采用腹腔注射鲑肾杆菌的方式进行攻毒，与未接种疫苗相比，接种疫苗的大西洋鲑幼鱼的相对保护率在 72%～91%（Salonius 等，2005）。海水网箱养殖的大西洋鲑暴露于天然条件下的鲑肾杆菌后，Renogen© 疫苗与其他病原疫苗共同接种的实验组鲑存活率显著高于未接种的对照组（Salonius 等，2005；Burnley 等，2010）。Salonius 等（2005）报道，BKD 暴发的 2 个月时间里，与对照组鲑相比，Renogen© 接种组的相对保护率达到 80%；此外，Burnley 等（2010）研究表明，BKD 暴发的 240d 内，与对照组相比，接种 Renogen© 鲑的死亡风险显著降低（危险系数= 0.68，$P=0.018$）。相反，Rhodes 等（2004b）

的研究表明，接种 Renogen©疫苗的大鳞大麻哈幼鱼，经腹腔注射鲑肾杆菌急性攻毒，幼鱼的相对保护率却很低（0.4%～15.0%），与对照组相比，疫苗对鲑幼鱼产生了一定的免疫保护效果，幼鱼平均存活时间（4～13d，$P=0$～0.04）并没有显著增加。Alcorn 等（2005）报道，注射免疫 Renogen©疫苗的大鳞大麻哈鱼与肾杆菌感染的鱼一起混合养殖 285d，攻毒后结果显示，注射疫苗组与对照组相比存活率差异不显著（$P>0.05$）。

细菌性肾病疫苗的保护时效性研究很少。据报道，在野外研究中，接种 Renogen©的大西洋鲑经历 BKD 自然暴发期后，其存活率比对照组显著提高，且疫苗的保护期约为 23 个月（Salonius 等，2005）或 27 个月（Burnley 等，2010），但是很有可能是大西洋鲑长期接触鲑肾杆菌，才分别在接种后的 21 个月和 19 个月出现 BKD 感染个体死亡。Burnley 等（2010）研究发现，鲑在免疫接种 4 个月后转移至海水，可能在短时间内接触到鲑肾杆菌而引起感染，这是由于商品规格的鲑也养殖在刚入海的鲑养殖网箱内（养殖的不连续性）。在同一地点养殖一年以上的鱼，可能会通过大鱼将鲑肾杆菌持续传播给小鱼，从而导致临床疾病的发生。

在可能感染细菌性肾病的地区，虽然商业疫苗可以加强大西洋鲑养殖者对该病的防控，但事实上养殖户通常不特别依赖于疫苗。该病的预防措施主要是隔离亲鱼与海水养殖阶段的鱼，以阻断鲑肾杆菌的垂直传播，并且反复检测亲鱼，或将亲鱼置于高级别的生物安全系统中，以预防可能的感染。当大部分鲑处在养殖阶段时，定期检测养殖群体中的 BKD 并在疾病的早期口服抗生素是应对疾病的有效策略。虽然疫苗接种在疾病临床暴发期间有利于降低鲑的死亡率（Burnley 等，2010），但养殖场管理人员认为，与流行地区的其他预防和控制措施相比，疫苗的使用在养殖成本控制方面是不划算的。

细菌性肾病疫苗对已经接触病原的个体，以及通过垂直传播的无症状的鱼是否能够产生保护力尚不清楚。研究发现，当种群中鲑肾杆菌自然感染率（通过细菌培养检测）为 30%～40%时，对大西洋鲑幼鱼接种 Renogen©疫苗，不能成功预防该病（Salonius 等，2005）。然而，用自然感染鲑肾杆菌的大鳞大麻哈幼鱼进行的实验中，设置单独接种 Renogen©疫苗、混合接种 Renogen©疫苗与缺失 MSA 活疫苗以及对照组，结果表明，实验组存活率显著高于对照组，联合疫苗免疫的治疗效果最好（Rhodes 等，2004b）。

通常认为当疫苗对 BKD 等慢性病提供一定的保护水平后，会显著提高鲑的生长率，但这一假设并无定论。Salonius 等（2005）报道，接种 Renogen©疫苗 23 个月后的大西洋鲑比接种其他病原菌疫苗，鱼体体重增加了 16%，但 Burnley 等（2010）的研究结果，在接种 27 个月后，接种 Renogen©疫苗的大西洋鲑与接种其他病原菌疫苗的鲑相比，在体重上无显著差异。然而在上述两项现场生产实验中，所有组的实验鱼均出现了自然感染 BKD 而死亡。

Renogen©疫苗除了有效预防大西洋鲑 BKD 的报道外，也有研究发现 Renogen©疫苗能预防由革兰氏阴性胞内寄生的鲑鱼立克次氏体（*Piscirickettsia salmonis*）引起的鲑鱼立克次氏体败血病（SRS）。在实验室和野外实验显示，接种 Renogen©疫苗的银大麻哈鱼，感染 SRS 的死亡率显著低于对照组（Salonius 等，2005）。

鱼类疫苗的副作用主要是引起腹腔粘连和生长率降低，这与使用油类佐剂有关（Midtlyng，1996；Midtlyng 和 Lillehaug，1998）。一些试验阶段的 BKD 疫苗含有油类佐剂，但商品活疫苗 Renogen©是一种不含佐剂的注射型疫苗。

第九节 管理条例

商品疫苗 Renogen©已经在加拿大、美国和智利获得上市许可，用于鲑科鱼类免疫接种预防 BKD，但在欧洲和日本还没有获得许可（Sommerset 等，2005）。也许是 BKD 疫苗市场的规模以及疫苗许可费用等经济方面的问题，阻碍了该疫苗在一些国家的商业应用。苏格兰等一些地方颁布了 BKD 根除计划，因此无法使用该疫苗（Murray 等，2012）。由于 Renogen©疫苗中含有一种活的非致病性环境细菌，且已证明这种细菌不会发生毒力恢复，因此该疫苗较鲑肾杆菌减毒菌苗更容易获得申请许可（Salonius 等，2005）。

第十节 发展方向

更有效的预防细菌性肾病的疫苗是控制该病在多种鲑科鱼类中发生的有力工具。前几节已经介绍了疫苗的特性和与疫苗性能有关的宿主反应，下面主要讨论 BKD 疫苗改进的研究方向。

一、DNA 疫苗和蛋白亚单位疫苗

鲑肾杆菌基因组的已知序列使得筛选 DNA 疫苗候选基因成为可能。有限的研究显示，虹鳟（*Oncorhynchus mykiss*）肌肉注射编码 MSA 的 DNA 表达载体或鲑肾杆菌基因组中随机混合的 DNA 片段的表达载体后，再进行注射攻毒，能够获得显著的免疫保护（Gomez-Chiarri 等，1996）。许多胞外蛋白或外膜蛋白可能会被 MSA 掩盖，但是可在 DNA 疫苗结构中单独表达或联合表达。超过 440 个开放阅读框（ORFs）似乎具有 Sec 依赖分泌途径的先导肽序列特征，并且已经鉴定出类似转肽酶基因和潜在的转肽酶底物基因（Wiens 等，2008）。转肽酶及其表面蛋白底物是革兰氏阳性菌重要的毒力决定因子，具有黏附宿主细胞、定植以及免疫逃逸的功能（Sudheesh 等，2007；Maresso 和 Schneewind，2008）。研究表明，小鼠免疫接种重组的转肽酶（Gianfaldoni 等，2009）或重组转肽酶底物蛋白（Stranger-Jones 等，2006），可以分别产生对肺炎链球菌和金黄葡萄球菌毒株感染的免疫保护。

二、佐剂

弗氏佐剂和油类佐剂会造成注射部位的炎症和粘连等病理变化（Mutoloki 等，2006），避免这些副作用对于保证鱼类的健康和满足市场实际需求均具有重要意义。细菌（和病毒）DNA 的未甲基化的 CpG 二核苷酸序列在脊椎动物中具有较强的免疫刺激性，这是 DNA 佐剂的基本作用机制（Higgins 等，2007）。当细菌的侵入突破宿主的皮肤或胃肠黏膜等物理屏障时，宿主细胞 Toll 样受体（TLRs）会识别保守的细菌 CpG 序列，这在先天性免疫反应的启动中起着关键作用，并影响随后针对特异性抗原的获得性免疫应答，这种刺激应答是序列依赖型的，在鱼类中相关研究很少（Rhodes 等，2004b；Pedersen 等，2006）。探究对鱼类具有免

疫刺激作用的 DNA 序列，可能会发现具有增强 DNA 疫苗或全菌疫苗预防 BKD 保护效果的佐剂。鳟和鲑中克隆出的 TLRs（Rebl 等，2010）有助于系统筛选其配体，如细菌未甲基化的 CpG 群，既可以结合鱼类特有的 TLRs，也可以结合脊椎动物保守的 TLRs。

三、候选疫苗的体外检测

在利用鲑肾杆菌基因组筛选潜在的候选疫苗时，需要开发高通量的疫苗评价方法和/或能预测临床疾病进展的免疫力指标。鲑科鱼类的原代头肾细胞能够在感染后表现出促炎性反应和免疫基因表达（Grayson 等，2002），建立可长期传代培养和可转染的鲑巨噬细胞系，用于从预测的开放阅读框（ORFs）中筛选候选疫苗，可提高筛选的能力和速度。目前已建立金鱼巨噬细胞系并对其特性进行了分析（*Carassius auratus*；Wang 等，1995），在鲑科鱼类也建立了具有巨噬细胞特征的细胞系（Ganassin 和 Bols，1998；Collet 和 Collins，2009）。

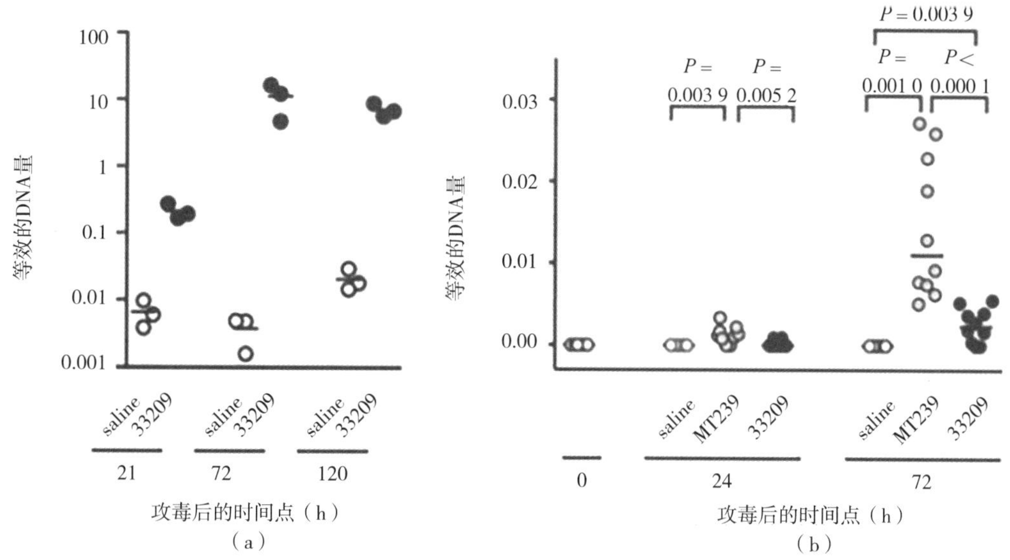

图 22-2　鲑肾杆菌实验攻毒后，细胞免疫反应相关的细胞因子——干扰素-γ（IFN-γ）的基因表达

（a）与对照组相比，ATCC 33209 毒性菌株攻毒后，虹鳟幼鱼［平均体重为（24±6）g］肾脏中 IFN-γ 表达升高。每条横标尺是 3 尾鱼的中值，引物和探针序列参照 Wiens 和 Vallejo，2010；（b）与 33209 毒性菌株或生理盐水对照组相比，减毒 MT239 菌株攻毒 3d 后，大鳞大麻哈幼鱼肾脏中 IFN-γ 的表达显著提高。每个横标尺是 10 尾鱼的中值，*P* 值是 Kruskal-Wallis 检验，正向引物：5′-CACCTGCAGAACCTGTGGG-3′；反向引物：5′-AACTCGGACAGAGCCTTCCC-3′；探针：5′ 6-FAM-CATC-GAGACCAGTGACACCACAGTCCA-BHQ1-3′

来源：（a）G. D. Wiens，未发表；（b）L. D. Rhodes，未发表

四、鱼类疫苗筛选的目标

细菌性肾病疫苗的测试应该简便合理，可通过检测能够预测疾病结果的生物标志物，筛选更多的候选疫苗，而不是以死亡率作为疫苗测试的最终结果。慢性疾病如 BKD，需要长期的攻毒试验（通常需要持续 3 个月）来评估死亡率。有指示作用的生物标记物在疾病发生的较早期就能被检测到，这样可以缩短疫苗评估的时间。例如，某些免疫基因的表达上调，包括肿瘤坏死因子（TNF-alpha）（Grayson 等，2002）、诱导型一氧化氮（iNOS）（Metzger 等，2010）、干扰素诱导基因（Rhodes 等，2009）和 γ-干扰素（图 22-2），是和宿主对鲑肾杆菌早期应答有

关，疫苗接种后，这些基因的表达可能有助于检测攻毒实验中鱼体的有效反应。由于宿主-病原动态变化在感染过程中的复杂性，早期反应的生物标志物可能缺乏特异性，如果通过快速的分子方法检测鱼体内的细菌载量，将这些标志物与细菌载量结合并标准化校正，能够提高标志物的预测准确度（Suzuki 和 Sakai，2007；Metzger 等，2010）。

参考文献

Alcorn，S.，Murray，A. L.，Pascho，R. J. and Varney，J. 2005. A cohabitation challenge to compare the efficacies of vaccines for bacterial kidney disease（BKD） in Chinook salmon *Oncorhynchus tshawytscha*. *Dis Aquat Org* 63：151-160.

Austin，B. 1985. Evaluation of antimicrobial compounds for the control of bacterial kidney disease in rainbow trout，*Salmo gairdneri* Richardson. *J Fish Dis* 8：209-220.

Austin，B. and Austin，D. A. 1987. *Bacterial Fish Pathogens*：*Disease in Farmed and Wild Fish*. Chichester，Ellis Horwood.

Balfry，S. K.，Albright，L. J. and Evelyn，T. P. T. 1996. Horizontal transfer of *Renibacterium salmoninarum* among farmed salmonids via the fecal-oral route. *Dis Aquat Org* 25：63-69.

Bandín，I.，Ellis，A. E.，Barja，J. L. and Secombes，C. J. 1993. Interaction between rainbow trout macrophages and *Renibacterium salmoninarum in vitro*. *Fish Shellfish Immunol* 3：25-33.

Banner，C. R.，Long，J. J.，Fryer，J. L. and Rohovec，J. S. 1986. Occurrence of salmonid fish infected with *Renibacterium salmoninarum* in the Pacific Ocean. *J Fish Dis* 9：273-275.

Beacham，T. D. and Evelyn，T. P. T. 1992. Population and genetic variation in resistance of Chinook salmonto vibriosis，furunculosis，and bacterial kidney disease. *J Aquat Anim Health* 4：153-167.

Belding，D. L. and Merrill，B. 1935. A preliminary report upon a hatchery disease of the Salmonidae. *Trans Am Fish Soc* 65：76-84.

Bell，G. R.，Higgs，D. A. and Traxler，G. S. 1984. The effect of dietary ascorbate，zinc，and manganese on the development of experimentally induced bacterial kidney disease in sockeye salmon（*Oncorhynchus nerka*）. *Aquaculture* 36：293-311.

Bell，G. R.，Traxler，G. S. and Dworschak，C. 1988. Development *in vitro* and pathogenicity of an erythromycin-resistant strain of *Renibacterium salmoninarum*，the causative agent of bacterial kidney disease in salmonids. *Dis Aquat Org* 4：19-25.

Bravo，S. and Midtlyng，P. J. 2007. The use of fish vaccines in the Chilean salmon industry 1999-2003. *Aquaculture* 270：36-42.

Brown，L. L.，Albright，L. J. and Evelyn，T. P. T. 1990. Control of vertical transmission of *Renibacterium salmoninarum* by injection of antibiotics into maturing female coho salmon *Oncorhynchus kisutch*. *Dis Aquat Org* 9：127-131.

Brown，L. L.，Iwama，G. K. and Evelyn，T. P. T. 1996. The effect of early exposure of coho salmon（*Oncorhynchus kisutch*） eggs to the p57 protein of *Renibacterium salmoninarum* on the development of immunity to the pathogen. *Fish Shellfish Immunol* 6：149-165.

Bruno，D. W. 1986. Histopathology of bacterial kidney disease in laboratory infected rainbow trout，*Salmo gairdneri* Richardson，and Atlantic salmon，*Salmo salar* L.，with reference to naturally infected fish. *J Fish Dis* 9：523-537.

Bruno, D. W. and Munro, A. L. S. 1986a. Uniformity in the biochemical properties of *Renibacterium salmoninarum* isolates obtained from several sources. *FEMS Microbiol Lett* 33: 247-250.

Bruno, D. W. and Munro, A. L. S. 1986b. Observations on *Renibacterium salmoninarum* and the salmonid egg. *Dis Aquat Org* 1: 83-87.

Bruno, D., Collet, B., Turnbull, A., *et al.*, 2007. Evaluation and development of diagnostic methods for *Renibacterium salmoninarum* causing bacterial kidney disease (BKD) in the UK. *Aquaculture* 269: 114-122.

Bullock, G. L., Stuckey, H. M. and Chen, P. K. 1974. Corynebacterial kidney disease of salmonids: growth and serological studies on the causative bacterium. *Appl Microbiol* 28: 811-814.

Burnley, T. A., Stryhn, H., Burnley, H. J. and Hammell, K. L. 2010. Randomized clinical field trial of a bacterial kidney disease vaccine in Atlantic salmon, *Salmo salar* L. *J Fish Dis* 33: 545-557.

Campos-Perez, J. J., Ellis, A. E. and Secombes, C. J. 1997. Investigation of factors influencing the ability of *Renibacterium salmoninarum* to stimulate rainbow trout macrophage respiratory burst activity. *Fish Shellfish Immunol* 7: 555-566.

Chambers, E., Gardiner, R. and Peeler, E. J. 2008. An investigation into the prevalence of *Renibacterium salmoninarum* in farmed rainbow trout, *Oncorhynchus mykiss* (Walbaum), and wild fish populations in selected river catchments in England and Wales between 1998 and 2000. *J Fish Dis* 31: 89-96.

Coady, A. M., Murray, A. L., Elliott, D. G. and Rhodes, L. D. 2006. Both msa genes in *Renibacterium salmoninarum* are needed for full virulence in bacterial kidney disease. *Appl Environ Microbiol* 72: 2672-2678.

Collet, B. and Collins, M. D. 2009. Comparative gene expression profile in two Atlantic salmon cell lines TO and SHK-1. *Vet Immunol Immunopathol* 130: 92-95.

Dale, O. B., Gutenberger, S. K. and Rohovec, J. S. 1997. Estimation of variation of virulence of *Renibacterium salmoninarum* by survival analysis of experimental infection of salmonid fish. *J Fish Dis* 20: 177-183.

Daly, J. G. and Stevenson, R. M. 1985. Charcoal agar, a new growth medium for the fish disease bacterium *Renibacterium salmoninarum*. *Appl Environ Microbiol* 50: 868-871.

Daly, J. G., Griffiths, S. G., Kew, A. K., *et al.*, 2001. Characterization of attenuated *Renibacterium salmoninarum* strains and their use as live vaccines. *Dis Aquat Org* 44: 121-126.

Eissa, A. E., Elsayed, E. E., McDonald, R. and Faisal, M. 2006. First record of *Renibacterium salmoninarum* in the sea lamprey (*Petromyzon marinus*). *J Wildl Dis* 42: 556-560.

Elliott, D. G. and Pascho, R. J. 2001. Evidence that coded-wire-tagging procedures can enhance transmission of *Renibacterium salmoninarum* in Chinook salmon. *J Aquat Anim Health* 13: 181-193.

Elliott, D. G., Pascho, R. J. and Bullock, G. L. 1989. Developments in the control of bacterial kidney disease of salmonid fishes. *Dis Aquat Org* 6: 201-215.

Elliott, D. G., Pascho, R. J., Jackson, L. M., *et al.*, 1997. *Renibacterium salmoninarum* in spring-summer Chinook salmon smolts at dams on the Columbia and Snake Rivers. *J Aquat Anim Health* 9: 114-126.

Ellis, A. E. 1999. Immunity to bacteria in fish. *Fish Shellfish Immunol* 9: 291-308.

Evelyn, T. P. T. 1971. The agglutinin response in sockeye salmon vaccinated intraperitoneally with a heat-killed preparation of the bacterium responsible for salmonid kidney disease. *J Wildl Dis* 7: 328-335.

Evelyn, T. P. T. 1988. Bacterial kidney disease in British Columbia, Canada: comments on its epizootiology and methods for its control on fish farms. *AQUA NOR 87 Trondheim International Conference*. Trondheim, Norske Fiskeoppdretternes Salgslag A/L, 51-57.

Evelyn, T. P. T. 1993. Bacterial kidney disease—BKD, in *Bacterial Diseases of Fish* (eds V. Inglis, R. J. Roberts and N. R. Bromage) . New York, Halsted Press, 177-195.

Evelyn, T. P. T., Hoskins, G. E. and Bell, G. R. 1973. First record of bacterial kidney disease in an apparently wild salmonid in British Columbia. *J Fish Res Board Can* 30: 1578-1580.

Evelyn, T. P. T., Ketcheson, J. E. and Prosperi-Porta, L. 1984. Further evidence for the presence of *Renibacterium salmoninarum* in salmonid eggs and for the failure of povidone-iodine to reduce the intra-ovum infection rate in water-hardened eggs. *J Fish Dis* 7: 173-182.

Evelyn, T. P. T., Prosperi-Porta, L. and Ketcheson, J. E. 1986. Experimental intra-ovum infection of salmonid eggs with *Renibacterium salmoninarum* and vertical transmission of the pathogen with such eggs despite their treatment with erythromycin. *Dis Aquat Org* 1: 197-202.

Evenden, A. J., Grayson, T. H., Gilpin, M. L. and Munn, C. B. 1993. *Renibacterium salmoninarum* and bacterial kidney disease - the unfinished jigsaw. *Annu Rev Fish Dis* 3: 87-104.

Fairgrieve, W. T., Masada, C. L., McAuley, W. C., *et al.*, 2005. Accumulation and clearance of orally administered erythromycin and its derivative, azithromycin, in juvenile fall Chinook salmon *Oncorhynchus tshawytscha*. *Dis Aquat Org* 64: 99-106.

Faisal, M., Loch, T. P., Brenden, T. O., *et al.*, 2010. Assessment of *Renibacterium salmoninarum* infections in four lake whitefish (*Coregonus clupeaformis*) stocks from northern Lakes Huron and Michigan. *J Great Lakes Res* 36: 29-37.

Fenichel, E. P., Tsao, J. I. and Jones, M. L. 2009. Modeling fish health to inform research and management: *Renibacterium salmoninarum* dynamics in Lake Michigan. *Ecol Appl* 19: 747-760.

Ferguson, H. W. 2006. *Systemic Pathology of Fish: a Text and Atlas of Normal Tissues in Teleosts and their Responses in Disease*. 2nd edn. London, Scotian Press.

Fiedler, F. and Draxl, R. 1986. Biochemical and immunochemical properties of the cell surface of *Renibacterium salmoninarum*. *J Bacteriol* 168: 799-804.

Flaño, E., Kaattari, S. L., Razquin, B. and Villena, A. J. 1996. Histopathology of the thymus of coho salmon *Oncorhynchus kisutch* experimentally infected with *Renibacterium salmoninarum*. *Dis Aquat Org* 26: 11-18.

Fredriksen, A., Endresen, C. and Wergeland, H. I. 1997. Immunosuppressive effect of a low molecular weight surface protein from *Renibacterium salmoninarum* on lymphocytes from Atlantic salmon (*Salmo salar* L) . *Fish Shellfish Immunol* 7: 273-282.

Fryer, J. L. and Sanders, J. E. 1981. Bacterial kidney disease of salmonid fish. *Annu Rev Microbiol* 35: 273-298.

Fryer, J. L. and Lannan, C. N. 1993. The history and current status of *Renibacterium salmoninarum*, the causative agent of bacterial kidney disease in Pacific salmon. *Fish Res* 17: 15-33.

Ganassin, R. C. and Bols, N. C. 1998. Development of a monocyte-macrophage-like cell line, RTS11, from rainbow trout spleen. *Fish Shellfish Immunol* 8: 457-476.

Getchell, R. G., Rohovec, J. S. and Fryer, J. L. 1985. Comparison of *Renibacterium salmoninarum* isolates by antigenic analysis. *Fish Pathol* 20: 149-159.

Gianfaldoni, C., Maccari, S., Pancotto, L., *et al.*, 2009. Sortase A confers protection against *Streptococcus pneumoniae* in mice. *Infect Immun* 77: 2957-2961.

Gomez-Chiarri, M., Brown, L. L. and Levine, R. P. 1996. Protection against *Renibacterium salmoninarum* infection by DNA-based immunization, in *Aquaculture Biotechnology* (eds E. Donaldson and D. McKinlay) . Proceedings, 2nd International Congress on the Biology of Fish, San Francisco, 155-7. Available at http://www.fishbiologycongress.org/1996.html.

Grayson, T. H., Evenden, A. J., Gilpin, M. L., *et al.*, 1995. A gene from *Renibacterium salmoninarum* encoding a product which shows homology to bacterial zinc-metalloproteases. *Microbiology* 141: 1331-1341.

Grayson, T. H., Gilpin, M. L., Evenden, A. J. and Munn, C. B. 2001. Evidence for the immune recognition of two haemolysins of *Renibacterium salmoninarum* by fish displaying clinical symptoms of bacterial kidney disease (BKD). *Fish Shellfish Immunol* 11: 367-370.

Grayson, T. H., Cooper, L. F., Wrathmell, A. B., *et al.*, 2002. Host responses to *Renibacterium salmoninarum* and specific components of the pathogen reveal the mechanisms of immune suppression and activation. *Immunology* 106: 273-283.

Griffiths, S. G., Melville, K. J. and Salonius, K. 1998. Reduction of *Renibacterium salmoninarum* culture activity in Atlantic salmon following vaccination with avirulent strains. *Fish Shellfish Immunol* 8: 607-619.

Guðmundsdóttir, S., Helgason, S., Sigurjónsdóttir, H., *et al.*, 2000. Measures applied to control *Renibacterium salmoninarum* infection in Atlantic salmon: a retrospective study of two sea ranches in Iceland. *Aquaculture* 186: 193-203.

Hard, J. J., Elliott, D. G., Pascho, R. J., *et al.*, 2006. Genetic effects of ELISA-based segregation for control of bacterial kidney disease in Chinook salmon (*Oncorhynchus tshawytscha*). *Can J Fish Aquat Sci* 63: 2793-2808.

Hardie, L. J., Ellis, A. E. and Secombes, C. J. 1996. *In vitro* activation of rainbow trout macrophages stimulates inhibition of *Renibacterium salmoninarum* growth concomitant with augmented generation of respiratory burst products. *Dis Aquat Org* 25: 175-183.

Hayakawa, Y., Harada, T., Hatai, K., *et al.*, 1989. Histopathology of BKD (bacterial kidney disease) occurred in sea-cultured coho salmon (*Oncorhynchus kisutch*). *Fish Pathol* 24: 17-21.

Hendricks, J. D. and Leek, S. L. 1975. Kidney disease postorbital lesions in Chinook salmon (*Oncorhynchus tshawytscha*). *Trans Am Fish Soc* 104: 805-807.

Higgins, D., Marshall, J. D., Traquina, P., *et al.*, 2007. Immunostimulatory DNA as a vaccine adjuvant. *Expert Rev Vaccines* 6: 747-759.

Hoffmann, R., Popp, W. and van de Graaff, S. 1984. Atypical BKD predominantly causing ocular and skin lesions. *Bull Eur Assoc Fish Pathol* 4: 7-9.

Holey, M. F., Elliott, R. F., Marcquenski, S. V., *et al.*, 1998. Chinook salmon epizootics in Lake Michigan: possible contributing factors and management implications. *J Aquat Anim Health* 10: 202-210.

Jansson, E., Hongslo, T., Johannisson, A., *et al.*, 2003. Bacterial kidney disease as a model for studies of cell mediated immunity in rainbow trout (*Oncorhynchus mykiss*). *Fish Shellfish Immunol* 14: 347-362.

Johnson, K. A., Flynn, J. K. and Amend, D. F. 1982. Duration of immunity in salmonids vaccinated by direct immersion with *Yersinia ruckeri* and *Vibrio anguillarum* bacterins. *J Fish Dis* 5: 207-213.

Johnson, R. M., Bryden, C. A. and Heath, D. D. 2003. Utility of genetically based health indicators for selection purposes in captive-reared Chinook salmon, *Oncorhynchus tshawytscha*, Walbaum. *Aquacult Res* 34: 1029-1036.

Jones, D. T., Moffitt, C. M. and Peters, K. K. 2007. Temperature-mediated differences in bacterial kidney disease expression and survival in *Renibacterium salmoninarum*—challenged bull trout and other salmonids. *N Am J Fish Manag* 27: 695-706.

Kaattari, S. L. and Piganelli, J. D. 1997. Immunization with bacterial antigens: bacterial kidney disease. *Dev Biol Stand* 90: 145-152.

Kawamura, H., Awakura, T., Watanabe, K. and Matsumoto, H. 1977. Therapeutic effects of erythromycin and the sensitivity of four salmonid fishes to bacterial kidney disease. *Sci Rep Hokkaido Fish Hatch* 32: 21-36.

Kent, M. L., Traxler, G. S., Kieser, D., *et al.*, 1998. Survey of salmonid pathogens in ocean-caught fishes in

British Columbia, Canada. *J Aquat Anim Health* 10: 211-219.

Kurath, G., Garver, K. A., Corbeil, S., *et al.*, 2006. Protective immunity and lack of histopathological damage two years after DNA vaccination against infectious hematopoietic necrosis virus in trout. *Vaccine* 24: 345-354.

Lee, E. G. H. and Evelyn, T. P. T. 1994. Prevention of vertical transmission of the bacterial kidney disease agent *Renibacterium salmoninarum* by broodstock injection with erythromycin. *Dis Aquat Org* 18: 1-4.

Lovely, J. E., Cabo, C., Griffiths, S. G. and Lynch, W. H. 1994. Detection of *Renibacterium salmoninarum* infection in asymptomatic Atlantic salmon. *J Aquat Anim Health* 6: 126-132.

Lumsden, J. S., Russell, S., Huber, P., *et al.*, 2008. An immune-complex glomerulonephritis of Chinook salmon, *Oncorhynchus tshawytscha* (Walbaum). *J Fish Dis* 31: 889-898.

MacLean, D. G. and Yoder, W. G. 1970. Kidney disease among Michigan salmon in 1967. *Prog Fish-Cult* 32: 26-30.

Maresso, A. W. and Schneewind, O. 2008. Sortase as a target of anti-infective therapy. *Pharmacol Rev* 60: 128-141.

Maule, A. G., Rondorf, D. W., Beeman, J. and Haner, P. 1996. Incidence of *Renibacterium salmoninarum* infections in juvenile hatchery spring Chinook salmon in the Columbia and Snake rivers. *J Aquat Anim Health* 8: 37-46.

McCarthy, D. H., Croy, T. R. and Amend, D. F. 1984. Immunization of rainbow trout, *Salmo gairdneri* Richardson, against bacterial kidney disease: preliminary efficacy evaluation. *J Fish Dis* 7: 65-71.

McIntosh, D., Austin, B., Flaño, E., *et al.*, 2000. Lack of uptake of *Renibacterium salmoninarum* by gill epithelia of rainbow trout. *J Fish Biol* 56: 1053-1061.

Mesa, M. G., Poe, T. P., Maule, A. G. and Schreck, C. B. 1998. Vulnerability to predation and physiological stress responses in juvenile Chinook salmon (*Oncorhynchus tshawytscha*) experimentally infected with *Renibacterium salmoninarum*. *Can J Fish Aquat Sci* 55: 1599-1606.

Mesa, M. G., Maule, A. G., Poe, T. P. and Schreck, C. B. 1999. Influence of bacterial kidney disease on smoltification in salmonids: is it a case of double jeopardy? *Aquaculture* 174: 25-41.

Metzger, D. C., Elliott, D. G., Wargo, A., *et al.*, 2010. Pathological and immunological responses associated with differential survival of Chinook salmon following *Renibacterium salmoninarum* challenge. *Dis Aquat Org* 90: 31-41.

Meyers, T. R., Korn, D., Glass, K., *et al.*, 2003. Retrospective analysis of antigen prevalences of *Renibacterium salmoninarum* (Rs) detected by enzyme-linked immunosorbent assay in Alaskan Pacific salmon and trout from 1988 to 2000 and management of Rs in hatchery Chinook and coho salmon. *J Aquat Anim Health* 15: 101-110.

Midtlyng, P. J. 1996. A field study on intraperitoneal vaccination of Atlantic salmon (*Salmo salar* L.) against furunculosis. *Fish Shellfish Immunol* 6: 553-565.

Midtlyng, P. J. and Lillehaug, A. 1998. Growth of Atlantic salmon *Salmo salar* after intraperitoneal administration of vaccines containing adjuvants. *Dis Aquat Org* 32: 91-97.

Mitchum, D. L. and Sherman, L. E. 1981. Transmission of bacterial kidney disease from wild to stocked hatchery trout. *Can J Fish Aquat Sci* 38: 547-551.

Mitchum, D. L., Sherman, L. E. and Baxter, G. T. 1979. Bacterial kidney disease in feral populations of brook trout (*Salvelinus fontinalis*), brown trout (*Salmo trutta*) and rainbow trout (*Salmo gairdneri*). *J Fish Res Board Can* 36: 1370-1376.

Moffitt, C. M. 1991. Oral and injectable applications of erythromycin in salmonid fish culture. *Vet Hum Toxicol* 33 (Suppl. 1): 49-53.

Moffitt, C. M. 1992. Survival of juvenile Chinook salmon challenged with *Renibacterium salmoninarum* and administered oral doses of erythromycin thiocyanate for different durations. *J Aquat Anim Health* 4：119-125.

Moffitt，C. M. and Kiryu，Y. 1999. Toxicity，teratogenesis，and efficacy of injectable erythromycin（Erythro-200）administered repeatedly to adult spring Chinook salmon. *J Aquat Anim Health* 11：1-11.

Munson，A. D.，Elliott，D. G. and Johnson，K. 2010. Management of bacterial kidney disease in Chinook salmon hatcheries based on broodstock testing by enzyme-linked immunosorbent assay：a multiyear study. *N Am J Fish Manag* 30：940-955.

Murray，A. G.，Hall，M.，Munro，L. A. and Wallace，I. S. 2011. Modelling management strategies for a disease including undetected sub-clinical infection：Bacterial kidney disease in Scottish salmon and trout farms. *Epidemics* 3：171-182.

Murray，A. G.，Munro，L. A. and Wallace，I. S.，*et al.*，2012. Epidemiology of *Renibacterium salmoninarum* in Scotland and the potential for compartmentalised management of salmon and trout farming areas. *Aquaculture* 324-325：1-13.

Murray，C. B.，Evelyn，T. P. T.，Beacham，T. D.，*et al.*，1992. Experimental induction of bacterial kidney disease in Chinook salmon by immersion and cohabitation challenges. *Dis Aquat Org* 12：91-96.

Mutoloki，S.，Reite，O. B.，Brudeseth，B.，*et al.*，2006. A comparative immunopathological study of injection site reactions in salmonids following intraperitoneal injection with oil-adjuvanted vaccines. *Vaccine* 24：578-588.

O'Farrell，C. L.，Elliott，D. G. and Landolt，M. L. 2000. Mortality and kidney histopathology of Chinook salmon *Oncorhynchus tshawytscha* exposed to virulent and attenuated *Renibacterium salmoninarum* strains. *Dis Aquat Org* 43：199-209.

Ordal，E. J. and Earp，B. J. 1956. Cultivation and transmission of the etiological agent of bacterial kidney disease. *Proc Soc Exp Biol Med* 92：85-88.

Pascho，R. J.，Elliott，D. G. and Streufert，J. M. 1991. Brood stock segregation of spring Chinook salmon *Oncorhynchus tshawytscha* by use of the enzyme-linked immunosorbent assay（ELISA）and the fluorescent antibody technique（FAT）affects the prevalence and levels of *Renibacterium salmoninarum* infection in progeny. *Dis Aquat Org* 12：25-40.

Pascho，R. J.，Goodrich，T. D. and McKibben，C. L. 1997. Evaluation by enzyme-linked immunosorbent assay（ELISA）of *Renibacterium salmoninarum* bacterins affected by persistence of bacterial antigens. *J Aquat Anim Health* 9：99-107.

Pascho，R. J.，Elliott，D. G. and Chase，D. M. 2002. Comparison of traditional and molecular methods for detection of *Renibacterium salmoninarum*，in *Molecular Diagnosis of Salmonid Diseases*（ed. C. O. Cunningham）. Dordrecht，Kluwer Academic Publishers，157-209.

Paterson，W. D.，Desautels，D. and Weber，J. M. 1981. Immune response of Atlantic salmon，*Salmo salar* L.，to the causative agent of bacterial kidney disease，*Renibacterium salmoninarum*. *J Fish Dis* 4：99-111.

Pedersen，G. M.，Johansen，A.，Olsen，R. L. and Jørgensen，J. B. 2006. Stimulation of type I IFN activity in Atlantic salmon（*Salmo salar* L.）leukocytes：synergistic effects of cationic proteins and CpG ODN. *Fish Shellfish Immunol* 20：503-518.

Piganelli，J. D.，Wiens，G. D. and Kaattari，S. L. 1999a. Elevated temperature treatment as a novel method for decreasing p57 on the cell surface of *Renibacterium salmoninarum*. *Dis Aquat Org* 36：29-35.

Piganelli，J. D.，Wiens，G. D.，Zhang，J. A.，*et al.*，1999b. Evaluation of a whole cell，p57-vaccine against *Renibacterium salmoninarum*. *Dis Aquat Org* 36：37-44.

Pippy，J. H. C. 1969. Kidney disease in juvenile Atlantic salmon（*Salmo salar*）in the Margaree River. *J Fish Res*

Board Can 26: 2535-2537.

Polinski, M. P., Fehringer, T. R., Johnson, K. A., *et al.*, 2010. Characterization of susceptibility and carrier status of burbot, *Lota lota* (L.), to IHNV, IPNV, *Flavobacterium psychrophilum*, *Aeromonas salmonicida* and *Renibacterium salmoninarum*. *J Fish Dis* 33: 559-570.

Purcell, M. K., Murray, A. L., Elz, A., *et al.*, 2008. Decreased mortality of Lake Michigan Chinook salmon after bacterial kidney disease challenge: evidence for pathogen-driven selection? *J Aquat Anim Health* 20: 225-235.

Rebl, A., Goldammer, T. and Seyfert, H.-M. 2010. Toll-like receptor signaling in bony fish. *Vet Immunol Immunopathol* 134: 139-150.

Rhodes, L. D., Coady, A. M. and Deinhard, R. K. 2004a. Identification of a third msa gene in *Renibacterium salmoninarum* and the associated virulence phenotype. *Appl Environ Microbiol* 70: 6488-6494.

Rhodes, L. D., Rathbone, C. K., Corbett, S. C., *et al.*, 2004b. Efficacy of cellular vaccines and genetic adjuvants against bacterial kidney disease in Chinook salmon (*Oncorhynchus tshawytscha*). *Fish Shellfish Immunol* 16: 461-474.

Rhodes, L. D., Nguyen, O. T., Deinhard, R. K., *et al.*, 2008. Characterization of *Renibacterium salmoninarum* with reduced susceptibility to macrolide antibiotics by a standard antibiotic susceptibility test. *Dis Aquat Org* 80: 173-180.

Rhodes, L. D., Wallis, S. and Demlow, S. E. 2009. Genes associated with an effective host response by Chinook salmon to *Renibacterium salmoninarum*. *Dev Comp Immunol* 33: 176-186.

Rhodes, L. D., Rice, C. A., Greene, C. M., *et al.*, 2011. Nearshore ecosystem predictors of a bacterial infection in juvenile Chinook salmon. *Mar Ecol Prog Ser* 432: 161-172.

Rockey, D. D., Turaga, P. S., Wiens, G. D., *et al.*, 1991. Serine proteinase of *Renibacterium salmoninarum* digests a major autologous extracellular and cell-surface protein. *Can J Microbiol* 37: 758-763.

Sakai, M. and Kobayashi, M. 1992. Detection of *Renibacterium salmoninarum*, the causative agent of bacterial kidney disease in salmonid fish, from pen-cultured coho salmon. *Appl Environ Microbiol* 58: 1061-1063.

Sakai, M., Ogasawara, F., Atsuta, S. and Kobayashi, M. 1989. Comparative sensitivity of carp *Cyprinus carpio* L and rainbow trout *Salmo gairdneri* Richardson, to *Renibacterium salmoninarum*. *J Fish Dis* 12: 367-372.

Sakai, M., Atsuta, S. and Kobayashi, M. 1991. Susceptibility of five salmonid fishes to *Renibacterium salmoninarum*. *Fish Pathol* 26: 159-160.

Sakai, M., Atsuta, S. and Kobayashi, M. 1993. The immune response of rainbow trout (*Oncorhynchus mykiss*) injected with five *Renibacterium salmoninarum* bacterins. *Aquaculture* 113: 11-18.

Salonius, K., Siderakis, C., MacKinnon, A. M. and Griffiths, S. G. 2005. Use of *Arthrobacter davidanieli* as a live vaccine against *Renibacterium salmoninarum* and *Piscirickettsia salmonis* in salmonids. *Dev Biol* (*Basel*) 121: 189-197.

Sami, S., Fischer-Scherl, T., Hoffmann, R. W. and Pfeil-Putzien, C. 1992. Immune complex-mediated glomerulonephritis associated with bacterial kidney disease in the rainbow trout (*Oncorhynchus mykiss*). *Vet Pathol* 29: 169-174.

Sandell, T. A. and Jacobson, K. C. 2011. Comparison and evaluation of *Renibacterium salmoninarum* quantitative PCR diagnostic assays using field samples of Chinook and coho salmon. *Dis Aquat Org* 93: 129-139.

Sanders, J. E. and Fryer, J. L. 1980. *Renibacterium salmoninarum* gen. nov., sp. nov., the causative agent of bacterial kidney disease in salmonid fishes. *Int J Syst Bacteriol* 30: 496-502.

Sanders, J. E., Pilcher, K. S. and Fryer, J. L. 1978. Relation of water temperature to bacterial kidney disease in coho salmon (*Oncorhynchus kisutch*), sockeye salmon (*O. nerka*), and steelhead trout (*Salmo gairdneri*). *J Fish Res Board Can* 35: 8-11.

Senson, P. R. and Stevenson, R. M. 1999. Production of the 57 kDa major surface antigen by a non-agglutinating strain of the fish pathogen *Renibacterium salmoninarum*. *Dis Aquat Org* 38: 23-31.

Smith, I. W. 1964. The occurrence and pathology of Dee disease. *Freshw Salmon Fish Res* 34: 1-12.

Sommerset, I., Krossøy, B., Biering, E. and Frost, P. 2005. Vaccines for fish in aquaculture. *Expert Rev Vaccines* 4: 89-101.

Souter, B. W., Dwilow, A. G. and Knight, K. 1987. *Renibacterium salmoninarum* in wild Arctic charr *Salvelinus alpinus* and lake trout *S. namaycush* from the Northwest Territories, Canada. *Dis Aquat Org* 3: 151-154.

Speare, D. J. 1997. Differences in patterns of meningoencephalitis due to bacterial kidney disease in farmed Atlantic and Chinook salmon. *Res Vet Sci* 62: 79-80.

Starliper, C. E. and Morrison, P. 2000. Bacterial pathogen contagion studies among freshwater bivalves and salmonid fishes. *J Shellfish Res* 19: 251-258.

Starliper, C. E., Smith, D. R. and Shatzer, T. 1997. Virulence of *Renibacterium salmoninarum* to salmonids. *J Aquat Anim Health* 9: 1-7.

Stranger-Jones, Y. K., Bae, T. and Schneewind, O. 2006. Vaccine assembly from surface proteins of *Staphylococcus aureus*. *Proc Natl Acad Sci USA* 103: 16942-16947.

Sudheesh, P. S., Crane, S., Cain, K. D. and Strom, M. S. 2007. Sortase inhibitor phenyl vinyl sulfone inhibits *Renibacterium salmoninarum* adherence and invasion of host cells. *Dis Aquat Org* 78: 115-127.

Suzuki, K. and Sakai, D. K. 2007. Real-time PCR for quantification of viable *Renibacterium salmoninarum* in chum salmon *Oncorhynchus keta*. *Dis Aquat Org* 74: 209-223.

Suzumoto, B., Schreck, C. and McIntyre, J. 1977. Relative resistances of three transferrin genotypes of coho salmon (*Oncorhynchus kisutch*) and their hematological responses to bacterial kidney disease. *J Fish Res Board Can* 34: 1-8.

Turaga, P., Wiens, G. and Kaattari, S. 1987. Bacterial kidney disease: the potential role of soluble protein antigen (s). *J Fish Biol* 31: 191-194.

Wang, R., Neumann, N. F., Shen, Q. and Belosevic, M. 1995. Establishment and characterization of a macrophage cell line from goldfish. *Fish Shellfish Immunol* 5: 329-346.

Wiens, G. D. and Vallejo, R. L. 2010. Temporal and pathogen-load dependent changes in rainbow trout (*Oncorhynchus mykiss*) immune response traits following challenge with biotype 2 *Yersinia ruckeri*. *Fish Shellfish Immunol* 29: 639-647.

Wiens, G. D., Pascho, R. and Winton, J. R. 2002. A single Ala139-to-Glu substitution in the *Renibacterium salmoninarum* virulence-associated protein p57 results in antigenic variation and is associated with enhanced p57 binding to Chinook salmon leukocytes. *Appl Environ Microbiol* 68: 3969-3977.

Wiens, G. D., Rockey, D. D., Wu, Z., *et al.*, 2008. Genome sequence of the fish pathogen *Renibacterium salmoninarum* suggests reductive evolution away from an environmental *Arthrobacter* ancestor. *J Bacteriol* 190: 6970-6982.

Wolf, K. and Dunbar, C. E. 1959. Test of 34 therapeutic agents for control of kidney disease in trout. *Trans Am Fish Soc* 88: 117-124.

Wolke, R. E. 1975. Pathology of bacterial and fungal diseases affecting fish, in *The Pathology of Fishes* (ed. W. E. Ribelin and G. Migaki). Madison, University of Wisconsin Press, 33-116.

Wood, J. W. and Wallis, J. 1955. Kidney disease in adult Chinook salmon and its transmission by feeding to young Chinook salmon. *Oreg Fish Comm Res Brief* 6: 32-40.

Wood, P. A. and Kaattari, S. L. 1996. Enhanced immunogenicity of *Renibacterium salmoninarum* in Chinook salmon after removal of the bacterial cell surface-associated 57 kDa protein. *Dis Aquat Org* 25: 71-79.

Young, C. L. and Chapman, G. B. 1978. Ultrastructural aspects of the causative agent and renal histopathology of bacterial kidney disease in brook trout (*Salvelinus fontinalis*). *J Fish Res Board Can* 35: 1234-1248.

第二十三章　黄杆菌病的疫苗接种

Krister Sundell, Eva Högfors-Rönnholm and Tom Wiklund

本章概要　SUMMARY

近年来发现多种黄杆菌科（Flavobacteriaceae）细菌可以引发鱼类疾病。已开展广泛研究的有嗜鳃黄杆菌（*Flavobacterium branchiophilum*）（细菌性鳃病）、柱状黄杆菌（*F. columnare*）（柱形病）、嗜冷黄杆菌（*F. psychrophilum*）（细菌性冷水病）和海洋屈挠杆菌（*Tenacibaculum maritimum*）（屈挠杆菌病），这些病原菌均能对水产养殖业造成重大的经济影响。传统细菌疫苗免疫预防黄杆菌科细菌引起的疾病效果不够稳定，现在开展较多的工作是重组蛋白及其制备的亚单位疫苗的免疫原性评价研究。到目前为止，只有柱形病的疫苗已经商业化，而细菌性冷水病的免疫保护仍依靠自体疫苗。自从美国推出柱形病的减毒活疫苗以来，已开始了类似疫苗在预防嗜冷黄杆菌感染中的评估研究。

第一节　引　言

黄杆菌科（Flavobacteriaceae）有嗜鳃黄杆菌（*Flavobacterium branchiophilum*）、柱状黄杆菌（*F. columnare*）、嗜冷黄杆菌（*F. psychrophilum*）和海洋屈挠杆菌（*Tenacibaculum maritimum*）等几种重要的鱼类致病菌（Bernardet 等，1996；Avendaño-Herrera 等，2006a；Barnes 等，2011；Starliper 等，2011）。此外，还有很多黄杆菌科的细菌与鱼类的偶发性死亡有关，如大菱鲆金黄杆菌（*Chryseobacterium scophthalmu*；Mudarris 等，1994；Vandamme 等，1994）、鱼害金黄杆菌（*C. piscicola*；Ilardi 等，2009）、双鱼金黄杆菌（*C. piscium*；de Beer 等，2006）、约氏黄杆菌（*Flavobacterium johnsoniae*；Bernardet 和 Bowman，2006；Flemming 等，2007）和解卵屈挠杆菌（*Flexibacter ovolyticus*；Hansen 等，1992）。感染鱼类的黄杆菌科细菌种类不断被发现，在智利的患病鲑鳟中分离的智利黄杆菌（*Flavobacterium chilense*）和阿劳科黄杆菌（*F. araucananum*），以及在西班牙患病虹鳟中分离的虹鳟黄杆菌（*F. oncorhynchi*）是黄杆菌科的3个新种（Kämpfer 等，2012；Zamora 等，2012）。

本章将聚焦嗜鳃黄杆菌、柱状黄杆菌、嗜冷黄杆菌和海洋屈挠杆菌疫苗的免疫接种研究现状，这几种黄杆菌在世界范围内可感染多种鱼类，给海水和淡水养殖均造成严重的危害。

第二节 细菌性鳃病（嗜鳃黄杆菌 *Flavobacterium branchiophilum*）

最初，嗜鳃黄杆菌是引起鲑和其他淡水鱼如六须鲶（*Silurus glanis*）和鲢（*Hypophthalmichtys molitrix*）鳃部感染的一种病原菌（Wakabayashi 等，1980；Farkas，1985；Ostland 等，1994）。育苗场的溪红点鲑（*Salvelinus fontinalis*）和虹鳟更易受该菌感染（Starliper 等，2011）。在加拿大、匈牙利、日本、韩国和美国已有该菌的报道（Farkas，1985；Wakabayashi 等，1989；Ferguson 等，1991；Ostland 等，1994；Ko 和 Heo，1997），在荷兰也可能存在该病原（Farkas，1985）。

嗜鳃黄杆菌主要感染鱼鳃并大量定植于鱼鳃表面（Wakabayashi 等，1980；Speare 等，1991），因此，由其引起的疾病被称为细菌性鳃病（BGD）。虽然有其他嗜纤维菌样的细菌病原（包括柱状黄杆菌）也在感染的鱼鳃组织中被分离出来（Farkas 和 Oláh，1986），但是嗜鳃黄杆菌被认为是 BGD 最主要的病原菌（Speare 等，1995）。该病原几乎不能从感染鱼的内脏组织器官中分离到（Bernardet，1997），因此被认为不会引起鱼类的全身感染（Ototake 等，1985；Starliper 等，2011）。

细菌性鳃病的最初病症包括层状上皮水样变性和坏死以及层状水肿（Speare 等，1991）；鱼体感染嗜鳃黄杆菌后期会发生鳃丝融合（Ototake 和 Wakabayashi，1985）和上皮增生等亚急性或慢性反应（Speare 等，1991）。

鱼体接触嗜鳃黄杆菌后，细菌会迅速附着并定植于鳃上。研究表明，嗜鳃黄杆菌的毒力与其表面的菌毛黏附作用（Heo 等，1990b）、胞外产物的凝血活力以及细菌凝集效应有关（Ototake 等，1985）。嗜鳃黄杆菌主要是引起鳃的气体交换不足（Wakabayashi 和 Iwado，1985）和血液循环障碍（Byrne 等，1991），从而造成鱼的死亡。

鲑嗜鳃黄杆菌可通过水体或与已感染的种内或种间鱼类接触进行水平传播（Ostland 等，1995；Ferguson 等，1991）。致病性和非致病性的嗜鳃黄杆菌分离株都可以附着在鱼鳃组织上，但只有致病性的菌株能够导致 BGD（Ostland 等，1995）。嗜鳃黄杆菌的菌毛有利于细菌附着于鳃上皮（Heo 等，1990b）。Ostland 等（1997）报道，虹鳟在含有嗜鳃黄杆菌菌毛粗提物的水体中浸泡后，嗜鳃黄杆菌的黏附减少，但没有完全抑制其黏附，细菌黏附仅减少22%～33%。以上结果表明，除了菌毛，其他菌体结构也可能参与该菌的黏附。

不同区域来源的嗜鳃黄杆菌分离株具有相同的抗原，但有研究发现不同区域的分离株也具有抗原多样性。例如，日本分离株与北美和匈牙利分离株的血清型完全不同（Wakabayashi 等，1980；Huh 等，1989），同一个地理区域的分离株也具有抗原差异（Ostland 等，1994）。Heo 等（1990a）发现，在外周循环系统中不产生该病原的特异性抗体，可能是因为自然感染嗜鳃黄杆菌的鱼体没有系统性感染的缘故。但溪红点鲑通过水平感染，经嗜鳃黄杆菌浸浴攻毒后，血清和鳃中均有轻微的抗体反应（Lumsden 等，1993）。后来，Lumsden 等（1994）报道，以未接触嗜鳃黄杆菌为对照，攻毒实验中幸存的虹鳟在第二次攻毒后存活率增加。

实验室条件下，嗜鳃黄杆菌的免疫防治取得了很好的效果。Lumsden 等（1994）使用丙酮灭活的嗜鳃黄杆菌进行反复的浸浴或腹腔注射免疫虹鳟，结果显示，虹鳟经 3 次浸浴免疫，嗜鳃黄杆菌实验攻毒后产生了良好的的保护作用。另一项研究中，浸泡免疫接种丙酮灭活的嗜鳃黄杆菌 2 次，同样取得了良好的保护效果（Lumsden 等，1995）。腹腔注射丙酮灭活的嗜鳃黄杆菌也具有免疫保护性，但其保护效果不及浸泡免疫（Lumsden 等，1994，1995），这表明具有保护性的血清抗体可以转移到鳃的表面。随后的研究发现，虹鳟腹腔注射丙酮灭活的细菌后，鳃黏膜中存在嗜鳃黄杆菌的特异性抗体（Lumsden 等，1995）。然而实验表明，相同抗原免疫，与注射免疫相比，浸泡免疫的虹鳟获得了更好的保护（Lumsden 等，1995）。细菌攻毒后，与未接种疫苗的对照组相比，免疫组鱼鳃中的嗜鳃黄杆菌数量明显减少，可以认为免疫组鳃部检测到的嗜鳃黄杆菌的显著减少是由于局部免疫反应的作用。因此建议，未来的研究应针对该菌的水平感染，评估疫苗免疫保护效果。

第三节 柱形病（柱状黄杆菌 *Flavobacterium columnare*）

柱状黄杆菌是一种感染淡水鱼类，并引起柱形病的细菌病原。数十年来，柱状黄杆菌由于其地理分布广和宿主范围广而一直备受关注。如今，柱形病对一些重要的水产养殖经济鱼类构成了巨大威胁，如斑点叉尾鮰（*Ictalurus punctatus*）、鲤（*Cyprinus carpio*）和尼罗罗非鱼（*Oreochromis niloticus*）。近年来，在欧洲部分地区（Kubilay 等，2008；Pulkkinen 等，2010）和智利（Avendaño-Herrera 等，2011），柱状黄杆菌已经对大西洋鲑（*Salmo salar*）、褐鳟（*S. trutta*）和虹鳟幼鱼生产构成威胁，在美国柱状黄杆菌感染给斑点叉尾鮰产业造成了非常严重的经济损失，估计每年可达3 000万美元（Wagner 等，2002）。

柱形病全年均可发病，但典型发病水温高于 20℃，主要与水质差、高温、高密度和养殖操作管理不当导致的应激有关。实验证明，饵料变化和饥饿可以增加鱼体对柱状黄杆菌感染的敏感性（Shoemaker 等，2003；Bebak 等，2009）。感染柱状黄杆菌的鱼体通常出现严重的鳃和皮肤组织坏死，但偶尔也会发生无明显外部症状的全身感染。该菌可通过鱼体间的接触直接传播，也可通过水体间接传播（Welker 等，2005）。尽管菌株的致病性决定了柱形病暴发的严重性，但是高温和高密度能加速疾病的进程（Suomalainen 等，2005；Noga，2010）。

虽然对柱状黄杆菌已进行了广泛的研究，但柱状黄杆菌的毒力机制仍然不明晰。通过血清学研究发现了该病原菌的 4 种主要血清型，但还未明确血清型和毒力之间的关联性（Anacker 等，1959；Noga，2010）。Triyanto 和 Wakabayashi 发现柱状黄杆菌有 3 种不同的遗传基因型，其在毒力（Decostere 等，1998）和宿主特异性（Olivares-Fuster 等，2007）上存在差异。组织溶解性软骨素 AC 裂解酶（Tissue-degrading chondroitin AC lyase）活性及其对鱼类黏液的趋化作用和柱状黄杆菌毒力呈正相关（Suomalainen 等，2006；Klesius 等，2008）。实验室条件下，从鱼类中分离出来的柱状黄杆菌具有不同的菌体形态（Thomas-Jinu 等，2004）和菌落形态（Song 等，1988）。此外，实验证明，根据不同的黏附能力和组织降解特性，柱状黄杆菌可以分成 4 种菌落类型（Kunttu 等，2009，2011）。尽管形态学上的差异与柱状黄杆菌的毒力没有直接关系，但是它们可能在柱形病的发病过程中起着重要的作用。

在一些重要经济鱼类中发现了针对柱状黄杆菌的体液抗体和黏液抗体反应（Fujihar 等，1971；Schachte 等，1973；Grabowski 等，2004；Shoemaker 等，2005），这为预防柱形病疫苗的研发提供了理论基础。建立确定鱼类柱状黄杆菌疫苗功效的标准化感染模型，将使疫苗研发的速度进一步加快。如今，在鮰（Klesius 等，1999；Shoemaker 等，2003）和虹鳟（LaFrentz 等，2012）中已建立可重复浸泡感染程序，使用斑马鱼（*Danio rerio*）的替代实验模型也已经建立（Moyer 等，2007）。

总的来说，通过任何类型的疫苗接种控制柱形病都比较困难，因为鱼类在孵化早期对柱状黄杆菌的敏感性就很高，此阶段免疫记忆还没有完全发育成熟，不可能形成较长时间的免疫保护，鱼体的抗病力主要依赖非特异性免疫反应。柱状黄杆菌抗血清的被动免疫实验表明，抗血清仅对斑点叉尾鮰幼鱼有部分保护作用，这表明除了抗体的反应，细胞介导的免疫反应可能在柱形病的获得性免疫方面起更重要作用（Shelby 等，2007）。Leal 等（2010）也有相似的发现，尼罗罗非鱼幼鱼注射接种柱状黄杆菌疫苗后，抗体滴度与疾病保护之间无相关性。与成鱼相比，柱状黄杆菌更易感染幼鱼，因此注射接种不是控制柱形病的最佳免疫策略，这就有必要大力研发更具实用性的疫苗。尽管银鲑口服柱状黄杆菌疫苗可以产生一定的保护作用（Fujihara 等，1971；Ransom，1975），但由于抗原可能在胃肠道中降解，这种接种途径可能没有效果。在少数报道的预防柱形病的口服疫苗实验中，Leal 等（2010）将菌苗包被在藻酸盐微粒中以保护抗原不受降解，但经过免疫，对尼罗罗非鱼并没有产生保护效果。

用福尔马林灭活的菌苗进行浸泡免疫，效果也不理想（Moore 等，1990；Mano 等，1996），因此，有人认为甲醛会破坏细菌重要的抗原，不适合用于制备灭活柱状黄杆菌疫苗（Bader 等，1997）。尽管如此，在北美和智利仍有一些商品化的疫苗预防柱形病，在美国和加拿大，不含佐剂的浸泡疫苗 Fryvacc 1 可用于鲑科鱼类的免疫接种；在智利除了 Fryvacc 1，还有柱状黄杆菌和鲁克氏耶尔森菌的二联疫苗 Fryvacc 2，自 1999 年以来，一直用于鲑科鱼类的浸泡免疫接种（Bravo 等，2007）。

为了克服疫苗制备和疫苗接种过程中的抗原破坏等相关问题，研发和使用减毒浸泡活疫苗已变得更有吸引力，但是活疫苗的生物安全性可能还存在一定的问题。2005 年，美国许可减毒活疫苗 Aquavac-Col® 用于预防鮰柱形病，在执业兽医的指导下，也允许用于预防其他鱼类的柱形病（Yanong，2008）。尽管在实验室条件下，疫苗 Aquavac-Col® 已经显示出能预防柱状黄杆菌感染，提高斑点叉尾鮰发眼卵和鱼苗的存活率（Shoemaker 等，2007，2011），但在养殖场环境下的有效性研究仍然有限。然而，在佛罗里达大口黑鲈（*Micropterus salmoides floridanus*）饲料驯化期鱼苗的野外试验中，接种疫苗 Aquavac-Col® 可以显著降低大口黑鲈鱼苗柱形病死亡的风险（Bebak 等，2009），但疫苗 Aquavac-Col® 在鲑科鱼类中的使用效果还不清楚。

鱼类接种疫苗 Aquavac-Col® 产生保护的免疫机制尚不完全清晰。研究证明，活疫苗免疫斑点叉尾鮰后刺激鱼体产生的抗体，可与柱状黄杆菌的脂多糖和蛋白抗原反应（Shoemaker 等，2007）。免疫基因表达研究显示，斑点叉尾鮰接种疫苗 Aquavac-Col® 10 后，共有 28 个基因上调表达，在这些基因中，近一半是免疫应答相关基因，主要参与鱼体感染后的先天性免疫或特异性免疫（Pridgeon 等，2010）。疫苗 Aquavac-Col® 源自一个无毒的抗利福平的柱状黄杆菌基因Ⅰ型突变体，然而已证明，基因Ⅱ型比基因Ⅰ型对斑点叉尾鮰的致病力更强，因此已经开始研制基因Ⅱ型分离株的柱状黄杆菌活疫苗（Olivares-Fuster 等，2011）。

二代重组疫苗避免了使用活菌疫苗可能产生的风险，因此更易被推广应用。到目前为止，一些柱状黄杆菌蛋白已被确定为新型重组疫苗的潜在抗原（Olivares-Fuster 等，2010；Liu 等，2012）。然而，一些蛋白的基因在不同基因型的菌株中序列差异显著，这些差异性可以认为是蛋白免疫原性的差异（Olivares-Fuster 等，2010）。首次（也是迄今唯一）柱形病亚单位疫苗的报道，是用高免疫原性的热休克蛋白 DnaJ 亚单位疫苗对斑点叉尾鮰进行免疫接种，但是并未产生针对柱形病的保护（Olivares-Fuster 等，2010）。由于保护性抗原的鉴定对研发有效的亚单位疫苗至关重要，因此需要深入开展柱状黄杆菌的免疫原性的研究。

第四节　细菌性冷水病（嗜冷黄杆菌 *Flavobacterium psychrophilum*）

嗜冷黄杆菌（Bernardet 等，1996）是细菌性冷水病（BCWD）（Holt 等，1993）的病原，可感染多种冷水鱼类。除了大多数养殖和野生的鲑科鱼类以外，该病原菌还会严重感染多种非鲑科鱼类（Starliper，2011）。自 1948 年在美国首次分离出嗜冷黄杆菌以来，该病原在世界范围内迅速传播，造成全球淡水鲑鱼养殖业的严重经济损失（Barnes 和 Brown，2011；Starliper，2011）。

细菌性冷水病通常在水温低于 15℃ 时暴发，临床症状因鱼体大小而不同。鱼苗感染后，病程发展迅速，有时甚至没有出现临床症状，就已出现全身性急性感染，造成脾脏肥大，引发高的死亡率（Dalsgaard，1993；Nematollahi 等，2003b）。鱼苗中的嗜冷黄杆菌感染曾被称为虹鳟鱼苗综合征（RTFS）（Lorenzen，1994），现在统称为细菌性冷水病。被感染的鱼苗主要症状为嗜睡、贫血、眼球突出、皮肤颜色变深，有时呈现皮肤损伤。幼鱼的临床症状通常是贫血、眼球突出和较大的皮肤坏死损伤与溃疡。鱼种和幼鱼中的嗜冷黄杆菌感染相对较轻，死亡率也比鱼苗低（Dalsgaard，1993；Nematollahi 等，2003b）。

尽管对 BCDW 进行了较多的研究，但仍不清楚嗜冷黄杆菌的侵染机制。从亲鱼到后代的垂直传播方式已有报道（Starliper，2011），但由水体经鳃、皮肤和/或消化道感染鱼体的水平传播方式可能是该病暴发的主要原因（Dalsgaard，1993；Starliper，2011）。

2007 年，致病性嗜冷黄杆菌分离株的全基因组序列测序完成，同时报道了几种潜在的编码毒力因子的基因（Duchaud 等，2007）。关于致病机制也有一些相关报道，例如一些外分泌蛋白酶可以降解宿主组织（Bertolini 等，1994）利于菌体获取营养（Pérez-Pascual 等，2011）；溶血素对细胞的破坏也利于菌体的营养获得（Högfors-Rönnholm 等，2010b）；菌体分泌的黏附素对宿主组织（Kondo 等，2002；Nematollahi 等，2003a）和细胞（Møller 等，2003；Wiklund 等，2003；Högfors-Rönnholm 等，2010b）的黏附以及生物膜的形成有重要作用（Álvarez 等，2006）。由于嗜冷黄杆菌既有抗原变异（Wakabayashi 等，1994；Lorenzen 等，1997）又有形态变异（Högfors-Rönnholm 等，2010a），有可能表现出不同的毒力机制，这对确定该病原体的毒力机制较为困难。研究表明，尽管所有的嗜冷黄杆菌分离株具有一些相同的抗原，但分为 3 种（Wakabayashi 等，1994；Lorenzen 等，1997）或者 7 种（Mata 等，2002）血清型。虽然两种形态变异体具有相似的代谢特征并且对虹鳟都具有毒性，但它们的生理特征、酶活力、血凝活力以及溶血活力（Högfors-Rönnholm 等，2010a，2010b）均不同。因此，对于一种有效预防嗜冷黄杆菌的疫苗来说，可能需要来自不同血清型和变异型的抗原。

到目前为止，还没有预防嗜冷黄杆菌的商品通用疫苗。其中一个主要的问题是，嗜冷黄杆菌在鱼苗中感染最为严重。新孵化的鱼苗获得性免疫系统尚未成熟，注射接种既不可行也不经济。因此，考虑到被感染的鱼类和鱼体大小，理想的疫苗接种方法是口服或浸泡免疫。另一个问题是缺乏自然的和可重复的攻毒模型用于测试疫苗的功效，在香鱼（*Plecoglossus altivelis*）中建立起一种浸泡攻毒模型（Kondo 等，2003）；但对于鲑科鱼类，只有注射嗜冷黄杆菌的攻毒模型才能引起鱼的死亡（Lorenzen 等，2010）。

尽管如此，疫苗接种实验已经证明，虹鳟（LaFrentz 等，2002；Madetoja 等，2006；Högfors 等，2008）和香鱼（Rahman 等，2000，2003）腹腔注射含佐剂或不含佐剂的嗜冷黄杆菌全菌灭活疫苗后，可以产生特异性抗体反应和良好的针对嗜冷黄杆菌感染的免疫保护。虽然受免鱼产生的特异性抗体与嗜冷黄杆菌的免疫保护性相关，但要全面了解保护鱼体免受 BCWD 感染，仍需进一步研究免疫机制。

到目前为止，通过浸浴或浸泡免疫预防嗜冷黄杆菌的研究报道还较少。使用全菌灭活制剂浸泡免疫虹鳟未能产生特异性抗体反应，也未使鱼体产生对嗜冷黄杆菌感染的保护（LaFrentz 等，2002；Lorenzen 等，2010)。使用嗜冷黄杆菌抗原制剂对香鱼（Kondo 等，2003）和虹鳟（Aoki 等，2007）进行口服免疫的实验效果较好，两种鱼在免疫后浸泡攻毒的存活率都较高，但是抗原制备和攻毒菌株都是从银鲑中分离出来的嗜冷黄杆菌 NCIMB 1947^T，该菌株已被证明对虹鳟没有感染性（Madsen 等，2000)。

因为嗜冷黄杆菌生长缓慢，因此全菌疫苗对于商品化应用来说可能不是最经济的选择。嗜冷黄杆菌疫苗研究的重点聚焦于鉴定菌体表面的蛋白质或肽段分子，作为重组亚单位疫苗的抗原（LaFrentz 等，2004；Crump 等，2005；Gliniewicz 等，2012)。在无可用的油类佐剂的情况下，在鱼类生长的早期阶段，亚单位疫苗通过口服或浸泡接种方式具有应用的潜力。迄今为止，大多数亚单位疫苗的接种免疫试验，在没有使用常规佐剂的情况下，对嗜冷黄杆菌感染的保护率都较低（Rahman 等，2000；Högfors 等，2008；Plant 等，2011)。Crump 等（2007）给虹鳟腹腔注射重组蛋白配以佐剂，然后用嗜冷黄杆菌进行注射攻毒，可获得 82% 的相对保护率。LaFrentz 等（2004）观察到，虹鳟腹腔注射 70～100 ku 的蛋白片段的佐剂疫苗，然后皮下注射嗜冷黄杆菌攻毒，产生的特异性抗体效价和存活率都较高。虹鳟的疫苗接种实验中，Plant 等（2009）从 70～100 ku 的片段中提取出两种高免疫原性的热休克蛋白——Hsp60 和 Hsp70，腹腔注射免疫虹鳟后，诱导了抗体的产生，但无论是单独还是混合免疫，都未能使虹鳟获得对嗜冷黄杆菌感染/攻毒的保护效果。克隆这两种热休克蛋白基因作为 DNA 疫苗肌肉注射免疫虹鳟，Hsp60 和 Hsp70 都产生保护效果，但只检测到较低效价的特异性抗体（Plant 等，2009)。另一项虹鳟免疫实验中，测试了其他三种重组高分子蛋白抗原（延伸因子 Tu、SufB Fe-S 组装蛋白以及 ATP 合酶 β)，这些抗原也未能诱导抗体的产生，也无对 BCWD 的保护效果（Plant 等，2011)。

嗜冷黄杆菌的外膜蛋白片段是亚单位疫苗的较好的候选抗原（Rahman 等，2002；Dumetz 等，2006，2007)。Rahman 等（2002）指出，鱼体腹腔注射不加佐剂的外膜蛋白片段，嗜冷黄杆菌攻毒后，虹鳟幼鱼（93%～95%）和香鱼（64%～71%）的存活率都很高。虹鳟腹腔注射免疫 OmpH 样蛋白（富含 P18 片段）配以油类佐剂，嗜冷黄杆菌攻毒后，虹鳟的存活率也较高（89%），且存活率与特异性血清抗体效价密切相关。虽然单独的富含 P18 片段免疫接种并没有显著降低虹鳟死亡率，但是虹鳟体内的特异性抗体滴度明显增加（Dumetz 等，2006），这些研究结果均有助于浸泡疫苗的研发。

预防嗜冷黄杆菌感染的减毒活疫苗的研究也有报道（Álvarez 等，2008；LaFrentz 等，2008）。Álvarez 等（2008）发现，虹鳟肌肉注射因缺铁而存在生长缺陷的嗜冷黄杆菌突变体，经具有侵染性的本源嗜冷黄杆菌菌株攻毒后，虹鳟的存活率可以达 82%。虹鳟腹腔注射利福平减毒的嗜冷黄杆菌疫苗（2010 年获得专利）免疫实验中，经本源性毒力菌株注射攻毒，虹鳟的存活率仅为 45%，但虹鳟的特异性抗体滴度明显提高（LaFrentz 等，2008）。在相应的浸泡免疫实验中，免疫前去除或不去除脂鳍，虹鳟的存活率分别为 28% 和 45%（LaFrentz 等，2008）。Lorenzen 等（2010）发现，虹鳟鱼苗浸泡野生型强毒嗜冷黄杆菌菌株并没有导致仔鱼死亡，并且之后注射攻毒嗜冷黄杆菌，虹鳟的存活率也较高（60%～88%），虽然存活率升高以及特异性抗体滴度增加，但强毒嗜冷黄杆菌野生型在野外条件下用于疫苗接种并不可行。这些研究结果表明嗜冷黄杆菌浸泡活疫苗用于鱼苗具有应用潜力，因为毒力衰减的菌体可以刺激鱼体的免疫系统而不会致病。

如今，虽然没有商品化的嗜冷黄杆菌疫苗，但是一些国家已经使用自体疫苗，使该病的暴发明显减少（Fredriksen 等，2013）。预防嗜冷黄杆菌的自体疫苗已在智利（Bravo 等，2007）、挪威（Johansen 等，2011）以及英国养殖的鲑鳟中应用，且疫苗接种大大减少了这些国家的鲑和虹鳟孵化场中 BCDW 的暴发（Wallis 等，2009）。

第五节　屈挠杆菌病（海洋屈挠杆菌 *Tenacibaculum maritimum*）

海洋屈挠杆菌（*Tenacibaculum maritimum*，Suzuki 等，2001），曾称为 *Cytophaga marina*、*Flexibacter marinus* 和 *F. maritimus*，是溯河产卵和海产鱼类屈挠杆菌病的病原。由海洋屈挠杆菌引起的疾病有几个不同的命名，比如"黑斑坏死病（black patch necrosis）""海水柱形病（saltwater columnaris disease）""海洋屈挠杆菌病 marine flexibacteriosis""口蚀综合征（eroded mouth syndrome）""海鱼滑菌病（gliding bacterial disease of sea fish）"（Bernardet 等，1990；Santos 等，1999）。鱼类被感染后的主要症状有皮肤溃疡、侵蚀性皮肤损伤、鳞片脱落、鱼鳍破损、坏死性鳃病变、口腔病变（口腔腐烂）以及偶发性角膜病变和眼破裂（Chen 等，1995；Handlinger 等，1997；Ostland 等，1999）。从感染鱼体的溃疡和病灶处（Chen 等，1995；Handlinger 等，1997；Ostland 等，1999）可以分离出该病原菌，该病原菌可入侵内脏各器官，导致全身性感染（Alsina 等，1993；Romalde 等，2005）。海洋屈挠杆菌没有严格的宿主特异性，已从澳大利亚、欧洲、日本、北美和南美的多种养殖鱼类中分离到（Avendaño-Herrera 等，2006a；Plumb 等，2011）。

海洋屈挠杆菌中至少有三个主要宿主相关的血清型（O1、O2 和 O3）。自金头鲷（*Sparus aurata*）的分离菌株属于血清型 O1，大菱鲆（*Scophthalmus maximus*）的分离菌株定义为血清型 O2 或 O3，塞内加尔鳎（*Solea senegalensis*）的分离株为血清型 O1 或 O3（Avendaño-Herrera 等，2004；2005；Castro 等，2007）。这些血清型仅限于伊比利亚半岛的分离株，从其他宿主和/或地理区域获得的分离菌株，可能会发现其他血清型（Avendaño-Herrera 等，2006a；van Gelderen 等，2010a）。

海洋屈挠杆菌的感染实验研究结果差异很大，取决于实验鱼的种类、感染途径和感染剂量。腹腔注射和皮下注射海洋屈挠杆菌通常不能引起感染（Wakabayashi 等，1984；Alsina 等，1993；Avendaño-Herrera 等，2006b）。最初，研究测试真鲷（*Pagrus major*）和黑鲷（*Acanthopagrus schlegeli*）的浸泡感染，发现感染效果不稳定（Wakabayashi 等，1984；Baxa 等，1987）。Avendaño-Herrera 等（2006b）报道了一种以海洋屈挠杆菌血清型 O2 和 O3 长时间（18h）浸泡大菱鲆的有效感染模型；Van Gelderen 等（2010b）报道了海洋屈挠杆菌强毒株，1h 攻毒大西洋鲑的可重复感染模型。以上两种模型均可用于疫苗效果的评价。

鱼体接种海洋屈挠杆菌后的抗体反应研究较少。Salati 等（2005）发现，海鲈（*Dicentrarchus labrax*）接种福尔马林灭活的海洋屈挠杆菌菌体和脂多糖合剂，与未接种的鱼相比，抗体反应显著增强。

目前已经对多种鱼类进行了海洋屈挠杆菌的浸泡免疫和注射免疫的研究。与未接种疫苗的对照组相比，福尔马林灭活的浸泡疫苗诱导了真鲷显著的免疫保护作用（$P<0.05$）（Kato 等，2007）。海洋屈挠杆菌疫苗浸泡免疫体重为 1～2 g 的大菱鲆，在体重为 20～30g 时进行注射加强免疫，细菌攻毒后，大菱鲆获得的相对保护率为 85%（Toranzo 等，2009），该疫苗已经申请相关专利。实验室条件下，塞内加尔舌鳎腹腔注射疫苗，保护性良好（相对保护率为 94%）（Romalde 等，2005）。此外，研究了养殖牙鲆（*Paralichthys olivaceus*）多联疫苗注射免疫的效果，该疫苗包含海洋屈挠杆菌在内的 5 种全菌抗原，海洋屈挠杆菌攻毒后，相对保护率为 85%（Kang，2011）。早在 20 世纪 90 年代，大西洋鲑屈挠杆菌病疫苗就已经有报道，但其结果与现在的不一致（Carson 等，1994）。在目前的研究中，大西洋鲑腹腔注射含佐剂的海洋屈挠杆菌疫苗后，鱼体获得了有效的保护（相对保护率为 80%）（Van Gelderen 等，2009）。

海洋屈挠杆菌是多种海洋鱼类养殖的潜在制约因素（Santos 等，1999），由于分离病原比较困难，屈挠杆菌病对鱼类养殖业的危害性可能被低估（Labrie 等，2005）。但有迹象表明，优化养殖条件和管理措施可以减少海洋屈挠杆菌对水产养殖业的不利影响（J. Carson，个人交流）。

参考文献

Alsina，M. and Blanch，A. R. 1993. First isolation of *Flexibacter maritimus* from cultivated turbot（*Scophthalmus maximus*）. *Bull Eur Assoc Fish Pathol* 13：157-60.

Álvarez，B.，Secades，P.，Prieto，M.，*et al.*，2006. A mutation in *Flavobacterium psychrophilum tlpB* inhibits gliding motility and induces biofilm formation. *Appl Environ Microbiol* 72：4044-4053.

Álvarez，B.，Álvarez，J.，Menéndez，A. and Guijarro，J. A. 2008. A mutant in one of two *exbD* loci of a TonB system in *Flavobacterium psychrophilum* shows attenuated virulence and confers protection against cold water disease. *Microbiology* 154：1144-1151.

Anacker，R. L. and Ordal，E. J. 1959. Studies on the myxobacterium *Chondrococcus columnaris*. I. Serological *typing*. *J Bacteriol* 78：25-32.

Aoki，M.，Kondo，M.，Nakatsuka，Y.，*et al.*，2007. Stationary phase culture supernatant containing membrane vesicles induced immunity to rainbow trout *Oncorhynchus mykiss* fry syndrome. *Vaccine* 25：561-569.

Avendaño-Herrera，R.，Magariños，B.，López-Romalde，S.，*et al.*，2004. Phenotypic characterization and description of two major O-serotypes in *Tenacibaculum maritimum* strains from marine fishes. *Dis Aquat Org* 58：1-8.

Avendaño-Herrera，R.，Magariños，B.，Moriñigo，M. A.，*et al.*，2005. A novel O-serotype in *Tenacibaculum maritimum* strains isolated from cultured sole（*Solea senegalensis*）. *Bull Eur Assoc Fish Pathol* 25：70-74.

Avendaño-Herrera，R.，Toranzo，A. E. and Magariños，B. 2006a. Tenacibaculosis infection in marine fish caused by *Tenacibaculum maritimum*：a review. *Dis Aquat Org* 71：255-266.

Avendaño-Herrera，R.，Toranzo，A. E. and Magariños，B. 2006b. A challenge model for *Tenacibaculum maritimum* infection in turbot，*Scophthalmus maximus*（L.）*J Fish Dis* 29：371-374.

Avendaño-Herrera，R.，Gherardelli，V.，Olmos，P.，*et al.*，2011. *Flavobacterium columnare* associated with mortality of salmonids farmed in Chile：a case report of two outbreaks. *Bull Eur Assoc Fish Pathol* 31：36-44.

Bader，J. A.，Klesius，P. H. and Vinitnantharat，S. 1997. Comparison of whole-cell antigens of pressure and formalin-killed *Flexibacter columnaris* from channel catfish（*Ictalurus punctatus*）. *Am J Vet Res* 58：185-188.

Barnes，M. E. and Brown，M. L. 2011. A review of *Flavobacterium psychrophilum* biology，clinical signs，and bacterial cold water disease prevention and treatment. *Open Fish Sci J* 4：40-48.

Baxa，D. V.，Kawai，K. and Kusuda，R. 1987. Experimental infection of *Flexibacter maritimus* in black sea bream（*Acanthopagrus schlegeli*）fry. *Fish Pathol* 22：105-109.

Bebak，J.，Matthews，M. and Shoemaker，C. 2009. Survival of vaccinated，feed-trained largemouth bass fry（*Micropterus salmoides floridanus*）during natural exposure to *Flavobacterium columnare*. *Vaccine* 27：4297-4301.

Bernardet，J.-F. 1997. Immunization with bacterial antigens. *Flavobacterium* and *Flexibacter* infections. *Biol Stand*（*Basel*）90：179-188.

Bernardet，J.-F. and Bowman，J. P. 2006. The genus *Flavobacterium*，in *The Prokaryotes*（eds M. Dworkin，S. Falkow，E. Rosenberg，*et al.*，），Vol. 7，3rd edn. New York，Springer，481-531.

Bernardet，J.-F.，Campbell，A. C. and Buswell，J. A. 1990. *Flexibacter maritimus* is the agent of "black patchnecrosis" in Dover sole in Scotland. *Dis Aquat Org* 8：233-237.

Bernardet，J.-F.，Segers，P.，Vancanneyt，M.，*et al.*，1996. Cutting a Gordian knot：emended classification and description of the genus *Flavobacterium*，emended description of the family *Flavobacteriaceae*，and proposal of *Flavobacterium hydatis* nom. nov.（Basonym，*Cytophaga aquatilis* Strohl and Tait 1978）. *Int J Syst Bacteriol* 46：128-148.

Bertolini，J. M.，Wakabayashi，H.，Watral，V. G.，*et al.*，1994. Electrophoretic detection of proteases from selected strains of *Flexibacter psychrophilus* and assessment of their variability. *J Aquat Anim Health* 6：224-233.

Bravo，S. and Midtlyng，P. J. 2007. The use of fish vaccines in the Chilean salmon industry 1999—2003. *Aquaculture* 270：36-42.

Byrne，P.，Ferguson，H. W.，Lumsden，J. S. and Ostland，V. E. 1991. Blood chemistry of bacterial gill disease in brook trout *Salvelinus fontinalis*. *Dis Aquat Org* 10：1-6.

Carson，J.，Schmidtke，L. and Lewis，T. 1994. Development of a vaccine against disease：results of efficacy testing of three types of vaccine，in *Proceedings of the SALTAS Research Review Seminar*（ed. P. Valentine）. Hobart，Tasmania，SALTAS，149-158.

Castro, N., Magariños, B., Núñez, S. and Toranzo, A. E. 2007. Reassessment of the *Tenacibaculum maritimum* serotypes causing mortalities in cultured marine fish. *Bull Eur Assoc Fish Pathol* 27: 229-233.

Chen, M. F., Henry-Ford, D. and Groff, J. M. 1995. Isolation and characterization of *Flexibacter maritimus* from marine fishes of California. *J Aquat Anim Health* 7: 318-326.

Crump, E. M., Burian, J., Allen, P. D. and Kay, W. W. 2005. Identification and expression of a host-recognized antigen, FspA, from *Flavobacterium psychrophilum*. *Microbiology* 151: 3127-3135.

Crump, E. M., Burian, J., Allen, P. D., *et al.*, 2007. Identification of a ribosomal L10-like protein from *Flavobacterium psychrophilum* as a recombinant vaccine candidate for rainbow trout fry syndrome. *J Mol Microbiol Biotechnol* 13: 55-64.

Dalsgaard, I. 1993. Virulence mechanisms in *Cytophaga psychrophila* and other *Cytophaga*-like bacteria pathogenic for fish. *Annu Rev Fish Dis* 3: 127-144.

De Beer, H., Hugo, C. J., Jooste, P. J., *et al.*, 2006. *Chryseobacterium piscium* sp. nov., isolated from fish of the South Atlantic Ocean off South Africa. *Int J Syst Evol Microbiol* 56: 1317-1322.

Decostere, A., Haesebrouck, F. and Devriese, L. A. 1998. Characterization of four *Flavobacterium columnare* (*Flexibacter columnaris*) strains isolated from tropical fish. *Vet Microbiol* 62: 35-45.

Duchaud, E., Boussaha, M., Loux, V., *et al.*, 2007. Complete genome sequence of the fish pathogen *Flavobacterium psychrophilum*. *Nat Biotechnol* 25: 763-769.

Dumetz, F., Duchaud, E., LaPatra, S. E., *et al.*, 2006. A protective immune response is generated in rainbow trout by an OmpH-like surface antigen (P18) of *Flavobacterium psychrophilum*. *Appl Environ Microbiol* 72: 4845-4852.

Dumetz, F., LaPatra, S. E., Duchaud, E., *et al.*, 2007. The *Flavobacterium psychrophilum* OmpA, an outer membrane glycoprotein, induces humoral response in rainbow trout. *J Appl Microbiol* 103: 1461-1470.

Farkas, J. 1985. Filamentous *Flavobacterium* sp. isolated from fish with gill diseases in cold water. *Aquaculture* 44: 1-10.

Farkas, J. and Oláh, J. 1986. Gill necrosis-a complex disease of carp. *Aquaculture* 58: 17-26.

Ferguson, H. W., Ostland, V. E., Byrne, P. and Lumsden, J. S. 1991. Experimental production of bacterial gill disease in trout by horizontal transmission and by bath challenge. *J Aquat Anim Health* 3: 118-123.

Flemming, L., Rawlings, D. and Chenia, H. 2007. Phenotypic and molecular characterisation of fish-borne *Flavobacterium johnsoniae*-like isolates from aquaculture systems in South Africa. *Res Microbiol* 158: 18-30.

Fredriksen, N. B., Olsen, R. H., Furevik, A., *et al.*, 2013. Efficacy of a divalent and a multivalent water-in-oil formulated vaccine against a highly virulent strain of *Flavobacterium psychrophilum* after intramuscular challenge of rainbow trout (*Oncorhynchus mykiss*). *Vaccine* 31: 1994-1998.

Fujihara, M. P. and Nakatani, R. E. 1971. Antibody production and immune responses of rainbow trout and coho salmon to *Chondrococcus columnaris*. *J Fish Res Board Can* 28: 1253-1258.

Gliniewicz, K., LaPatra, S. E., LaFrentz, B. R., *et al.*, 2012. Comparative proteomic analysis of virulent and rifampicin-attenuated *Flavobacterium psychrophilum*. *J Fish Dis* 35: 529-539.

Grabowski, L. D., LaPatra, S. E. and Cain, K. D. 2004. Systemic and mucosal antibody response in tilapia, *Oreochromis niloticus* (L.) following immunization with *Flavobacterium columnare*. *J Fish Dis* 27: 573-581.

Handlinger, J., Soltani, M. and Percival, S. 1997. The pathology of *Flexibacter maritimus* in aquaculture species in Tasmania, Australia. *J Fish Dis* 20: 159-168.

Hansen, G. H., Bergh, Ø., Michaelsen, J. and Knappskog, D. 1992. *Flexibacter ovolyticus* sp. nov., a pathogen of eggs and larvae of Atlantic halibut, *Hippoglossus hippoglossus* L. *Int J Syst Bacteriol* 42: 451-458.

Heo, G.-J., Kasai, K. and Wakabayashi, H. 1990a. Occurrence of *Flavobacterium branchiophila* associated with bacterial gill disease at a trout hatchery. *Fish Pathol* 25: 99-105.

Heo, G.-J., Wakabayashi, H. and Watabe, S. 1990b. Purification and characterization of pili from *Flavobacterium branchiophila*. *Fish Pathol* 25: 21-27.

Högfors, E., Pullinen, K. R., Madetoja, J. and Wiklund, T. 2008. Immunization of rainbow trout, *Oncorhynchus mykiss* (Walbaum), with a low molecular mass fraction isolated from *Flavobacterium psychrophilum*. *J Fish Dis* 31: 899-911.

Högfors-Rönnholm, E. and Wiklund, T. 2010a. Phase variation in *Flavobacterium psychrophilum*: characterization of two distinct colony phenotypes. *Dis Aquat Org* 90: 43-53.

Högfors-Rönnholm, E. and Wiklund, T. 2010b. Hemolytic activity in *Flavobacterium psychrophilum* is a contact-dependent two-step mechanism and differently expressed in smooth and rough phenotypes. *Microb Pathogen* 49: 369-375.

Holt, R. A., Rohovec, J. S. and Fryer, J. L. 1993. Bacterial cold-water disease, in Bacterial Diseases of Fish (eds V. Inglis, R. J. Roberts, N. R. Bromage). Oxford, Blackwell Scientific Publications, 3-23.

Huh, G.-J. and Wakabayashi, H. 1989. Serological characteristics of *Flavobacterium branchiophila* isolated from gill diseases of freshwater fishes in Japan, USA and Hungary. *J Aquat Anim Health* 1: 142-147.

Ilardi, P., Fernández, J. and Avendaño-Herrera, R. 2009. *Chryseobacterium piscicola* sp. nov., isolated from diseased salmonid fish. *Int J Syst Evol Microbiol* 59: 3001-3005.

Johansen, L. H., Jensen, I., Mikkelsen, H., *et al.*, 2011. Disease interaction and pathogens exchange between wild and farmed fish populations with special reference to Norway. *Aquaculture* 315: 167-186.

Kämpfer, P., Lodders, N., Martin, K. and Avendaño-Herrera, R. 2012. *Flavobacterium chilense* sp. nov. and *Flavobacterium araucananum* sp. nov., isolated from farmed salmonid fish. *Int J Syst Evol Microbiol* 62: 1402-1408.

Kang, B. 2011. Efficacy of *Edwardsiella tarda*, *Streptococcus iniae*, *Streptococcus parauberis*, *Vibrio anguillarum*, and *Tenacibaculum maritimum*. International Conference & Exhibition on Vaccines & Vaccination. Philadelphia, US (abstract).

Kato, F., Ishimaru, K., Murata, O and Kumai, H. 2007. The efficacy of immersion-vaccination against gliding bacterial disease in red sea bream *Pagrus major*. *Nippon Suisan Gakkaishi* 73: 51-54.

Klesius, P., Lim, C. and Shoemaker, C. 1999. Effect of feed deprivation on innate resistance and antibody response to *Flavobacterium columnare* in channel catfish, *Ictalurus punctatus*. *Bull Eur Assoc Fish Pathol* 19: 156-158.

Klesius, P. H., Shoemaker, C. A. and Evans, J. J. 2008. *Flavobacterium columnare* chemotaxis to channel catfish mucus. *FEMS Microbiol Lett* 288: 216-220.

Ko, Y.-M. and Heo, G.-J. 1997. Characteristics of *Flavobacterium branchiophilum* isolated from rainbow trout in Korea. *Fish Pathol* 32: 97-102.

Kondo, M., Kawai, K., Kurohara, K. and Oshima, S. 2002. Adherence of *Flavobacterium psychrophilum* on the body surface of the ayu *Plecoglossus altivelis*. *Microbes Infect* 4: 279-283.

Kondo, M., Kawai, K., Okabe, M., *et al.*, 2003. Efficacy of oral vaccine against bacterial coldwater disease in ayu *Plecoglossus altivelis*. *Dis Aquat Org* 55: 261-264.

Kubilay, A., Altun, S., Diler, Ö. and Ekici, S. 2008. Isolation of *Flavobacterium columnare* from cultured rainbow trout (*Oncorhynchus mykiss*) fry in Turkey. *Turkish J Fish Aquat Sci* 8: 165-169.

Kunttu, H. M., Suomalainen, L. R., Jokinen, E. I. and Valtonen, E. T. 2009. *Flavobacterium columnare* colony types: connection to adhesion and virulence. *Microb Pathog* 46: 21-27.

Kunttu, H. M., Jokinen, E. I., Valtonen, E. T. and Sundberg, L. R. 2011. Virulent and nonvirulent *Flavobacterium columnare* colony morphologies: characterization of chondroitin AC lyase activity and adhesion to polystyrene. *J Appl Microbiol* 111: 1319-1326.

Labrie, L., Grisez, L., Komar, C. and Tan, Z. 2005. *Tenacibaculum maritimum* an underestimated fish pathogen in Asian marine fish culture, in *International Peace and Development through Aquaculture*. World Aquaculture Society, Bali, Indonesia (abstract).

LaFrentz, B. R., LaPatra, S. E., Jones, G. R., *et al.*, 2002. Characterization of serum and mucosal antibody responses and relative per cent survival in rainbow trout, *Oncorhynchus mykiss* (Walbaum), following immunization and challenge with *Flavobacterium psychrophilum*. *J Fish Dis* 25: 703-713.

LaFrentz, B. R., LaPatra, S. E., Jones, G. R. and Cain, K. D. 2003. Passive immunization of rainbow trout, *Oncorhynchus mykiss* (Walbaum), against *Flavobacterium psychrophilum*, the causative agent of bacterial coldwater disease and rainbow trout fry syndrome. *J Fish Dis* 26: 377-384.

LaFrentz, B. R., LaPatra, S. E., Jones, G. R. and Cain, K. D. 2004. Protective immunity in rainbow trout *Oncorhynchus mykiss* following immunization with distinct molecular mass fractions isolated from *Flavobacterium psychrophilum*. *Dis Aquat Org* 59: 17-26.

LaFrentz, B. R., LaPatra, S. E., Call, D. R. and Cain, K. D. 2008. Isolation of rifampicin resistant *Flavobacterium psychrophilum* strains and their potential as live attenuated vaccine candidates. *Vaccine* 26: 5582-5589.

LaFrentz, B. R., LaPatra, S. E., Shoemaker, C. A. and Klesius, P. H. 2012. Reproducible challenge model to investigate the virulence of *Flavobacterium columnare* genomovars in rainbow trout (*Oncorhynchus mykiss*). *Dis Aquat Org* 101: 115-122.

Leal, C. A. G., Carvalho-Castro, G. A., Sacchetin, P. S. C., *et al.*, 2010. Oral and parenteral vaccines against *Flavobacterium columnare*: evaluation of humoral immune response by ELISA and *in vivo* efficiency in Nile tilapia (*Oreochromis niloticus*). *Aquac Int* 18: 657-666.

Liu, Z. X., Liu, G. Y., Li, N., *et al.*, 2012. Identification of immunogenic proteins of *Flavobacterium columnare* by two-dimensional electrophoresis immunoblotting with antibacterial sera from grass carp, *Ctenopharyngodon idella* (Valenciennes). *J Fish Dis* 35: 255-263.

Lorenzen, E. 1994. *Studies on Flexibacter psychrophilus in relation to rainbow trout fry syndrome*. PhD thesis, Royal Veterinary and Agricultural School, Copenhagen.

Lorenzen, E. and Olesen, N. J. 1997. Characterization of isolates of *Flavobacterium psychrophilum* associated with coldwater disease or rainbow trout fry syndrome II: serological studies. *Dis Aquat Org* 31: 209-220.

Lorenzen, E., Brudeseth, B. E., Wiklund, T. and Lorenzen, N. 2010. Immersion exposure of rainbow trout (*Oncorhynchus mykiss*) fry to wildtype *Flavobacterium psychrophilum* induces no mortality, but protects against later intraperitoneal challenge. *Fish Shellfish Immunol* 28: 440-444.

Lumsden, J. S., Ostland, V. E., Byrne, P. J. and Ferguson, H. W. 1993. Detection of a distinct gill-surface antibody response following horizontal infection and bath challenge of brook trout *Salvelinus fontinalis* with *Flavobacterium branchiophilum*, the causative agent of bacterial gill disease. *Dis Aquat Org* 16: 21-27.

Lumsden, J. S., Ostland, V. E., MacPhee, D. D., *et al.*, 1994. Protection of rainbow trout from experimentally induced bacterial gill disease caused by *Flavobacterium branchiophilum*. *J Aquatic Anim Health* 6: 292-302.

Lumsden, J. S., Ostland, V. E., MacPhee, D. D. and Ferguson, H. W. 1995. Production of gill-associated and serum antibody by rainbow trout (*Oncorhynchus mykiss*) following immersion immunization with acetone-killed *Flavobacterium branchiophilum* and the relationship to protection from experimental challenge. *Fish Shellfish Immunol* 5：151-165.

Madetoja，J.，Lönnström，L. G.，Björkblom，C.，*et al.*，2006. Efficacy of injection vaccines against *Flavobacterium psychrophilum* in rainbow trout，*Oncorhynchus mykiss* (Walbaum). *J Fish Dis* 29：9-20.

Madsen，L. and Dalsgaard，I. 2000. Comparative studies of Danish *Flavobacterium psychrophilum* isolates：ribotypes，plasmid profiles，serotypes and virulence. *J Fish Dis* 23：211-218.

Mano，N.，Inui，T.，Arai，D.，*et al.*，1996. Immune response in the skin of eel against *Cytophaga columnaris*. *Fish Pathol* 31：65-70.

Mata，M.，Skarmeta，A. and Santos，Y. 2002. A proposed serotyping system for *Flavobacterium psychrophilum*. *Lett Appl Microbiol* 35：166-170.

Møller，D. M.，Larsen，J. L.，Madsen，L. and Dalsgaard，I. 2003. Involvement of sialic acid-binding lectin with hemagglutination and hydrophobicity of *Flavobacterium psychrophilum*. *Appl Environ Microbiol* 69：5275-5280.

Moore，A. A.，Eimers，M. E. and Cardella，N. A. 1990. Attempts to control *Flexibacter columnaris* epizootics in pond-reared channel catfish by vaccination. *J Aquat Anim Health* 2：109-111.

Moyer，T. R. and Hunnicutt，D. W. 2007. Susceptibility of zebra fish *Danio rerio* to infection by *Flavobacterium columnare* and *F. johnsoniae*. *Dis Aquat Org* 76：39-44.

Mudarris，M.，Austin，B.，Segers，P.，*et al.*，1994. *Flavobacterium scophthalmum* sp. nov.，a pathogen of turbot (*Scophthalmus maximus* L.). *Int J Syst Bacteriol* 44：447-453.

Nematollahi，A.，Decostere，A.，Pasmans，F.，*et al.*，2003a. Adhesion of high and low virulence *Flavobacterium psychrophilum* strains to isolated gill arches of rainbow trout *Oncorhynchus mykiss*. *Dis Aquat Org* 55：101-107.

Nematollahi，A.，Decostere，A.，Pasmans，F. and Haesebrouck，F. 2003b. *Flavobacterium psychrophilum* infections in salmonid fish. *J Fish Dis* 26：563-574.

Noga，E. J. 2010. Problem 37. Columnaris infection (myxobacterial disease，saddleback，fin rot，cotton wool disease，black patch necrosis)，in *Fish Disease：Diagnosis and Treatment*，2nd edn. Ames，Iowa，Wiley-Blackwell，166-169.

Olivares-Fuster，O. and Arias，C. R. 2011. Development and characterization of rifampicin-resistant mutants from high virulent strains of *Flavobacterium columnare*. *J Fish Dis* 34：385-394.

Olivares-Fuster，O.，Baker，J. L.，Terhune，J. S.，*et al.*，2007. Host-specific association between *Flavobacterium columnare* genomovars and fish species. *Syst Appl Microbiol* 30：624-633.

Olivares-Fuster，O.，Terhune，J. S.，Shoemaker，C. A. and Arias，C. R. 2010. Cloning，expression，and immunogenicity of *Flavobacterium columnare* heat shock protein DnaJ. *J Aquat Anim Health* 22：78-86.

Ostland，V. E.，Lumsden，J. S.，MacPhee，D. D. and Ferguson，H. W. 1994. Characteristics of *Flavobacterium branchiophilum*，the cause of salmonid bacterial gill disease in Ontario. *J Aquat Anim Health* 6：13-26.

Ostland，V. E.，MacPhee，D. D.，Lumsden，J. S. and Ferguson，H. W. 1995. Virulence of *Flavobacterium branchiophilum* in experimentally infected salmonids. *J Fish Dis* 18：249-262.

Ostland，V. E.，Lumsden，J. S.，MacPhee，D. D.，*et al.*，1997. Inhibition of the attachment of *Flavobacterium branchiophilum* to the gills of rainbow trout，*Oncorhynchus mykiss* (Walbaum). *J Fish Dis* 20：109-117.

Ostland，V. E.，LaTrace，C.，Morrison，D. and Ferguson，H. W. 1999. *Flexibacter maritimus* associated with a bacterial stomatitis in Atlantic salmon smolts reared in net-pens in British Columbia. *J Aquat Anim Health* 11：35-44.

Ototake, M. and Wakabayashi, H. 1985. Characteristics of extracellular products of *Flavobacterium* sp., a pathogen of bacterial gill disease. *Fish Pathol* 20: 167-171.

Pérez-Pascual, D., Gómez, E., Álvarez, B., *et al.*, 2011. Comparative analysis and mutation effects of fpp2-fpp1 tandem genes encoding proteolytic extracellular enzymes of *Flavobacterium psychrophilum*. *Microbiology* 157: 1196-1204.

Plant, K. P., LaPatra, S. E. and Cain, K. D. 2009. Vaccination of rainbow trout, *Oncorhynchus mykiss* (Walbaum), with recombinant and DNA vaccines produced to *Flavobacterium psychrophilum* heat shock proteins 60 and 70. *J Fish Dis* 32: 521-534.

Plant, K. P., LaPatra, S. E., Call, D. R. and Cain, K. D. 2011. Immunization of rainbow trout, *Oncorhynchus mykiss* (Walbaum), with *Flavobacterium psychrophilum* proteins elongation factor-Tu, SufB Fe-S assembly protein and ATP synthaseβ. *J Fish Dis* 34: 247-250.

Plumb, J. A. and Hanson, L. A. 2011. *Health Maintenance and Principal Microbial Diseases of Cultured Fishes* (3rd edn). Ames, Iowa, Wiley-Blackwell.

Pridgeon, J. W. and Klesius, P. H. 2010. Identification and expression profile of multiple genes in channel catfish fry 10 min after modified live *Flavobacterium columnare* vaccination. *Vet Immunol Immunopathol* 138: 25-33.

Pulkkinen, K., Suomalainen, L. R., Read, A. F., *et al.*, 2010. Intensive fish farming and the evolution of pathogen virulence: the case of columnaris disease in Finland. *Proc Biol Sci* 277: 593-600.

Rahman, M. H., Ototake, M., Iida, Y., *et al.*, 2000. Efficacy of oil-adjuvanted vaccine for coldwater disease in ayu *Plecoglossus altivelis*. *Fish Pathol* 35: 199-203.

Rahman, M. H., Kuroda, A., Dijkstra, J. M., *et al.*, 2002. The outer membrane fraction of *Flavobacterium psychrophilum* induces protective immunity in rainbow trout and ayu. *Fish Shellfish Immunol* 12: 169-179.

Rahman, M. H., Ototake, M. and Nakanishi, T. 2003. Water-soluble adjuvants enhance the protective effect of *Flavobacterium psychrophilum* vaccines in ayu *Plecoglossus altivelis*. *Fish Pathol* 38: 171-176.

Ransom, D. P. 1975. *Immune responses of salmonids: A) Oral immunization against Flexibacter columnaris. B) Effects of combining antigens in parenterally administered polyvalent vaccines*. MA thesis, Oregon State University, Corvallis.

Romalde, J. L., Ravelo, C., López-Romalde, S., *et al.*, 2005. Vaccination strategies to prevent emerging diseases for Spanish aquaculture. *Dev Biol (Basel)* 121: 85-95.

Salati, F., Cubadda, C., Viale, I. and Kusuda, R. 2005. Immune response of sea bass *Dicentrarchus labrax* to *Tenacibaculum maritimum* antigens. *Fish Sci* 71: 563-567.

Santos, Y., Pazos, F. and Barja, J. L. 1999. *Flexibacter maritimus*, causal agent of flexibacteriosis in marine fish. *ICES identification leaflets for diseases and parasites of fish and shellfish*. Copenhagen, ICES, Leaflet No. 55. Schachte, J. H. Jr and Mora, E. C. 1973. Production of agglutinating antibodies in the channel catfish (*Ictalurus punctatus*) against *Chondrococcus columnaris*. *J Fish Res Board Can* 30: 116-118.

Shelby, R. A., Shoemaker, C. A. andKlesius, P. H. 2007. Passive immunization of channel catfish *Ictalurus punctatus* with anti-*Flavobacterium columnare* sera. *Dis Aquat Org* 77: 143-147.

Shoemaker, C. A., Klesius, P. H., Lim, C. and Yildirim, M. 2003. Feed deprivation of channel catfish, *Ictalurus punctatus* (Rafinesque), influences organosomatic indices, chemical composition and susceptibility to *Flavobacterium columnare*. *J Fish Dis* 26: 553-561.

Shoemaker, C. A., Xu, D., Shelby, R. A. and Klesius, P. H. 2005. Detection of cutaneous antibodies against *Flavobacterium columnare* in channel catfish, *Ictalurus punctatus* (Rafinesque). *Aquac Res* 36: 813-818.

Shoemaker, C. A., Klesius, P. H. and Evans, J. J. 2007. Immunization of eyed channel catfish, *Ictalurus punctatus*, eggs with monovalent *Flavobacterium columnare* vaccine and bivalent *F. columnare* and *Edwardsiella ictaluri* vaccine. *Vaccine* 25: 1126-1131.

Shoemaker, C. A., Klesius, P. H., Drennan, J. D. and Evans, J. J. 2011. Efficacy of a modified live *Flavobacterium columnare* vaccine in fish. *Fish Shellfish Immunol* 30: 304-308.

Song, Y. L., Fryer, J. L. and Rohovec, J. S. 1988. Comparison of gliding bacteria isolated from fish in North America and other areas of the Pacific Rim. *Fish Pathol* 23: 197-202.

Speare, D. J., Ferguson, H. W., Beamish, F. W. M., *et al.*, 1991. Pathology of bacterial gill disease: sequential development of lesions during natural outbreaks of disease. *J Fish Dis* 14: 21-32.

Speare, D. J., Markham, R. J. F., Despres, B., *et al.*, 1995. Examination of gills from salmonids with bacterial gill disease using monoclonal antibody probes for *Flavobacterium branchiophilum* and *Cytophaga columnaris*. *J Vet Diagn Invest* 7: 500-505.

Starliper, C. E. 2011. Bacterial coldwater disease of fishes caused by *Flavobacterium psychrophilum*. *J Adv Res* 2: 97-108.

Starliper, C. E. and Schill, W. B. 2011. Flavobacterial diseases: Columnaris disease, coldwater disease and bacterial gill disease, in *Fish Diseases and Disorders*, *Vol*. 3: *Viral*, *Bacterial and Fungal Infections* (eds P. T. K. Woo and D. W. Bruno). 2nd edn. Wallingford, CABI, 606-631.

Suomalainen, L. R., Tiirola, M. and Valtonen, E. T. 2005. Influence of rearing conditions on *Flavobacterium columnare* infection of rainbow trout, *Oncorhynchus mykiss* (Walbaum). *J Fish Dis* 28: 271-277.

Suomalainen, L. R., Tiirola, M. and Valtonen, E. T. 2006. Chondroitin AC lyase activity is related to virulence in fish pathogenic *Flavobacterium columnare*. *J Fish Dis* 29: 757-763.

Suzuki, M., Nakagawa, Y., Harayama, S. and Yamamoto, S. 2001. Phylogenetic analysis and taxonomic study of marine *Cytophaga*-like bacteria: proposal for *Tenacibaculum* gen. nov. with *Tenacibaculum maritimum* comb. nov. and *Tenacibaculum ovolyticum* comb. nov., and description of *Tenacibaculum mesophilum* sp. nov. and *Tenacibaculum amylolyticum* sp. nov. *Int J Syst Evol Microbiol* 51: 1639-1652.

Thomas-Jinu, S. and Goodwin, A. E. 2004. Morphologic and genetic characteristics of *Flavobacterium columnare* isolates: Correlations with virulence in fish. *J Fish Dis* 27: 29-35.

Toranzo, A. E., Romalde, J. L., Magariños, B. and Barja, J. L. 2009. Present and future of aquaculture vaccines against fish bacterial diseases, in *The Use of Veterinary Drugs and Vaccines in Mediterranean Aquaculture* (eds C. Rodgers and B. Basurko). Op Méditerr, A/No 86, 155-176.

Triyanto, K. and Wakabayashi, H. 1999. Genotypic diversity of strains of *Flavobacterium columnare* from diseased fishes. *Fish Pathol* 34: 65-71.

Vandamme, P., Bernardet, J.-F., Segers, P., *et al.*, 1994. New perspectives in the classification of the Flavobacteria: description of *Chryseobacterium* gen. nov., *Bergeyella* gen. nov., and *Empedobacter* nom. rev. *Int J Syst Bacteriol* 44: 827-831.

Van Gelderen, R., Carson, J. and Nowak, B. 2009. Experimental vaccination of Atlantic salmon (*Salmo salar* L.) against marine flexibacteriosis. *Aquaculture* 288: 7-13.

Van Gelderen, R., Carson, J., Gudkovs, N. and Nowak, B. 2010a. Physical characterisation of *Tenacibaculum maritimum* for vaccine development. *J Appl Microbiol* 109: 1668-1676.

Van Gelderen, R., Carson, J. and Nowak, B. 2010b. Experimentally induced marine flexibacteriosis in Atlantic salmon smolts *Salmo salar*. I. Pathogenicity. *Dis Aquat Org* 91: 121-128.

Wagner, B. A., Wise, D. J., Khoo, L. H. and Terhune, J. S. 2002. The epidemiology of bacterial diseases in food-size channel catfish. *J Aquat Anim Health* 14: 263-272.

Wakabayashi, H. and Iwado, T. 1985. Changes in glycogen, pyruvate and lactate in rainbow trout with bacterial gill disease. *Fish Pathol* 20: 161-165.

Wakabayashi, H., Egusa, S. and Fryer, J. L. 1980. Characteristics of filamentous bacteria isolated from a gill disease of salmonids. *Can J Fish Aquat Sci* 37: 1499-1504.

Wakabayashi, H., Hikida, M. and Masumura, K. 1984. *Flexibacter* infection in cultured marine fish in Japan. *Helgoländer Meeresunters* 37: 587-593.

Wakabayashi, H., Huh, G. J. and Kimura, N. 1989. *Flavobacterium branchiophila* sp. nov., a causative agent of bacterial gill disease of freshwater fishes. *Int J System Bacteriol* 39: 213-216.

Wakabayashi, H., Toyama, T. and Iida, T. 1994. A study on serotyping of *Cytophaga psychrophila* isolated from fishes in Japan. *Fish Pathol* 29: 101-104.

Wallis, T., Dalsgaard, I., Hopewell, R. and Kardos, G. 2009. The potential of autogenous vaccines based on *Flavobacterium psychrophilum* for the control of rainbow trout fry syndrome and red mark syndrome in salmonids: Typing of *F. psychrophilum* isolates from salmon and trout in the UK. *Proceedings of the 2nd Conference on Members of the Genus Flavobacterium*, MGEN, Paris.

Welker, T. L., Shoemaker, C. A., Arias, C. R. and Klesius, P. H. 2005. Transmission and detection of *Flavobacterium columnare* in channel catfish, *Ictalurus punctatus*. *Dis Aquat Org* 63: 129-138.

Wiklund, T. and Dalsgaard, I. 2003. Association of *Flavobacterium psychrophilum* with rainbow trout (*Oncorhynchus mykiss*) kidney phagocytosis *in vitro*. *Fish Shellfish Immunol* 15: 387-395.

Yanong, R. P. E. 2008. *Use of vaccines in finfish aquaculture*. Gainesville: University of Florida. Available at http: //edis. ifas. ufl. edu/fa156.

Zamora, L., Fernández-Garayzábal, J. F., Svensson-Stadler, L. A., *et al.*, 2012. *Flavobacterium oncorhynchi* sp. nov., new species isolated from rainbow trout (*Oncorhynchus mykiss*). *System Appl Microbiol* 35: 86-91.

第二十四章　病毒性出血性败血症和传染性造血器官坏死症的疫苗接种

Stéphane Biacchesi and Michel Brémont

本章概要　SUMMARY

病毒性出血性败血症（VHS）和传染性造血器官坏死症（IHN）是感染性极强的病毒性疾病，主要引起鲑幼鱼（但不仅限于幼鱼）的高死亡率。疫苗研发工作尽管已经进行了40多年，但目前还没有针对这两种疾病的大规模疫苗接种。基于灭活病毒、重组G蛋白、减毒毒株和DNA载体的几种候选疫苗，能够有效地诱导鱼体产生持久的特异性免疫保护。最近，一种传染性造血器官坏死病毒（IHNV）的DNA疫苗已在加拿大获得许可，并在大西洋鲑中进行商业化应用。然而，该疫苗在欧洲的商品化并没有得到批准，并且也缺乏鱼苗的大规模疫苗接种系统。与此同时，反向遗传学的发展使研发多价减毒活疫苗成为可能，但是，需要进行深入的研究来提高此类疫苗的安全性。

第一节　发生流行与危害

病毒性出血性败血症（VHS）和传染性造血器官坏死症（IHN）是感染性极强的病毒性疾病，导致世界范围内养殖和野生鱼类幼鱼的高死亡率，被世界动物卫生组织（OIE）列为必须上报的病毒性疾病（Skall等，2005；Walker和Winton，2010）。VHS和IHN是OIE收录的7种鱼类病毒病中的2种，它们严重危害鱼类养殖生产，影响社会经济发展（OIE，2013），可对自然资源和濒危鱼种造成巨大威胁。在2001年至2004年期间，IHN给加拿大水产养殖业造成了2亿多加元（CAD）的经济损失（Salonius等，2007）。

有关VHS和IHN在欧洲的流行病学数据显示，受VHS和IHN感染影响的养殖场分别有36%和38%，但存在严重的漏报现象（Olesen和Nicolajsen，2011）。对于这两种疾病而言，传染源是疫情暴发受感染后存活下来的养殖或野生鱼类，这些鱼都可能成为长期的病毒携带者

(Bootland 和 Leong，2011)。病毒传播途径主要为水平传播，即通过患病鱼释放病毒粒子污染水体引起鱼类患病，因此，所有患病群体都必须彻底销毁。此外，碘伏溶液对受精卵表面的消毒可减少病毒的垂直传播（OIE，2013)。生活在传染性造血器官坏死病毒（IHNV）和病毒性出血性败血症病毒（VHSV）流行区域的鱼类，可能同时感染这两种病毒（Brudeseth 等，2002；Biacchesi 等，2010)，这与共栖感染试验中观察到的结果相似。由于这两种病毒之间不存在交叉保护，因此，研发同时预防 VHSV 和 IHNV 的高效疫苗非常重要（Hattenberger-Baudouy 等，1995)。在丹麦，通过一个长期的监测和控制政策，已经彻底清除了 VHS，并且之后两年多再没暴发新的 VHS 疫情（Mellergaard，2011)。而其他国家在没有任何治疗措施的情况下，则是通过预防病原输入和对患病群体的彻底销毁来控制 VHSV 和 IHNV。

病毒性出血性败血症是由病毒性出血性败血症病毒（VHSV，又称艾特韦病毒）引起养殖虹鳟（*Oncorhynchus mykiss*)、大菱鲆（*Scophthalmus maximus*)、牙鲆（*Paralichthys olivaceus*）等多种淡水和海水鱼类的主要疾病。迄今为止，VHSV 已经从北半球，包括北美、亚洲和欧洲，多达 82 种不同的淡水和海水物种中分离到（Skall 等，2005；Jonstrup 等，2009；Kim 和 Faisal，2011；OIE，2013)。自 1938 年以来，普遍认为 VHSV 引起的疾病仅对欧洲大陆的虹鳟养殖业构成威胁（Wolf，1988；Bootland 和 Leong，2011)，但在 20 世纪 80 年代以后，VHSV 已蔓延到整个北半球，并对许多淡水和海水鱼类造成巨大威胁（Skall 等，2005)。在越来越多的野生无症状海水鱼类中发现 VHSV，这意味着该病毒存在更加广泛的宿主范围，这给 VHS 的防控带来新的挑战，并可能导致养殖鱼类中出现高毒性的新型毒株（Snow，2011；Kurath 和 Winton，2011)。

有两个案例有力地证明了上述情况。2007 年，挪威某虹鳟海水养殖场暴发了 VHS 疫情，分离出 VHSV 毒株受到质疑（Dale 等，2009)，因为海水分离株通常不会感染虹鳟，但这个分离株与海水分离株密切相关，这意味着 VHSV 具有很强的适应性。21 世纪初，北美五大湖地区出现了鱼群大规模死亡现象，涉及至少 31 种淡水鱼类（Kim 和 Faisal，2011)，这是证明该病毒具有较强适应性的另一个案例。事实上，2003 年从圣克莱尔湖捕获的梭鱼（*Esox masquinongy*）中第一次分离到了 VHSV（Elsayed 等，2006)，此后 VHSV 就在五大湖传播，此前在该区域没有发现弹状病毒。

传染性造血器官坏死症是一种由传染性造血器官坏死病毒（IHNV）引起的病毒性疾病，主要感染鲑科鱼类（OIE，2013)。IHNV 主要发生在太平洋的鲑科鱼类，并造成损失，包括红大麻哈鱼（*Oncorhynchus nerka*)、大鳞大麻哈鱼（*O. tshawytscha*)、细鳞大麻哈鱼（*O. gurbuscha*)、大麻哈鱼（*O. keta*)、马苏大麻哈鱼（*O. masou*)、大西洋鲑（*Salmo salar*）和虹鳟。20 世纪 50 年代，从北美西部地区养殖场的濒死红大麻哈鱼上首次分离出 IHNV，20 世纪 70 年代，IHNV 开始肆虐并成为养殖虹鳟中一种危害严重的病毒。早在 20 世纪 50 年代到 60 年代，美国西部的鲑孵化场广泛使用未处理的野杂鱼投喂鱼苗，以及用常规方式进行鱼苗的运输，导致 IHNV 在该地迅速传播。起初，其他地区并无 IHNV，人们认为该病毒是通过鱼和/或鱼卵运输传播到欧洲和亚洲的（Wolf，1988；Bootland 和 Leong，2011)。迄今为止，在北半球，被 IHNV 感染或在实验条件下易受感染的淡水和海水鱼类多达 23 种（OIE，2013)。从野生和养殖的海水及淡水鱼类中均分离到 IHNV，这表明 IHNV 在感染周期内对宿主种类和宿主环境具有高度适应性。此外，非鲑科的海水和淡水鱼类可能是潜在的病毒源，在 IHNV 及其宿主的生命周期中起重要作用。事实上，溯河产卵鱼类成年后返回产卵水域时，最初体内不携带病毒，随着生长到产卵年龄，它们的患病率随之增加。但是，不能排除鱼体本身可能携带少量 IHNV，由于溯河时造成的应激使得鱼体自身免疫受到抑制，进而病毒扩增并产生传染性 (Meyers，1998；Kim 等，1999)。

第二节　病原学

传染性造血器官坏死病毒和病毒性出血性败血症病毒均属于单股负链 RNA 病毒目 (Mononegavirales)、弹状病毒科（Rhabdoviridae）（Hoffmann 等，2005；Tordo 等，2005)。病毒粒子均具有囊膜，呈典型的弹状（长约 180 nm，直径约 70 nm)，基因组由一个大约 11 000 bp 不分节段的单股负链 RNA 组成，以 3′-N-P-M-G-NV-L-5′ 为顺序编码 6 种蛋白质（Morzunov 等，1995；Schutze 等，1995，1999)。病毒 RNA 被核蛋白（N)、聚合酶相关磷酸蛋白（P）以及 RNA 依赖的 RNA 聚合酶（L）紧密包裹，形成螺旋状核糖核蛋白复合物（RNP)；基质蛋白 (M）与 RNP 及病毒囊膜相互作用，共同参与病毒的出芽阶段；最后，病毒表面糖蛋白（G）介导病毒入胞，因此 G 蛋白是诱导中和抗体和保护性抗体产生的唯一靶蛋白（Lorenzen 等，1990，2000；Bearzotti 等，1995)。与其他弹状病毒相比，IHNV 和 VHSV 基因组具有一个位于 *G* 和 *L* 基因之间的编码非病毒结构蛋白 *NV* 基因（Kurath 等，1985；Kurath 和 Leong，1985；Schutze 等，1996)。由于存在 *NV* 基因，IHNV 和 VHSV 被归为粒外弹状病毒属（*Novirhabdovirus*）（Tordo 等，2005)。反向遗传学的研究证明，NV 蛋白不是病毒的结构蛋白，但对病毒的胞内有效复制必不可少（Biacchesi 等，2000，2010；Ammayappan 等，2011；Kim 等，2011)，在虹鳟（Thoulouze 等，2004；Harmache 等，2006a)、黄鲈（*Perca flavescens*）（Ammayappan 等，2011）和牙鲆（Kim 和 Kim，2011）上的研究表明，NV 蛋白对该病毒的致病性至关重要。

对从世界范围内众多养殖和野生鱼类体内分离到的 VHSV 和 IHNV，开展了大量的分子流行病学研究，结果有助于对 VHSV 和 IHNV 分离株进行基因分型，从而更好地了解病毒的自然分布、起源和宿主范围（Bandin 和 Dopazo，2011；Snow，2011)。迄今为止，根据 *N* 基因和 *G* 基因的序列，已经鉴定出 4 种 VHSV 的主要基因型（Ⅰ～Ⅳ）（Einer-Jensen 等，2004；Snow 等，2004)。系统发育分析显示，VHSV 主要是在地理分布上存在差异，并不是在宿主上存在差异。此外，源于海洋鱼类的 VHSV 分离株，对大菱鲆具有一定的致病性，但对虹鳟和大西洋鲑的致病力较低。基于全基因组的序列分析表明，源自海洋和淡水的 VHSV 分离株对虹鳟的致病能力各不相同，这可能只与各分离株之间少数氨基酸残基的差异相关（Betts 和 Stone，2000；Campbell 等，2009)。有趣的是，分子生物学分析表明，VHSV 起源于海水环境，病毒在传播过程中，主要是从海水鱼类到鳟，可能发生过几次物种跨越（Kurath 和 Winton，2011)。目前，VHSV 分离株是分为 3 个血清型，还是 1 个血清型内的 3 个亚型尚无定论（Olesen 等，1993)，因此，虽然 VHSV 的血清型与基因型并不相关，但不同分离株具有相同的几个抗原表位。关于 IHNV，根据几株北美分离株的 *G* 基因部分核苷酸序列的系统发育分析发现，IHNV 主要有3种基因型（U、M、L）（Kurath等，2003），但只有1种主要的血清型（Engelking等，1991）；欧洲分离株似乎仅具有单一起源，与M基因型密切相关（Enzmann等，2005）；相比之下，亚洲分离株与U基因型密切相关，但在某种程度上已趋于独立分化（Nishizawa等，2006）。虽然，某些U基因型和M基因型的分离株在宿主特异性上存在差异，如U基因型感染红鲑而不感染虹鳟，M基因型感染虹鳟而不感染红鲑（Penaranda等，2009，2011b），但分离株之间的系统发育关系主要与地理分布相关。

第三节 致病机理

传染性造血器官坏死症和VHS是急性全身性疾病（Wolf，1988；Bootland和Leong，2011），任何年龄段的鱼都容易感染，其中鱼苗和稚鱼最为敏感。死亡率因各种环境因素和生理因素有所不同，如鱼的年龄、饲养条件、应激、鱼的种类、病毒株和水温等。水温10℃左右时，发病最为严重，而高于15℃时，几乎不发病。病毒通过污染的水体直接接触感染鱼，或投喂污染的饲料传播。利用非侵入性生物发光成像设备和表达荧光素酶的重组病毒证明，鱼鳍是IHNV入侵虹鳟的途径（Harmache等，2006a）。经口传播也已被证实是IHNV和VHSV的另一种感染途径（Helmick等，1995；Schonherz等，2012）。实验条件下，VHSV或IHNV感染虹鳟鱼苗（体重0.3～2 g），4～6d后鱼苗迅速发病，15～20d后的累积死亡率达80%～100%。VHS和IHN的临床症状和病灶没有明显的典型特征，与其他常见的全身性病毒病基本相同（Wolf，1988）。感染鱼首先出现嗜睡，随后剧烈游动并伴随异常的游泳行为，如螺旋式游动。由于渗透压平衡受损，病鱼表现为眼球突出并伴随贫血，如鳃丝颜色变浅、苍白。眼周组织、骨骼肌和内脏有明显的点状出血。在急性期，肾脏呈暗红色，脾脏肿大，肝脏苍白。幼鱼肾脏和脾脏的造血组织感染最为严重，最早表现出大面积坏死。

病毒性出血性败血症可分为急性和慢性两种类型。急性型的特点是死亡率高，前肾和脾脏的内皮细胞是病毒的主要靶细胞；肝血窦充血，肝细胞发生局灶性到大面积的变性和坏死；脾脏以及肾脏的造血组织和肾单位也有类似的病变，此外，VHSV在肾脏和脾脏中的滴度较高。相比之下，慢性型的特点是死亡率低且持续时间长，VHSV在脑中的滴度较高。

传染性造血器官坏死症也可以分为两种类型：一种是“造血”型，另一种是“嗜神经”型（LaPatra等，1995）。患有“造血”型IHN的虹鳟表现出典型的疾病症状是病毒在肾脏和脾脏中的滴度最高；组织病理学检查显示，肾脏的黑色素巨噬细胞、脾脏红髓和肝实质等造血组织严重坏死，肠黏膜下层嗜酸性粒细胞坏死被认为是IHN的特征性病症，胰腺的腺泡细胞以及胰岛细胞的细胞核内和细胞质内偶尔会观察到包涵体。与之相反，患有“嗜神经”型IHN的鱼表现出行为异常，死亡率低且持续时间长，且IHNV在脑中的滴度最高。虽然有报道称糖蛋白G的突变可能是产生“嗜神经”型IHN的主要原因，但这种病毒趋向性改变的机制尚不清楚（Kim等，1994）。

第四节 疫　苗

尽管开展VHSV和IHNV的疫苗研发已有40多年，最近也有IHNV的一种DNA疫苗在加拿大获批用于大西洋鲑，但目前还没有大规模接种这两种病毒的疫苗。研发的灭活病毒、原核和真核系统表达的重组 G 蛋白、减毒病毒株和 DNA 载体等候选疫苗均能有效诱导鱼体产生持久的特异性免疫保护，但没有一种候选疫苗符合以下全部要求：具有良好的免疫原性、能诱导产生免疫保护、对多种鱼类有效、生产简单、成本较低、安全并且便于在幼鱼中大规模接种。以下将详述 VHSV 和 IHNV 疫苗的研发进展。

一、非复制型疫苗

最初，用灭活的全病毒制备预防 VHSV 和 IHNV 的疫苗。通过化学或物理处理很容易实现病毒灭活，最关键是在不改变病毒粒子的免疫原性和抗原性的前提下，实现病毒的大规模生产和完全灭活（无残余感染性）。研究表明，灭活病毒疫苗，尤其是由 β-丙内酯处理的灭活病毒疫苗，腹腔注射接种后能够有效诱导鱼体产生长期保护（Amend，1976；de Kinkelin 等，1995；Anderson 等，2008）。相比之下，疫苗通过浸泡免疫产生的保护效果并不明显。因此，非复制型疫苗不适合在幼鱼中大规模使用，但仍是高经济价值鱼类的潜在疫苗，比如对转移到开放水域养殖前的鲑实施疫苗接种。此外，即使不添加佐剂，灭活病毒疫苗也能够对鱼体产生高效的保护，还能减少疫苗的副作用（Anderson 等，2008）。

二、亚单位疫苗

糖蛋白 G 是保护性抗体和中和抗体的唯一靶蛋白，重组亚单位疫苗就是基于该蛋白制备的。用几种原核和真核系统表达全部或部分 G 蛋白的研究表明，由大肠杆菌（*Escherichia coli*）（Gilmore 等，1988；Xu 等，1991；Lorenzen 等，1993）、新月柄杆菌（*Caulobacter crescentus*）（Simon 等，2001）、鲁克氏耶尔森菌（*Yersina ruckeri*）（Estepa 等，1994）、酵母（Estepa 等，1994）和杆状病毒/昆虫细胞（Lecocq-Xhonneux 等，1994；Cain 等，1999；Biering 等，2005）表达的 VHSV 和 IHNV 重组 G 蛋白均未能诱导鱼体产生高水平保护，尽管某些情况下可以在被免疫鱼体的血清中检测到中和抗体。重组 G 蛋白只有与佐剂混合后通过腹腔注射免疫时，才能产生较低的保护水平（Biering 等，2005；Lecocq-Xhonneux 等，1994）。同样，根据预测的 IHNV G 蛋白抗原决定簇合成的多肽，其免疫原性也较低（Emmenegger 等，1997）。由于生产和接种成本较高，重组 G 蛋白似乎不适合鱼苗的大规模接种。为了解决这些问题，应用减毒杀鲑气单胞菌（*Aeromonas salmonicida*）表达 VHSV 和 IHNV 的 G 蛋白片段（Noonan 等，1995），结果发现，浸泡或喷雾免疫后，这些 G 蛋白片段为鱼体提供了一定的保护。总之，上述研究表明，表达抗原结构正确的 G 蛋白作为疫苗并能诱导鱼体产生保护性免疫是非常复杂的。

三、DNA 疫苗

Anderson 等（1996）首次报道了成功使用 DNA 疫苗保护虹鳟免受 IHNV 感染。随后，在 VHSV 方面也有类似报道（Lorenzen 等，1998）。这些疫苗由编码 IHNV 或 VHSV G 蛋白的 DNA 质粒组成，DNA 质粒由人类巨细胞病毒强大结构的启动子调控，通过肌肉注射进行免疫接种。自从 DNA 疫苗报道以后，多个实验室开始全面研究这两种病毒的 DNA 疫苗，结果表明，在很多情况下，这些疫苗在有些海水和淡水鱼中都能高效诱导产生具有保护效果的特异性免疫应答

（详见第五章）。此外，VHSV 和 IHNV 的 DNA 疫苗的单剂量同时接种的实验结果表明，鱼体能产生针对这两种病毒的特异性中和抗体（Boudinot 等，1998），从而抵御这两种病毒的共同感染（Einer-Jensen 等，2009）。单次肌肉注射 1～10 ng 的疫苗就可以有效保护虹鳟幼鱼（Corbeil 等，2000b；McLauchlan 等，2003）。生产实际应用中，疫苗剂量在 0.1～1μg 便可对稚鱼（体重介于 1～10g）产生较高水平的保护，而体型较大的个体（体重＞100g）则需要 0.5～1μg。通过基因枪肌内接种疫苗可以诱导幼鱼产生保护性免疫，但腹腔注射接种疫苗提供的保护水平却比较低（Corbeil 等，2000a；McLauchlan 等，2003）。多数研究发现，DNA 疫苗可快速激活鱼体的非特异性免疫，足以在接种后 4～7d 内保护鱼体免受任何病毒的感染（LaPatra 等，2001；Lorenzen 等，2002）。这种早期抗病毒应答是Ⅰ型干扰素信号通路上调表达的结果，随后产生特异性免疫反应，诱导机体产生中和抗体（Purcell 等，2006）。这种特异性免疫反应可为鱼体提供长达 2 年的保护效果（Kurath 等，2006）。最重要的是，DNA 疫苗能够针对异源病毒株产生广泛、有效的交叉保护，但当 *G* 基因与攻毒的毒株基因型不同时，疫苗的效果则有所降低（Lorenzen 等，1999；Penaranda 等，2011a）。

2005 年，由于这些疫苗有一定的功效，预防大西洋鲑 IHNV 的 DNA 疫苗在加拿大获得许可并商业化使用（APEX-IHN；加拿大诺华动物保健公司 Novartis Animal Health Canada Inc.）（详见第五章以及 Salonius 等，2007）。然而，这种疫苗并没有在欧洲获得批准，主要原因是 DNA 疫苗接种的鱼被认为是转基因生物（GMOs），这可能会带来严重的经济影响。此外，DNA 疫苗的某些方面还有待改进，首先，肌肉注射 DNA 疫苗需要大量劳动力，也不可能对幼鱼进行大规模接种，而其他免疫途径，例如皮肤划伤、口服胶囊型 DNA 疫苗、浸泡微球包裹的 DNA 疫苗以及超声加强浸泡免疫等，产生的免疫效果都非常有限（Corbeil 等，2000a；Fernandez-Alonso 等，2001；Adomako 等，2012）。其次，还需要更多的研究来明确不同基因型分离株间的交叉保护水平，以及是否需要利用编码遗传分支较远的 G 蛋白的 DNA 载体混合接种，尽可能诱导广泛的免疫保护。尽管鱼体对 DNA 疫苗既耐受又安全（Kurath 等，2006），但仍需解决一些安全问题，包括质粒 DNA 在免疫动物基因组中的存留和/或整合，人类巨细胞病毒启动子的使用以及 DNA 疫苗在环境中的释放等安全问题。事实上，质粒 DNA 会在体内长期存留（疫苗接种后可达 728d）（Salonius 等，2007；Tonheim 等，2007），但这一问题可以通过使用编码细胞死亡因子的自杀性 DNA 载体，诱导并调控启动子得以解决（Alonso 等，2011）。另外，人类巨细胞病毒启动子也可以替换为鱼源的启动子（Alonso 等，2003；Chico 等，2009）。

四、复制型疫苗

减毒活疫苗最初是通过传统方法制备获得：细胞传代培养（Jørgensen，1982；Ristow 等，2000；Adelmann 等，2008）、中和逃避突变体（Kim 等，1994；Roberti 等，1998 Gaudin 等，1999）和抗热病毒（de Kinkelin 等，1980；Bernard 等，1983）。在这些研究中，浸泡和口服免疫减毒活疫苗可以对鱼体产生高水平的保护，然而，在大多数情况下，鱼体内会残留病毒。反向遗传学的发展使编辑 VHSV 和 IHNV 的 RNA 基因组成为可能，进而实现定点突变，包括核苷酸替换、基因敲除、异源基因交换和插入，从而降低病毒毒力并将其作为基因载体（Biacchesi，2011），该技术降低了常规方法获得减毒病毒的随机性和不确定性。此外，反向遗传学创造的重组病毒可用来制备“DIVA”（区分感染态和免疫态动物）疫苗，这使得在养殖环境中很容易区分免疫的和自然感染的个体（Meeusen 等，2007）。为了获得可以制备安全减毒活疫苗的完全减毒病毒，已开展了多种研究方法的探索。

非结构蛋白基因缺失（Δ*NV*）的重组 IHNV 和 VHSV 在虹鳟（Thoulouze 等，2004；Harmache 等，2006a）、黄鲈（Ammayappan 等，2011）以及牙鲆（Kim 和 Kim，2011）中表现出不可逆的毒力衰减。这些病毒已被高度减毒，当腹腔注射感染鱼体时，引起的死亡率较低（感染 4 周后，累计死亡率为 20%～25%，而野生型病毒的累计死亡率是 90%～100%）。免疫 30d 后用高致病性病毒进行攻毒，Δ*NV* 病毒免疫的鱼体获得了全面保护（Thoulouze 等，2004），但若在免疫后较长时间进行攻毒，仍可以观察到死亡个体（A. Harmache 和 M. Brémont，数据未发表）。Harmache 等用荧光素酶基因替代病毒的 *NV* 基因，并利用非侵入性生物发光成像技术进行研究，结果仅在鱼鳍中发现了少量的病毒复制，病毒可以在鱼鳍内存留 3 周而不进一步扩散（Harmache 等，2006a）。上述结果可以解释 RT-PCR 检测不到鱼体内脏中 IHNV-Δ*NV*，以及为何机体在某种程度上不能产生抗体（Thoulouze 等，2004；Romero 等，2008）。这些结果表明，这种短期保护是鱼鳍内的病毒突变体与野生型病毒间互相干涉的结果。Δ*NV* 病毒也可以被用作疫苗载体（Novoa 等，2006；Romero 等，2008；Kim 和 Kim，2012a；A. Harmache 和 M. Brémont，数据未发表），虽然这些病毒毒力极低，但它们在培养的细胞中仍能进行有效复制，高效表达外源蛋白，并且鱼体在注射后可以诱导产生中和抗体。然而，鱼体浸泡 Δ*NV* 病毒后，该病毒无感染性，因此，不能作为浸泡疫苗载体。

Harmache 等开展了 IHNV 作为疫苗载体预防其他鱼类病毒性疾病的可行性研究（Harmache 等，2006b；A. Harmache 和 M. Brémont，数据未发表）。研究构建了重组 IHNV 载体，其中携带了对鱼类危害最大的 4 种病毒的 3 个外来抗原片段：VHSV 的糖蛋白 G、传染性胰腺坏死病毒的 VP2、昏睡病病毒的 C-E3-E2、传染性鲑贫血性病毒（ISAV）的血凝素-酯酶糖蛋白和融合蛋白。结果发现，虽然重组病毒的基因组大小比野生型病毒增加了 50% 以上（外来序列约 5.5 kb），但仍可高效、稳定地表达这 3 种外来抗原。这种重组 IHNV 在鳟中的毒性很低，引起的累计死亡率只有 10%。然而，存活的鳟在随后的攻毒实验中，不仅可以免受 IHNV 的侵害，还可以抵御其他 3 种病毒的感染。因此，基因组大小的增加能有效降低病毒毒力，并且由于弹状病毒基因组的重组率很低，该载体相对稳定。IHNV 基因组中引入额外的定向突变可以降低病毒毒力，从而使其更加安全，这种突变之前已有研究报道。事实上，IHNV 基因组中存在 4 个随机突变就足以使重组病毒毒力大幅降低。IHNV 的 N 蛋白中替换两个氨基酸（P 蛋白替换一个，M 蛋白替换一个），鳟浸泡免疫该疫苗后累计死亡率低于 10%，并且强力诱导产生抗 IHNV 的特异性抗体（Biacchesi，2002；Romero 等，2005）。

重组病毒是通过不同的、远缘的、抗原性存在差异的弹状病毒独特的表面糖蛋白交换构建的（Biacchesi 等，2002；Kim 和 Kim，2012b）。研究发现，将 IHNV 的 *G* 基因替换为 VHSV 的 *G* 基因（两者 G 蛋白同源性为 51%）或鲤春病毒（SVCV；鱼类疱疹病毒，两者 G 蛋白同源性为 29%）的 *G* 基因，对 IHNV 基因组进行改造，同样，VHSV 的 *G* 基因用牙鲆弹状病毒（另一种鱼类弹状病毒，两者 *G* 基因同源性为 51%）的 *G* 基因替换，成功构建的嵌合病毒与母本病毒一样能在细胞中进行高效的传代与复制。此外，在虹鳟体内，IHNV- Gvhsv 与它的母本

(IHNV 和 VHSV) 毒力相同 (Biacchesi, 2002; Romero 等, 2005)。有趣的是，虽然 SVCV 不能感染虹鳟，但是 IHNV-Gsvcv 浸泡虹鳟后，累积死亡率高达 93% (母本病毒 IHNV 的累计死亡率为 95%)，这说明 SVCV 的 G 蛋白能在 IHNV 中充分发挥功能，并且 SVCV 中应该存在另一种与宿主特异性有关的蛋白 (S. Biacchesi 和 M. Brémont，未发表的数据)。因此，*G* 基因交换构建的嵌合病毒的这些优势，使其能够作为疫苗载体用于再次免疫。

第五节 结　　语

综上所述，尽管在过去的 40 年里发表了大量研究成果，但目前仍没有高免疫原性、能提供长期保护以及易于操作的 VHSV 和 IHNV 疫苗。利用反向遗传学可对 VHSV 和 IHNV 进行基因改造，从而研制有效的多价减毒活疫苗。从成本、保护效果以及接种简便等方面，与灭活疫苗和 DNA 疫苗相比，复制型疫苗可能是最理想的鱼类疫苗之一，并且能够对幼鱼实施大规模接种，弥补了灭活疫苗和 DNA 疫苗的这一不足。然而，仍需更多的研究去除减毒活疫苗的残余毒力并提高病毒的持续性，避免减毒株毒力逆转为野生型。虽然减毒活疫苗在人和动物病毒性疾病预防方面已被证明有效，但监管的限制仍是这类疫苗面临的主要问题 (Lauring 等，2010)。水产养殖的显著特点是池塘或网箱中的养殖群体与野生群体之间没有物理屏障，因此，即使减毒活疫苗对接种鱼体没有致病性，但也必须保证其对水环境中所有物种是安全的。一种解决方法是，在封闭的环境中进行疫苗接种，随后再将免疫的群体转移到开放的水环境中。另外，反向遗传学技术系统可以制备单一复制周期的病毒疫苗载体，如复制子 (Biacchesi，2011)。虽然这类疫苗已被证明用于注射免疫哺乳动物是安全有效的，但在鱼类浸泡免疫中，其免疫效果仍需验证。总之，基因免疫似乎对 VHSV 和 IHNV 的免疫接种都是高效的，但在推广应用之前，仍需对其免疫接种程序进行改进。目前，正在积极开展新的口服和浸泡免疫策略的研究 (Plant 和 Lapatra，2011)，虽然尚未达到肌肉注射 DNA 疫苗的保护水平，但其中一些策略仍有很大的发展潜力。

第六节 致　　谢

作者获得了法国国家农业研究所 (INRA) 的资助，感谢 INRA 鱼类分子病毒学小组 (FMVG) 中所有成员为这项工作所做的贡献。

参考文献

Adelmann, M., Kollner, B., Bergmann, S. M., *et al.*, 2008. Development of an oral vaccine for immunisation of rainbow trout (*Oncorhynchus mykiss*) against viral hemorrhagic septicaemia. *Vaccine* 26: 837-844.

Adomako, M., St-Hilaire, S., Zheng, Y., *et al.*, 2012. Oral DNA vaccination of rainbow trout, *Oncorhynchus mykiss* (Walbaum), against infectious hematopoietic necrosis virus using PLGA [Poly (D, L-Lactic-CoGlycolic Acid)] nanoparticles. *J Fish Dis* 35: 203-214.

Alonso, M., Johnson, M., Simon, B. and Leong, J. A. 2003. A fish specific expression vector containing the interferon regulatory factor 1A (IRF1A) promoter for genetic immunization of fish. *Vaccine* 21: 1591-1600.

Alonso, M., Chiou, P. P. and Leong, J. A. 2011. Development of a suicidal DNA vaccine for infectious hematopoietic necrosis virus (IHNV). *Fish Shellfish Immunol* 30: 815-823.

Amend, D. F. 1976. Prevention and control of viral disease of salmonids. *J Fish Res Board Can* 33: 1059-1066.

Ammayappan, A., Kurath, G., Thompson, T. M. and Vakharia, V. N. 2011. A reverse genetics system for the Great Lakes strain of viral hemorrhagic septicemia virus: the NV gene is required for pathogenicity. *Mar Biotechnol (NY)* 13: 672-683.

Anderson, E. D., Mourich, D. V., Fahrenkrug, S. C., *et al.*, 1996. Genetic immunization of rainbow trout (*Oncorhynchus mykiss*) against infectious hematopoietic necrosis virus. *Mol Mar Biol Biotechnol* 5: 114-122.

Anderson, E., Clouthier, S., Shewmaker, W., *et al.*, 2008. Inactivated infectious hematopoietic necrosis virus (IHNV) vaccines. *J Fish Dis* 31: 729-745.

Bandin, I. and Dopazo, C. P. 2011. Host range, host specificity and hypothesized host shift events among viruses of lower vertebrates. *Vet Res* 42: 67.

Bearzotti, M., Monnier, A. F., Vende, P., *et al.*, 1995. The glycoprotein of viral hemorrhagic septicemia virus (VHSV): antigenicity and role in virulence. *Vet Res* 26: 413-422.

Bernard, J., de Kinkelin, P. and Bearzotti-LeBerre, M. 1983. Viral hemorrhagic septicemia of rainbow trout: relation between the G polypeptide and antibody production in protection of the fish after infection with the F25 attenuated variant. *Infect Immun* 39: 7-14.

Betts, A. M. and Stone, D. M. 2000. Nucleotide sequence analysis of the entire coding regions of virulent and avirulent strains of viral hemorrhagic septicaemia virus. *Virus Genes* 20: 259-262.

Biacchesi, S. 2002. *Recovery of recombinant piscine rhabdovirus by reverse genetics*. Paris XI Orsay, Orsay.

Biacchesi, S. 2011. The reverse genetics applied to fish RNA viruses. *Vet Res* 42: 12.

Biacchesi, S., Thoulouze, M. I., Bearzotti, M., *et al.*, 2000. Recovery of NV knockout infectious hematopoietic necrosis virus expressing foreign genes. *J Virol* 74: 11247-11253.

Biacchesi, S., Bearzotti, M., Bouguyon, E. and Brémont, M. 2002. Heterologous exchanges of the glycoprotein and the matrix protein in a *Novirhabdovirus*. *J Virol* 76: 2881-2819.

Biacchesi, S., Lamoureux, A., Mérour, E., *et al.*, 2010. Limited interference at the early stage of infection between two recombinant novirhabdoviruses: viral hemorrhagic septicemia virus and infectious hematopoietic necrosis virus. *J Virol* 84: 10038-10050.

Biering, E., Villoing, S., Sommerset, I. and Christie, K. E. 2005. Update on viral vaccines for fish. *Dev Biol (Basel)* 121: 97-113.

Bootland, L. M. and Leong, J. C. 2011. Infectious haematopoietic necrosis virus, in *Fish Diseases and Disorders, Vol. 3: Viral, Bacterial and Fungal Infections*, 2nd edn (eds P. T. K. Woo and D. W. Bruno). Wallingford, CABI, 66-109.

Boudinot, P., Blanco, M., de Kinkelin, P. and Benmansour, A. 1998. Combined DNA immunization with the glycoprotein gene of viral hemorrhagic septicemia virus and infectious hematopoietic necrosis virus induces double-specific protective immunity and nonspecific response in rainbow trout. *Virology* 249: 297-306.

Brudeseth, B. E., Castric, J. and Evensen, Ø. 2002. Studies on pathogenesis following single and double infection with viral hemorrhagic septicemia virus and infectious hematopoietic necrosis virus in rainbow trout (*Oncorhynchus mykiss*). *Vet Pathol* 39: 180-189.

Cain, K. D., LaPatra, S. E., Shewmaker, B., *et al.*, 1999. Immunogenicity of a recombinant infectious hematopoietic necrosis virus glycoprotein produced in insect cells. *Dis Aquat Org* 36: 67-72.

Campbell, S., Collet, B., Einer-Jensen, K., *et al.*, 2009. Identifying potential virulence determinants in viral haemorrhagic septicaemia virus (VHSV) for rainbow trout. *Dis Aquat Org* 86: 205-212.

Chico, V., Ortega-Villaizan, M., Falco, A., *et al.*, 2009. The immunogenicity of viral haemorragic septicaemia rhabdovirus (VHSV) DNA vaccines can depend on plasmid regulatory sequences. *Vaccine* 27: 1938-1948.

Corbeil, S., Kurath, G. and LaPatra, S. E. 2000a. Fish DNA vaccine against infectious hematopoietic necrosis virus: efficacy of various routes of immunisation. *Fish Shellfish Immunol* 10: 711-723.

Corbeil, S., LaPatra, S. E., Anderson, E. D. and Kurath, G. 2000b. Nanogram quantities of a DNA vaccine protect rainbow trout fry against heterologous strains of infectious hematopoietic necrosis virus. *Vaccine* 18: 2817-2824.

Dale, O. B., Ørpetveit, I., Lyngstad, T. M., *et al.*, 2009. Outbreak of viral haemorrhagic septicaemia (VHS) in seawater-farmed rainbow trout in Norway caused by VHS virus Genotype III. *Dis Aquat Org* 85: 93-103.

de Kinkelin, P., Bearzotti-LeBerre, M. and Bernard, J. 1980. Viral hemorrhagic septicemia of rainbow trout: selection of a thermoresistant virus variant and comparison of polypeptide synthesis with the wild-type virus strain. *J Virol* 36: 652-658.

de Kinkelin, P., Bearzotti, M., Castric, J., *et al.*, 1995. Eighteen years of vaccination against viral haemorrhagic septicaemia in France. *Vet Res* 26: 379-387.

Einer-Jensen, K., Ahrens, P., Forsberg, R. and Lorenzen, N. 2004. Evolution of the fish rhabdovirus viral haemorrhagic septicaemia virus. *J Gen Virol* 85: 1167-1179.

Einer-Jensen, K., Delgado, L., Lorenzen, E., *et al.*, 2009. Dual DNA vaccination of rainbow trout (*Oncorhynchus mykiss*) against two different rhabdoviruses, VHSV and IHNV, induces specific divalent protection. *Vaccine* 27: 1248-1253.

Elsayed, E., Faisal, M., Thomas, M., *et al.*, 2006. Isolation of viral haemorrhagic septicaemia virus from muskellunge, *Esox masquinongy* (Mitchill), in Lake St Clair, Michigan, USA reveals a new sublineage of the North American genotype. *J Fish Dis* 29: 611-619.

Emmenegger, E., Landolt, M., LaPatra, S. and Winton, J. 1997. Immunogenicity of synthetic peptides representing antigenic determinants on the infectious hematopoietic necrosis virus glycoprotein. *Dis Aquat Org* 28: 175-184.

Engelking, H. M., Harry, J. B. and Leong, J. A. 1991. Comparison of representative strains of infectious hematopoietic necrosis virus by serological neutralization and cross-protection assays. *Appl Environ Microbiol* 57: 1372-1378.

Enzmann, P. J., Kurath, G., Fichtner, D. and Bergmann, S. M. 2005. Infectious hematopoietic necrosis virus: monophyletic origin of European isolates from North American genogroup M. *Dis Aquat Org* 66: 187-195.

Estepa, A., Thiry, M. and Coll, J. M. 1994. Recombinant protein fragments from haemorrhagic septicaemia rhabdovirus stimulate trout leukocyte anamnestic responses *in vitro*. *J Gen Virol* 75: 1329-1338.

Fernandez-Alonso, M., Rocha, A. and Coll, J. M. 2001. DNA vaccination by immersion and ultrasound to trout viral haemorrhagic septicaemia virus. *Vaccine* 19: 3067-3075.

Gaudin, Y., de Kinkelin, P. and Benmansour, A. 1999. Mutations in the glycoprotein of viral haemorrhagic septicaemia virus that affect virulence for fish and the pH threshold for membrane fusion. *J Gen Virol* 80: 1221-1229.

Gilmore, R. D., Engelking, H. M., Manning, D. S. and Leong, J. A. 1988. Expression in *Escherichia coli* of an epitope of the glycoprotein of infectious hematopoietic necrosis virus protects against viral challenge. *Bio/Technology* 6: 295-300.

Harmache, A., LeBerre, M., Droineau, S., *et al.*, 2006a. Bioluminescence imaging of live infected salmonids reveals that the fin bases are the major portal of entry for *Novirhabdovirus*. *J Virol* 80: 3655-3659.

Harmache, A., LeBerre, M., Lamoureux, A. and Bremont, M. 2006b. Recombinant IHNV expressing three foreign expression units an efficient bath immersion vaccine in salmonids, in *XIII International Conference on Negative Strand Viruses*. Salamanca, Spain.

Hattenberger-Baudouy, A. M., Danton, M., Merle, G. and de Kinkelin, P. 1995. Serum neutralization test for epidemiological studies of salmonid rhabdoviroses in France. *Vet Res* 26: 512-520.

Helmick, C. M., Bailey, J. F., LaPatra, S. and Ristow, S. 1995. The esophagus/cardiac stomach region: site of attachment and internalization of infectious hematopoietic necrosis virus in challenged juvenile rainbow trout *Oncorhynchus mykiss* and coho salmon *O. kisutch*. *Dis Aquat Org* 23: 189-199.

Hoffmann, B., Beer, M., Schutze, H. and Mettenleiter, T. C. 2005. Fish rhabdoviruses: molecular epidemiology and evolution. *Curr Top Microbiol Immunol* 292: 81-117.

Jonstrup, S. P., Gray, T., Kahns, S., *et al.*, 2009. FishPathogens. eu/vhsv: a user-friendly viral haemorrhagic septicaemia virus isolate and sequence database. *J Fish Dis* 32: 925-929.

Jørgensen, P. E. V. 1982. Egtved virus: temperature-dependent immune response of trout to infection with low-virulence virus. *J Fish Dis* 5: 47-55.

Kim, C. H., Winton, J. R. and Leong, J. C. 1994. Neutralization-resistant variants of infectious hematopoietic necrosis virus have altered virulence and tissue tropism. *J Virol* 68: 8447-8453.

Kim, C. H., Dummer, D. M., Chiou, P. P. and Leong, J. A. 1999. Truncated particles produced in fish surviving infectious hematopoietic necrosis virus infection: mediators of persistence? *J Virol* 73: 843-849.

Kim, M. S. and Kim, K. H. 2011. Protection of olive flounder, *Paralichthys olivaceus*, against viral hemorrhagic septicemia virus (VHSV) by immunization with NV gene-knockout recombinant VHSV. *Aquaculture* 314: 39-43.

Kim, M. S. and Kim, K. H. 2012a. Effects of NV gene knock-out recombinant viral hemorrhagic septicemia virus (VHSV) on Mx gene expression in *Epithelioma papulosum cyprini* (EPC) cells and olive flounder (*Paralichthys olivaceus*). *Fish Shellfish Immunol* 32: 459-463.

Kim, M. S. and Kim, K. H. 2012b. Generation of recombinant viral hemorrhagic septicemia viruses (VHSVs) harboring G gene of hirame rhabdovirus (HIRRV) and their ability to induce serum neutralization activity in olive flounder (*Paralichthys olivaceus*). *Aquaculture* 330-333: 37-41.

Kim, M. S., Kim, D. S. and Kim, K. H. 2011. Generation and characterization of NV gene-knockout recombinant viral hemorrhagic septicemia virus (VHSV) genotype IVa. *Dis Aquat Org* 97: 25-35.

Kim, R. and Faisal, M. 2011. Emergence and resurgence of the viral hemorrhagic septicemia virus (*Novirhabdovirus*, *Rhabdoviridae*, *Mononegavirales*). *J Adv Res* 2: 9-21.

Kurath, G. and Leong, J. C. 1985. Characterization of infectious hematopoietic necrosis virus mRNA species reveals a nonvirion rhabdovirus protein. *J Virol* 53: 462-468.

Kurath, G. and Winton, J. 2011. Complex dynamics at the interface between wild and domestic viruses of finfish. *Curr Opin Virol* 1: 73-80.

Kurath, G., Ahern, K. G., Pearson, G. D. and Leong, J. C. 1985. Molecular cloning of the six mRNA species of infectious hematopoietic necrosis virus, a fish rhabdovirus, and gene order determination by R-loop mapping. *J Virol* 53: 469-476.

Kurath, G., Garver, K. A., Troyer, R. M., *et al.*, 2003. Phylogeography of infectious haematopoietic necrosis virus in North America. *J Gen Virol* 84: 803-814.

Kurath, G., Garver, K. A., Corbeil, S., *et al.*, 2006. Protective immunity and lack of histopathological damage two years after DNA vaccination against infectious hematopoietic necrosis virus in trout. *Vaccine* 24: 345-354.

LaPatra, S. E., Lauda, K. A., Jones, G. R., *et al.*, 1995. Characterization of IHNV isolates associated with neurotropism. *Vet Res* 26: 433-437.

LaPatra, S. E., Corbeil, S., Jones, G. R., *et al.*, 2001. Protection of rainbow trout against infectious hematopoietic necrosis virus four days after specific or semi-specific DNA vaccination. *Vaccine* 19: 4011-4019.

Lauring, A. S., Jones, J. O. and Andino, R. 2010. Rationalizing the development of live attenuated virus vaccines. *Nat Biotechnol* 28: 573-579.

Lecocq-Xhonneux, F., Thiry, M., Dheur, I., *et al.*, 1994. A recombinant viral haemorrhagic septicaemia virus glycoprotein expressed in insect cells induces protective immunity in rainbow trout. *J Gen Virol* 75: 1579-1587.

Lorenzen, N., Olesen, N. J. and Jørgensen, P. E. 1990. Neutralization of Egtved virus pathogenicity to cell cultures and fish by monoclonal antibodies to the viral G protein. *J Gen Virol* 71: 561-567.

Lorenzen, N., Olesen, N. J., Vestergård-Jørgensen, P. E., *et al.*, 1993. Molecular cloning and expression in *Escherichia coli* of the glycoprotein gene of VHS virus, and immunization of rainbow trout with the recombinant protein. *J Gen Virol* 74: 623-630.

Lorenzen, N., Lorenzen, E., Einer-Jensen, K., *et al.*, 1998. Protective immunity to VHS in rainbow trout (*Oncorhynchus mykiss*, Walbaum) following DNA vaccination. *Fish Shellfish Immunol* 8: 261-270.

Lorenzen, N., Lorenzen, E., Einer-Jensen, K., *et al.*, 1999. Genetic vaccination of rainbow trout against viral haemorrhagic septicaemia virus: small amounts of plasmid DNA protect against a heterologous serotype. *Virus Res* 63: 19-25.

Lorenzen, N., Cupit, P. M., Einer-Jensen, K., *et al.*, 2000. Immunoprophylaxis in fish by injection of mouse antibody genes. *Nat Biotechnol* 18: 1177-1180.

Lorenzen, N., Lorenzen, E., Einer-Jensen, K. and LaPatra, S. E. 2002. Immunity induced shortly after DNA vaccination of rainbow trout against rhabdoviruses protects against heterologous virus but not against bacterial pathogens. *Dev Comp Immunol* 26: 173-179.

McLauchlan, P. E., Collet, B., Ingerslev, E., *et al.*, 2003. DNA vaccination against viral haemorrhagic septicaemia (VHS) in rainbow trout: size, dose, route of injection and duration of protection: early protection correlates with Mx expression. *Fish Shellfish Immunol* 15: 39-50.

Meeusen, E. N., Walker, J., Peters, A., *et al.*, 2007. Current status of veterinary vaccines. *Clin Microbiol Rev* 20: 489-510.

Mellergaard S. 2011. Surveillance and eradication of VHS in Denmark, in *Report on the 15th Annual Meeting of the National Reference Laboratories for Fish Diseases*, Copenhagen: 18. Available at http: //www. crl-fish. eu/upload/sites/eurl-fish/activities/annual _ meetings/15 _ am _ report. pdf.

Meyers, T. R. 1998. Healthy juvenile sockeye salmon reared in virus-free hatchery water return as adults infected with infectious hematopoietic necrosis virus (IHNV): a case report and review of controversial issues in the epizootiology of IHNV. *J Aquat Anim Health* 10: 172-181.

Morzunov, S. P., Winton, J. R. and Nichol, S. T. 1995. The complete genome structure and phylogenetic relationship of infectious hematopoietic necrosis virus. *Virus Res* 38: 175-192.

Nishizawa, T., Kinoshita, S., Kim, W. S., *et al.*, 2006. Nucleotide diversity of Japanese isolates of infectious hematopoietic necrosis virus (IHNV) based on the glycoprotein gene. *Dis Aquat Org* 71: 267-272.

Noonan, B., Enzmann, P. J. and Trust, T. J. 1995. Recombinant infectious hematopoietic necrosis virus and viral hemorrhagic septicemia virus glycoprotein epitopes expressed in *Aeromonas salmonicida* induce protective immunity in rainbow trout (*Oncorhynchus mykiss*). *Appl Environ Microbiol* 61: 3586-3591.

Novoa, B., Romero, A., Mulero, V., *et al.*, 2006. Zebrafish (*Danio rerio*) as a model for the study of vaccination against viral haemorrhagic septicemia virus (VHSV). *Vaccine* 24: 5806-5816.

OIE. 2013. *Manual of diagnostic tests for aquatic animals*. Available at http://www.oie.int/en/internationalstandard-setting/aquatic-manual/access-online/.

Olesen, N. J. and Nicolajsen, N. 2011. Overview of the disease situation and surveillance in Europe in 2010, in *Report on the 15th Annual Meeting of the National Reference Laboratories for Fish Diseases*, Copenhagen: 9-11. Available at http://www.crl-fish.eu/upload/sites/eurl-fish/activities/annual_meetings/15_am_report.pdf.

Olesen, N. J., Lorenzen, N. and Jørgensen, P. E. 1993. Serological differences among isolates of viral haemorrhagic septicaemia virus detected by neutralizing monoclonal and polyclonal antibodies. *Dis Aquat Org* 16: 163-170.

Penaranda, M. M., Purcell, M. K. and Kurath, G. 2009. Differential virulence mechanisms of infectious hematopoietic necrosis virus in rainbow trout (*Oncorhynchus mykiss*) include host entry and virus replication kinetics. *J Gen Virol* 90: 2172-2182.

Penaranda, M. M., LaPatra, S. E. and Kurath, G. 2011a. Specificity of DNA vaccines against the U and M genogroups of infectious hematopoietic necrosis virus (IHNV) in rainbow trout (*Oncorhynchus mykiss*). *Fish Shellfish Immunol* 31: 43-51.

Penaranda, M. M., Wargo, A. R. and Kurath, G. 2011b. *In vivo* fitness correlates with host-specific virulence of *Infectious hematopoietic necrosis virus* (IHNV) in sockeye salmon and rainbow trout. *Virology* 417: 312-319.

Plant, K. P. and LaPatra, S. E. 2011. Advances in fish vaccine delivery. *Dev Comp Immunol* 35: 1256-1262.

Purcell, M. K., Nichols, K. M., Winton, J. R., *et al.*, 2006. Comprehensive gene expression profiling following DNA vaccination of rainbow trout against infectious hematopoietic necrosis virus. *Mol Immunol* 43: 2089-2106.

Ristow, S. S., LaPatra, S. E., Dixon, R., *et al.*, 2000. Responses of cloned rainbow trout *Oncorhynchus mykiss* to an attenuated strain of infectious hematopoietic necrosis virus. *Dis Aquat Org* 42: 163-172.

Roberti, K. A., Rohovec, J. S. and Winton, J. R. 1998. Vaccination of rainbow trout against infectious hematopoietic necrosis (IHN) by using attenuated mutants selected by neutralizing monoclonal antibodies. *J Aquat Anim Health* 10: 328-337.

Romero, A., Figueras, A., Tafalla, C., *et al.*, 2005. Histological, serological and virulence studies on rainbow trout experimentally infected with recombinant infectious hematopoietic necrosis viruses. *Dis Aquat Org* 68: 17-28.

Romero, A., Figueras, A., Thoulouze, M. I., *et al.*, 2008. Recombinant infectious hematopoietic necrosis viruses induce protection for rainbow trout *Oncorhynchus mykiss*. *Dis Aquat Organ* 80: 123-135.

Salonius, K., Simard, N., Harland, R. and Ulmer, J. B. 2007. The road to licensure of a DNA vaccine. *Curr Opin Investig Drugs* 8: 635-641.

Schonherz, A. A., Hansen, M. H., Jørgensen, H. B., *et al.*, 2012. Oral transmission as a route of infection for viral haemorrhagic septicaemia virus in rainbow trout, *Oncorhynchus mykiss* (Walbaum). *J Fish Dis* 35: 395-406.

Schutze, H., Enzmann, P. J., Kuchling, R., *et al.*, 1995. Complete genomic sequence of the fish rhabdovirus infectious haematopoietic necrosis virus. *J Gen Virol* 76: 2519-2527.

Schutze, H., Enzmann, P. J., Mundt, E. and Mettenleiter, T. C. 1996. Identification of the non-virion (NV) protein of fish rhabdoviruses viral haemorrhagic septicaemia virus and infectious haematopoietic necrosis virus. *J Gen Virol* 77: 1259-1263.

Schutze, H., Mundt, E. and Mettenleiter, T. C. 1999. Complete genomic sequence of viral hemorrhagic septicemia virus, a fish rhabdovirus. *Virus Genes* 19: 59-65.

Simon, B., Nomellini, J., Chiou, P., *et al.*, 2001. Recombinant vaccines against infectious hematopoietic necrosis virus: production by the *Caulobacter crescentus* S-layer protein secretion system and evaluation in laboratory trials. *Dis Aquat Org* 44: 17-27.

Skall, H. F., Olesen, N. J. and Mellergaard, S. 2005. Viral haemorrhagic septicaemia virus in marine fish and its implications for fish farming: a review. *J Fish Dis* 28: 509-529.

Snow, M. 2011. The contribution of molecular epidemiology to the understanding and control of viral diseases of salmonid aquaculture. *Vet Res* 42: 56.

Snow, M., Bain, N., Black, J., *et al.*, 2004. Genetic population structure of marine viral haemorrhagic septicaemia virus (VHSV). *Dis Aquat Org* 61: 11-21.

Thoulouze, M. I., Bouguyon, E., Carpentier, C. and Bremont, M. 2004. Essential role of the NV protein of *Novirhabdovirus* for pathogenicity in rainbow trout. *J Virol* 78: 4098-4107.

Tonheim, T. C., Leirvik, J., Løvoll, M., *et al.*, 2007. Detection of supercoiled plasmid DNA and luciferase expression in Atlantic salmon (*Salmo salar* L.) 535 days after injection. *Fish Shellfish Immunol* 23: 867-876.

Tordo, N., Benmansour, A., Calisher, C., *et al.*, 2005. Rhabdovidae, in *Virus Taxonomy: Classification and Nomenclature of Viruses* (eds C. M. Fauquet, M. A. Mayo, J. Maniloff, *et al.*,). 8th Report of the International Committee on the Taxonomy of Viruses. London, Elsevier, 623-644.

Walker, P. J. and Winton, J. R. 2010. Emerging viral diseases of fish and shrimp. *Vet Res* 41: 51.

Wolf, K. 1988. *Fish Viruses and Fish Viral Diseases*. Ithaca, NY, Comstock.

Xu, L., Mourich, D. V., Engelking, H. M., *et al.*, 1991. Epitope mapping and characterization of the infectious hematopoietic necrosis virus glycoprotein, using fusion proteins synthesized in *Escherichia coli*. *J Virol* 65: 1611-1615.

第二十五章　传染性胰脏坏死症的疫苗接种

Espen Rimstad

本章概要　SUMMARY

目前，在鲑养殖的国家中还没有接种传染性胰脏坏死症（IPN）疫苗的成功案例。接种过 IPN 疫苗的养殖场中，疾病暴发仍然普遍存在。起初，IPN 在鱼苗中较为常见，但随着鲑海水养殖模式普及，该病已经出现在鲑生命周期（包括银化降海后）的各个阶段。制备 IPN 疫苗的抗原种类有很多，包括来自细胞培养的病毒，或重组表达的病毒蛋白。虽然有多种口服和浸泡的疫苗接种策略，但 IPN 疫苗通常在鲑降海前通过腹腔注射接种，以保护其海水养成阶段。疫苗接种后，在鱼体内尚未发现与 IPN 保护效果相关的特异性抗体反应。

第一节　发生流行与危害

传染性胰脏坏死病毒（IPNV）于 1957 年从溪红点鲑（*Salvelinus fontinalis*）中分离到，这是首个从鱼类中分离获得的病毒（Wolf 等，1960）。早在 20 世纪 40 年代，就有鲑幼鱼急性黏膜肠炎的报道，后来被称为传染性胰脏坏死症（IPN）（M'Gonigle，1940）。该病是对养殖幼鲑具有高度传染性的病毒性疾病（Wolf 等，1960）。

与 IPNV 抗原性和遗传相似的病毒也从世界各地许多非鲑科鱼类中分离到，但这些患病鱼类的临床症状与 IPN 不完全相同（Castric 等，1987；Nakajima 和 Sorimachi，1994）。例如，日本鳗鲡（*Anguilla japonica*）表现出肾炎和层状鳃（Hsu 等，1993；Lee 等，1999），五条鰤（*Seriola quinqueradiata*）幼鱼表现出腹水和颅内出血（Nakajima 和 Sorimachi，1994），大菱鲆（*Scophthalmus maximus*）表现出造血器官和肾脏坏死（Castric 等，1987）。这些全球性病毒的存在，是鱼类疾病暴发和养殖经济损失的巨大隐患。

鲑 IPN 的临床症状表现为皮肤黑色素沉积、腹部肿胀和螺旋游动（Wolf 等，1960）。根据毒株类型（McAllister 和 Owens，1995）、感染程度（Bebak 和 McAllister，2009）、宿主种类、管理措施和水温等环境因素（Bebak-Williams 等，2002），患病群体的累积死亡率从 10% 至

90%不等。

鲑传染性胰脏坏死症的地理分布很广，常发生于北美洲、南美洲、欧洲、亚洲和大洋洲以鲑养殖为主的大多数国家（Ahne，1980）。IPN 造成的经济损失，还包括该病引发的其他继发性疾病。IPN 暴发过后，存活的鲑可能会携带病毒，造成持续感染（Yamamoto，1975a）。

传染性胰脏坏死病毒可以通过鱼卵、养殖设施、各种动物和水体流通进入育苗场，而病毒通过不同方式入侵造成的影响也有所不同。当大西洋鲑（*Salmo salar*）转移至海水养殖时，不同育苗场的鱼苗混养、转移时的运输方式以及小规格的银化苗都会增加 IPN 暴发的风险（Jarp，1999）。

IPNV 可以通过水体以及鱼和鱼卵的运输进行水平传播。在疾病暴发期间，病毒大量扩散，水平传播是主要的传播方式。亚临床感染的个体可以间歇地释放病毒，这可能是主要的病毒来源（Urquhart 等，2008）。

该病毒还可以通过受精卵从亲代到子代进行垂直传播，这已在溪红点鲑（Wolf 等，1968；Bullock 等，1976）、虹鳟（*Oncorhynchus mykiss*）（Dorson 和 Torchy，1985）以及北极红点鲑（*Salvelinus alpinus*）（Ahne 和 Negele，1985）中得到证实。有研究表明，IPNV 在大西洋鲑中也存在垂直传播（Smail 和 Munro，1989），传播的概率取决于卵巢液或精液中的病毒载量（Bootland 等，1991，1995）。对溪红点鲑的研究发现，卵巢液中分离到 IPNV 的滴度比卵子匀浆液中分离到的病毒滴度高 10^4 $TCID_{50}$/mL（Wolf 等，1968，$TCID_{50}$ 为半数组织培养感染剂量——译者注）。尽管 IPNV 不会感染精子或卵子，但可以附着在精子（Mulcahy 和 Pascho，1984；Dorson 和 Torchy，1985）以及卵膜上（Ahne 和 Negele，1985），从而导致病毒传播，而使用碘伏对卵子表面进行消毒并不能完全阻止 IPNV 的垂直传播（Bullock 等，1976）。

从持续暴发 IPN 的大西洋鲑养殖场附近海域的很多海水鱼类中都分离出了 IPNV（Wallace 等，2008）。由此可见，病毒对不同鱼类的侵染力非常复杂，研究还发现，距暴发地越远，疫情过后的时间越久，感染 IPNV 个体的数量越少（Wallace 等，2008）。

据估计，大西洋鲑对 IPN 的抵抗力存在很大差异，很大程度上取决于宿主的遗传多态性，无论在淡水还是海水阶段，关键的数量性状位点（QTL）都会影响其对 IPN 的抵抗力（Wetten 等，2007；Moen 等，2009；Houston 等，2010）。降低感染应激、减少带毒个体数量以及降低病毒的垂直传播概率，可以达到一定的保护效果。然而，这不是一个关于选择抗病育种还是疫苗接种策略的问题，而是这两种方案都要采取。

第二节　病原学

传染性胰脏坏死病毒是双 RNA 病毒科（Birnaviridae）水生双 RNA 病毒属（*Aquabirnavirus*）的代表种（Delmas 等，2011）。水生双 RNA 病毒分布广泛，通常分离的病毒与疾病并不是直接相关，这就使水生双 RNA 病毒属病毒的命名非常困难。目前，公认的水生双 RNA 病毒属有 3 种：传染性胰脏坏死病毒、樱蛤病毒（Tellina virus）和黄尾鰤腹水病毒（Yellowtail ascites virus），而海洋双 RNA 病毒还未被列为单独的种（Delmas 等，2011）。水生双 RNA 病毒之间亲缘关系非常密切，但具有相当复杂的抗原多样性（Okamoto 等，1983；

Melby 等，1994；Hill 和 Way，1995），中和实验结果显示，两个不同的血清型之间没有任何交叉反应（Underwood 等，1977；Olesen 等，1988）。大多数分离株属于血清型 A，该血清型有 9 种能发生交叉反应，病毒衣壳蛋白（VP2）的氨基酸序列与血清型和地理分布密切相关（Hill 和 Way，1995）。根据核苷酸序列的相似性，水生双 RNA 病毒分为 6 种基因型，随着在世界各地的不同鱼类中发现新的水生双 RNA 病毒，这个数目预计还会增加（Blake 等，2001）。

双 RNA 病毒的基因组由一个双节段（片段 A 和 B）、双链（ds）RNA 组成，中等大小，由单层二十面体的衣壳包裹，IPNV 是一种没有囊膜的裸病毒。

基因组片段 A 包含两个开放阅读框（ORFs），其中较大的开放阅读框编码一个多聚蛋白，其顺序排列为 NH_2-pVP2-VP4-VP3-COOH。在蛋白翻译过程中，多聚蛋白会利用病毒蛋白酶 VP4 在 pVP2-VP4 和 VP4-VP3 连接处发生自我剪切（Duncan 等，1987）。VP2 前体蛋白（pVP2）在羧基末端进行多次剪切，加工修饰形成成熟的 VP2，这一过程非常缓慢，可能由宿主细胞蛋白酶介导完成，而不是 VP4 蛋白的水解（Dobos，1995）。此外，VP2 还可能发生 O 链接的糖基化（Estay 等，1990，Hjalmarsson 等，1999）。VP3 是一种 dsRNA 内部的结合蛋白（Pedersen 等，2007），可诱导细胞凋亡（Chiu 等，2010），空衣壳是没有 RNA 的病毒样粒子（VLPs），不存在 VP3（Dobos，1995），因此，空衣壳中 VP3 的缺失表明 VP2 能够独立形成 VLPs。

基因组片段 A 还包含一个小的 ORF，与多聚蛋白 ORF 的氨基末端重叠。这个小 ORF 编码一种辅助性非结构蛋白 VP5，但其功能目前尚不清楚（Hong 等，2002；Santi 等，2005a，2005b；Chiu 等，2010）。基因组片段 B 编码 VP1，即病毒 RNA 依赖的 RNA 聚合酶（RdRp）（Duncan 等，1991）。

传染性胰脏坏死病毒在水环境中的稳定性对其传播非常重要，一般来说，病毒易于在低温环境中生存。有报道称，在 15℃ 未经处理的淡水、半咸水和海水中，IPNV 的失活率大约为每 3 周 $3\log_{10}$（Toranzo 和 Hetrick，1982），相当于每周减少 90%，IPNV 感染力的持续时间最终取决于当地的实际情况。

第三节 致病机理

起初发现鲑对 IPN 的敏感性与年龄相关（Snieszko 等，1959；Dorson 和 Torchy，1981），通常随着年龄的增长而降低（Wolf 和 Quimby，1971）。然而，随着鲑的海水养殖发展，IPN 不仅在幼鱼阶段发生，也在银化转入海水后发生（Smail 等，1992；Jarp 等，1995；Bowden 等，2002）。水温在 10～16°C 时，鲑的死亡率最高（Dorson 和 Torchy，1981）。对幼鱼进行 IPNV 攻毒后，检测其中和抗体，结果发现感染后存活下来的鱼均未产生对 IPNV 的免疫耐受（Dorson 和 Torchy，1981）。研究表明，无症状感染的鱼通过精液和卵巢液传播病毒（Bootland 等，1995），应激会增加病毒经粪便的释放（Yamamoto，1975b；Reno 等，1978；Yamamoto 和 Kilistoff，1979）。银化后的鲑腹腔注射 IPNV，IPNV 的释放率在攻毒后第 11 天达到峰值，约为 6.8×10^3 $TCID_{50}$/（h·kg）（Urquhart 等，2008）。因此，同一种群或不同种群，如不同育苗场银化后的鲑混合在同一海域，携带 IPNV 的个体对其他鱼都是危险因素。

目前，尚未发现抗 IPNV 抗体对病毒的持续感染具有保护作用，并且中和抗体也不能清除 IPNV（Bootland 等，1991，1995）。在大西洋鲑中，IPNV 携带个体体内的抗 IPNV 抗体水平差异很大，并且组织内的 IPNV 水平和血清中的抗体水平并不相关（Melby 和 Falk，1995）。成年的溪红点鲑接种弗氏完全佐剂的灭活 IPNV 疫苗后，体液免疫反应显著增强，但受免鱼仍然是 IPNV 携带个体，并且鱼体内存在的中和抗体并没有降低病毒感染白细胞的比例（Bootland 等，1995）。

鲑银化后浸浴 4h 感染 IPNV 所需的最小病毒量少于 10^{-1} $TCID_{50}$/mL（Urquhart 等，2008），研究结果表明，体液免疫既不能改变鲑的带毒状态，也不能阻止体内的病毒释放，且 IPNV 感染所需的病毒量很低。

第四节　疫苗与疫苗效果

1995 年，IPNV 的商业疫苗在挪威上市，通常是与细菌疫苗混合的多联制剂。这些疫苗由 IPNV 的多种抗原组分制成，包括免疫原性最强的蛋白 VP2、细胞培养的灭活病毒和利用不同表达系统获得的重组蛋白（rVP2），然而证明 IPNV 疫苗功效的报道较少。目前，有 4 家公司生产用于腹腔注射的油类佐剂 IPNV 疫苗，其中挪威法玛克（Pharmaq AS）、诺华动物保健公司（Novartis Animal Health）和 Centrovet 公司生产的疫苗主要成分是灭活的 IPNV，而 MSD 动物保健公司（MSD Animal Health）生产的疫苗则是大肠杆菌表达的 VP2 蛋白片段。

大多数 IPNV 商品疫苗是用于银化后海水养殖期的鲑。然而，即使疫苗有效，接种疫苗前的幼鱼阶段还是很容易感染 IPN。据记载，挪威 IPN 疫情大多数发生在已接种疫苗的海水养殖阶段的鲑，截至 2010 年，IPN 的暴发仍然没有下降的趋势。2011 年出现的疫情下降趋势可能是由于引入了 QTL 选择育种的新品种鲑（Hjeltnes 等，2012）。由此可见，疫苗在目前养殖条件下（如大西洋鲑的商业化养殖）的效果仍非常有限。

疫苗在实际养殖中效果不佳，还会产生副作用，如腹腔注射油类佐剂，虽然会对某些病原菌起到积极的保护作用，但引起组织粘连和黑化，因此，急需可替代的抗原并探索不同的抗原接种策略。

建立适合 IPN 感染的实验模型存在一定的困难（Bowden 等，2002；Ramstad 等，2007），由于缺乏与野外生产环境效果紧密相关的攻毒模型，因此研发 IPN 高效疫苗较难。

然而，在实验条件下的攻毒模型中，疫苗效果很好。鲑接种含佐剂的灭活 IPNV 疫苗后，可诱导产生中和抗体，但血清中的抗体水平与产生的保护力不一定相关。在养殖实验中，两组接种 IPN 疫苗的大西洋鲑在转入海水 6 周后，在 IPN 暴发时，其相对保护率大致相同，分别为 50.6% 和 53.2%，其中一组针对 IPNV 特异性抗体水平显著升高，而另一组却不显著。

一、抗原疫苗

含有大肠杆菌表达的 IPNV 重组 VP2（rVP2）多价疫苗与油类佐剂混合后免疫大西洋鲑幼鱼，攻毒 12 周后发现，鱼的 IPNV 感染率与对照组相比显著降低（Frost 和 Ness，1997）。使用含有 rVP2 的商品疫苗免疫鲑，随后进行浸泡攻毒和共栖攻毒，结果显示，与不含 rVP2 的疫苗相比，该疫苗产生的保护率约为70%（Ramstad等，2007）。

以大肠杆菌表达的rVP2或从细胞培养的IPNV中纯化的VP2免疫兔，只有后者的血清能够中和IPNV，使病毒不侵染培养的细胞。该实验结果表明，rVP2和天然VP2的空间构象表位存在差异，或者大肠杆菌表达的rVP2缺少一些在诱导中和免疫反应中发挥重要作用的糖蛋白，并且这些糖蛋白可被兔的免疫系统识别（Fridholm等，2007）。

用不同的疫苗（全病毒疫苗、纳米颗粒包裹的全病毒疫苗、大肠杆菌表达的亚单位抗原与铜绿假单胞菌外毒素的假定转位结构域融合制备的疫苗、编码片段A的DNA疫苗）免疫大西洋鲑的比较研究发现，虽然各组间的保护效果差别不大，但全病毒疫苗的保护效果最好（Munang'andu等，2012）。这类实验设计的一个复杂问题是，很难确定每组免疫的抗原都是等量的。例如，全病毒疫苗是以$TCID_{50}$/mL的病毒量测定，而亚单位疫苗则是以g/mL的纯化蛋白量测定。尽管如此，当免疫原和免疫方式相同时，含有VP2蛋白毒性或非毒性功能域的IPNV制备的疫苗，在攻毒实验后具有相同的保护效果，这说明相同血清型的IPNV分离株之间具有较强的交叉保护（Munang'andu等，2012）。

除了前述的大肠杆菌表达系统外，表达VP2的系统还有很多。以干酪乳杆菌（*Lactobacillus casei*）系统表达的VP2-VP3融合蛋白，虹鳟口服免疫后，与对照组相比，体内的病毒载量减少了90%（Zhao等，2012）。利用酵母系统表达VP2产生rVP2-VLPs，将其注射或投喂免疫虹鳟，均产生抗IPNV的抗体，用异源IPNV分离株攻毒，检测鱼体的病毒载量，结果发现，无论是注射免疫还是口服免疫，实验组虹鳟体内的病毒载量明显低于未免疫组或对照组（Allnutt等，2007）。利用杆状病毒表达系统在昆虫细胞或幼虫中表达IPNV的所有结构蛋白，可产生VLPs，用这种VLPs浸泡免疫虹鳟幼鱼，却无保护效果；对银化前的大西洋鲑腹腔注射油类佐剂VLPs，保护效果也较弱（Shivappa等，2005）。利用塞姆利基森林甲病毒表达系统在哺乳动物细胞系中表达IPNV片段A，也可产生VLPs。有研究利用抗IPNV构像表位的单克隆抗体，识别选择表达正确的多聚蛋白作为候选疫苗，但尚无免疫效果的报道（McKenna等，2001）。

二、DNA疫苗

DNA疫苗将抗原的体外表达系统转移到接种疫苗鱼体的细胞上，这种内源表达的抗原会激活宿主整体免疫系统的应答。已发现，DNA疫苗对鲑的弹状病毒病非常有效，能诱导快速、持久的免疫保护，然而，目前IPNV的DNA疫苗还没有产生这么好的效果。虽然，油类佐剂是腹腔注射多联抗原疫苗的必备成分，但不适合DNA疫苗。

质粒表达由巨细胞病毒启动子调控，编码IPNV基因组片段A的多聚蛋白ORF的DNA载体疫苗，肌肉注射免疫大西洋鲑后可获得84%的相对保护率，而对照组的死亡率仅为33%；而仅表达VP2蛋白的DNA疫苗或表达部分VP2蛋白的DNA疫苗均未获得如此的保护效果（Mikalsen等，2004）。

利用海藻酸盐微球包裹编码VP2的DNA疫苗口服免疫褐鳟和虹鳟，免疫15d后，鱼体内均可检测到中和抗体的产生；分别在免疫后第15天和第30天进行攻毒实验，两种鱼均获得约80%的相对保护率，并且实验组中存活的鱼体内病毒载量显著低于对照组（de Las Heras等，

2009，2010)。

第五节 疫苗诱导的免疫反应

硬骨鱼的抗体不存在类型转换（Kaattari 等，2002)，且抗体的亲和力较低。疫苗接种后，尚未发现诱导鱼体产生的特异性抗体滴度或中和抗体与抗 IPN 感染的保护效果存在相关性(Shivappa 等，2005；Cuesta 等，2010)，即体液免疫应答和保护效果无关，但这只是一个大胆的推测。另外，体液免疫保护的缺乏也可能与病毒增殖的特性有关。研究表明，体液免疫不足以防止 IPNV 引发疾病，但体液免疫能多大程度上预防 IPNV 的感染，以及减少病毒在宿主体内的扩散仍然不清楚，因此，以体液免疫测定评估 IPN 疫苗效果就受到限制。完善评价指标，如抵抗病毒感染的能力（相对保护率)、组织病理学变化或减少 IPNV 载量的能力等，将能更好地判定疫苗的保护效果。理想的疫苗应该可以预防感染的发生，然而 IPN 疫苗的保护效果很大程度上取决于实验鱼类的遗传特性（Ramstad 和 Midtlyng，2008)。此外，毒株致病力的差异并不是大西洋鲑在养殖环境中暴发疾病差异的主要原因，两项研究结果表明，大西洋鲑银化入海后的疾病暴发在很大程度上由非病毒因素引起（Julin 等，2013)。

最近在减毒、可复制的亚病毒粒子上的研究表明，接种虹鳟幼鱼后，同时检测中和抗体的滴度和非特异性细胞毒性的活力，比仅检测体液免疫反应能更好地评估疫苗效果（Rivas-Aravena 等，2012)。

第六节 管理条例

总的来说，鱼类 DNA 疫苗监管的相关问题有很多，但这些问题不仅仅局限于 IPN 的 DNA 疫苗。从本质上讲，DNA 疫苗带来的风险不比目前使用的鱼类疫苗更大，但主要涉及 DNA 疫苗在鱼体内的持续性，长期的环境和食品安全问题，以及消费者对其接受的程度（Lorenzen 和 LaPatra，2005)。

同样，目前普遍关注的环境安全问题也阻碍了鱼类活病毒疫苗的研发和应用。

参考文献

Ahne，W. 1980. Occurrence of infectious pancreatic necrosis virus（IPN）in different fish species. *Berliner Munch Tierarztl Wochenschr* 93：14-16.

Ahne，W. and Negele，R. 1985. Studies on the transmission of infectious pancreatic necrosis virus via eyed eggs and sexual products of salmonid fish，in *Fish and Shellfish Pathology*（ed. A. Ellis）. London，Academic Press，262-270.

Allnutt，F. C. T.，Bowers，R. M.，Rowe，C. G.，*et al.*，2007. Antigenicity of infectious pancreatic necrosis virus VP2 subviral particles expressed in yeast. *Vaccine* 25：4880-4888.

Bebak，J. and McAllister，P. E. 2009. Continuous exposure to infectious pancreatic necrosis virus during early life stages of rainbow trout，*Oncorhynchus mykiss*（Walbaum）. *J Fish Dis* 32：173-181.

Bebak-Williams, J., McAllister, P. E., Smith, G. and Boston, R. 2002. Effect of fish density and number of infectious fish on the survival of rainbow trout fry, *Oncorhynchus mykiss* (Walbaum), during epidemics of infectious pancreatic necrosis. *J Fish Dis* 25: 715-726.

Blake, S., Ma, J. Y., Caporale, D. A., *et al.*, 2001. Phylogenetic relationships of aquatic birnaviruses based on deduced amino acid sequences of genome segment A cDNA. *Dis Aquat Org* 45: 89-102.

Bootland, L. M., Dobos, P. and Stevenson, R. M. W. 1991. The IPNV carrier state and demonstration of vertical transmission in experimentally infected brook trout. *Dis Aquat Org* 10: 13-21.

Bootland, L. M., Dobos, P. and Stevenson, R. M. W. 1995. Immunization of adult brook trout, *Salvelinus fontinalis* (Mitchill), fails to prevent the infectious pancreatic necrosis virus (IPNV) carrier state. *J Fish Dis* 18: 449-458.

Bowden, T. J., Smail, D. A. and Ellis, A. E. 2002. Development of a reproducible infectious pancreatic necrosis virus challenge model for Atlantic salmon, *Salmo salar* L. *J Fish Dis* 25: 555-563.

Bullock, G. L., Rucker, R. R., Amend, D., *et al.*, 1976. Infectious pancreatic necrosis-transmission with iodine-treated and nontreated eggs of brook trout (*Salvelinus fontinalis*). *J Fish Res Board Can* 33: 1197-1198.

Castric, J., Baudin-Laurencin, F., Coustans, M. F. and Auffret, M. 1987. Isolation of infectious pancreatic necrosis virus, Ab serotype, from an epizootic in farmed turbot, *Scophthalmus maximus*. *Aquaculture* 67: 117-126.

Chiu, C. L., Wu, J. L., Her, G. M., *et al.*, 2010. Aquatic birnavirus capsid protein, VP3, induces apoptosis via the Bad-mediated mitochondria pathway in fish and mouse cells. *Apoptosis* 15: 653-668.

Cuesta, A., Chaves-Pozo, E., de Las Heras, A. I., *et al.*, 2010. An active DNA vaccine against infectious pancreatic necrosis virus (IPNV) with a different mode of action than fish *Rhabdovirus* DNA vaccines. *Vaccine* 28: 3291-3300.

de Las Heras, A. I., Prieto, S. I. P. and Saint-Jean, S. R. 2009. *In vitro* and *in vivo* immune responses induced by a DNA vaccine encoding the VP2 gene of the infectious pancreatic necrosis virus. *Fish Shellfish Immunol* 27: 120-129.

de Las Heras, A. I., Saint-Jean, S. R. and Perez-Prieto, S. I. 2010. Immunogenic and protective effects of an oral DNA vaccine against infectious pancreatic necrosis virus in fish. *Fish Shellfish Immunol* 28: 562-570.

Delmas, B., Mundt, E., Vakharia, V. N. and Wu, J. L. 2011. Birnaviridae, in *Virus Taxonomy Classification and Nomenclature of Viruses* (eds A. M. Q. King, M. J. Adams. E. B. Carstens and E. J. Lefkowitz). Amsterdam, Elsevier, 499-507.

Dobos, P. 1995. The molecular biology of infectious pancreatic necrosis virus. *Annu Review Fish Dis* 5: 25-54.

Dorson, M. and Torchy, C. 1981. The influence of fish age and water temperature on mortalities of rainbow trout, *Salmo gairdneri* Richardson, caused by a European strain of infectious Pancreatic Necrosis Virus. *J Fish Dis* 4: 213-221.

Dorson, M. and Torchy, C. 1985. Experimental transmission of infectious pancreatic necrosis virus via the sexual products, in *Fish and Shellfish Pathology* (ed. A. Ellis). London, Academic Press, 251-261.

Duncan, R., Nagy, E., Krell, P. J. and Dobos, P. 1987. Synthesis of the infectious pancreatic necrosis virus polyprotein, detection of a virus-encoded protease, and fine-structure mapping of genome segment-a coding regions. *J Virol* 61: 3655-3664.

Duncan, R., Mason, C. L., Nagy, E., *et al.*, 1991. Sequence analysis of infectious pancreatic necrosis virus genome segment-B and its encoded VP1 protein: a putative RNA-dependent RNA polymerase lacking the Gly-Asp-

Asp Motif. *Virology* 181：541-552.

Erdal，J. I.，Skjelstad，B. and Soleim，K. B. 2003. Evaluation of protection from vaccination against infectious pancreatic necrosis (IPN) in a GCP field trial in Norway，in *Abstracts from the 11th International Conference of EAFP：Diseases of Fish and Shellfish*. Saint Julians，Malta，2003.

Estay，A.，Farias，G.，Soler，M. and Kuznar，J. 1990. Further analysis on the structural proteins of infectious Pancreatic Necrosis Virus. *Virus Res* 15：85-96.

Fridholm，H.，Eliasson，L. and Everitt，E. 2007. Immunogenicity properties of authentic and heterologously synthesized structural protein VP2 of infectious pancreatic necrosis virus. *Viral Immunol* 20：635-648.

Frost，P. and Ness，A. 1997. Vaccination of Atlantic salmon with recombinant VP2 of infectious pancreatic necrosis virus (IPNV)，added to a multivalent vaccine，suppresses viral replication following IPNV challenge. *Fish Shellfish Immunol* 7：609-619.

Hill，B. J. and Way，K. 1995. Serological classification of infectious pancreatic necrosis (IPN) virus and other aquaticbirnaviruses. *Annu Rev Fish Dis* 5：55-77.

Hjalmarsson，A.，Carlemalm，E. and Everitt，E. 1999. Infectious pancreatic necrosis virus：Identification of a VP3-containing ribonucleoprotein core structure and evidence for O-linked glycosylation of the capsid protein VP2. *J Virol* 73：3484-3490.

Hjeltnes，B.，Olsen，A. B. and Hellberg，H. (eds) 2012. *Fiskehelserapporten 2011* [Farmed fish health report 2011]. Oslo，Norwegian Veterinary Institute.

Hong，J. R.，Gong，H. Y. and Wu，J. L. 2002. IPNV VP5，a novel anti-apoptosis gene of the Bcl-2 family，regulates Mcl-1 and viral protein expression. *Virology* 295：217-229.

Houston，R. D.，Haley，C. S.，Hamilton，A.，*et al.*，2010. The susceptibility of Atlantic salmon fry to freshwater infectious pancreatic necrosis is largely explained by a major QTL. *Heredity* 105：318-327.

Hsu，Y. L.，Chen，B. S. and Wu，J. L. 1993. Demonstration of infectious pancreatic necrosis virus-Strain Vr-299 in Japanese eel，*Anguilla japonica* Temminck and Schlegel. *J Fish Dis* 16：123-129.

Jarp，J. 1999. Epidemiological aspects of viral diseases in the Norwegian farmed Atlantic salmon (*Salmo salar* L.). *Bull Eur Assoc Fish Pathol* 19：240-244.

Jarp，J.，Gjevre，A. G.，Olsen，A. B. and Bruheim，T. 1995. Risk-factors for furunculosis，infectious pancreatic necrosis and mortality in post-smolt of Atlantic salmon，*Salmo salar* L. *J Fish Dis* 18：67-78.

Julin，K.，Mennen，S. and Sommer，A. I. 2013. Study of virulence in field isolates of infectious pancreatic necrosis virus obtained from the northern part of Norway. *J Fish Dis* 36：89-102.

Kaattari，S. L.，Zhang，H. L. L.，Khor，I. W.，*et al.*，2002. Affinity maturation in trout：clonal dominance of high affinity antibodies late in the immune response. *Dev Comp Immunol* 26：191-200.

Lee，N. S.，Nomura，Y. and Miyazaki，T. 1999. Gill lamellar pillar cell necrosis，a new birnavirus disease in Japanese eels. *Dis Aquat Org* 37：13-21.

Lorenzen，N. and LaPatra，S. E. 2005. DNA vaccines for aquacultured fish. *Rev Sci Tech* 24：201-213.

McAllister，P. E. and Owens，W. J. 1995. Assessment of the virulence of fish and molluscan isolates of infectious pancreatic necrosis virus for salmonid fish by challenge of brook trout，*Salvelinus fontinalis* (Mitchill). *J Fish Dis* 18：97-103.

McKenna，B. M.，Fitzpatrick，R. M.，Phenix，K. V.，*et al.*，2001. Formation of infectious pancreatic necrosis virus-like particles following expression of segment A by recombinant semliki forest virus. *Mar Biotechnol (NY)* 3：103-110.

Melby, H. P., Caswell-Reno, P. and Falk, K. 1994. Antigenic analysis of Norwegian aquatic birnavirus isolates using monoclonal antibodies-with special reference to fish species, age and health status. *J Fish Dis* 17: 85-91.

Melby, H. P. and Falk, K. 1995. Study of the interaction between a persistent infectious pancreatic necrosis virus (IPNV) infection and experimental infectious salmon anaemia (ISA) in Atlantic salmon, *Salmo salar* L. *J Fish Dis* 18: 579-586.

M'Gonigle, R. H. 1940. Acute catarrhal enteritis of salmonid fingerlings. *Trans Am Fish Soc* 70: 297-302.

Mikalsen, A. B., Torgersen, J., Aleström, P., *et al.*, 2004. Protection of Atlantic salmon *Salmo salar* against infectious pancreatic necrosis after DNA vaccination. *Dis Aquat Org* 60: 11-20.

Moen, T., Baranski, M., Sonesson, A. K. and Kjøglum, S. 2009. Confirmation and fine-mapping of a major QTL for resistance to infectious pancreatic necrosis in Atlantic salmon (*Salmo salar*): population-level associations between markers and trait. *BMC Genomics* 10: 368

Mulcahy, D. and Pascho, R. J. 1984. Adsorption to fish sperm of vertically transmitted fish viruses. *Science* 225: 333-335.

Munang'andu, H. M., Fredriksen, B. N., Mutoloki, S., *et al.*, 2012. Comparison of vaccine efficacy for different antigen delivery systems for infectious pancreatic necrosis virus vaccines in Atlantic salmon (*Salmo salar* L.) in a cohabitation challenge model. *Vaccine* 30: 4007-4016.

Nakajima, K. and Sorimachi, M. 1994. Serological and biochemical characterization of 2 birnaviruses-Vdv and Yav isolated from cultured yellowtail. *Fish Pathol* 29: 183-186.

Okamoto, N., Sano, T., Hedrick, R. P. and Fryer, J. L. 1983. Antigenic relationships of selected strains of infectious pancreatic necrosis virus and European eel virus. *J Fish Dis* 6: 19-25.

Olesen, N. J., Jørgensen, P. E. V., Bloch, B. and Mellergaard, S. 1988. Isolation of an IPN-like virus belonging to the serogroup II of the aquatic birnaviruses from dab, *Limanda limanda* L. *J Fish Dis* 11: 449-451.

Pedersen, T., Skjesol, A. and Jørgensen, J. B. 2007. VP3, a structural protein of infectious pancreatic necrosis virus, interacts with RNA-dependent RNA polymerase VP1 and with double-stranded RNA. *J Virol* 81: 6652-6663.

Ramstad, A. and Midtlyng, P. J. 2008. Strong genetic influence on IPN vaccination-and-challenge trials in Atlantic salmon, *Salmo salar* L. *J Fish Dis* 31: 567-578.

Ramstad, A., Romstad, A. B., Knappskog, D. H. and Midtlyng, P. J. 2007. Field validation of experimental challenge models for IPN vaccines. *J Fish Dis* 30: 723-731.

Reno, P. W., Darley, S. and Savan, M. 1978. Infectious pancreatic necrosis-experimental induction of a carrier state in trout. *J Fish Res Board Can* 35: 1451-1456.

Rivas-Aravena, A., Martin, M. C. S., Galaz, J., *et al.*, 2012. Evaluation of the immune response against immature viral particles of infectious pancreatic necrosis virus (IPNV): a new model to develop an attenuated vaccine. *Vaccine* 30: 5110-5117.

Santi, N., Sandtrø, A., Sindre, H., *et al.*, 2005a. Infectious pancreatic necrosis virus induces apoptosis *in vitro* and *in vivo* independent of VP5 expression. *Virology* 342: 13-25.

Santi, N., Song, H., Vakharia, V. N. and Evensen, Ø. 2005b. Infectious pancreatic necrosis virus VP5 is dispensable for virulence and persistence. *J Virol* 79: 9206-9216.

Shivappa, R. B., McAllister, P. E., Edwards, G. H., *et al.*, 2005. Development of a subunit vaccine for Infectious pancreatic necrosis virus using a baculovirus insect/larvae system. *Dev Biol* (*Basel*) 121: 165-174.

Smail, D. A. and Munro, A. L. S. 1989. Infectious pancreatic necrosis virus in Atlantic salmon: transmission via the sexual products? in *Viruses of Lower Vertebrates* (eds W. Ahne and E. Kurstak). Berlin, Springer Verlag,

292-301.

Smail, D. A., Bruno, D. W., Dear, G., *et al.*, 1992. Infectious pancreatic necrosis (IPN) virus Sp serotype in farmed Atlantic salmon, *Salmo salar* L, post-smolts associated with mortality and clinical disease. *J Fish Dis* 15: 77-83.

Snieszko, S. F., Wolf, K., Amper, J. E. and Pettijohn, L. L. 1959. Infectious nature of pancreatic necrosis. *Trans Am Fish Soc* 88: 289-293.

Toranzo, A. E. and Hetrick, F. M. 1982. Comparative stability of 2 salmonid viruses and poliovirus in fresh, estuarine and marine waters. *J Fish Dis* 5: 223-231.

Underwood, B. O., Smale, C. J., Brown, F. and Hill, B. J. 1977. Relationship of a virus from *Tellina tenuis* to infectious pancreatic necrosis virus. *J Gen Virol* 36: 93-109.

Urquhart, K., Murray, A. G., Gregory, A., *et al.*, 2008. Estimation of infectious dose and viral shedding rates for infectious pancreatic necrosis virus in Atlantic salmon, *Salmo salar* L., post-smolts. *J Fish Dis* 31: 879-887.

Wallace, I. S., Gregory, A., Murray, A. G., *et al.*, 2008. Distribution of infectious pancreatic necrosis virus (IPNV) in wild marine fish from Scottish waters with respect to clinically infected aquaculture sites producing Atlantic salmon, *Salmo salar* L. *J Fish Dis* 31: 177-186.

Wetten, M., Aasmundstad, T., Kjøglum, S. and Storset, A. 2007. Genetic analysis of resistance to infectious pancreatic necrosis in Atlantic salmon (*Salmo salar* L.). *Aquaculture* 272: 111-117.

Wolf, K. and Quimby, M. C. 1971. Salmonid viruses-infectious pancreatic necrosis virus-morphology, pathology, and serology of first European isolations. *Arch Gesamte Virusforsch* 34: 144-156.

Wolf, K., Snieszko, S. F., Dunbar, C. E. and Pyle, E. 1960. Virus nature of infectious pancreatic necrosis in trout. *Proc Soc Exp Biol Med* 104: 105-108.

Wolf, K., Bradford, A. D. and Quimby, M. C. 1968. Egg-associated transmission of IPN virus of trouts. *Virology* 21: 317-321.

Yamamoto, T. 1975a. Frequency of detection and survival of infectious pancreatic necrosis virus in a carrier population of brook trout (*Salvelinus fontinalis*) in a lake. *J Fish Res Board Canada* 32: 568-570.

Yamamoto, T. 1975b. Infectious pancreatic necrosis (IPN) virus carriers and antibody-production in a population of rainbow-trout (*Salmo gairdneri*). *Can J Microbiol* 21: 1343-1347.

Yamamoto, T. and Kilistoff, J. 1979. Infectious pancreatic necrosis virus-quantification of carriers in lake populations during a 6-year period. *J Fish Res Board Canada* 36: 562-567.

Zhao, L. L., Liu, M., Ge, J. W., *et al.*, 2012. Expression of infectious pancreatic necrosis virus (IPNV) VP2-VP3 fusion protein in *Lactobacillus casei* and immunogenicity in rainbow trouts. *Vaccine* 30: 1823-1829.

第二十六章　传染性鲑贫血症的疫苗接种

Knut Falk

本章概要　SUMMARY

传染性鲑贫血症（Infectious salmon anemia，ISA）是海水养殖大西洋鲑（*Salmo salar*）的一种危害严重的疾病，由 ISA 病毒（ISAV，属正黏病毒）引起。最常见的 ISA 疫苗是由细胞培养制备的灭活全病毒油类佐剂疫苗，对银化鲑在移到海水网箱前进行腹腔注射免疫。据报道，在实验条件下，该疫苗的相对保护率（RPS）为 84%～95%，但在鲑免疫群体中暴发了几次 ISA 疫情，这使自然养殖条件下疫苗的效果备受质疑。最近，智利一家公司推出了基于重组表达 ISAV 蛋白的疫苗，可用于鱼的注射和口服免疫，然而疫苗效果的研究却未见报道。1999 年，加拿大将 ISA 的疫苗接种纳入该病的防控计划，2005 年和 2010 年法罗群岛和智利也分别制定了相应的防控计划。2009 年，挪威在多次暴发 ISA 疫情的地区也将疫苗接种纳入该疫病防控计划。

第一节　发生流行与危害

传染性鲑贫血症（ISA）是养殖大西洋鲑中一种危害严重的传染性病毒病，于 1984 年在挪威首次报道（Thorud 和 Djupvik，1988）。ISA 是世界动物卫生组织（OIE）规定必须通报的一种疾病，在欧盟被列为 2 类疫病。目前，该病在加拿大、美国、苏格兰、法罗群岛和智利等国家和地区的养殖大西洋鲑中均有报道（Rimstad 等，2011）。

该病曾在挪威水产养殖业中广泛传播，仅 1990 年就暴发 90 余次，达到高峰（Hastein 等，1999）。虽然 1995 年才鉴定出该病的病原（Dannevig 等，1995），但养殖业者早已采取了一系列生物安全措施预防和控制该病。这些措施以常规的卫生准则为基础，以降低感染风险和切断传播途径为重点，包括对患病群体的早期发现和扑杀，加强运输管控，对加工产生的内脏和废弃物进行消毒，按鱼的年龄分养以及加强健康管理和认证。这些措施取得了显著的成效，例如，在 1994 年挪威仅有两起疫情出现，这表明 ISA 可以在不使用药物和疫苗，甚至是不了解其病原、流行病学和致病机制的情况下得以控制。目前，ISA在挪威的发病率很低，之前受该病影响的其他国家也控制住了该病，只有偶尔出现的几起疾病暴发案例。关于ISA的流行病学信息可以在OIE的世界动物健康信息数据库（http://www.oie.int/wahis_2/public/wahid.php/Wahidhome/Home）中查阅。

人工养殖的大西洋鲑是ISA自然发病的唯一物种，而且大多发生在鲑的海水养殖阶段。然而已有实验证明，ISAV可在褐鳟、虹鳟、北极鲑、马苏大麻哈鱼、银大麻哈鱼、鲱和大西洋鳕等中增殖而不引起鱼发病（Rimstad等，2011）。此外，研究人员在健康的野生大西洋鲑和海鳟中也检测到ISAV（Raynard等，2001a；Plarre等，2005），因此，这些鱼类种群可能是病毒的潜在携带者或病毒来源。

在养殖鲑中，ISA是一个发展缓慢的系统性疾病。在发病初期，通常仅在养殖场的少数网箱中发现患病个体，每日死亡率为0.05%～0.1%。然而，若不采取任何措施防止该病的蔓延，病毒则会传播到其他网箱，累计死亡率会在短短的几个月内达到80%以上。

第二节 病原学

传染性鲑贫血症病毒（ISAV）属于正黏病毒科（Orthomyxoviridae），传染性鲑贫血病毒属（*Isavirus*），具有正黏病毒科的主要功能、生化和形态学特征（Palese等，2007；Rimstad等，2011）。病毒基因组由一个分节段的、单股负链RNA组成，编码10种蛋白（Mjaaland等，1997；Clouthier等，2002；Falk等，2004）。病毒粒子主要由4种结构蛋白组成：核蛋白（NP）、基质（M）蛋白、血凝素酯酶（HE）蛋白和融合（F）蛋白（Mjaaland等，1997；Clouthier等，2002；Falk等，2004）。其中，HE蛋白和F蛋白是镶嵌在病毒囊膜中的糖蛋白，共同组成病毒粒子特有的突起（Falk等，1997）。这两种蛋白对病毒的毒力很重要，是关键的保护性抗原。HE蛋白由病毒基因的第6个片段编码，介导病毒与受体的结合并促使宿主细胞释放（通过受体破坏酶，RDE）新形成的病毒粒子（Falk等，2004；Hellebø等，2004）。

该病毒基因组高度保守，其中，变异程度最高的两个基因片段分别编码HE蛋白和F蛋白（即片段6和片段5）。基于*HE*基因的系统发育分析表明，该病毒有欧洲和北美两个主要的进化分支（Devold等，2001）。此外，HE蛋白跨膜结构域上游存在一个由11～35个氨基酸组成的多态区域（HPR），该区域是区分ISAV特性和亚型的依据（Rimstad等，2011）。HPR突变体可能是由全长原始序列（HPR0）中HPR缺失造成的（Mjaaland等，2002），最早在苏格兰的野生鲑中发现（Cunningham等，2002）。从暴发疾病的鲑中获得的ISAV分离株均缺失HPR，而HPR0亚型与ISA的临床症状和病理特征无关联（Cunningham等，2002；Cook-Versloot等，2004；Nylund等，2006；McBeath等，2009；Christiansen等，2011），且应用RT-PCR方法常在鲑鳃中检测到HPR0（Christiansen等，2011；Lyngstad等，2012）。HPR缺失的ISAV很容易在体外易感细胞中培养（Dannevig等，1995；Devold等，2001），与之相反，HPR0亚型病毒却无法进行细胞培养，这就限制了对该病毒亚型特性的深入研究。

ISAV各分离株的毒力存在差异，从野外发病和实验条件致病群体的病程发展、死亡率以及临床症状不同也得到确认（Mjaaland等，2005；Kibenge等，2006；Ritchie等，2009）。在分子水平上，与其他正黏病毒相似，ISAV 毒力由多种基因控制。2003 年人结膜炎和荷兰高致病性禽流感（HPAI）暴发期间，对从人致死病例中分离出的 H7N7 流感病毒进行了分析，证实两种 H7N7 病毒的基因片段 5 上存在 15 个氨基酸位点的差异（de Witt 等，2010）。然而，迄今为止，只有两个推定的 ISAV 毒力标志的差异位点被鉴定出来，一种是上面提到的 HPR0 亚型；另一种也与 HPR0 亚型病毒相关，其 F 蛋白氨基酸序列上靠近蛋白水解酶激活位点的第 266 个氨基酸——谷氨酰胺被亮氨酸替换（Markussen 等，2008）。

第三节　致病机理

传染性鲑贫血症是一种致死的系统性疾病，病鱼的临床特征表现为循环系统衰竭，并伴有严重的贫血、充血、腹水以及肝脾肿大（Evensen 等，1991）。病鱼解剖后常见脏器不同程度的出血和坏死。ISAV 主要损害循环系统，其主要靶细胞是所有器官的血管内皮细胞，包括血窦、心内膜和清道夫内皮细胞（Scavenger endothelial cells）（Aamelfot 等，2012），这在一定程度上解释了该病的临床症状和病理特征。

该病自然发病的潜伏期很难确定，但估计有几周甚至几个月（Vågsholm 等，1994；Jarp 和 Karlsen，1997）。实验条件下，通过腹腔注射、共栖感染或浸泡攻毒（Raynard 等，2001b），ISAV 在大西洋鲑中的潜伏期一般为 10～20d（Rimstad 等，2011）。

研究表明，大西洋鲑可产生针对 ISAV 的非特异性细胞免疫和体液免疫反应（Falk 等，1995；Jørgensen 等，2007；Ritchie 等，2009；LeBlanc 等，2010；Lauscher 等，2011）。将 ISAV 感染存活个体的血清注射到鱼体内（即被动免疫），被动免疫的鱼在攻毒后产生一定的保护效果，这表明体液免疫起重要作用（Falk 等，1995）。通过编码 HE 蛋白的 DNA 疫苗证明了 HE 蛋白抗原在免疫保护上的重要性（Mikalsen 等，2005；Wolf 等，2013）。然而，4 个大西洋鲑养殖场在暴发 ISA 前的 1～4 个月，通过 qPCR 在鱼鳃中检测到 HPR0 型 ISAV，这表明感染 HPR0 型 ISAV 并不能对鱼体产生显著的保护作用（Lyngstad 等，2011）。

第四节　疫　　苗

有关 ISA 疫苗及其效果的文献很少，Jones 等（1999）研究表明，由细胞培养的 ISAV 制备的灭活佐剂疫苗的相对保护率为 84%～95%，有些会议摘要和网络出版的文献支持该结果。Lauscher 等（2011）的研究证实，细胞培养的 ISAV 灭活疫苗相对保护率为 86%，且疫苗的保护效果与疫苗注射的剂量和检测到的抗 ISAV 抗体密切相关。

ISAV 商品疫苗的研发虽然主要集中于灭活疫苗，但利用分子生物学技术制备疫苗的研究也有报道。Mikalsen 等（2005）以编码 HE 蛋白的载体注射大西洋鲑 3 次，每次间隔 3 周，产生的保护效果不高，各组的相对保护率为 40%～60%，此实验的攻毒方式是腹腔注射，这可能是导致保护效果不理想的原因。Wolf 等（2013）以表达 ISAV 凝血素-酯酶的鲑甲病毒复制子肌肉注射免疫鲑，共栖攻毒后，结果发现，相对保护率在 65%～69%，而免疫灭活 ISAV 疫苗相对保护率为 80%（Wolf 等，2013）。

目前有两种 ISAV 商品疫苗，为诺华动物保健公司（Novartis Animal Health）和法玛克公司（Pharmaq）生产的由细胞培养 ISAV 制备的油类佐剂灭活疫苗，通过腹腔注射接种（Finne-Fridell 等，2011）。这些灭活疫苗可以作为单价疫苗使用，也可以作为多联疫苗的一个组分使用，通常在鲑转移至海水前注射免疫接种（表 26-1）。除此之外，智利制药公司（Centrovet）目前正在推广一种 ISAV 的油类佐剂注射疫苗和一种口服疫苗，均由酵母表达 ISAV 的 HE 蛋白制备而成。在口服疫苗的制备过程中，使用了 Advanced BioNutrition Corporation 的 MicroMatrix 专利技术，包被表达的 HE 蛋白，目前还没有关于 Centrovet 公司的这两种疫苗的文献报道，仅在一个会议摘要中报道了其中的口服疫苗，据称疫苗的效果良好（Tobar 等，2010）。

1999 年，ISAV 疫苗接种在加拿大东海岸进行了首次尝试，但是在疫苗接种的第一年，免疫群体中就发生 ISA 疫情，使疫苗的效果受到质疑。目前为止，还没有在养殖环境下 ISAV 疫苗效果的文献报道。在加拿大，疫苗接种与防控计划相结合，包括养殖和运输中的预防措施、鲑健康状况的持续监测以及疾病暴发时的控制措施等。通过这些措施，2006 年到 2012 年间，加拿大没有暴发过 ISA 疫情，仅 2012 年在加拿大的新斯科舍暴发了一次 ISA 疫情。另一个实例是，2000—2005 年在法罗群岛出现了多次 ISA 暴发，通过疫苗接种和成功的防控计划控制了 ISA 疫情。正如 Hastein 等（1999）的综述，这些成功案例是基于常规的卫生准则、病毒疫苗的免疫接种和严格的防控管理体制相结合的结果。自 2005 年以来，虽然广泛的疫情监测经常发现 HPR0 型 ISAV，但在法罗群岛再未暴发 ISA（Christiansen 等，2011）。在智利也采取了包括疫苗接种在内的类似措施，自 2010 年 9 月以来，除了 2012 年报道过一例 RT-PCR 检测 ISAV 阳性外，再也没有暴发过 ISA。在挪威，为预防当地 ISA 流行，对鲑实施了接种疫苗，然而在 2010 年接种疫苗的群体中却暴发了 3 起 ISA 疫情。

表 26-1　传染性鲑贫血症的商品疫苗

产品名称	疫苗类型	抗原	佐剂	免疫方式	疫苗公司
ILAvacc	灭活的细胞培养病毒疫苗，单联	ISAV，北美生物型	矿物油	腹腔注射	诺华动物保健公司，加拿大
Pentium Forte Plus ILA Pentium Ⅶ	灭活的细胞培养病毒疫苗，多联	ISAV，北美生物型	矿物油	腹腔注射	诺华动物保健公司，加拿大
Alpha Ject micro 1 ISA	灭活的细胞培养病毒疫苗，单联	ISAV，欧洲生物型	矿物油	腹腔注射	法玛克，挪威
Alpha Ject micro 7 ILA Alpha Ject 5-1	灭活的细胞培养病毒疫苗，多联	ISAV，欧洲生物型	矿物油	腹腔注射	法玛克，挪威
传染性鲑贫血症的单价亚单位疫苗，或与鲑鱼立克次氏体、传染性胰腺坏死病毒和奥达弧菌的多联疫苗	亚单位疫苗酵母表达	ISAV 重组 HE 蛋白	矿物油	腹腔注射	Centrovet，智利
传染性鲑贫血症的单价亚单位疫苗，或与鲑鱼立克次氏体的二联疫苗	亚单位疫苗酵母表达	ISAV 重组 HE 蛋白		口服	Centrovet，智利

第五节　监管问题

加拿大、法罗群岛和智利分别自 1999 年、2005 年以及 2010 年起将 ISA 的疫苗接种纳入

ISAV 防控计划。2006 年之前，ISA 一直是欧盟包括挪威的“1 类疫病”，这意味着禁止使用疫苗预防 ISA。然而，苏格兰不实施疫苗接种，通过采取预防措施成功控制了该病。2006 年，ISA 被降级至“2 类疫病”，这使得 ISA 疫苗接种可在尚未确定 ISA 流行的区域（即“三级”区域）作为防控计划的一部分。由于这一变化，2007 年至 2010 年期间，挪威北部地区使用 ISA 疫苗预防 ISA 流行，但在最后一个流行的年份，接种过疫苗的大西洋鲑仍暴发了 3 次 ISA。因此，ISAV 疫苗接种结合良好的管理制度、生物安全防控措施等会带来预期的成效，但 ISAV 疫苗本身的实际应用效果问题仍有待探索。

参考文献

Aamelfot, M., Dale, O. B., Weli, S. C., *et al.*, 2012. Expression of the infectious salmon anemia virus receptor on Atlantic salmon endothelial cells correlates with the cell tropism of the virus. *J Virol* 86: 10571-10578.

Aspehaug, V., Falk, K., Krossøy, B, *et al.*, 2004. Infectious salmon anemia virus (ISAV) genomic segment 3 encodes the viral nucleoprotein (NP), an RNA binding protein with two monopartite nuclear localization signals (NLS). *Virus Res* 106: 51-60.

Aspehaug, V., Mikalsen, A. B., Snow, M., *et al.*, 2005. Characterization of the infectious salmon anemia virus fusion protein. *J Virol* 79: 12544-12553.

Christiansen, D. H., Østergaard, P. S., Snow, M., *et al.*, 2011. A low-pathogenic variant of infectious salmon anemia virus (ISAV-HPR0) is highly prevalent and causes a non-clinical transient infection in farmed Atlantic salmon (*Salmo salar* L.) in the Faroe Islands. *J Gen Virol* 92: 909-918.

Clouthier, S. C., Rector, T., Brown, N. E. C. and Anderson, E. D. 2002. Genomic organization of infectious salmon anaemia virus. *J Gen Virol* 83: 421-428.

Cook-Versloot, M., Griffiths, S., Cusack, R., *et al.*, 2004. Identification and characterization of infectious salmon anaemia virus (ISAV) haemagglutinin gene highly polymorphic region (HPR) type 0 in North America. *Bull Eur Assoc Fish Pathol* 24: 203-208.

Cunningham, C. O., Gregory, A., Black, J., *et al.*, 2002. A novel variant of the infectious salmon anaemia virus (ISAV) haemagglutinin gene suggests mechanisms for virus diversity. *Bull Eur Assoc Fish Pathol* 22: 366-374.

Dannevig, B. H., Falk, K. and Namork, E. 1995. Isolation of the causal virus of infectious salmon anaemia (ISA) in a long-term cell line from Atlantic salmon head kidney. *J Gen Virol* 76: 1353-1359.

Devold, M., Falk, K., Dale, O. B., *et al.*, 2001. Strain variation, based on the hemagglutinin gene, in Norwegian ISA virus isolates collected from 1987 to 2001: indications of recombination. *Dis Aquat Org* 47: 119-128.

de Witt, E., Munster, V. J., van Riel, D., *et al.*, 2010. Molecular determinants of adaptation of highly pathogenic avian influenza H7N7 viruses to efficient replication in the human host. *J Virol* 84: 1597-1606.

Evensen, Ø., Thorud, K. E. and Olsen, Y. A. 1991. A morphological study of the gross and light microscopic lesions of infectious anaemia in Atlantic salmon (*Salmo salar*). *Res Vet Sci* 51: 215-222.

Falk, K. and Dannevig, B. H. 1995. Demonstration of infectious salmon anaemia (ISA) viral antigens in cell cultures and tissue sections. *Vet Res* 26: 499-504.

Falk, K., Namork, E., Rimstad, E., *et al.*, 1997. Characterization of infectious salmon anemia virus, an orthomyxo-like virus isolated from Atlantic salmon (*Salmo salar* L.). *J Virol* 71: 9016-9023.

Falk, K., Aspehaug, V., Vlasak, R. and Endresen, C. 2004. Identification and characterization of viral structural proteins of infectious salmon anemia virus. *J Virol* 78：3063-3071.

Finne-Fridell，F.，Aas-Eng，A.，Fyrand，K.，*et al.*，2011. Infeksiøs lakseanemi-vaksineutvikling. *Nor Fiskeoppdr* 36（5）：54-56.

Håstein，T.，Hill，B. J. and Winton，J. R. 1999. Successful aquatic animal disease emergency programmes. *Rev Sci Tech* 18：214-227.

Hellebø，A.，Vilas，U.，Falk，K. and Vlasak，R. 2004. Infectious salmon anemia virus specifically binds to and hydrolyzes 4-*O*-acetylated sialic acids. *J Virol* 78：3055-3062.

Jarp，J. and Karlsen，E. 1997. Infectious salmon anaemia（ISA）risk factors in sea-cultured Atlantic salmon *Salmo salar*. *Dis Aquat Org* 28：79-86.

Jones，S. R. M.，MacKinnon，A. M. and Salonius，K. 1999. Vaccination of fresh-water-reared Atlantic salmon reduces mortality associated with infectious salmon anaemia virus. *Bull Eur Assoc Fish Pathol* 19：98-101.

Jørgensen，S. M.，Hetland，D. L.，Press，C. McL.，*et al.*，2007. Effect of early infectious salmon anaemia virus（ISAV）infection on expression of MHC pathway genes and type Ⅰ and Ⅱ interferon in Atlantic salmon（*Salmo salar* L.）tissues. *Fish Shellfish Immunol* 23：576-588.

Kibenge，F. S. B.，Kibenge，M. J. T.，Groman，D. and McGeachy，S. 2006. *In vivo* correlates of infectious salmon anemia virus pathogenesis in fish. *J Gen Virol* 87：2645-2652.

Krossøy，B.，Devold，M.，Sanders，L.，*et al.*，2001. Cloning and identification of the infectious salmon anaemia virus haemagglutinin. *J Gen Virol* 82：1757-1765.

Lauscher，A.，Krossøy，B.，Frost，P.，*et al.*，2011. Immune responses in Atlantic salmon（*Salmo salar*）following protective vaccination against infectious salmon anemia（ISA）and subsequent ISA virus infection. *Vaccine* 29：6392-6401.

LeBlanc，F.，Laflamme，M. and Gagné，N. 2010. Genetic markers of the immune response of Atlantic salmon（*Salmo salar*）to infectious salmon anemia virus（ISAV）. *Fish Shellfish Immunol* 29：217-232.

Lyngstad，T. M.，Hjortaas，M. J.，Kristoffersen，A. B.，*et al.*，2011. Use of molecular epidemiology to trace transmission pathways for infectious salmon anaemia virus（ISAV）in Norwegian salmon farming. *Epidemics* 3：1-11.

Lyngstad，T. M.，Kristoffersen，A. B.，Hjortaas，M. J.，*et al.*，2012. Low virulent infectious salmon anaemia virus（ISAV-HPR0）is prevalent and geographically structured in Norwegian salmon farming. *Dis Aquat Org* 101：197-206.

Markussen，T.，Jonassen，C. M.，Numanovic，S.，*et al.*，2008. Evolutionary mechanisms involved in the virulence of infectious salmon anaemia virus（ISAV），a piscine orthomyxovirus. *Virology* 374：515-527.

McBeath，A. J. A.，Bain，N. and Snow，M. 2009. Surveillance for infectious salmon anaemia virus HPR0 in marine Atlantic salmon farms across Scotland. *Dis Aquat Org* 87：161-169.

Mikalsen，A. B.，Sindre，H.，Mjaaland，S.，Rimstad，E. 2005. Expression，antigenicity and studies on cell receptor binding of the hemagglutinin of infectious salmon anemia virus. *Arch Virol* 150：1621-1637.

Mjaaland，S.，Rimstad，E.，Falk，K. and Dannevig，B. H. 1997. Genomic characterization of the virus causing infectious salmon anemia in Atlantic salmon（*Salmo salar* L.）：an orthomyxo-like virus in a teleost. *J Virol* 71：7681-7686.

Mjaaland，S.，Hungnes，O.，Teig，A.，*et al.*，2002. Polymorphism in the infectious salmon anemia virus hemagglutinin gene：importance and possible implications for evolution and ecology of infectious salmon anemia disease. *Virology* 304：379-391.

Mjaaland, S., Markussen, T., Sindre, H., *et al.*, 2005. Susceptibility and immune responses following experimental infection of MHC compatible Atlantic salmon (*Salmo salar* L.) with different infectious salmon anaemia virus isolates. *Arch Virol* 8: 1621-1637.

Nylund, A., Plarre, H., Karlsen, M., *et al.*, 2006. Transmission of infectious salmon anaemia virus (ISAV) in farmed populations of Atlantic salmon (*Salmo salar*). *Arch Virol* 152: 151-179.

Palese, P. and Shaw, M. L. 2007. Orthomyxoviridae: the viruses and their replication, in *Fields Virology* (5th edn) (eds D. M. Knipe and P. M. Howley). Philadelphia, Lippincott Williams & Wilkins, 1647-1689.

Plarre, H., Devold, M., Snow, M. and Nylund, A. 2005. Prevalence of infectious salmon anaemia virus (ISAV) in wild salmonids in western Norway. *Dis Aquat Org* 66: 71-79.

Raynard, R. S., Murray, A. G. and Gregory, A. 2001a. Infectious salmon anaemia virus in wild fish from Scotland. *Dis Aquat Org* 46: 93-100.

Raynard, R. S., Snow, M. and Bruno, D. W. 2001b. Experimental infection models and susceptibility of Atlantic salmon *Salmo salar* to a Scottish isolate of infectious salmon anaemia virus. *Dis Aquat Org* 47: 169-174.

Rimstad, E., Mjaaland, S., Snow, M., *et al.*, 2001. Characterization of the infectious salmon anemia virus genomic segment that encodes the putative hemagglutinin. *J Virol* 75: 5352-5356.

Rimstad, E., Dale, O. B., Dannevig, B. H. and Falk, K. 2011. Infectious salmon anaemia, in *Fish Diseases and Disorders: Volume 3: Viral, Bacterial and Fungal Infections* (2nd edn) (eds P. T. K. Woo and D. W. Bruno). Oxfordshire, CAB International, 143-165.

Ritchie, R. J., McDonald, J. T., Glebe, B., *et al.*, 2009. Comparative virulence of infectious salmon anaemia virus isolates in Atlantic salmon, *Salmo salar* L. *J Fish Dis* 32: 157-171.

Thorud, K. and Djupvik, H. O. 1988. Infectious salmon anaemia in Atlantic salmon (*Salmo salar*). *Bull Eur Assoc Fish Pathol* 8: 109-111.

Tobar, J. A., Caruffo, M., Betz, Y., *et al.*, 2010. Oral immunity against infectious salmon anaemia in Atlantic salmon (*Salmo salar*). XVIII *International Conference on Bioencapsulation*, Porto, Portugal. Available at http://impascience.eu/bioencapsulation/340_contribution_texts/2010-10-01_P-088.pdf? PHPSESSID=9ae917c5d251 14d8dd4a81ebbfddac16.

Vågsholm, I., Djupvik, H. O., Willumsen, F. V., *et al.*, 1994. Infectious salmon anemia (ISA) epidemiology in Norway. *Prev Vet Med* 19: 277-290.

Wolf, A., Hodneland, K., Frost, P., *et al.*, 2013. A hemagglutinin-esterase-expressing salmonid alphavirus replicon protects Atlantic salmon (*Salmo salar*) against infectious salmon anemia (ISA). *Vaccine* 31: 661-669.

第二十七章　锦鲤疱疹病毒病的疫苗接种

Arnon Dishon，Ofer Ashoulin，E. Scott Weber III and Moshe Kotler

本章概要　SUMMARY

十多年来，锦鲤疱疹病毒病一直严重威胁鲤和观赏锦鲤产业。该病的病原是一种基因组较大的双链 DNA 病毒，被鉴定为鲤疱疹病毒 3 型。该病为暴发性疫病且死亡率高，受感染鱼的死亡率可达 100%。 集约化养殖、病毒的高传染性以及全球范围的锦鲤贸易流通，导致了该病的快速传播。锦鲤疱疹病毒病在一些国家的野生和养殖鲤中持续流行发生，并在世界范围内传播，尽管采取限制进口、扑杀感染群体等措施控制疾病传播，但收效甚微。2005 年，该病的减毒活疫苗在以色列开始使用，为以色列锦鲤以及鲤养殖业的恢复和可持续发展起到了重要作用。作为控制该致死性疾病的预防措施，该病疫苗已在印度尼西亚和美国获得商业许可。

第一节　发生流行与危害

鲤（*Cyprinus carpio carpio*）是一种广泛养殖的重要经济鱼类，作为传统食物并提供蛋白源，每年产量为 370 万 t，主要产自中国及亚洲其他国家以及东欧国家（FAO，2013）。锦鲤（*Cyprinus carpio koi*）是鲤的亚种，是观赏鱼爱好者养殖在庭院池塘或水族馆的一种观赏鱼类。

20 世纪 90 年代中期以来，鲤和锦鲤养殖业的发展受到锦鲤疱疹病毒病（Koi herpesvirus disease，KHVD）的严重影响。1998 年，KHVD 首次在以色列沿岸以及美国大西洋沿岸中部地区发生，引起锦鲤和鲤的大规模死亡（Ariav 等，1999；Hedrick 等，2000）。

Hedrick 等（2000）从患病鱼中分离出疱疹病毒，将其命名为锦鲤疱疹病毒（Koi herpesvirus，KHV）。该病对以色列水产养殖业造成毁灭性的打击，在几年内蔓延到 90% 的鲤养殖场，每年的损失估计达 300 万美元（Perelberg 等，2003）。后来，德国（Bretzinger 等，1999）和英国（Walster，1999）报道了相似的锦鲤死亡情况。德国（Neukirch 和 Kunz，2001）和以色列

(Perelberg 等，2003；Ronen 等，2003) 的其他研究者也相继分离到 KHV。PCR 技术追溯分析了 1996 年英国收集的不明原因大量死亡的锦鲤样本，检测出 KHV 的 DNA (Haenen 等，2004)。在欧洲超过 15 个国家存在 KHV，这不仅对锦鲤市场而且对食源鲤养殖的大型企业也构成了威胁 (Haenen 等，2004；Haenen 和 Olesen，2009)。

2002 年 3 月，在印度尼西亚发生了毁灭性的 KHVD 大暴发 (Sunarto 等，2005)。经香港从中国进口锦鲤后，在东爪哇发生了 KHVD 的首次暴发，养殖锦鲤大量死亡，此后，由于带病鱼的流通，该病快速传播到西爪哇，到 2003 年初，暴发性流行已波及南苏门答腊。该病对印度尼西亚造成了长期的社会经济影响，据鱼类健康与环境局估算，截至 2002 年 12 月和 2003 年 12 月，造成的经济损失分别达1 000万美元和1 500万美元 (Bondad-Reantaso 等，2007)。

2003 年，KHVD 在日本首次暴发；2004 年和 2005 年，引发了日本全国范围内的锦鲤大规模死亡 (Sano 等，2004)，威胁到 7 500 万美元的日本观赏锦鲤产业 (Haenen 等，2004)。日本当局出台了新的管理办法，努力控制疾病的传播，尽管如此，全国范围的调查显示，这种病毒仍普遍存在于日本的自然水域 (Minamoto 等，2009，2012；Uchii 等，2009)。

2002 年 4 月，中国广东省的一个养殖场发生了进口锦鲤的疾病暴发，死亡率很高，PCR 分析证实鱼体感染了 KHV (Lih 等，2002)，此后，该疾病的发生和病毒的分离鉴定在中国均有报道 (Dong 等，2011，2013)。在亚洲出现 KHVD 暴发流行的国家或地区还有泰国 (Pikulkaew 等，2009)、菲律宾 (Somga 等，2010)、新加坡 (Lio-Po，2011)、韩国 (Oh 等，2001)、马来西亚 (Haenen 等，2004；Rathore 等，2012) 等。除了南美、北非以及澳大利亚之外，全球其他地区均有 KHVD 的发生和流行，并造成严重的经济损失。

第二节 病原学

由于该病的病原在形态特性上与疱疹病毒目 (Herpesvirales) 相似，最初被命名为锦鲤疱疹病毒 (Koi herpesvirus) (Hedrick 等，2000)。病毒具有囊膜，有典型二十面体的衣壳，衣壳由包被在双链 DNA 周围的蛋白构成 (Hedrick 等，2000；Hutoran 等，2005)。根据疾病的病理特点，这种病毒也曾被称为鲤间质性肾炎与鳃坏死病毒 (CNGV) (Pikarsky 等，2004；Ronen 等，2005)。基因组测序显示，该病毒含有一个非常大 (295 kb) 的基因组，与已知的哺乳类、鸟类和爬行类的疱疹病毒的序列同源性低 (Hutoran 等，2005；Waltzek 等，2005；Aoki 等，2007)。根据国际病毒分类委员会 (ICTV) 对疱疹病毒分类的最新修订，并比对与其他鲤疱疹病毒的同源性，该病毒归为异疱疹病毒科 (Alloherpesviridae)、鲤病毒属 (*Cyprinivirus*)、鲤疱疹病毒 3 型 (Cyprinid herpesvirus 3，CyHV-3) (King 等，2012)。

该病毒是迄今为止疱疹病毒目中基因组最大的病毒，由在中心大的核蛋白及其左右两侧 22 kbp 重复的序列组成。对日本、美国和以色列的 3 个病毒分离株的全基因序列比较显示，相似性达 99%；以色列分离株和美国分离株的关系更近，二者与日本分离株关系较远 (Aoki 等，2007)。对采自野生和养殖患病鲤、锦鲤样本 (主要来自亚洲和欧洲，美国和以色列各一个样本) 的 DNA 进行比较，结果显示，CyHV-3 基因组内的 3 个区域中有 10 个可变区。基于 3 个位点的多态性观察分析，分为亚洲基因型和欧洲基因型，分别包含 7 个和 2 个突变位点 (Kurita

等，2009）。串联重复序列（VNTR）的数量差异分析表明，印度尼西亚分离株与日本分离株关系相距较远（Avarre等，2011），此外，研究还表明，印度尼西亚分离株可能是欧洲基因型与亚洲基因型之间的中间株型（Sunarto等，2011）。

基因组分析发现，锦鲤疱疹病毒包含编码156个蛋白质的开放阅读框（ORFs）（Aoki等，2007），共40个结构蛋白，包括3个衣壳蛋白、13个囊膜蛋白、2个内膜蛋白和22个未分类的蛋白（Rosenkranz等，2008；Michel等，2010b）。所有156个ORFs的体外转录研究表明，如其他疱疹病毒一样，锦鲤疱疹病毒分为即早期型（IE或α）、早期型（E或β）和后期型（L或γ）（Ilouze等，2012a）。研究证明，CyHV-3基因组的复制和转录受温度调控，这有助于探究变温动物病毒潜伏期进化的机制（Dishon等，2007；Ilouze等，2012b）。

第三节 致病机理

锦鲤疱疹病毒病在露天池塘水温18～28℃时的春季和秋季发生，具有明显的季节性。CyHV-3具有水平传播、高传染性的特点，在同一池塘内由感染个体传染至健康个体，致死率达80%～100%。CyHV-3可感染各种年龄和大小的鲤，在感染后的6～22d发生死亡，8～12d为死亡峰值期。尽管该病具有高传染性和致死率，但仅发生在锦鲤和鲤中（*Cyprinus carpio*）（Bretzinger等，1999；Walster，1999；Perelberg等，2003），并不引起其他鱼类发病。研究表明，鲤科鱼类金鱼（*Carrassius auratus*）可能是CyHV-3的携带者（Soliman和El-Matbouli，2005；Davidovich等，2007；Sadler等，2008）。患病鱼厌食、昏睡，在水面或进/出水口处聚集，表现为缺氧状态。在感染后的第3天，患病鱼鳃出现坏死，伴有鱼体外部寄生虫和细菌的数量增加（Ariav等，1999）。患病鱼的皮肤苍白、失去光泽，表皮呈现微凸斑块以及黏液分泌增加（Hedrick等，2000；Perelberg等，2003，2005；Haenen等，2004）。

皮肤是CyHV-3侵入鱼体的主要入口（Costes等，2009b），一旦病毒在表皮中复制，就会迅速扩散，引起全身感染。感染24h后进行qPCR检测，在肝脏、胃肠道、肾脏、脑、脾脏等几乎所有内脏器官均能检测到较多的病毒（Gilad等，2004）。

当水温处于18～28℃的“发病温度”时，锦鲤疱疹病毒病就会暴发。在“发病温度”将健康的鲤鱼放于病鱼池中，或者放于含有细胞培养病毒的水中，均会导致75%～95%的高死亡率（Hedrick等，2000；Gilad等，2002，2003；Perelberg等，2003，2005；Hutoran等，2005）。然而，实验条件下，水温在29℃和15℃时感染，实验鱼无死亡，且没有表现出任何临床症状（Gilad等，2003；Ronen等，2003）。这些实验表明，与其他鱼类病毒病一样，锦鲤疱疹病毒病的发生与温度有关（Gilad等，2003；Ronen等，2003）。

体外研究表明，30℃时病毒能在培养细胞中存活30d，CyHV-3感染的细胞呈现变形空泡，培养温度上升后，转为正常且空斑消失；当温度调节至“发病温度”后，细胞再度呈现变形，出现空斑。这些结果表明，CyHV-3能在鱼体内存在很长时间，一旦水温转为“发病温度”，能使鱼重新感染，并出现锦鲤疱疹病毒病的临床症状（Dishon等，2007）。

体内实验也证实了温度与CyHV-3的复活、长期携带以及潜伏期有关。锦鲤感染病毒后的第65天，qPCR检测存活的锦鲤，结果呈现CyHV-3阳性（Gilad等，2004）。有研究显示，曾接触过 CyHV-3、临床表现健康的鱼，受到温度应激胁迫会发病（St-Hilaire 等，2005；Eide 等，2011），但也有研究报道称该结果无法重复（Gilad 等，2003）。

第四节　疫苗与疫苗接种

甲醛灭活 CyHV-3 疫苗的研究显示，将灭活的病毒包裹在脂质囊内，然后投喂鲤 3d。21d 后，攻毒检测疫苗的功效，该口服疫苗的保护率最高达 70%（Yasumoto 等，2006）。细菌人工克隆（BAC）制备缺失 *TK* 基因的重组 CyHV-3 减毒活疫苗，实验结果表明该疫苗只是部分减毒，尚有很强的致病力，然而存活下来的鱼获得了免疫保护（Costes 等，2009a）。未见这些疫苗在野外大规模接种试验和商业化应用的进一步报道。

减毒活疫苗在水产养殖中应用有许多优势，是最适合于鲤大规模接种的疫苗（Benmansour 和 de Kinkelin，1997；Shoemaker 和 Klesius，2014）。一般来说，活疫苗可刺激各个阶段的机体免疫，可实现体液免疫和细胞免疫反应的系统性与局部性的统一。由于热灭活病毒的免疫原性极低，且需要大量的抗原才能达到理想的免疫效果和免疫持续时间（Marsden 等，1996，1998），减毒活疫苗在鱼类中应用的优势尤其明显。此外，突变的 DNA 病毒出现复壮并威胁到接种群体的可能性很小。

自 2003 年起，以色列就开展了旨在获得一种无致病性的减毒 CyHV-3 的研究（Ronen 等，2003，2005；Perelberg 等，2005）。CyHV-3 以色列分离株，在锦鲤鳍细胞系中多次传代培养后，分离获得减毒株。第 20 代细胞培养收集的病毒注射或浸浴健康的幼鱼后，仅引起一小部分的鱼发病（图 27-1）（Ronen 等，2003，2005；Perelberg 等，2005）。体外细胞培养缺乏机体免疫系统，病毒基因会发生改变，这种改变在多次继代细胞培养过程中不断积累，产生病毒基因突变，使病毒更适宜体外细胞培养，因没经历鱼体免疫系统的筛选，因此病毒在鱼体内的致病力较弱。传代后，克隆病毒经紫外线照射，然后再克隆，以便获得在病毒基因组中插入突变

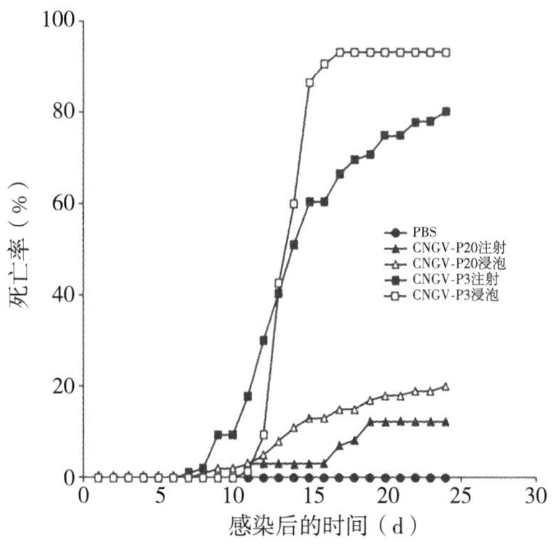

图 27-1　鲤感染（腹腔注射或浸泡）CNGV-P20 或 CNGV-P3 的死亡率

每组鱼 100 尾，平均体重 50g，阴性对照为注射 PBS

（引自 Ronen 等，2003）

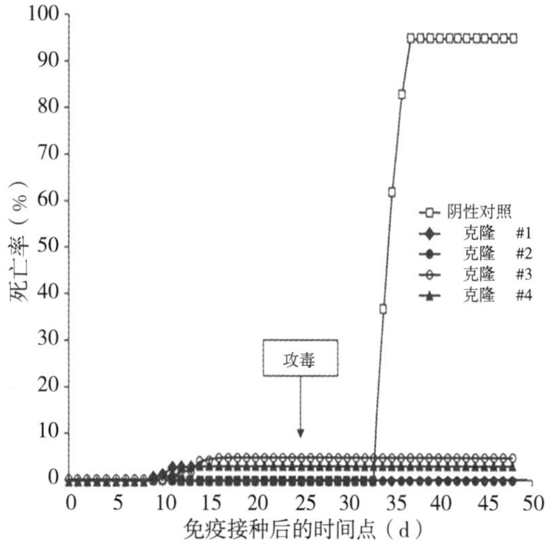

图 27-2　免疫接种的鲤攻毒后的死亡率

每组鱼 100 尾，平均体重 50g，腹腔分别注射免疫 CNGV-P26 的 4 个克隆，25d 后，与病鱼共养。注射 PBS 作为阴性对照

（引自 Ronen 等，2003）

基因的突变株（图 27-2）（Freese，1963；Drake，1969，1976；Ronen 等，2003）。对减毒病毒株的序列分析表明，超过 60%的毒株基因发生多个突变，包括预测开放阅读框内的 4 个缺失和插入（Ilouze 等，2008）。根据减毒病毒株的基因缺失建立的 PCR 检测方法，可以区分疫苗株与野生型的 CyHV-3（未发表）。

筛选出的减毒病毒株不会诱发疾病，能有效地保护鱼体在相当长的时间内不受感染(Ronen 等，2003；Perelberg 等，2005)。鲤对致病性和减毒性病毒都非常敏感，在含有病毒的水中，短暂的浸泡就足以使其感染。减毒病毒的免疫接种效果与温度相关，如果鱼免疫接种后，立即在非“发病温度”的水环境中养殖，这些鱼不能获得对锦鲤疱疹病毒的免疫保护。减毒病毒必须在鱼体内增殖，才能诱导鱼体产生对病毒感染的抵抗力。与致病性病毒只在“发病温度”时才会发病一样，减毒病毒也需要适当的温度才能诱发免疫保护。将鲤浸泡在含有减毒病毒的水中 40～60min 后，在“发病温度”(18～28℃）饲养 5d，可获得有效的保护（Perelberg 等，2005)。

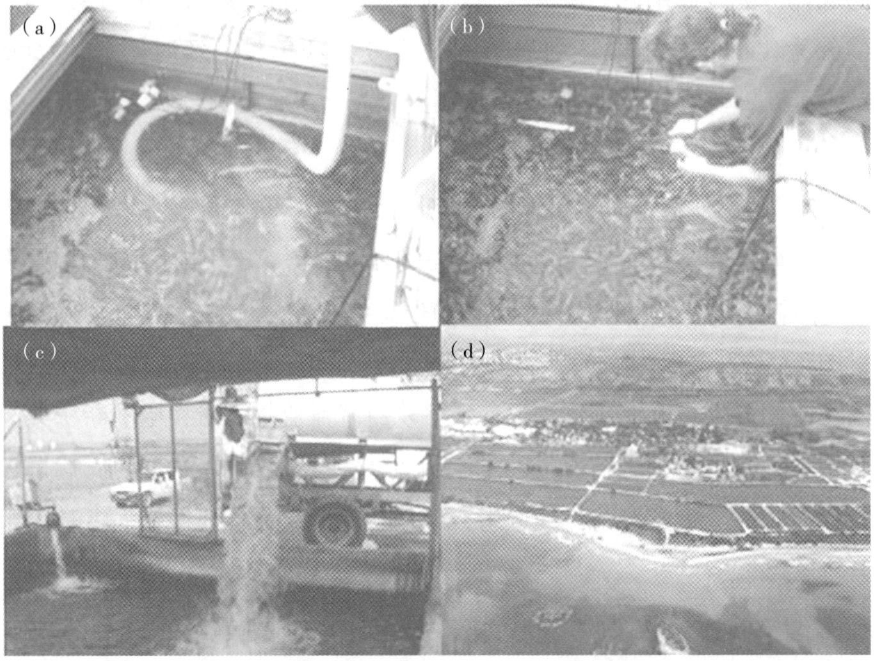

图 27-3　地中海地区 Maagan Michael 鱼类养殖场内锦鲤的免疫接种

(a）锦鲤被集中于接种箱中，监测水温并充氧；(b）疫苗化冻并倒入接种箱中；(c）锦鲤浸泡接种疫苗 45～60min 后，被转移至无病毒的环境中（至少）5d；(d）锦鲤被重新放入土池中

（由 Dagon 鱼类养殖场惠赠，Madan-Ma'agan Michael 同意许可）

在以色列，采用浸泡方式，对数以万计的最小为 3 月龄的幼鱼进行野外现场免疫（Lillehaug，2014）（Ashoulin，2008)。从池塘中捕捞锦鲤，称重后以 20%～30%（*w/v*）的密度放入一个大的充氧的免疫接种箱中。疫苗按生物量的比例倒入免疫水箱，浸泡免疫 40～60min。然后将已免疫的锦鲤转移至“发病温度”的池塘中至少 5d，最后放回土池（图 27-3)。

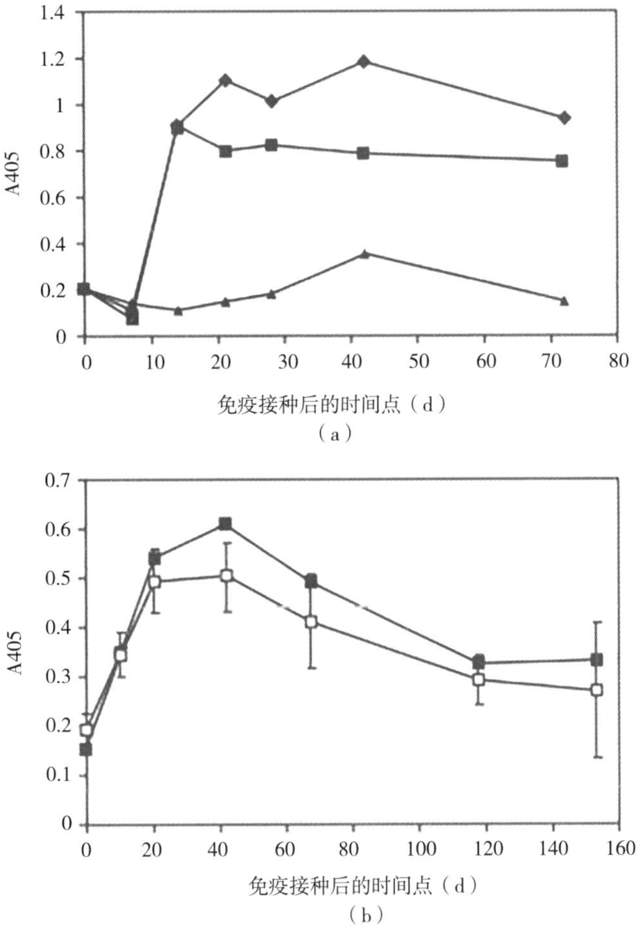

图 27-4　鱼免疫接种 CyHV-3 后体内抗病毒抗体变化的结果

（a）5～7 尾鱼的混合血清稀释 150 倍，ELISA 结果。（◆）野生株病毒，（■）减毒病毒免疫鱼，（▲）注射 PBS 鱼为阴性对照。每个时间点取样 5～7 尾鱼。（b）免疫接种减毒病毒后鱼体内抗体的变化。（■）5～7 尾鱼的混合血清 ELISA 结果（□）单尾鱼抗病毒抗体，ELISA 结果取平均值。每个时间点的 OD_{405} 值为从病毒免疫鱼的血清 OD_{405} 值中减去未免疫鱼血清的 OD_{405} 值。结果以标准差表示，指的是血清样品稀释 150 倍的值

（引自 Perelberg 等，2008）

第五节　效　　果

当温度和接种方案适当时，减毒疫苗是有效的，在实验室条件下和野外养殖的鲤体内均产生持久的保护，相对保护率大于 80%（Perelberg 等，2005，2008；Ronen 等，2005；Zak 等，2007；Ashoulin，2008；Weber 等，2011）。鱼对疱疹病毒病的免疫保护与抗野生型病毒特异性抗体水平的提高密切相关。鱼初次接种疫苗后，体内抗 CyHV-3 特异性抗体滴度在免疫后的第 7 天开始升高，在免疫后的 20～40d 达到峰值（图 27-4）。在不进行再次免疫的情况下，鱼体内的抗 CyHV-3 抗体水平在 6～9 个月内逐渐降低到略高于对照水平。在初次免疫后，鱼的存活率与抗 CyHV-3 抗体滴度的增加存在相关性，但在鱼的长期保护性实验中，即使没有检测到鱼体内的抗体，免疫过的鱼对攻毒感染仍具有显著的保护作用。这可能是由于病毒诱导活化了记忆细胞，然后引起快速的体液和细胞保护反应从而获得免疫保护（图 27-5）（Perelberg 等，2008）。虽然疫苗的保护作用是显著的，但免疫接种 13

个月后与9个月时相比，保护作用已下降，因此应该考虑每年进行加强免疫（Perelberg等，2008）。

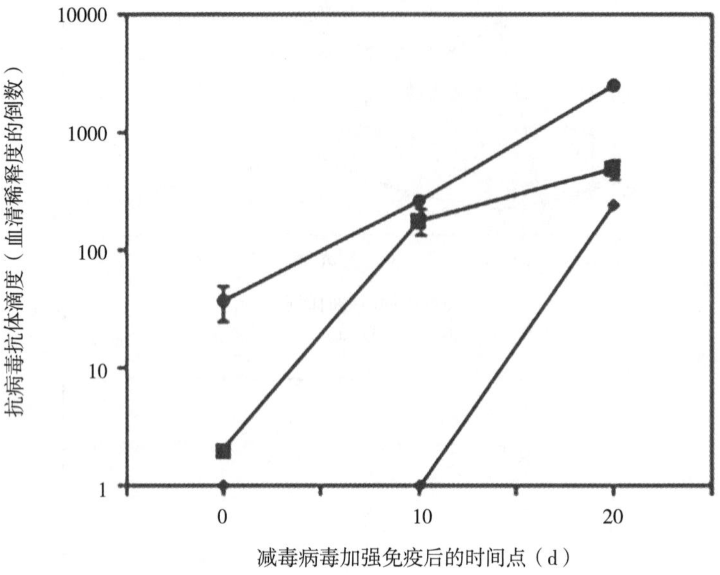

图 27-5　鱼体经减毒疫苗加强免疫后体内抗病毒抗体的水平

具有 CyHV-3 不同感染史的鱼减毒病毒加强免疫后，ELISA 检测梯度稀释的血清中抗病毒抗体水平。抗体活性以 405nm 处光学吸收值为 0.8 时对应的稀释度的倒数表示。（●）加强免疫前，4 龄的鲤接触过野生型 CyHV-3 3～5 次。（■）加强免疫前，免疫过减毒病毒 6 个月。（◆）从未接触过病毒的鱼。每个时间点取样 3 尾鱼的混合血清

（引自 Perelberg 等，2008）

第六节　安 全 性

为了监管，在以色列、印度尼西亚和美国，通过野外实地研究，已确定了该疫苗的良好安全性。良好的卫生条件、减少胁迫、合理的养殖管理是预防传染病的关键（Sommerset 等，2005）。在接种疫苗的鲤中观察到的副作用与疫苗接种之前鱼的健康状况有关（Perelberg 等，2008），水质差、养殖密度过大、低溶氧水平加上疫苗接种的应激，受免鱼可能出现不良的副作用，如果不及时治疗，可能会导致一部分接种鱼死亡。受胁迫的应激鱼容易被寄生虫（例如鱼波豆虫 *Ichthyobodo* sp.、三代虫 *Gyrodactylus* sp.）和条件致病菌（气单胞菌属 *Aeromonas* sp.）感染，某些情况下，疫苗接种过程和免疫应激使鱼增加了感染病原的概率。在美国对养殖锦鲤的免疫接种显示，大规格锦鲤（体重>80g）对疫苗的耐受性高于小规格鱼（Weber 等，2011）。

对减毒疫苗的耐受性可能会由于鱼的品系不同而有差异。研究表明，进口到以色列的匈牙利鲤（Szarvas-22 和 Dinnyes）接种疫苗后死亡率高于当地种或者匈牙利杂交品种（Zak 等，2007）。养殖人员总结原因认为，匈牙利鲤水土不服，对新的气候和水质适应性差，因为在匈牙利鲤群体中明显存在与疫苗无关的其他生理问题。

使用减毒活疫苗的主要安全问题之一是，减毒株再突变或逆转为致病性表型（Michel 等，2010a；Rathore 等，2012），为了避免不必要的重组和互补，用作疫苗的减毒株会在培养细胞中克隆两次。此外，通过“恢复实验”（VICH-GL41，2007；VS-MEMO800.201，2008）验证安

全性，即将减毒病毒疫苗接种于易感宿主，经过适当的孵育时间，从宿主获取减毒病毒，再接种到易感宿主动物，经 5 次传代，减毒病毒不会转为致病性表型即视为减毒成功（Dishon，2009）。

活疫苗还面临生态安全问题，与接种目标动物相比，活疫苗制剂可能导致更敏感的物种发病，因而限制了这类疫苗的使用。不像水生 RNA 病毒对多种鱼类具有致病性，人们普遍认为锦鲤疱疹病毒仅感染鲤和锦鲤（Ilouze 等，2006；Michel 等，2010a）。野生致病性病毒通过粪便（Dishon 等，2005）和分泌黏液将病毒颗粒（Perelberg 等，2003）释放到水环境中。接种减毒疫苗后有一个最短释放时间，因此，在免疫后 2～3 周内无法使用 PCR 和 ELISA 技术在受免鱼的排泄物中检测到减毒病毒（Dishon 等，2005；Dishon，2009）。对受免鱼的组织提取物 qPCR 分析显示，免疫后第 28 天，60％的鱼没有检测到疫苗病毒，其余的鱼检测到的病毒载量也非常低（Weber 等，2011）。此外，CyHV-3 攻毒后的检测结果表明，野生型病毒不能对接种疫苗鱼持续感染，因为攻毒后的第 21 天受免鱼的 qPCR 检测呈 CyHV-3 阴性（O'connor 等，未发表）。

第七节　监管问题

目前，CyHV- 3 的减毒活疫苗（由 Kovax Ltd. 生产）是唯一获得许可的商业化疫苗。该疫苗 2005 年在以色列获得许可，在锦鲤和食用鲤中广泛使用。经野外生产应用研究后，2010 年在印度尼西亚获准在鲤幼鱼上使用。2012 年初，美国农业部签发了该疫苗供应和销售的正式许可。

参考文献

Aoki，T.，Hirono，I.，Kurokawa，K.，*et al.*，2007. Genome sequences of three koi herpesvirus isolates representing the expanding distribution of an emerging disease threatening koi and common carp worldwide. *J Virol* 81：5058-5065.

Ariav，R.，Tinman，S.，Paperna，I. and Bejerano，I. 1999. First report of newly emerging viral disease of *Cyprinus carpio* species in Israel，in *Proceedings of the 9th International Conference of the EAFP*，Rhodes，Greece.

Ashoulin，O. 2008. *KV3 In Practice-A Clinician Perspective*. International Workshop on CyHV-3（KHV）Cyprinid Herpes Viruses-basic and applied aspects，Caesaria，Israel.

Avarre，J. C.，Madeira，J. P.，Santika，A.，*et al.*，2011. Investigation of Cyprinid herpesvirus-3 genetic diversity by a multi-locus variable number of tandem repeats analysis. *J Virol Methods* 173：320-327.

Benmansour，A. and de Kinkelin，P. 1997. Live fish vaccines：history and perspectives. *Dev Biol Stand* 90：279-289.

Bondad-Reantaso，M. G.，Sunarto，A. and Subasinghe，R. P. 2007. Managing the koi herpesvirus disease outbreak in Indonesia and the lessons learned. *Dev Biol*（*Basel*）129：21-28.

Bretzinger，A.，Fischer-Scherl，T.，Oumouma，M.，*et al.*，1999. Mass mortalities in koi，*Cyprinus carpio*，associated with gill and skin disease. *Bull Eur Assoc Fish Pathol* 19：182-185.

Costes, B., Lieffrig, F. and Vanderplasschen, A. 2009a. A recombinant koi herpesvirus (KHV) or Cyprinid herpesvirus 3 (CyHV-3) and a vaccine for the prevention of a disease caused by KHV/CyHV-3 in *Cyprinus carpio carpio* or *Cyprinus carpio koi*. European Patent 2195021 A1.

Costes, B., Raj, V. S., Michel, B., *et al.*, 2009b. The major portal of entry of koi herpesvirus in *Cyprinus carpio* is the skin. *J Virol* 83: 2819-2830.

Davidovich, M., Dishon, A., Ilouze, M. and Kotler, M. 2007. Susceptibility of cyprinid cultured cells to cyprinid herpesvirus 3. *Arch Virol* 152: 1541-1546.

Dishon, A. 2009. *Back passage/shed assay for Cyprinid Herpes Virus Type* 3 *modified live virus according to VS Memorandum No*. 800. 201. Kovax Ltd.

Dishon, A., Perelberg, A., Bishara-Shieban, J., *et al.*, 2005. Detection of carp interstitial nephritis and gill necrosis virus in fish droppings. *Appl Environment Microbiol* 71: 7285-7291.

Dishon, A., Davidovich, M., Ilouze, M. and Kotler, M. 2007. Persistence of cyprinid herpesvirus 3 in infected cultured carp cells. *J Virol* 81: 4828-4836.

Dong, C., Li, X., Weng, S., *et al.*, 2013. Emergence of fatal European genotype CyHV-3/KHV in mainland China. *Vet Microbiol* 162: 239-244.

Dong, C., Weng, S., Li, X., *et al.*, 2011. Characterization of a new cell line from caudal fin of koi, *Cyprinus carpio koi*, and first isolation of cyprinid herpesvirus 3 in China. *Virus Res* 161: 140-149.

Drake, J. W. 1969. Mutagenic mechanisms. *Annu Rev Genetics* 3: 247-268.

Drake, J. W. 1976. The biochemistry of mutagenesis. *Annu Rev Biochem* 45: 11-37.

Eide, K. E., Miller-Morgan, T., Heidel, J. R., *et al.*, 2011. Investigation of koi herpesvirus (KHV) latency in koi. *J Virol* 85: 4954-4962.

FAO. 2013. FAO-Fisheries and Aquaculture Information and Statistics Service 10/04/2013.

Freese, E. 1963. Molecular mechanisms of mutations, in *Molecular Genetics* (ed. J. H. Taylor). New York, Academic Press, p. 207.

Gilad, O., Yun, S., Andree, K. B., *et al.*, 2002. Initial characteristics of koi herpesvirus and development of a polymerase chain reaction assay to detect the virus in koi, *Cyprinus carpio koi*. *Dis Aquat Org* 48: 101-108.

Gilad, O., Yun, S., Adkison, M. A., *et al.*, 2003. Molecular comparison of isolates of an emerging fish pathogen, koi herpesvirus, and the effect of water temperature on mortality of experimentally infected koi. *J Gen Virol* 84: 2661-2668.

Gilad, O., Yun, S., Zagmutt-Vergara, F. J., *et al.*, 2004. Concentrations of a Koi herpesvirus (KHV) in tissues of experimentally infected *Cyprinus carpio koi* as assessed by real-time TaqMan PCR. *Dis Aquat Org* 60: 179-187.

Haenen, O. and Olesen, N. J. 2009. Koi herpes virus world wide: results on global koi herpesvirus questionnaire, in 14*th EAFP International Conference on Diseases of Fish and Shellfish*, Prague, Czech Republic (poster).

Haenen, O. L. M., Way, K., Bergmann, S. M. and Ariel, E. 2004. The emergence of koi herpesvirus and its significance to European aquaculture. *Bull Eur Ass Fish Pathol* 24: 293-307.

Hedrick, R. P., Gilad, O., Yun, S., *et al.*, 2000. A herpesvirus associated with mass mortality of juvenile and adult koi, a strain of common carp. *J Aquat Animal Health* 12: 44-55.

Hutoran, M., Ronen, A., Perelberg, A., *et al.*, 2005. Description of an as yet unclassified DNA virus from diseased *Cyprinus carpio* species. *J Virol* 79: 1983-1991.

Ilouze, M., Dishon, A. and Kotler, M. 2006. Characterization of a novel virus causing a lethal disease in carp and koi. *Microbiol Mol Biol Rev* 70：147-156.

Ilouze，M.，Dishon，A.，Davidovich，M.，*et al.*，2008. Characterization of attenuated CyHV-3 genes essential for virus multiplication and for inducing a mortal disease in carp，in *International Workshop on CyHV-3（KHV）Cyprinid Herpes Viruses-Basic and Applied Aspects*. Caesarea，Israel.

Ilouze，M.，Dishon，A. and Kotler，M. 2012a. Coordinated and sequential transcription of the cyprinid herpesvirus-3 annotated genes. *Virus Res* 169：98-106.

Ilouze，M.，Dishon，A. and Kotler，M. 2012b. Down-regulation of the cyprinid herpesvirus-3 annotated genes in cultured cells maintained at restrictive high temperature. *Virus Res* 169：289-295.

King，A. M. Q.，Lefkowitz，E.，Adams，M. J. and Carstens，E. B.（eds）2012. *Virus Taxonomy. Ninth Report of the International Committee on Taxonomy of Viruses*. San Diego，Elsevier Academic Press.

Kurita，J.，Yuasa，K.，Ito，T.，*et al.*，2009. Molecular epidemiology of Koi herpesvirus. *Fish Pathol* 44：59-66.

Lih，H.，Shi，X.，Gao，L. and Jiang，Y. 2002. Study on the aetiology of koi epizootic disease using method of nested-polymerase chain reaction assay（nested-PCR）. *J Huazhong Agricult Univ* 21：414-418.

Lillehaug，A. 2014. Vaccination strategies and procedures，in *Fish Vaccination*（eds R. Gudding，A. Lillehaug and Ø. Evensen）. Oxford，John Wiley & Sons，Ltd.，140-152.

Lio-Po，G. D. 2011. Recent developments in the study and surveillance of koi herpesvirus（KHV）in Asia，in *Diseases in Asian Aquaculture VII. Taipei* 2008.（eds M. G. Bondad-Reantaso，J. B. Jones，F. Corsin and T. Aoki）. Selangor，Malaysia，13-28.

Marsden，M. J.，Vaughan，L. M.，Foster，T. J. and Secombes，C. J. 1996. A live（delta aroA）*Aeromonas salmonicida* vaccine for furunculosis preferentially stimulates T-cell responses relative to B-cell responses in rainbow trout（*Oncorhynchus mykiss*）. *Infect Immun* 64：3863-3869.

Marsden，M. J.，Vaughan，L. M.，Fitzpatrick，R. M.，*et al.*，1998. Potency testing of a live，genetically attenuated vaccine for salmonids. *Vaccine* 16：1087-1094.

Michel，B.，Fournier，G.，Lieffrig，F.，*et al.*，2010a. Cyprinid herpesvirus 3. *Emerging Infect Dis* 16：1835-43.

Michel，B.，Leroy，B.，Stalin Raj，V.，*et al.*，2010b. The genome of cyprinid herpesvirus 3 encodes 40 proteins incorporated in mature virions. *J Gen Virol* 91：452-462.

Minamoto，T.，Honjo，M. N.，Uchii，K.，*et al.*，2009. Detection of cyprinid herpesvirus 3 DNA in river water during and after an outbreak. *Vet Microbiol* 135：261-266.

Minamoto，T.，Honjo，M. N.，Yamanaka，H.，*et al.*，2012. Nationwide Cyprinid herpesvirus 3 contamination in natural rivers of Japan. *Res Vet Sci* 93：508-514.

Neukirch，M. and Kunz，U. 2001. Isolation and preliminary characterization of several viruses from koi（*Cyprinus carpio*）suffering gill necrosis and mortality. *Dis Aquat Org* 21：125-135.

Oh，M. J.，Jung，S. J.，Choi，T. J.，*et al.*，2001. A viral disease occurring in cultured carp *Cyprinus carpio* in Korea. *Fish Pathol* 36：147-151.

Perelberg，A.，Smirnov，M.，Hutoran，M.，*et al.*，2003. Epidemiological description of a new viral disease afflicting cultured *Cyprinus carpio* in Israel. *Isr J Aquac-Bamidgeh* 55：5-12.

Perelberg，A.，Ronen，A.，Hutoran，M.，*et al.*，2005. Protection of cultured *Cyprinus carpio* against a lethal viral disease by an attenuated virus vaccine. *Vaccine* 23：3396-3403.

Perelberg，A.，Ilouze，M.，Kotler，M. and Steinitz，M. 2008. Antibody response and resistance of *Cyprinus carpio* immunized with cyprinid herpes virus 3 (CyHV-3). *Vaccine* 26: 3750-3756.

Pikarsky, E., Ronen, A., Abramowitz, J., *et al.*, 2004. Pathogenesis of acute viral disease induced in fish by carp interstitial nephritis and gill necrosis virus. *J Virol* 78: 9544-9551.

Pikulkaew, S., Meeyam, T. and Banlunara, W. 2009. The outbreak of koi herpesvirus (KHV) in koi (*Cyprinus carpio koi*) from Chiang Mai Province, Thailand. *Thai J Vet Med* 39: 53-58.

Rathore, G., Kumar, G., Raja Swaminathan, T. and Swain, P. 2012. Koi herpes virus: a review and risk assessment of Indian aquaculture. *Indian J Virol* 23: 124-133.

Ronen, A., Perelberg, A., Abramowitz, J., *et al.*, 2003. Efficient vaccine against the virus causing a lethal disease in cultured *Cyprinus carpio*. *Vaccine* 21: 4677-4684.

Ronen, A., Perelberg, A., Hutoran, M., *et al.*, 2005. Prevention of a mortal disease of carps induced by the carp interstitial and gill necrosis virus (CNGV) in Israel. *Bull Fish Res Agency Suppl* 2: 9-11.

Rosenkranz, D., Klupp, B. G., Teifke, J. P., *et al.*, 2008. Identification of envelope protein pORF81 of koi herpesvirus. *J Gen Virol* 89: 896-900.

Sadler, J., Marecaux, E. and Goodwin, A. E. 2008. Detection of koi herpes virus (CyHV-3) in goldfish, *Carassius auratus* (L.), exposed to infected koi. *J Fish Dis* 31: 71-72.

Sano, M., Takafumi, I., Kurita, J., *et al.*, 2004. First detection of Koi herpesvirus in cultured common carp *Cyprinus carpio* in Japan. *Fish Pathol* 39: 165-176.

Shoemaker, C. and Klesius, P. 2014. Replicating vaccines, in *Fish Vaccination* (eds R. Gudding, A. Lillehaug and Ø. Evensen). Oxford, John Wiley & Sons, Ltd., 33-46.

Soliman, H. and El-Matbouli, M. 2005. An inexpensive and rapid diagnostic method of Koi herpesvirus (KHV) infection by loop-mediated isothermal amplification. *Virol J* 2: 83.

Somga, J., de la Pena, L. D. and Sombito, C. D. 2010. Koi herpesvirus associated mortalities in quarantined koi carp in the Philippines. *Bull Eur Assoc Fish Pathol* 30: 2-7.

Sommerset, I., Krossøy, B., Biering, E. and Frost, P. 2005. Vaccines for fish in aquaculture. *Exp Rev Vaccines* 4: 89-101.

St-Hilaire, S., Beevers, N., Way, K., *et al.*, 2005. Reactivation of koi herpesvirus infections in common carp *Cyprinus carpio*. *Dis Aquat Org* 67: 15-23.

Sunarto, A., Rukyani, A. and Itami, T. 2005. Indonesian experience on the outbreak of koi herpesvirus in koi and carp (*Cyprinus carpio*). *Bull Fish Res Agency Suppl* 2: 15-21.

Sunarto, A., McColl, K. A., Crane, M. S., *et al.*, 2011. Isolation and characterization of koi herpesvirus (KHV) from Indonesia: identification of a new genetic lineage. *J Fish Dis* 34: 87-101.

Tu, C., Weng, M., Shiau, J. and Lin. S. 2004. Detection of Koi herpesvirus in koi *Cyprinus carpio* in Taiwan. *Fish Pathol* 39: 109-110.

Uchii, K., Matsui, K., Iida, T. and Kawabata, Z. 2009. Distribution of the introduced cyprinid herpesvirus 3 in a wild population of common carp, *Cyprinus carpio* L. *J Fish Dis* 32: 857-864.

VICH-GL41. 2007. Target animal safety: examination of live veterinary vaccines in target animals for absence of reversion to virulence, in *International Cooperation on Harmonisation of Technical Requirements for Registration of Veterinary Medicinal Products*, Bruxelles, Belgium.

VS-MEMO800. 201. 2008. *Veterinary Services Memorandum* 800. 201. Washington, DC.

Walster, C. 1999. Clinical observations of severe mortalities in koi carp, *Cyprinus carpio*, with gill disease. *Fish Vet J* 3: 54-58.

Waltzek, T. B., Kelley, G. O., Stone, D. M., *et al.*, 2005. Koi herpesvirus represents a third cyprinid herpesvirus (CyHV-3) in the family Herpesviridae. *J Gen Virol* 86: 1659-1667.

Weber, E. S., Hedrick, R., Dishon, A. and Salonius, K. 2011. Detailed study of the kinetics of infection following vaccination with a modified live CyHV3 vaccine and challenge with wild-type CyHV3 virus in koi carp (*Cyprinis carpio koi*). *Proceedings of the Eastern Fish Health Conference*, Lake Placid, New York.

Yasumoto, S., Kuzuya, Y., Yasuda, M., *et al.*, 2006. Oral immunization of common carp with a liposome vaccine fusing koi herpesvirus antigen. *Fish Pathol* 41: 141-145.

Zak, T., Perelberg, A., Magen, Y., *et al.*, 2007. Heterosis in the growth rate of Hungarian-Israeli common carp crossbreeds and evaluation of their sensitivity to Koi herpes virus (KHV) disease. *Isr J Aquac-Bamidgeh* 59: 63-72.

第二十八章 鲑甲病毒病的疫苗接种

Emilie Mérour and Michel Brémont

本章概要 SUMMARY

20 世纪 90 年代末，首次发现并鉴定出鲑甲病毒，包括鲑胰腺病病毒（Salmon pancreas disease virus，SPDV）和昏睡病病毒（Sleeping disease virus，SDV），并对其进行了分子特征研究，但仅有很少关于该病毒疫苗的研究。这些病毒对养殖和野生鱼类造成的影响日益严重。胰腺病在挪威已被列在必须上报的疾病名录中，对制定预防该病的策略更加重视。本章将介绍该病毒的分子生物学研究的最新进展、已研发出的诊断方法、最新的流行病学研究，以及已有的和未来可能用于预防这些病毒的疫苗。

第一节 发生流行与危害

1976 年，首次于苏格兰在大西洋鲑中发现胰腺病（PD）（Munro 等，1984），随后在爱尔兰、挪威、法国、美国和西班牙等地发现该病（Kent 和 Elston 1987；Raynard 等，1992），其中爱尔兰、挪威和苏格兰的大西洋鲑感染最为严重。患胰腺病的鲑表现为突然失去食欲、昏睡、粪便的数量增加、死亡率上升等，此外由于肌肉损伤，鱼体难以在水中保持平衡。

1985 年，法国首次在虹鳟中发现昏睡病，患病鱼的特征为行为异常，在养殖箱底部边缘静止不动，像是“睡眠状态”，所以俗称该病为昏睡病（Boucher 和 Baudin-Laurencin，1994）。推测这种行为主要由骨骼红肌大面积的坏死导致，该病病程缓慢，胰腺和心脏也会相继出现组织学病变（Boucher 和 Baudin-Laurencin，1996），虹鳟养殖过程中的各个阶段都可能感染该病。

目前鲑甲病毒（Salmonid alphavirus，SAV）分为 3 种亚型：SAV1 流行于爱尔兰和苏格兰，称为鲑胰腺病病毒，引起胰腺病；SAV2 称为昏睡病病毒，发生在淡水虹鳟中，引起昏睡病；SAV3 称为鲑胰腺病病毒，流行于挪威，在鲑和海水虹鳟中引起胰腺病，这些病毒在欧洲的地理分布如图 28-2 所示。近年来，SAV 的影响越来越严重，对大多数养殖场而言，根除 SAV 非常困难。研究表明，这些病毒能够直接通过水体快速水平传播（McLoughlin，1996）。目前，在挪威西部的大西洋鲑（*Salmo salar*）和虹鳟（*Oncorhynchus mykiss*）养殖系统中均能检测到 SAV3，且 SAV3 也出现在银化降海鲑的养殖区域。因此认为，几年前，通过西部大西洋鲑苗种的淡水主产地，经银化后鲑的运输，将 SAV3 从挪威西部传播到挪威北部（Bratland 和 Nylund，2009）。此外，通过自然或实验浸泡途径感染大西洋鲑，并未发现其精卵能够垂直传播 SAV3（Kongtorp 等，2010）。目前也不清楚在鱼类生存环境中是否存在 SAV 的载体或传播媒介。在法国淡水虹鳟中发现了独特的 SAV2 亚型，这可能是源于养殖场的一次单一感染并传播开的（Snow，2011）。通过实时 PCR 检测海水鱼类，虽然欧洲黄盖鲽（*Limanda limanda*）和鲽（*Pleuronectes platessa*）（Snow 等，2010）远离养殖场，但其体内也存在 SAV 的 RNA。因此，不能排除淡水虹鳟中的 SAV 可能来自海水物种。

第二节　病原学

该病病原在 20 世纪 90 年代就被怀疑是病毒（Boucher 等，1994；Castric 等，1997），后来经过研究确认病原属甲病毒（Villoing 等，2000）。昏睡病病毒（SDV）和鲑胰腺病病毒（SPDV）的基因组序列均已公布（Villoing 等，2000；Weston 等，2002），其基因组均由单股正链 RNA 分子组成，长度分别为 11 900 nt（核苷酸单位，译者注）和 11 919 nt，与所有甲病毒一样，SAV 基因组编码两个多聚蛋白（图 28-1），由蛋白水解产生非结构蛋白 nsP1、nsP2、nsP3、nsP4 共同构成复制酶复合物，以及主要结构蛋白衣壳（C）和两个外部糖蛋白（E2 和 E1）。

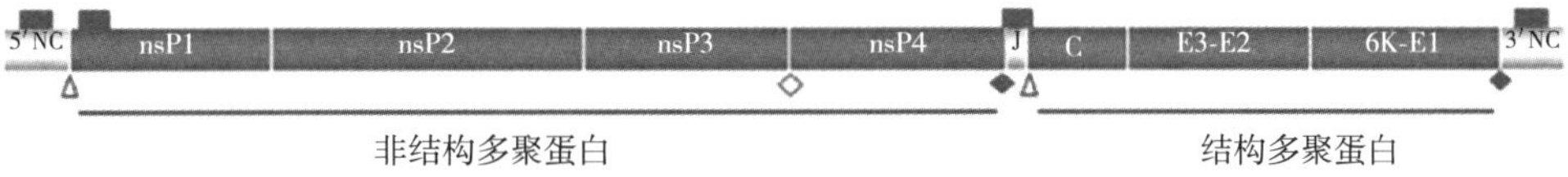

图 28-1　鲑甲病毒（SAV）RNA 基因组编码非结构多聚蛋白和结构多聚蛋白的示意图

昏睡病病毒（SDV）和鲑胰腺病病毒（SPDV）的非结构蛋白和结构蛋白的氨基酸相似性分别为 95%和 93.6%，并且存在血清交叉反应（Boucher 和 Baudin-Laurencin，1996），因此将 SDV 和 SPDV 同归为鲑甲病毒（SAV）（Weston 等，2002）。这种新的分类至少基于 3 个主要特点：①SDV 和 SPDV 非常相近，且不同于其他甲病毒；②非结构蛋白和结构蛋白均大于哺乳动物甲病毒的蛋白；③共栖感染实验证实病毒感染宿主不依赖于节肢动物（Boucher，1995），该现象在哺乳动物甲病毒中从未被记载。目前一些研究表明，SAV 可能共有 6 种亚型（Fringuelli 等，2008），只有 SAV2 可以感染鳟。

如上所述，不像哺乳动物典型的甲病毒，鲑甲病毒各亚型（SAVs）尽管基因组与其他甲病毒相似，但是基因组 5′和 3′的非编码区比典型的甲病毒要小得多（5′末端，SAVs 大约含有 20 个核苷酸，而典型的甲病毒含有 60～80 个核苷酸）并且 SAVs 的大多成熟蛋白比其他甲病毒的成熟蛋白大。SAV 蛋白的氨基酸序列与哺乳动物甲病毒的相似性较低，非结构蛋白具有 30%～35% 的相似性，结构蛋白具有 40%～45% 的相似性。另外，SAVs 的一个显著特征是：该病毒没有中间宿主（如节肢动物），因此认为其可以通过直接传播方式感染。已从象海豹身上的海兽虱（*Lepidophthirus macrorhini*）中分离出甲病毒（Linn 等，2001），这一发现说明，海兽虱可能是鲑甲病毒的载体或传播媒介，同时可能是野生病毒宿主，目前还没有确凿的证据证明上述观点，但发现在实验鱼类的养殖设施中，通过共栖感染，被感染的鱼可传播病毒至健康鱼。

像大多数感染鲑科鱼类的病毒一样，SAV 适宜在 10～14°C 低温条件下复制增殖。一些经典的鱼类细胞系，如 CHSE-214（大鳞大麻哈鱼胚胎细胞系）、BF2（蓝鳃太阳鱼成纤维细胞系）、RTG（虹鳟性腺细胞系）甚至 EPC（鲤上皮瘤细胞系）均可用于培养该病毒，细胞被感染后 7～10d 内，病毒诱导的细胞病理效应限于细胞变圆、折光增强，产生的病毒滴度可达到 10^8 pfu/mL。提纯 SDV 的 SDS 聚丙烯酰胺凝胶电泳，考马斯亮蓝染色结果，以及病毒粒子的电镜观察结果见图 28-3（MerouLepault，未发表）。

图 28-2　鲑甲病毒各亚型在欧洲的地理分布

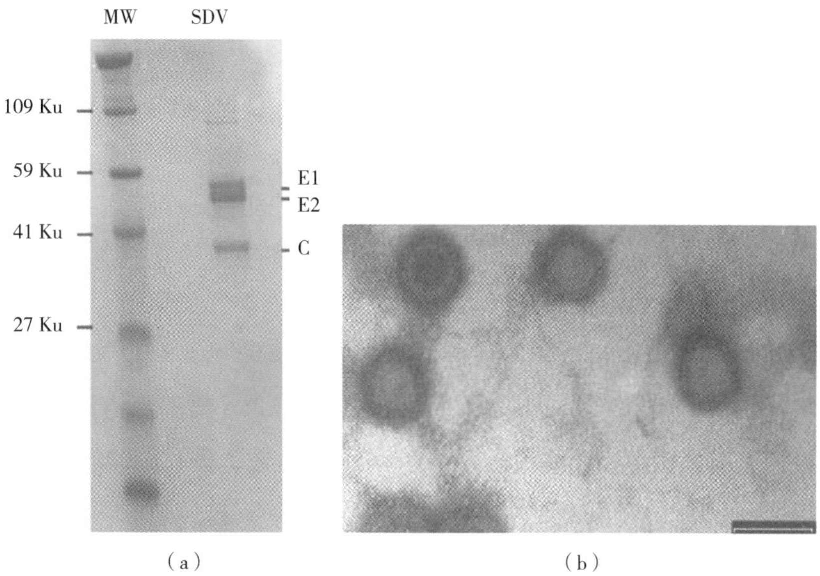

图 28-3　SDV 蛋白分析和电镜观察结果

(a) 蔗糖梯度纯化的 SDV，SDS-PAGE 蛋白分析。MW，分子量标记；SDV，纯化的昏睡病病毒。(b) 纯化 SDV 的电镜观察

第三节　致病机理

早期的研究证实，被感染鲑的血浆、白细胞和肾提取物可以传播鲑甲病毒病，发病的进程与温度相关，在 14°C 下的发病进程比 9°C 快（McLoughlin，未发表）。SAV1 和 SAV2 都会引起鲑和鳟的胰腺、心脏及肌肉发生病变，病变的严重程度取决于鱼的品系。对于鲑胰腺病，最初的组织学病变是胰腺的坏死，然而，随后的研究发现患病鲑还表现出严重的心肌和骨骼肌病变（McLoughlin 和 Graham，2007）。发病时间进程的研究结果表明，鲑急性发病会发生胰腺坏死，腺泡组织遭到破坏，但很少或无炎症反应，心肌的病变也不明显。骨骼肌病变出现在发病的晚期，往往在胰腺和心脏出现病变后约 4 周才发生，由于肌肉的病变，患病鱼表现出行为异常。

第四节　免疫力与疫苗研发

由血清被动免疫实验证实，体液免疫是鲑免受 SAV 感染死亡的主要方式（Houghton 和 Ellis，1996），体外实验也证明，Ⅰ型干扰素在大西洋鲑抵御 SAV 过程中发挥重要作用（Gahlawat 等，2009；Xu 等，2010）。感染鲑甲病毒的鱼康复后，抗病毒的能力至少可以持续 9 个月，自然发病存活的鲑，一般不会复发。感染 14～16d 后，在大多数被感染的大西洋鲑体内可检测到中和抗体，感染 28d 后，在 100% 的大西洋鲑体内可检测到中和抗体（McLoughlin 等，1996）。

迄今为止，大多数水产养殖用的病毒疫苗是灭活全病毒疫苗或重组蛋白疫苗，目前仅有 Intervet-Norbio（现为 MSD）公司研发的胰腺病的灭活病毒疫苗（NORVAX® Compact PD）获得商业许可（Bang Jensen 等，2012），用于大西洋鲑腹腔注射免疫，并对不同鲑甲病毒亚型都具有交叉保护作用，且保护期长，野外试验鱼体免疫后的保护时间可达 80 周，实验室试验达 12

个月。鲑接种疫苗的好处是降低了由 SAV 造成的死亡率，抑制 SAV 病毒的扩散以及减少了由 SAV 引起的病变。

基于 SAV 亚型 3 研发出了 ALV405，一种新灭活疫苗（Karlsen 等，2012），该疫苗已完成了实验室和野外条件下的有效性和安全性评估，结果证明，ALV405 可成为控制 SAV 流行的一种新的有效疫苗。通过 3 种攻毒模型进行的疫苗功效检测表明，腹腔注射攻毒，ALV405 可降低感染率和死亡率；共栖感染攻毒，ALV405 诱导的免疫保护可阻止胰腺病的发展扩散；自然感染环境中，ALV405 可降低 SAV 暴发时的死亡率。

基于 SAV3 的 E1、E2 刺突蛋白的亚单位疫苗以及 DNA 疫苗的免疫原性和保护力，已经在大西洋鲑上进行了研究，并与灭活全病毒疫苗进行了比较（Xu 等，2012）。DNA 疫苗是一种控制胞内病原感染的有前景的免疫接种策略（这也适用于硬骨鱼类），实验证明，这种方法在预防鱼类弹状病毒方面已获得了较好效果（Boudinot 等，1998）。然而，表达 SAV E1 和 E2 的 DNA 疫苗，在接种后并没有诱导明显的抗体反应，攻毒后受免鱼与未受免鱼比较，死亡率并没有降低。E2 亚单位疫苗和 SAV 灭活疫苗的效果对比研究显示，接种后鱼体产生了抗 SAV 的中和抗体，这是鱼体防止感染和死亡的主要机制。但 E1 亚单位疫苗免疫后，鱼体没有产生中和抗体。因此，E2 重组蛋白亚单位疫苗具有更好的应用前景。

减毒活疫苗对鱼类来说可能是最有前途的疫苗，因为与其他的疫苗相比，活疫苗目前的最大优点是可以通过浸泡接种。浸泡免疫是最简单的接种方法，也不需要佐剂，因此不会产生油类佐剂配方导致的副作用。SAV 唯一的完全减毒活疫苗是由鳟昏睡病病毒（SDV，SAV2）制备而成，是利用反向遗传学系统通过致病 SDV 株的 cDNA（SDV RNA 的全长 cDNA 拷贝）重组的一株可复制的 SDV 病毒（rSDV）(Moriette 等，2006)，经浸泡和注射感染虹鳟，均表明重组 rSDV 已完全减毒。rSDV 免疫鳟后，无论浸泡或注射野生型 SDV 攻毒，该疫苗均可产生长期保护（至少 7 个月），保护率可达 100%，且 rSDV 接种的鱼还可预防野生胰腺病病毒（SAV1）的感染。研究表明，该病毒毒性衰减主要是由于 E2 糖蛋白中单个氨基酸残基发生了变化（Mérour 等，2013）。虽然重组活病毒可能是减毒活疫苗研制的有效途径，但重组活病毒在水产养殖领域推广应用，仍存在一些安全性限制因素亟待研究。主要的限制因素包括消费者对安全的关注，和环境安全问题（Shoemaker 和 Klesius，2014）。未来，需要通过探索新的研制方法，制备出宜于浸泡接种、不复制、在环境中不释放且不扩散的重组减毒活疫苗。为此，需要进一步的持续研究。

参考文献

Bang Jensen，B.，Kristoffersen，A. B.，Myr，C. and Brun，E. 2012. Cohort study of effect of vaccination on pancreas disease in Norwegian salmon aquaculture. *Dis Aquat Org* 102：23-31.

Boucher，P. 1995. *Etiologie et pathogénie de la maladie du pancréas et de la maladie du sommeil chez les salmonidés d'élevage*. PhD thesis，University of Rennes 1.

Boucher，P. and Baudin-Laurencin，F. 1994. Sleeping disease of salmonids. *Bull Eur Assoc Fish Pathol* 14：179-180.

Boucher，P. and Baudin-Laurencin，F. 1996. Sleeping disease and pancreas disease comparative histopathology and acquired cross-protection. *J Fish Dis* 19：303-310.

Boucher, P., Castric, J. and Baudin-Laurencin F. 1994. Observation of virus-like particles in rainbow trout *Oncorhyncus mykiss* infected with sleeping disease. *Bull Eur Assoc Fish Pathol* 14: 215-216.

Boudinot, P., Blanco, M., deKinkelin, P. and Benmansour, A. 1998. Combined DNA immunization with the glycoprotein gene of viral hemorrhagic septicemia virus and infectious hematopoietic necrosis virus induces double-specific protective immunity and nonspecific response in rainbow trout. *Virology* 249: 297-306.

Bratland, A. and Nylund, A. 2009. Studies on the possibility of vertical transmission of Norwegian salmonid alphavirus in production of Atlantic salmon in Norway. *J Aquat Anim Health* 21: 173-178.

Castric, J., Baudin-Laurencin, F., Brémont, M., *et al.*, 1997. Isolation of the virus responsible for sleeping disease in experimentally infected rainbow trout (*Oncorhynchus mykiss*). *Bull Eur Assoc Fish Pathol* 17: 27-30.

Fringuelli, E., Rowley, H. M., Wilson, J. C., *et al.*, 2008. Phylogenetic analyses and molecular epidemiology of European salmonid alphaviruses (SAV) based on partial E2 and nsP3 gene nucleotide sequences. *J Fish Dis* 31: 811-823.

Gahlawat, S. K., Ellis, A. E. and Collet, B. 2009. Expression of interferon and interferon—induced genes in Atlantic salmon *Salmo salar* cell lines SHK-1 and TO following infection with salmon alphavirus SAV. *Fish Shellfish Immunol* 26: 672-675.

Houghton, G. and Ellis, A. E. 1996. Pancreas disease in Atlantic salmon: serum neutralisation and passive immunisation. *Fish Shellfish Immunol* 6: 465-472.

Karlsen, M., Tingbø, T., Solbakk, I. T., *et al.*, 2012. Efficacy and safety of an inactivated vaccine against salmonid alphavirus (family Togaviridae). *Vaccine* 30: 5688-5694.

Kent, M. L. and Elston, R. A. 1987. Pancreas disease in pen-reared Atlantic salmon in North America. *Bull Europ Assoc Fish Pathol* 7: 29-31.

Kongtorp, R. T., Stene, A., Andreassen, P. A., *et al.*, 2010. Lack of evidence for vertical transmission of SAV 3 using gametes of Atlantic salmon, *Salmo salar* L., exposed by natural and experimental routes. *J Fish Dis* 33: 879-888.

Linn, M. L., Gardner, J., Warrilow, D., *et al.*, 2001. Arbovirus of marine mammals: a new alphavirus isolated from the elephant seal louse, *Lepidophthirus macrorhini*. *J Virol* 75: 4103-4109.

McLoughlin, M. F. and Graham, D. A. 2007. Alphavirus infections in salmonids-a review. *J Fish Dis* 30: 511-531.

McLoughlin, M. F., Nelson, R. T., Rowley, H. M., *et al.*, 1996. Experimental pancreas disease in Atlantic salmon *Salmo salar* post-smolts induced by salmon pancreas disease virus (SPDV). *Dis Aquat Org* 26: 117-124.

Mérour, E., Lamoureux, A., Bernard, J., *et al.*, 2013. A fully attenuated recombinant salmonid alphavirus becomes pathogenic through a single amino acid change in the E2 glycoprotein. *J Virol* 87: 6027-6030.

Moriette, C., Leberre, M., Lamoureux, A., *et al.*, 2006. Recovery of a recombinant salmonid alphavirus fully attenuated and protective for rainbow trout. *J Virol* 80: 4088-4098.

Munro, A. L. S., Ellis, A. E., McVicar, A. H., *et al.*, 1984. An exocrine pancreas disease of farmed Atlantic salmon in Scotland. *Helgol Meeresunters* 37: 571-586.

Raynard, R., Houghton, G., Munro, A. L. S. 1992. *Pancreas Disease of Atlantic Salmon: Proceedings of a European Commission Workshop*. Edinburgh, Scottish Office Agriculture and Fisheries Department (Scottish Aquaculture Research Report No. 1).

Shoemaker, C. A. and Klesius, P. H. 2014. Replicating vaccines, in *Fish Vaccination* (eds R. Gudding, A. Lillehaug and Ø. Evensen). Oxford, John Wiley & Sons, Ltd., 33-46.

Snow, M. 2011. The contribution of molecular epidemiology to the understanding and control of viral diseases of salmonid aquaculture. *Vet Res* 42：56.

Snow，M.，Black，J. A.，Matejusova，I.，*et al.*，2010. Evidence for the detection of salmonid alphavirus（SAV）RNA in wild marine fish in areas remote from aquaculture activity：Implications for the origins of salmon pancreas disease（SPD）in aquaculture. *Dis Aquat Org* 91：189-200.

Villoing，S.，Béarzotti，M.，Chilmonczyk，S.，*et al.*，2000. Rainbow trout sleeping disease virus is an atypical alphavirus. *J Virol* 74：173-183.

Weston，J.，Villoing，S.，Brémont，M.，*et al.*，2002. Comparison of two aquatic alphaviruses，salmon pancreas disease virus and sleeping disease virus，by genome sequence analysis，monoclonal reactivity，and cross-infection. *J Virol* 76：6155-6163.

Xu，C.，Guo，T. C.，Mutoloki，S.，*et al.*，2010. Alpha interferon and not gamma interferon inhibits salmonid alphavirus subtype 3 replication *in vitro*. *J Virol* 84：8903-8912.

Xu，C.，Mutoloki，S. and Evensen，Ø. 2012. Superior protection conferred by inactivated whole virus vaccine over subunit and DNA vaccines against salmonid alphavirus infection in Atlantic salmon（*Salmo salar* L）. *Vaccine* 30：3918-3928.

第二十九章　野田病毒病的疫苗接种

Sonal Patel and Audun H. Nerland

本章概要　SUMMARY

β野田病毒（*Betanodavirus*）可感染世界范围内的多种鱼类，引起的鱼类疾病称为病毒性脑病和视网膜病（Viral encephalopathy and retinopathy，VER），或称为病毒性神经坏死（viral nervous necrosis，VNN），死亡率高达 100%。目前还没有预防 β野田病毒的商品疫苗。与其他胞内病原一样，β野田病毒疫苗的研发也面临着巨大的挑战。几种疫苗设计策略已在实验室和自然环境中进行了测试，其中重组病毒蛋白与有效佐剂的联合使用已获得良好的结果，具有较好的应用前景。然而，由于这种疾病的暴发主要发生在仔鱼和幼鱼早期阶段，此时鱼体免疫系统发育尚不完全，使得该病毒疫苗的研发尤其具有挑战性。虽然已开展了亲鱼疫苗接种的研究，仍需进一步探索。因此建议在仔鱼和幼鱼阶段使用免疫刺激剂，在幼鱼的获得性免疫系统形成后再进行疫苗接种，这将是解决该病高死亡率的最佳方法。

第一节　病毒性脑病和视网膜病

1990 年前后报道了一种感染多种鱼类神经系统的疾病，因为在鱼类病灶中观察到大量的病毒样颗粒，故认为病原是病毒，该病被称为病毒性神经坏死（VNN）（Yoshikoshi 和 Inoue，1990）、脑脊髓炎（Bloch 等，1991）或病毒性脑病和视网膜病（VER）（Munday 等，1992）。本章将统一称为病毒性脑病和视网膜病。

对该病毒的特性研究发现，其结构类似于之前从昆虫中分离到的野田病毒，因为首次在日本野田村的昆虫中发现，因此命名为野田病毒（Mori 等，1992）。后来感染昆虫的病毒命名为 α 野田病毒，感染鱼类的称为 β 野田病毒，其中 β 野田病毒也被称为神经坏死病毒（NNV）。接下来，在本章将简称 β 野田病毒为 NNV。

第二节 发生流行与危害

世界上有30多种鱼类被确认患有VER，患病鱼类主要是海水鱼类的仔鱼和幼鱼（Munday等，2002）。在仔鱼和幼鱼阶段的急性感染期，死亡率可达到100%，而易感的几种鱼类在生长期的死亡率也可达50%。在某些情况下，该病呈现亚急性，导致一部分鱼持续感染，出现无症状带毒的个体。NNV是很多地区温水性海水鱼类养殖面临的一大难题，例如地中海的海鲈和真鲷（Bovo等，1999；Ucko等，2004）、日本的鹦嘴鱼（Yoshikoshi和Inoue，1990）、挪威的大菱鲆（Bloch等，1991）以及澳大利亚的尖吻鲈（Glazebrook等，1990）。NNV的感染已成为限制冷水性大西洋庸鲽养殖产业发展的主要原因之一（Grotmol等，1997），同时也给大西洋鳕的养殖带来困扰（Patel等，2007）。虽然NNV易感鱼类大多为海水鱼，但在少数淡水鱼类中也发现了该病毒的感染。自2002年以来，感染NNV的淡水鱼类的种类数量持续增加（Munday等，2002；Athanassopoulou等，2004；Bigarré等，2009），有报道称条纹鲈和白鲈的杂交种（*Morone saxatilis* × *Morone chrysops*）以及大口黑鲈（*Micropterus salmoides*）均可感染NNV（Bovo等，2011）。

除少数鱼外，一般鱼的幼体阶段更易感染NNV。例如，NNV可导致大西洋庸鲽仔鱼的高死亡率，但当幼体完成变态发育（体重约2 g）后，鱼体对NNV似乎产生了很强的抗感染能力。

第三节 病原学

神经坏死病毒（NNV）属于野田病毒科（Nodaviridae），结构非常简单，粒子直径25～35nm、无囊膜。该病毒科成员首次分离自昆虫中，然后又从鱼体中分离出来，后来也从甲壳动物（Bonami和Widada，2011）和线虫（Félix等，2011）中分离出类似的病毒。该科病毒结构组成简单，因此能够在不同物种的细胞中增殖。NNV基因组由两个单股正链RNA组成：RNA1片段（3 100个核苷酸）编码RNA依赖的RNA聚合酶（RdRp）的催化部分，负责基因组的复制；RNA2片段（1 400个核苷酸）编码衣壳蛋白，包裹基因组构成二十面体结构；另外，RNA3片段是从RNA1中生成，可编码多肽B2，RNA3只存在于受感染的细胞，不被包入病毒颗粒。有迹象表明，B2与RNA干扰系统相互作用并抑制该系统的功能，破坏鱼体防御病毒侵染的免疫系统。

基于病毒RNA2片段的可变区域（T4区），将目前发现NNV病毒株分为4种主要的基因型（Nishizawa等，1997）：条纹石斑鱼神经坏死病毒（Striped jack nervous necrosis virus，SJNNV）、红斑石斑鱼神经坏死病毒（Red spotted grouper nervous necrosis virus，RGNNV）、红鳍东方鲀神经坏死病毒（Tiger puffer nervous necrosis virus，TPNNV）和条斑星鲽神经坏死病毒（Barfin flounder nervous necrosis virus，BFNNV）。而从大菱鲆中分离的NNV被认为是第5种基因型（Johansen等，2004b）。这些基因型以首次分离的鱼命名。之后的研究表明，这些病毒株并不是完全符合物种特异性，似乎与地理分布更相关，水温是地理分布的一个重要因素（Hata等，2010）。相关研究表明，NNV在极端环境条件下非常稳定（Frerichs等，2000）。

第四节 致病机理

神经坏死病毒的主要靶器官是中枢神经系统和视网膜，病鱼微观病变的特点是细胞空泡化和神经元变性。早期感染阶段可见免疫染色阳性的小病灶，随后，脑组织可见免疫检测阳性的巨噬样细胞且在胞内或胞外出现液泡。视网膜的光感受器出现退化的视锥细胞，可见黑色素的聚集。在病鱼的大脑和视网膜感光层中常见细胞结构破坏和细胞浸润，从而导致组织大量坏死。

研究认为，NNV 进入细胞依赖于内吞途径（Adachi 等，2007；Liu 等，2005），条纹鳢细胞（SSN-1）表面的唾液酸参与病毒颗粒的初始结合（Liu 等，2005）。NNV 可通过诱导细胞凋亡和继发性坏死而脱离宿主细胞（Guo 等，2003；Chen 等，2007；Su 等，2009）。

从世界各地感染鱼类的种类来看，盐度不是 NNV 的限制因素，这也增加了其他重要淡海水经济鱼类被感染的可能性。

患病鱼常见的临床症状有游泳行为异常，如翻滚、不协调地狂游以及食欲减退和色素沉积等（Munday 等，2002），偶尔可见肌肉组织的强直性痉挛导致鱼体暂时性屈曲。关于淡水鱼类患病的临床症状的报道较少，大多与海水鱼类的典型症状一致，包括活力差和消瘦。

神经坏死病毒的传播途径尚未完全明晰。病毒可以通过亲鱼的卵或精子垂直传播，也可以通过水或饲料进入养殖设施进行水平传播。病毒在亚临床感染的鱼中可以存活很长时间并且仍然具有传染性（Johansen 等，2004a），在性腺和卵子中可检测到 NNV。在对海鲈的研究中也证实了 NNV 的垂直传播途径（Breuil 等，2002；Kai 等，2010）。从养殖庸鲽幼鱼的感染个体内检测到高浓度 NNV（Nerland 等，2007），这表明病毒已发生水平传播。因此，可以推论，通常病毒是从亚感染的亲本到仔鱼进行垂直传播，如果其中的一些鱼苗被病毒感染，病毒就会随之进入水体，进行水平传播。总之，一旦病毒进入养殖环境或设施，可能就很难根除。

第五节 鱼体对神经坏死病毒的免疫反应

鱼类的生长阶段、协同感染、环境和生理应激、病毒的变异和宿主的遗传性状等因素都会影响疾病的结局以及疫苗效果。

大菱鲆 NNV 攻毒实验显示，感染死亡率与鱼的年龄有关，虽然差别很小，但观察到的高死亡率的鱼体重为 10g，中等死亡率的鱼体重约为 20g，低死亡率的鱼体重超过 40g。与此相反，如斑鳃棘鲈（*Plectropomus maculatus*）的成鱼更容易感染 NNV（Nopadon 等，2009）。

不同鱼类的免疫系统及其对某一病原免疫反应均存在差异，因此，揭示各种鱼类感染的 NNV 摄入和清除机制是极其困难的。鱼体免疫反应的不同，不仅是由于环境和生理上的应激，还与这些指标的波动有关。无论是感染还是疫苗接种，鱼体的体液反应和细胞反应对于获得有效的保护都非常重要。鱼在仔鱼/幼鱼阶段受源自母源成分的保护（如凝集素和抗体），且主要依赖于早期阶段就已存在的先天性免疫。对于大西洋庸鲽，其获得性免疫系统的发育可能需要几天至几个月的时间，这与温度直接相关。事实上，NNV 可感染不同养殖温度下的多种鱼类，这表明不可能通过某一种鱼、某一病毒基因型，对其他鱼和其他基因型的感染进程、免疫应答和疾病结局做出判断，需要有针对性地开展广泛的研究。因此，在制定 NNV 的疫苗策略时，必须注意这一点。

第六节 疫 苗

目前，虽然实验室条件下 NNV 疫苗通过多种接种策略已显示出较高的保护作用，但尚无商品化疫苗。现已开展疫苗测试的研究，实验鱼都是较大规格的鱼种，比易感的幼体阶段要大。疫苗在幼体阶段接种较难是由于：①免疫系统发育不完善；②鱼的个体小，注射免疫难以操作，浸泡或口服的大规模疫苗接种方案不成熟。有一种可行的方法是，对产卵前的亲鱼进行免疫接种，防止病毒垂直传播。

神经坏死病毒作为一种胞内病原体，鱼体对其产生的有效保护可能依赖于 B 细胞和 T 细胞的激活。病毒感染时，由 B 细胞产生的特异性抗体可与从感染细胞中释放出来的病毒粒子结合，但是对于细胞内部的病毒，除非细胞被细胞毒性 T 细胞（Tc-细胞）裂解，否则抗体无法与病毒接触，尤其是 NNV 可以隐藏在细胞内并持续很长一段时间。B 细胞和 Tc 细胞均依赖于辅助性 T 细胞（Th 细胞）的信号得以活化和增殖，因此，研发保护性的 NNV 疫苗，抗原的递呈必须同时刺激 B 细胞和 T 细胞。

除了特异性免疫外，先天性免疫系统也会对病毒感染给予免疫保护。病毒刺激后，先天性免疫系统可很快发生反应，并产生高保护，但这种保护作用不会持续很久。然而，对于疫苗设计来说，先天性免疫极有意义：首先，先天性免疫反应是炎症的一部分，而炎症反应通常是获得特异性免疫反应所必需的；其次，有效的特异性免疫产生有迟滞期，先天性免疫应答可在这段时间保护鱼体免受感染。

已开展了 NNV 疫苗的几种不同接种途径的实验研究，如注射（腹腔或肌肉）或大规模免疫（浸泡或口服），这些接种方案在保护水平、副作用、实用性和成本上各有利弊。通常，疫苗接种方式根据鱼体大小、病毒的种类和疫苗的配方来选择。由于 NNV 主要感染仔鱼和幼鱼，浸泡或口服的大规模免疫接种相对更实用且易于操作。当接种幼鱼和稍大的鱼时，免疫方式的选择更加灵活，如鲈、大西洋鳕的 NNV 接种。NNV 的垂直传播可能是卵子发育过程中病毒侵入卵子，也可能是受精过程中精子携带病毒进入卵子，这在养殖石斑鱼中得到证实。研究发现，NNV 不是在卵的表面而是存在于胚胎中，并且 NNV 可在胚胎发育过程中不断复制（Kuo 等，2012）。因此，亲鱼的疫苗接种对预防 NNV 传播非常重要。

在下面的章节和表 29-1 中，概述并列出了迄今各基因型 NNV 疫苗的实验研制方法。由于目前尚无 NNV 的商品疫苗，因此对这些疫苗的配方、导入途径、疫苗效果和副作用等不进行讨论，但会介绍实验室研制的疫苗和中试测试的疫苗。

表 29-1 部分β野田病毒不同基因型的实验疫苗的制备方法及其接种方式和攻毒检测

鱼类种类	NNV 基因型	抗原类型	剂量	平均体重	免疫方式	存活率		参考文献
						对照组	免疫组	
七带石斑	RGNNV	活病毒	$10^{4.3}$ $TCID_{50}$/mL 的 NNV，0.1mL/尾	52g、80g	IM	71～79.5	97.5～99.8	Oh 等(2012)

（续）

鱼类种类	NNV 基因型	抗原类型	剂量	平均体重	免疫方式	存活率		参考文献
						对照组	免疫组	
七带石斑	RGNNV	福尔马林灭活	$10^{9.1}TCID_{50}$/mL 的 NNV，0.1mL/尾	12.3 g	IP	4～40	67～96	Yamashita 等(2005)
七带石斑	RGNNV	重组蛋白	2 × 60μg/尾	28 g	IM	0～35	12～65	Tanaka 等(2001)
驼背鲈	RGNNV	3 株重组蛋白	3 × 0.21mg(0.7 mg/株)	27 g	IM	0	60～77.5	Yuasa 等(2002)
大菱鲆	BFNNV	重组蛋白	10μg/尾	2.2 g	IP	61	87	
大菱鲆	BFNNV	DNA 疫苗	20μg/尾	2.2 g	IM	57	57～59	Sommerset 等(2005)
大菱鲆	SJNNV	重组蛋白	15μg/尾	1.8 g	IP	52～56	68～92	Husgård 等(2001)
斜带石斑	RGNNV	重组蛋白	*	18～19 日龄	口服（卤虫包裹）	44～54	80～86	Lin 等(2007)

注：目前已有多篇重组蛋白作为抗原研发疫苗的论文，表中仅列举了不同基因型的 NNV 代表性的成果。对于每种实验疫苗，表中列出的是阴性对照组和免疫组的存活率而不是免疫组的相对存活率，这是由于大多数研究有多次重复试验，或多次疫苗浓度的试验，以及在不同的条件下的实验。

RGNNV? -从石斑鱼分离出来的 NNV，但基因型尚不清楚。

* 表达 NNV 衣壳蛋白的大肠杆菌，福尔马林灭活后喂食卤虫，以含大肠杆菌的活卤虫口服接种。

IM，肌肉注射；IP，腹腔注射。

鱼类的拉丁名称：七带石斑 *Epinephelus septemfasciatus*；驼背鲈 *Cromileptes altivelis*；大菱鲆 *Scophthalmus maximus*；斜带石斑 *Epinephelus coioides*。

第七节 可复制疫苗

鱼体被活病毒感染后，再注射相应病毒，能够产生特异性免疫保护，但无论是注射还是浸泡感染必须使用与致死相当的剂量。这种疫苗不需要相对昂贵的疫苗研发和制备过程。

活 NNV 与 polyI:C 共同免疫七带石斑鱼（*Epinehelus septemfasciatus*）可以预防 RGNNV 感染（Nishizawa 等，2009）。在实地研究中发现，免疫过的七带石斑鱼的存活率高达 97.5%～99.8%，未免疫组的存活率为 71%～79.5%，这取决于疫苗接种鱼体的大小和 VER 的暴发时间（Oh 等，2012），VER 暴发的 20d 前，对鱼类注射 polyI:C 可以预防 NNV 并提高存活率，这种保护作用可以持续 10 个月以上。类似的研究表明，使用不含特定抗原的免疫刺激物，可预防 NNV 感染发生，因此认为免疫刺激物是一类新的抗病毒制剂（Kuan 等，2012）。

研究表明，使用低剂量的活 NNV（每克鱼体 $10^{4.3}$ $TCID_{50}$ 或更少）感染七带石斑鱼，鱼体内 NNV 的增殖率与 NNV 毒力和鱼的死亡率密切相关（Nishizawa 等，2012）。在 17℃ 条件下，鱼体内 NNV 的增殖速度要比 26℃ 下低，峰值时的 NNV 滴度远低于临界值但大大高于阈值水平，这也许可以解释在 17℃ 水温养殖的鱼类在攻毒后可以存活，以及鱼体对 NNV 的特异性免疫保护反应增强的结果。由此可以推断，17℃ 水温养殖的鱼体内 NNV 扩增减少，这不仅可以控制 VER 的发生，还可以控制 NNV 活疫苗，即在低温下的 NNV 活疫苗接种具有很强的应用潜力。然而这种方法意味着具有感染性和致病性的病毒会被释放到环境中，因此，活疫苗是否会被批准值得怀疑。

第八节 灭活病毒疫苗

采用不同灭活方法制备 NNV 疫苗的研究取得了较好的效果。当七带石斑鱼腹腔注射福尔马林灭活的 RGNNV 疫苗后，无论是在实验室还是在网箱中自然感染的情况下，免疫组的死亡率都较对照组明显降低（Yamashita 等，2005）。在为期 9 周的实地试验中，七带石斑鱼免疫后养殖在自然感染的网箱内，相对保护率高达 85%。将福尔马林灭活疫苗经纳米微粒包裹或不包裹，浸泡免疫斜带石斑鱼苗（*Epinephelus coioides*）并进行浸泡攻毒，结果显示，纳米包裹的疫苗可以显著提升福尔马林灭活疫苗的功效（Kai 和 Chi，2008）。然而，必须严格遵守病毒灭活程序，通常 100% 的病毒失活是较难把握的，与细菌灭活相比，病毒灭活还受到分析方法上的局限。

第九节 重组蛋白/多肽疫苗

体外重组表达 NNV 衣壳蛋白是最常用的实验研发 NNV 疫苗的方法。在大肠杆菌中表达的重组衣壳蛋白（recCP）作为疫苗，已经在不同鱼类中取得了良好的结果。SJNNV 重组 C 片段油类佐剂疫苗，腹腔注射免疫大菱鲆幼鱼和其他鱼类，产生了显著的保护作用，并且在血清中检测到抗病毒的中和抗体（Husgard 等，2001）。研究发现，卤虫包裹的 recCP 可明显诱导斜带石斑鱼的免疫反应，抗原在后肠吸收，继而诱导产生特异性抗 recCP 的抗体（Yang 等，2007）。将点带石斑鱼 NNV 的 3 个日本分离株的重组 C 蛋白混合后，注射免疫驼背鲈（*Cromileptes altivelis*），用印度尼西亚的 NNV 分离株进行攻毒，驼背鲈也获得了较好的免疫保护效果（Yuasa 等，2002）。比较研究了大肠杆菌和鳗弧菌（*V. anguillarum*）作为 NNV 衣壳蛋白基因开放阅读框表达系统，通过卤虫包裹投喂免疫斜带石斑鱼，结果表明，基于鳗弧菌的疫苗诱导了较强的早期免疫反应，产生了较高的相对保护率（Chen 等，2011）。

大西洋庸鲽在接种油类佐剂 recCP 疫苗后，体内的病毒载量减少，8 尾鱼中有 5 尾呈 NNV 阴性，从而证实了 recCP 是有开发前景的候选疫苗的抗原（Øvergard 等，2013）。由于 VER 主要感染幼鱼，幼鱼在此阶段获得性免疫系统尚未发育完善，因此，使用卤虫包裹抗原对其进行口服免疫效果不明显。在大西洋庸鲽孵化 94d 后，对幼鱼进行卤虫包裹抗原疫苗实验，鱼体成功产生了有效的免疫保护（Patel，2009；Øvergård，2012）。另外，也已开展使用反向遗传学方法降低病毒的传染性，制备 NNV 减毒活疫苗的相关研究（Takizawa 等，2008）。

合成多肽或许可以替代传统疫苗或重组疫苗。根据热灭活 NNV 能够使鲈产生免疫保护，且其 N 端肽是一种潜在的保护性多肽（Coeurdacier 等，2003），或许能够找到多肽作为抗原制备疫苗的方法，这需要进一步的研究。迄今为止，合成多肽主要用于研究蛋白质结构，其用作疫苗的可能性研究尚未开展。

第十节 DNA 疫苗

DNA 疫苗是将编码抗原蛋白的序列克隆到一个强真核启动子的载体中制备的疫苗，然后经肌肉注射载体，对宿主实施接种。在细胞内，载体 DNA 表达抗原蛋白，模仿病毒复制感染，随后抗原被处理并递呈给免疫系统，刺激 Tc 细胞激活，作用于被感染细胞。借鉴弹状病毒 DNA 疫苗研制成功（Anderson 等，1996）的经验，编码 NNV 衣壳蛋白的基因被克隆到真核启动子的载体中（Sommerset 等，2005），然而该载体注射免疫大菱鲆并没有获得对 VER 的保护，其中的原因尚不清楚，可能是弹状病毒 VHSV G 蛋白是一种糖基化膜蛋白，具有独特的免疫学特性。有趣的发现是，编码 VHSV G 蛋白的 DNA 疫苗也能有效地预防 VER，但这种保护作用在接种后很短的时间内出现，且持续时间不长，这可能是先天性免疫的结果。

对使用 DNA 疫苗需要关注和考虑的问题是：DNA 疫苗在鱼体内的持续性，对水生环境的潜在影响以及消费者接受程度（Gomez-Casado 等，2011），以及接种过疫苗的鱼是否应该被视为转基因生物。

第十一节 前景和建议

通过疫苗接种方案有效可行地防控 VER，还需要掌握各种更加详细的指标数据，如接种途径、有效剂量和保护期限等。目前，多种疫苗的有效性数据是以注射免疫方式在幼鱼上得到的，而 VER 死亡率最高的是仔鱼，此时鱼太小，不适合注射免疫。因此，从幼体发育早期阶段到获得性免疫系统发育完善前，应使用免疫刺激剂增加幼体免疫力，使其渡过疾病易发期。此外，对产卵前亲鱼进行疫苗接种可能是保护幼体的一种有效手段，但相关研究尚少且效果也存在争议。为应对 NNV 感染的严峻形势，未来应聚焦于亲鱼疫苗接种策略的深入研究，以及研制对不同 NNV 株/基因型具有交叉保护的多价疫苗。考虑到地理分布和 NNV 的基因型，应将有效的预防与仔鱼、幼鱼和亲鱼阶段的免疫接种相结合，开展综合防控。

参考文献

Adachi, K., Ichinose, T., Takizawa, N., *et al.*, 2007. Inhibition of betanodavirus infection by inhibitors of endosomal acidification. *Arch Virol* 152: 2217-2224.

Anderson, E. D., Mourich, D. V., Fahrenkrug, S. C., *et al.*, 1996. Genetic immunization of rainbow trout (*Onchorhynchus mykiss*) against infectious hematopoietic necrosis virus. *Mol Mar Biol Biotech* 5: 114-122.

Athanassopoulou, F., Billinis, C. and Prapas, T. 2004. Important disease conditions of newly cultured species in intensive freshwater farms in Greece: first incidence of nodavirus infection in *Acipenser* sp. *Dis Aquat Org* 60: 247-252.

Bigarré, L., Cabon, J., Baud, M., *et al.*, 2009. Outbreak of betanodavirus infection in tilapia, *Oreochromis niloticus* (L.), in fresh water. *J Fish Dis* 32: 667-673.

Bloch, B., Gravningen, K. and Larsen, J. L. 1991. Encephalomyelitis among turbot associated with a picornavirus-like agent. *Dis Aquat Org* 10: 65-70.

Bonami, J. R. and Widada, J. S. 2011. Viral diseases of the giant fresh water prawn *Macrobrachium rosenbergii*: a review. *J Invertebr Pathol* 106: 131-142.

Bovo, G., Nishizawa, T., Maltese, C., *et al.*, 1999. Viral encephalopathy and retinopathy of farmed marine fish species in Italy. *Virus Research* 63: 143-146.

Bovo, G., Gustinelli, A., Quaglio, F., *et al.*, 2011. Viral encephalopathy and retinopathy outbreak in freshwater fish farmed in Italy. *Dis Aquat Org* 96: 45-54.

Breuil, G., Pepin, J. F. P., Boscher, S. and Thiery, R. 2002. Experimental vertical transmission of nodavirus from broodfish to eggs and larvae of the sea bass, *Dicentrarchus labrax* (L.). *J Fish Dis* 25: 697-702.

Chen, S. P., Wu, J. L., Su, Y. C. and Hong, J. R. 2007. Anti-Bcl-2 family members, zfBcl-x (L) and zfMcl-1a, prevent cytochrome c release from cells undergoing betanodavirus-induced secondary necrotic cell death. *Apoptosis* 12: 1043-1060.

Chen, Y. M., Shih, C. H., Liu, H. C., *et al.*, 2011. An oral nervous necrosis virus vaccine using *Vibrio anguillarum* as an expression host provides early protection. *Aquaculture* 321: 26-33.

Coeurdacier, J. L., Laporte, F. and Pepin, J. F. 2003. Preliminary approach to find synthetic peptides from nodavirus capsid potentially protective against sea bass viral encephalopathy and retinopathy. *Fish Shellfish Immunol* 14: 435-447.

Félix, M. A., Ashe, A., Piffaretti, J., *et al.*, 2011. Natural and experimental infection of *Caenorhabditis* nematodes by novel viruses related to nodaviruses. *PloS Biol* 9*e*1000586.

Frerichs, G. N., Tweedie, A., Starkey, W. G. and Richards, R. H. 2000. Temperature, pH and electrolyte sensitivity, and heat, UV and disinfectant inactivation of sea bass (*Dicentrarchus labrax*) neuropathy nodavirus. *Aquaculture* 185: 13-24.

Glazebrook, J. S., Heasman, M. P. and Debeer, S. W. 1990. Picorna-like viral particles associated with mass mortalities in larval barramundi, *Lates calcarifer* Bloch. *J Fish Dis* 13: 245-249.

Gomez-Casado, E., Estepa, A. and Coll, J. M. 2011. A comparative review on European-farmed finfish RNA viruses and their vaccines. *Vaccine* 29: 2657-2671.

Grotmol, S., Totland, G. K., Thorud, K. and Hjeltnes, B. K. 1997. Vacuolating encephalopathy and retinopathy associated with a nodavirus-like agent: a probable cause of mass mortality of cultured larval and juvenile Atlantic halibut *Hippoglossus hippoglossus*. *Dis Aquat Org* 29: 85-97.

Guo, Y. X., Wei, T., Dallmann, K. and Kwang, J. 2003. Induction of caspase-dependent apoptosis by betanodaviruses GGNNV and demonstration of protein alpha as an apoptosis inducer. *Virology* 308: 74-82.

Hata, N., Okinaka, Y., Iwamoto, T., *et al.*, 2010. Identification of RNA regions that determine temperature sensitivities in betanodaviruses. *Arch Virol* 155: 1597-1606.

Husgard, S., Grotmol, S., Hjeltnes, B. K., *et al.*, 2001. Immune response to a recombinant capsid protein of striped jack nervous necrosis virus (SJNNV) in turbot *Scophthalmus maximus* and Atlantic halibut *Hippoglossus hippoglossus*, and evaluation of a vaccine against SJNNV. *Dis Aquat Org* 45: 33-44.

Johansen, R., Grove, S., Svendsen, A. K., *et al.*, 2004a. A sequential study of pathological findings in Atlantic halibut, *Hippoglossus hippoglossus* (L), throughout one year after an acute outbreak of viral encephalopathy and retinopathy. *J Fish Dis* 27: 327-341.

Johansen, R., Sommerset, I., Tørud, B., *et al.*, 2004b. Characterization of nodavirus and viral encephalopathy and retinopathy in farmed turbot, *Scophthalmus maximus* (L.). *J Fish Dis* 27: 591-601.

Kai, Y. H. and Chi, S. C. 2008. Efficacies of inactivated vaccines against betanodavirus in grouper larvae (*Epinephelus coioides*) by bath immunization. *Vaccine* 26: 1450-1457.

Kai, Y. H., Su, H. M., Tai, K. T. and Chi, S. C. 2010. Vaccination of grouper broodfish (*Epinephelus tukula*) reduces the risk of vertical transmission by nervous necrosis virus. *Vaccine* 28: 996-1001.

Kuan, Y. C., Sheu, F., Lee, G. C., *et al.*, 2012. Administration of recombinant Reishi immunomodulatory protein (rLZ-8) diet enhances innate immune responses and elicits protection against nervous necrosis virus in grouper *Epinephelus coioides*. *Fish Shellfish Immunol* 32: 986-993.

Kuo, H. C., Wang, T. Y., Hsu, H. H., *et al.*, 2012. Nervous necrosis virus replicates following the embryo development and dual infection with iridovirus at juvenile stage in grouper. *PloS One* 7: e36183.

Lin, C. C., Liu, J. H. Y., Chen, M. S. and Yang, H. L. 2007. An oral nervous necrosis vaccine that induces protective immunity in larvae of grouper (*Epinephelus coioides*). *Aquaculture* 268: 265-273.

Liu, W. T., Hsu, C. H., Hong, Y. R., *et al.*, 2005. Early endocytosis pathways in SSN-1 cells infected by dragon grouper nervous necrosis virus. *J Gen Virol* 86: 2553-2561.

Mori, K. I., Nakai, T., Muroga, K., *et al.*, 1992. Properties of a new virus belonging to nodaviridae found in larval striped jack (*Pseudocaranx dentex*) with nervous necrosis. *Virology* 187: 368-371.

Munday, B. L., Langdon, J. S., Hyatt, A. and Humphrey, J. D. 1992. Mass mortality associated with a viral-induced vacuolating encephalopathy and retinopathy of larval and juvenile barramundi, *Lates calcarifer* Bloch. *Aquaculture* 103: 197-211.

Munday, B. L., Kwang, J. and Moody, N. 2002. Betanodavirus infections of teleost fish: a review. *J Fish Dis* 25: 127-142.

Nerland, A. H., Skår, C., Eriksen, T. B. and Bleie, H. 2007. Detection of nodavirus in seawater from rearing facilities for Atlantic halibut *Hippoglossus hippoglossus* larvae. *Dis Aquat Org* 73: 201-205.

Nishizawa, T., Furuhashi, M., Nagai, T., *et al.*, 1997. Genomic classification of fish nodaviruses by molecular phylogenetic analysis of the coat protein gene. *Appl Environ Microb* 63: 1633-1636.

Nishizawa, T., Takami, I., Kokawa, Y. and Yoshimizu, M. 2009. Fish immunization using a synthetic double-stranded RNA Poly (I: C), an interferon inducer, offers protection against RGNNV, a fish nodavirus. *Dis Aquat Org* 83: 115-122.

Nishizawa, T., Gye, H. J., Takami, I. and Oh, M. J. 2012. Potentiality of a live vaccine with nervous necrosis virus (NNV) for sevenband grouper *Epinephelus septemfasciatus* at a low rearing temperature. *Vaccine* 30: 1056-1063.

Nopadon, P., Aranya, P., Tipaporn, T., *et al.*, 2009. Nodavirus associated with pathological changes in adult spotted coralgroupers (*Plectropomus maculatus*) in Thailand with viral nervous necrosis. *Res Vet Sci* 87: 97-101.

Oh, M. J., Takami, I., Nishizawa, T., *et al.*, 2012. Field tests of Poly (I: C) immunization with nervous necrosis virus (NNV) in sevenband grouper, *Epinephelus septemfasciatus* (Thunberg). *J Fish Dis* 35: 187-191.

Patel, S. 2009. *Immune related genes in Atlantic halibut (Hippoglossus hippoglossus), Expression of B and T cell markers during ontogenesis*. PhD thesis, Molecular Biology Institute, University of Bergen.

Patel, S., Korsnes, K., Bergh, Ø., *et al.*, 2007. Nodavirus in farmed Atlantic cod *Gadus morhua* in Norway. *Dis Aquat Org* 77: 169-173.

Sommerset, I., Skern, R., Biering, E., *et al.*, 2005. Protection against Atlantic halibut nodavirus in turbot is induced by recombinant capsid protein vaccination but not following DNA vaccination. *Fish Shellfish Immunol* 18: 13-29.

Su, Y. C., Wu, J. L. and Hong, J. R. 2009. Betanodavirus non-structural protein B2: a novel necrotic death factor that induces mitochondria-mediated cell death in fish cells. *Virology* 385: 143-154.

Takizawa, N., Adachi, K. and Kobayashi, N. 2008. Establishment of reverse genetics system of Betanodavirus for the efficient recovery of infectious particles. *J Virol Methods* 151: 271-276.

Tanaka, S., Mori, K., Arimoto, M., *et al.*, 2001. Protective immunity of sevenband grouper *Epinephelus sep-*

temfasciatus Thunberg, against experimental viral nervous necrosis. *J Fish Dis* 24: 15-22.

Ucko, M., Colorni, A. and Diamant, A. 2004. Nodavirus infections in Israeli mariculture. *J Fish Dis* 27: 459-469.

Yamashita, H., Fujita, Y., Kawakami, H. and Nakai, T. 2005. The efficacy of inactivated virus vaccine against viral nervous necrosis (VNN). *Fish Pathol* 40: 15-21.

Yang, H. L., Lin, C. C., Lin, J. H. Y. and Chen, M. S. 2007. An oral nervous necrosis virus vaccine that induces protective immunity in larvae of grouper (*Epinephelus coioides*). *Aquaculture* 268: 265-273.

Yoshikoshi, K. and Inoue, K. 1990. Viral nervous necrosis in hatchery-reared larvae and juveniles of Japanese parrotfish, *Oplegnathus fasciatus* (Temminck and Schlegel). *J Fish Dis* 13: 69-77.

Yuasa, K., Koesharyani, I., Roza, D., *et al.*, 2002. Immune response of humpback grouper, *Cromileptes altivelis* (Valenciennes) injected with the recombinant coat protein of betanodavirus. *J Fish Dis* 25: 53-56.

Øvergård, A. C. 2012. *T-cells and cytokines of Atlantic halibut (Hippoglossus hippoglossus L.). Gene expression levels during ontogenesis, vaccination and experimental nodavirus infection*. PhD thesis, University of Bergen.

Øvergård, A. C., Patel, S., Nøstbakken, O. J. and Nerland, A. H. 2013. Halibut T-cell and cytokine response upon vaccination and challenge with nodavirus. *Vaccine* 31: 2395-2402.

第三十章　甲壳动物的免疫系统和免疫刺激

Indrani Karunasagar, Singaiah NaveenKumar,
Biswajit Maiti and Praveen Rai

本章概要　SUMMARY

甲壳动物免疫防御机制很强大，但是对其了解较少。甲壳动物主要依赖非特异性免疫防御机制，因此，能够激活非特异性免疫应答的刺激物在甲壳动物抵御疾病方面发挥着重要作用。目前多种化合物的免疫刺激作用已有报道，包括β-1，3-葡聚糖、细菌疫苗、肽聚糖、脂多糖、岩藻多糖等，它们均可增强甲壳动物的非特异性免疫防御系统中的体液和细胞免疫反应。通过注射、浸泡和口服等多种方式的刺激显示，免疫刺激物能够显著增强甲壳动物的非特异性免疫力，提高其在微生物病原感染后的存活率。

第一节　引　言

过去的10年间水产养殖产量迅速增长，截至2010年，世界水产养殖产量已达5 990万t，年均增长率6.3%，产值约1 194亿美元（FAO，2012）。对虾、蟹、龙虾和小龙虾等甲壳类的产量约占9.6%，约570万t（FAO，2012年）。水产养殖业在近20年发生了巨大变革，特别是20世纪90年代，养虾业以惊人的速度发展，表现在养殖区域不断扩张、集约化的养殖方式以及高投入高回报的养殖模式。然而，无论是纵向还是横向比较，水产养殖产量的增加与行业扩张并不成正比，这导致了甲壳类养殖行业的年增长率下降。自20世纪90年代以来，甲壳动物养殖总产量下降，主要原因是疾病的暴发。因为无脊椎动物易受病原微生物感染，所以养殖风险升高。在过去15年里，微生物感染导致的损失累计约为150亿美元，其中病毒性疾病约占全部损失的60%，细菌性疾病约为20%，其余的是寄生虫等其他病害造成的损失（Flegel，2012）。在养殖的对虾中，大约有20种病毒性疾病与对虾的大量死亡有关，白斑综合征病毒（WSSV）、黄头病毒（YHV）、传染性皮下和造血器官坏死病毒（IHHNV）和桃拉综合征病毒（TSV）均是危害十分严重的病毒（Lightner，2011）。哈维氏弧菌（*Vibrio harveyi*）、副溶血弧菌（*V. parahaemolyticus*）和其他弧菌等革兰氏阴性细菌会感染育苗期幼体和养殖期幼虾，造成重大经济损失。一些革兰氏阳性细菌，如气球菌（*Aerococcus* spp.）也能导致对虾死亡。寄生虫，尤其是纤毛虫、簇虫和微孢子虫等原生动物，也会感染虾类的各个发育阶段，导致虾养殖的病害问题。真菌是虾类的条件致病菌，链壶菌（*Lagenidium* spp.）和离壶菌（*Sirolpidium* spp.）是感染虾幼体的常见水生真菌。显然，各种各样的病原造成虾类养殖业的巨大损失，并威胁到水产养殖产业的健康可持续发展。因此，在世界范围内已开展了广泛的研究，以解决困扰该行业疾病管理的相关问题。

利用免疫学和分子生物学方法建立的灵敏、特异的诊断技术，有助于疾病的早期发现。水产养殖业中不建议使用抗生素，因为抗生素的大量使用会导致其在动物体内蓄积残留，并且诱导病原菌产生耐药性以及将耐药性转移到其他细菌，损害环境中对生物地球化学循环有利的/无害的生物，最终破坏生态平衡（Karunasagar 等，1994）。培育无特定病原的苗种以及在苗种放养前的病原筛查是已经实行的科学管理方法。尽管采用无特定病原苗种养殖，但疾病依旧会暴发，这说明幼体在培育过程中可能通过多种途径受到感染。循环养殖系统排除了外部污染，同时阻止了病原携带个体传播疾病，但对一些亚洲国家，在养殖面积较大的区域使用该方式从经济角度上是不可行的。在放养甲壳动物前，使用消毒剂和杀菌剂对养殖池塘进行处理，以及使用抗病毒药物、化学药物、益生菌和生物修复等防控措施，已取得一定的效果。开展甲壳动物免疫系统与免疫调节的研究将有助于甲壳动物养殖的管理和产业的可持续发展。

第二节 甲壳动物的免疫系统

甲壳动物主要依靠先天性免疫，该免疫系统在无脊椎动物中高度发达。甲壳动物的外骨骼覆盖整个身体，由多层壳质组成，成为抵御病原入侵的第一道防线。如果病原成功突破这一防线，甲壳动物将启动一系列细胞和体液的非特异性免疫应答，包括血细胞的透明细胞、半颗粒细胞和颗粒细胞，以及血淋巴和淋巴器官（只在对虾中发现）中的可溶性因子，均能积极参与机体防御反应（Rusaini 等，2010；Söderhäll 等，1992）。近年来，甲壳动物的免疫系统研究迅速发展。使用表达序列标签（EST）文库、微阵列研究和蛋白质组学等方法已经鉴定出许多保守蛋白序列，部分蛋白见表 30-1。抗菌肽（AMPs）、模式识别蛋白（PRPs）、凝集素、细胞因子样分子、水解酶等构成体液免疫防御的重要分子。通常，预防微生物入侵的细胞免疫反应包括吞噬作用、结节形成、创伤修复、细胞介导的细胞毒作用、包裹作用、凝集、黑化和酚氧化酶原（proPO）激活（Aguirre-Guzman 等，2009）等。

一、体液防御系统

1. 识别

先天性免疫反应的激活主要是宿主蛋白对微生物的特异性识别，这种蛋白称为模式识别蛋白（PRPs）。一些 PRPs，如 β-1, 3-葡聚糖结合蛋白（βGBP）、C 型凝集素、革兰氏阴性结合蛋白（GNBP）或脂多糖（LPS）结合蛋白（LBP）、含硫酯蛋白（TEPs）、肽聚糖结合蛋白（PGBP）、半乳凝素、清道夫受体（SRs）、纤维蛋白原相关蛋白（FREPs）、Toll 样受体（TLRs）

以及脂多糖和β-1,3-葡聚糖结合蛋白（LGBP），能识别入侵的微生物，并产生免疫应答（Wang等，2013年）。这些PRPs能够与各种微生物细胞壁成分结合，这些成分被称为病原相关分子模式（PAMPs）。PAMPs包括细菌和真菌细胞壁的结构组分，如肽聚糖（PG）、脂多糖（LPS）、脂磷壁酸（LTA）和β-1，3-葡聚糖。细菌和病毒未甲基化的CpG DNA以及来自病毒的单链RNA和双链RNA也可以作为PAMPs（Wang等，2013a）。一旦PRPs与PAMPs结合，会诱导机体体液和细胞的快速反应。

2. 凝集素

凝集素是一类糖蛋白，能与病原表面的特定碳水化合物基团（葡萄糖、甘露糖和果糖）相结合，因此凝集素被认为是甲壳动物的先天性免疫中的主要模式识别受体。动物体内已发现13种凝集素，其中对虾中发现7种，包括钙依赖型（C型）、L型、M型、P型、纤维蛋白原样结构域凝集素、半乳糖凝集素和钙联结蛋白/钙网蛋白（Wang等，2013b）。甲壳动物中研究较多和最详细的是C型凝集素，在C型凝集素中发现了碳水化合物识别域（CRD），CRD有助于识别病原，并通过促进吞噬、结节形成、包裹、酚氧化酶原激活和黑化等，在清除微生物中起着重要作用。对虾中C型凝集素的活性主要针对细菌和真菌，对病毒的作用有限。

3. Toll

Toll或TLRs在识别剪切后的Spätzle细胞因子中发挥关键作用。Toll和TLRs是进化上高度保守的跨膜糖蛋白，具有胞外和胞内结构域，这些结构域是由富含大量亮氨酸重复（LRR）序列和与白细胞介素-1受体（IL-1R）同源的胞质信号域组成（Bowie等，2000）。研究表明，Toll主要参与细菌感染引起的先天性免疫反应，而不针对病毒感染（Assavalapsakul等，2012）。

表30-1　已鉴定出的甲壳动物免疫系统重要的蛋白分子

序号	蛋白名称	种类	检索号	参考文献
1	α_2-巨球蛋白	斑节对虾（*Penaeus monodon*）	DQ355145	Lin等（2007）
2	β-肌动蛋白	斑节对虾 中国明对虾（*Fenneropenaeus chinens is*）， 凡纳滨对虾（*Litopenaeus vannamei*）	DQ205426， JN808449，JF288784	Zhang 等（2006）；Pon-prateep等（2012）
3	抗病毒样C型凝集素	印度明对虾（*Fenneropenaeus indicus*）	HM034320	未发表
4	抗脂多糖因子	北美螯虾（*Pacifastacus leniusculus*）	EF523760	Liu等（2006）
5	Argonuate1亚型A和B	斑节对虾	DQ663629，DQ663630	Dechklar等（2008）
6	造血激素	斑节对虾，北美螯虾	AY787657，AY787656	Söderhäll等（2005）
7	C型凝集素	斑节对虾	DQ078266	Luo等（2006）
8	半胱天冬酶	斑节对虾	DQ846887	Wongprasert等（2007）
9	半胱天冬酶3	斑节对虾	HM034317	未发表
10	几丁质酶	斑节对虾	AF157503.1	Tan等（2000）
11	丝氨酸蛋白酶同族体（c-SPH）	斑节对虾		Lin等（2006）
12	抗菌肽	史氏滨对虾（*Litopenaeus schmitti*）	EF182748	Rosa等（2007）
13	抗菌肽1，2和3	中国明对虾，北美螯虾	FJ853146 to FJ853148， EF523612 to EF523614	Sun等（2010）；Jiravan-ichpaisal等（2007）

（续）

序号	蛋白名称	种类	检索号	参考文献
14	亲环蛋白 A	斑节对虾	EU164775	Qiu 等（2009）
15	真核翻译起始因子 5（eIF5A）	斑节对虾	DQ851145	Phongdara 等（2007）
16	铁蛋白	斑节对虾、中国明对虾、凡纳滨对虾	EF523241，DQ205422，AY955373	Maiti 等（2010）；Zhang 等（2006）；Hsieh 等（2006）
17	谷胱甘肽过氧化物酶	凡纳滨对虾	AY973252	Liu 等（2007）
18	血蓝蛋白	斑节对虾、北美螯虾	AF431737，AF522504	Johnson（2002）；Söderhäll 等（2005）
19	热休克蛋白 70 同族体	斑节对虾	EF472918	Chuang 等（2007）
20	IFA-14 胞内脂肪酸结合蛋白	斑节对虾	EU287463	未发表
21	脂多糖和 β-1，3 葡聚糖结合蛋白	斑节对虾、北美螯虾，中华绒螯蟹（*Eriocheir sinensis*）	JN415536，AJ250128，FJ605172	Amparyup 等(2012)；Lee 等(2000)；Zhao 等(2009)
22	溶菌酶	斑节对虾	EU095851	Tyagi 等（2007）
23	甘露糖结合蛋白	北美螯虾	AY861653	Sricharoen 等（2005）
24	黑化作用相互作用蛋白	北美螯虾	EU308499	Söderhäll 等（2009）
25	蛋白酶制剂分子	北美螯虾	U81825	Liang 等（1997）
26	对虾抗菌肽（2a -2d；3a-3n，4a；4c & 4d）	白滨对虾（*Litopenaeus setiferus*）、凡纳滨对虾	Y14925-Y14928，AF390140-AF390147，AF390149,AF387660，AY039202-AY039207	Destoumieux 等(1997)；Cuthbertson 等(2002)
27	对虾抗菌肽-3	斑节对虾、印度明对虾	BI784459，HM535650	Supungul 等（2004）；Shanthi 等（2012）
28	对虾抗菌肽-5	斑节对虾	FJ686018	Tassanakajon 等（2010）
29	细胞黏附蛋白	斑节对虾、北美螯虾	AF188840，X91409	Sritunyalucksana 等(2001)；Johansson（1999）
30	吞噬作用激活蛋白（PAP）	斑节对虾		Deachamag 等（2006）
31	抗病毒蛋白	斑节对虾	DQ641258	Luo 等（2007）
32	酚氧化酶原（proPO）	斑节对虾、北美螯虾	AF099741，X83494	Sritunyalucksana 等(1999)；Aspan 等(1995)
33	Rab7	斑节对虾	DQ231062.1	Sritunyalucksana 等(2006)
34	活化蛋白激酶受体 c1(RACK1)	斑节对虾	EF569136	Tonganunt 等（2009）
35	核糖体蛋白 L26		EF523242	未发表
36	乳清酸单一结构域蛋白亚型 1,2,3	斑节对虾	EU623979,EU623980，EU623981	Amparyup 等（2008）
37	丝氨酸蛋白酶	中国明对虾、北美螯虾	FJ770383，AJ007668	Ren 等（2009）；Wang 等（2001b）
38	同线蛋白	斑节对虾	AF335106	未发表
39	谷氨酰胺转移酶	斑节对虾、北美螯虾 日本囊对虾(*Marsupenaeus japonicus*)，	AF469484，AF336805	Yeh 等（2006）；Wang 等（2001a）
40	原肌球蛋白Ⅱ	斑节对虾		Tan 等（2000）

二、酚氧化酶原（proPO）系统

酚氧化酶原（proPO）系统是无脊椎动物血淋巴清除异物的高效先天性免疫系统。甲壳动物的 proPO 系统参与吞噬、黑化、细胞毒性反应、细胞黏附、包裹和结节形成。在小龙虾中最先报道了 proPO cDNA 的基本结构（Aspan 等，1995）。随后，陆续从不同的无脊椎动物中鉴定出 proPO，包括棘皮动物、双壳类、腹足类、昆虫、海鞘、多足类和腕足类等（Cerenius 等，2008）。

1. proPO 系统的激活

对虾的 proPO 系统的激活是通过 PRPs（PGBP、LGBP 和 βGBP）识别微生物细胞壁成分（PAMPs）后，诱导黑色素的形成，是一个级联式反应过程（图 30-1）。这一过程通过连锁反应启动丝氨酸蛋白酶（SPs）的激活，从而导致 proPO 激活酶（PPAE）转化为活性蛋白酶，该活性蛋白酶将非活性酶前体 proPO 转化为酚氧化酶（PO）。PO 是参与黑色素合成的主要酶，是一种含铜酶和酪氨酸酶的双功能酶类，它能催化羟基化单酚并且将苯二酚氧化为邻醌（Sugumaran，1996；Söderhäll 等，1998）。这些醌类化合物和其他短期反应中间体共同在入侵微生物的周围聚合，形成黑色素，并增加对虾血细胞对细菌的黏附，从而通过结节形成而加速血细胞对异物的清除作用（da Silva，2002；Cerenius 等，2008）。对虾的 proPO 是在血细胞中合成的，存在于半颗粒细胞（SGCs）和颗粒细胞（GCs）中（Perazzolo 等，1997），proPO 的 mRNA 只在血细胞中表达（Sritunyalucksana 等，1999）。大多数甲壳类动物中仅具有一个 *proPO* 基因，而在斑节对虾（*Penaeus monodon*）和凡纳滨对虾（*Litopenaeus vannamei*）的血细胞中报道了 *PmproPO*1 和 *PmproPO*2 两个基因，发现了 PmPPAE1 和 PmPPAE2 两种 proPO 活化酶（Amparyup 等，2013）。

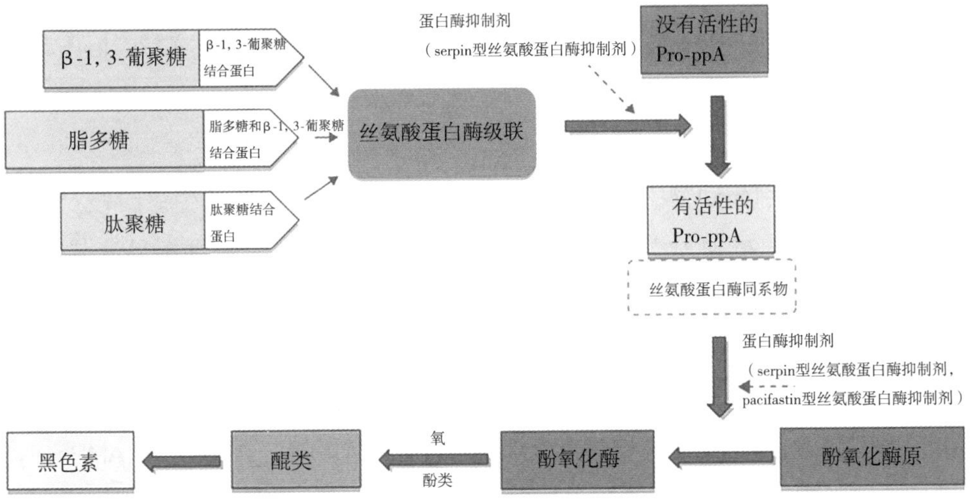

图 30-1　酚氧化酶原激活系统的概述

2. proPO 系统的调节

对虾机体对 proPO 系统具有精细的调节机制，以防止内源性级联反应的过度激活，避免该系统活性成分的有害影响。例如，PO 产生高活性且有毒的醌类中间体可造成宿主细胞的损伤和死亡。因此，必须具有正向和负向控制 proPO 系统合成黑色素的调节因子。黑色素的形成是由多种黑色素抑制分子在多个水平上控制和调节，主要的黑化作用抑制分子包括蛋白酶抑制剂（SP 级联的抑制）、酚氧化酶抑制剂（POI）和黑化作用抑制蛋白（MIP）（Daquing 等，1999；Sugumaran 等，2000），这些抑制分子大多数属于蛋白酶抑制剂家族，如 Kazal、Kunitz、丝氨酸蛋白酶抑制剂、α-巨球蛋白和金属蛋白酶抑制剂（Wedde 等，1998；Kanost，1999）。宿主蛋白酶抑制剂还能抑制入侵病原的蛋白酶。例如，对虾中已经报道3个丝氨酸蛋白酶抑制剂家族，命名为kazal型丝氨酸蛋白酶抑制剂（KPIs）、丝氨酸蛋白酶抑制剂和α_2-巨球蛋白（A2Ms）。

丝氨酸蛋白酶抑制剂（KPI）是富含半胱氨酸的蛋白质，包含一个或多个50～60个氨基酸残基的Kazal域，每个结构域包含由6个半胱氨酸残基在结构域内形成的3个二硫键（Rimphanitchayakit等，2010），KPIs包含不止一个Kazal域，因此每个KPI可能抑制多个蛋白酶。SPIPm2是斑节对虾血细胞中表达的Kazal型丝氨酸蛋白酶，除具有蛋白酶抑制活性外，还具有抑菌和抗病毒活性（Ponprateep等，2011）。

由于丝氨酸蛋白酶抑制剂是蛋白酶的抑制剂，因此也是自身杀伤性底物，能将反应中心环（RCL）插入蛋白酶的活性位点，使其构象发生变化，形成共价的丝氨酸蛋白酶抑制剂-蛋白酶复合物。据报道，中国明对虾和斑节对虾具有丝氨酸蛋白酶抑制剂，位于血细胞并在对病原体的应答反应过程中表达（Somnuk等，2012；Amparyup等，2013）。

α_2-巨球蛋白（A2Ms）是非特异性的广谱蛋白酶抑制剂，属于含硫酯蛋白质超家族，在进化过程中高度保守。A2Ms的反应机制很特别，目标蛋白酶在A2M分子的折叠内被捕获，该捕获过程主要是通过水解目标蛋白酶介导A2M分子的识别，导致蛋白酶分子在A2M分子中被包裹来实现。已在斑节对虾、凡纳滨对虾、中国明对虾和圣保罗明对虾（*F. paulensis*）的血细胞中鉴定出A2Ms（Wang，2013a）。对虾的A2Ms是同源二聚体，位于颗粒血细胞的囊泡中。微生物感染对虾时，虾体内的A2M表达水平增加。斑节对虾的A2M功能分析显示，A2M可以与线蛋白结合，该蛋白是具有多种生物学功能的胞质蛋白，可能有利于吞噬激活蛋白（PAP）进入吞噬细胞，能提高对虾WSSV感染的存活率（Tonganunt等，2005；Chotigeat等，2007）。

三、抗菌肽

抗菌肽（AMPs）是甲壳动物的重要防御分子，在甲壳动物较早的生命阶段产生，有助于甲壳动物抵御各种感染。抗菌肽为低分子量（＜10ku）的水脂两亲性多肽，具有抗菌功能（Boman，1995）。基于氨基酸序列、结构和功能，抗菌肽可分为：①α-螺旋蛋白；②β-折叠小蛋白质；③含硫醚环多肽；④含有一个或两个肽氨基酸过表达的多肽；⑤脂肽；⑥大环胱氨酸肽（Epand，Vogel，1999）。大多数AMPs通过在微生物膜中形成孔或破坏膜的完整性发挥作用（Tam等，2000）。对虾抗菌肽、甲壳抗菌肽和抗脂多糖因子（ALFs）等AMPs，具有较强的抗细菌和抗真菌活性并包含很多种类或亚型。对虾抗菌肽有PEN2、PEN3、PEN4和PEN5亚型；甲壳抗菌肽有3个亚型：crustin 1、crustin 2和crustin 3（表30-1）。对虾抗菌肽是一个独特的AMPs家族，最初从凡纳滨对虾的血浆和血细胞中分离到（Destoumieux等，1997；Tassanakajon等，2010），通常都是短的、分子量小的多肽（5.5～6.5ku）并主要作用于革兰氏阳性菌，具有杀菌或抑菌作用，而对革兰氏阴性细菌，如对虾弧菌病原未表现出抗菌活性；有趣的是抗菌肽能够抑制丝状真菌，如尖孢镰刀菌（*Fusarium oxysporum*）的生长。甲壳动物抗菌肽最先是在青蟹中报道的一种分子量为11.5ku、具有多种亚型的抗菌肽，且在凡纳滨对虾、白滨对虾、中国明对虾、斑节对虾和日本仿长额虾中均已报道。已经证明，无脊椎动物体内能够表达溶菌酶，参与抵御细菌感染。溶菌酶对革兰氏阴性和阳性菌均具有抗菌活性（Tyagi等，2007）。另一个重要的AMP是ALFs，对细菌、真菌、寄生虫和病毒均具有抗性（Tassanakajon

等，2010)。在罗氏沼虾（*Macrobrachium rosenbergii*）中已克隆、表达和鉴定出 ALFs，发现其具有广谱抗菌活性（Ren 等，2012)。在凡纳滨对虾中，其组蛋白（H2A、H2B、H3、H4）对几种革兰氏阳性细菌具有抗菌活性（Patat 等，2004)。

四、细胞免疫反应

1. 血淋巴细胞

甲壳动物循环系统拥有大量的血细胞，称为血淋巴细胞，在清除外来异物中发挥重要作用。通过形态学和染色反应可将甲壳动物中血淋巴细胞进行分类。在功能上可分为半颗粒细胞(SGs)、颗粒细胞（GCs）和透明细胞（Söderhäll 等，1992)。透明细胞缺乏颗粒，可能含有胞质包涵体，具有吞噬和诱导凝集功能。半颗粒细胞细胞质中包含不同数量的小颗粒，在β-1,3-葡聚糖或脂多糖等微生物多糖的存在下，这些颗粒可迅速分解，是参与包裹反应的主要细胞类型。颗粒细胞细胞质中含有较大的颗粒，并且含有酚氧化酶原、丝氨酸蛋白酶原、细胞黏附分子和抗菌肽等。

2. 吞噬作用

吞噬作用是最常见的细胞免疫反应，一旦病原体突破了角质层的理化屏障，吞噬作用就与体液因子一起构成首道防线。甲壳动物所有类型的血淋巴细胞都具有吞噬作用，但吞噬革兰氏阴性菌的活性实验表明，透明细胞和半颗粒细胞较颗粒细胞的吞噬作用更强。血淋巴细胞对细菌的有效吞噬依赖于血淋巴中各种因子的存在（Albores 等，1998)，有研究证明，酚氧化酶原系统作为调理素可协助吞噬作用（Smith 等，1983)。

3. 结节形成

当微生物的数量较多时，吞噬作用难以及时清除就会导致结节形成。结节形成是微生物被若干血细胞包裹的过程，由于宿主中 PO 活性，结节被黑化。第一个阶段是颗粒细胞和细菌间的接触，之后血细胞迅速脱颗粒并释放“黏性凝固物”，该过程须有细胞黏附蛋白和 βGBP 等内源性蛋白的存在。然后，proPO 的激活导致黑色素在细菌周围沉积（Söderhäll 等，1986)。第二阶段的特征是血细胞的黏附和扩散，在结节的周围不断堆积，结节内微生物迅速在血淋巴循环中被清除。结节在血细胞块中央，通常出现在鳃和血窦之间的肝胰腺小管中。

4. 包裹作用

包裹作用较容易观察，当被吞噬的病原较大时，几个循环的血细胞结合，一起封闭异物，这些血淋巴细胞变形并变黑。半颗粒细胞是对外来分子作出反应的首类血淋巴细胞，研究显示，血淋巴细胞脱颗粒后释放的具有多功能的细胞黏附蛋白能促进包裹作用和结节形成。血细胞进行包裹作用，首先需要识别非自身的外来物，之后非黏附细胞变为扁平的黏附细胞，逐一黏附外源异物（Johansson，1999)。

5. 细胞毒作用

颗粒细胞和半颗粒细胞能够分泌细胞毒性分子，破坏病变细胞和非病变细胞。尽管甲壳动物的细胞毒作用机制尚不清楚，但推测颗粒细胞一旦识别异物，就附着目标物并分泌细胞毒性分子，从而破坏外来异物或病变细胞。

五、甲壳动物的原始特异性免疫

无脊椎动物包括甲壳动物，被认为只有先天性免疫，没有特异性的免疫记忆（Janeway 等，2002）。最早的有关特异性免疫的研究来自桡足类动物——白色大剑水蚤（*Macrocyclops albidus*），研究结果显示，白色大剑水蚤接触过某种寄生虫后，再感染时获得了保护，而且该寄生虫的再次感染率比其他寄生虫低，这表明桡足类动物可能存在某种获得性免疫（Kurtz 等，2003）。WSSV 暴发时存活下来的日本对虾，再攻毒后，发现“类似免疫系统”抵御病毒感染（Venegas 等，2000）。还有研究证实了无脊椎动物免疫力的跨代转移，例如对虾（Little 等，2003）。上述的所有研究均表明无脊椎动物可能具有获得性免疫的替代系统，这些研究均基于几种无脊椎动物的定性研究（死亡率、相对存活率等），还没有分子调控机制的研究报道。然而，在黑腹果蝇（*Drosophila melanogaster*）和冈比亚按蚊（*Anopheles gambiae*）的研究中，发现唐氏综合征细胞黏附分子（DSCAM）可能是参与无脊椎动物特异性免疫应答的分子，该分子属于免疫球蛋白超家族（IgSF），机体攻毒实验表明，在主要的免疫细胞中表达病原特异性的 *DSCAM* 基因亚基。此外，冈比亚按蚊的 *Dscam* 基因沉默后对病原感染的抵抗力降低（Dong 等，2006）。*Dscam* 基因在养殖的凡纳滨对虾和斑节对虾中均已被鉴定出来（Chou 等，2009，2011）。

全面研究无脊椎动物的免疫系统，能更好地了解无脊椎动物的特异性免疫。上面提到的一些经典的实验研究，使疫苗接种防治养殖甲壳动物疾病成为一种可能。使用“疫苗接种”这一术语还为时过早，因为相关研究表明“疫苗”特异性保护的效果不稳定，特别是在对虾的 WSSV 预防方面，包括全病毒疫苗、亚单位重组疫苗和口服（亚单位）重组疫苗的研究。由于注射接种操作困难，现甲壳动物疫苗研究重点是口服疫苗（Mustaq 等，2011）以及其他接种策略，以获得最佳效益。

第三节 免疫刺激剂

约 40 年前，Snieszko（1974）提出，水生动物疾病的发生是由于破坏了宿主、病原与环境之间的微妙平衡，使用疫苗和免疫刺激剂改善宿主免疫因子是将疾病影响降到最低的策略之一。由于无脊椎动物主要依靠先天性免疫抵御传染性疾病，动物机体通过上调体液和细胞免疫进入警戒状态，抵御病原入侵感染，因此必须深入了解先天性免疫反应指标及其在免疫刺激后产生的保护作用。甲壳动物非特异性免疫系统的研究显示，葡聚糖、肽聚糖、菌苗、核苷酸等分子可以刺激机体产生免疫防御，获得对细菌和病毒感染的抵抗力。免疫刺激剂被定义为“提高非特异性或特异性免疫反应的化学制品、药物或应激物”（Anderson，1992）。随后，由于发现吞噬细胞的 PRRs，Bricknell 和 Dalmo（2005）重新定义免疫刺激剂为“自然产生的化合物，这些化合物主要通过调节宿主的免疫系统，增强宿主对相关病原的抵抗力”。有研究认为，免疫刺激剂可用于预防规律性暴发的疾病，以减少损失（Anderson，1992）。

免疫刺激剂在激活鱼类免疫系统中取得的效果较好，对虾等甲壳类动物中也借鉴了鱼类的相关研究成果。免疫刺激剂和佐剂已经在提高养殖对虾抗病性方面取得显著效果。对虾养殖中使用免疫刺激剂预防疾病是一个很有前景的发展方向，已经报道了一些具有免疫刺激特性的产品/分子。根据来源和作用方式，其可以分为细菌产品、藻类衍生产品、动物和植物源性衍生产品、营养因子、酶/激素/细胞因子（Sakai，1999）。已经从细菌（芽孢杆菌 *Bacillus* spp.）、真菌（酿酒酵母 *Saccharomyces cerevisia*；裂褶菌 *Schizophyllum commune*）、海藻（匍枝马尾藻 *Sargassum polycystum*）和草药中获得多种相关产品（Chang 等，2003；Liu 等，2006；Yogeeswaran 等，2012），并已在对虾养殖中应用。

一、免疫刺激剂诱导的血淋巴因子反应

许多血淋巴因子与甲壳动物的免疫系统有关，包括凝集素、沉淀素、杀菌蛋白、细胞溶解酶、溶菌酶、磷酸酶、磷脂酶、过氧化物酶、蛋白酶和其他抑菌和杀菌分子，并已经在甲壳动物中报道了这些血淋巴因子的作用。Evans 等（1969）报道，美洲龙虾（*Panulirus argus*）注射灭活细菌后，抗菌肽的表达上升。类似的报道还有很多，如螳虾蛄（*Squilla mantis*）注射弧菌菌苗后，体内凝集素和杀菌物质的含量上升（Danielli 等，1989）。从印度明对虾（*F. indicus*）体内分离并鉴定出一种天然凝集素，该凝集素能识别乙酰基并可选择性黏附多种细菌（Maheswari 等，2002）。斑节对虾注射热灭活的溶藻弧菌（*V. alginolyticus*）后 1d 内产生了较多抗菌肽和其他体液因子，并在第 2 天达到峰值且持续到第 5 天（Adams，1991）。较实用的口服实验方法是用含弧菌菌苗的饲料投喂对虾，进行口服免疫，结果发现免疫指标增强且存活率提高，证明菌苗有效（Devaraja 等，1998）。Sung 等（1996）证明，斑节对虾用免疫刺激剂如 β-1,3-葡聚糖、酵母聚糖或者弧菌菌苗处理后，血淋巴的杀菌活力增强。Devaraja 等（1998）报道，斑节对虾投喂含酵母 β-1,3-葡聚糖饲料后，血淋巴杀弧菌的活力增强；有趣的是，当弧菌菌苗和酵母 β-1,3-葡聚糖混合在饲料中后，与单一组分相比，混合组分诱导血淋巴产生了更强的杀弧菌活力，表明这些免疫刺激剂具有协同效应。Itami 等（1989）以注射或喷淋方式给对虾接种弧菌菌苗，研究菌苗诱导对虾的其他体液因子反应，结果表明，免疫对虾的血淋巴产生了趋化因子，能激活血细胞通过膜在博登腔内迁移。

从北美螯虾中，纯化鉴定出一种蛋白，部分表征表明其是血细胞上的 β-1,3-葡聚糖结合蛋白（βGBP）（Amparyup 等，2013；Lee 等，2000）。Zhao 等（2009）从中华绒螯蟹（*Eriocheir sinensis*）中克隆和鉴定出脂多糖和葡聚糖的结合蛋白，发现中华绒螯蟹注射细菌 1.5h 后，脂多糖和葡聚糖结合蛋白上调表达，在 12h 和 24h 逐步恢复。Lai 等（2011）研究了 WSSV 感染后，中国明对虾的 βGBP 在各组织中的差异表达。凡纳滨对虾注射溶藻弧菌 3h 和 6h 后 LGBP 的表达水平增加（Cheng 等，2005）。其他对虾中，如加州对虾（*P. californiensis*）、凡纳滨对虾（*L. vannamei*）和细角滨对虾（*L. stylirostris*），已纯化鉴定出 βGBP（Amparyup 等，2013）。βGBP 可以增强 PPAE 的活力，进而激活 PO（Aspan 等，1991；Amparyup 等，2012）。在斑节对虾中，浸泡或投喂免疫刺激剂 β-1,3-葡聚糖后，也会增强 PO 活性（Devaraja 等，1998）。

关于免疫刺激剂对 AMPs 调控的研究已有报道，斑节对虾投喂益生菌和肽聚糖后，以哈维氏弧菌攻毒，*crustin* 基因显著上调表达；然而，斑节对虾投喂免疫刺激物酵母、LPS 和葡聚糖后再攻毒，该基因却出现下调表达（Antony 等，2011）。Babu 等（2013）研究了水生假丝酵母（*Candida aquaetextoris* S527）作为免疫刺激剂添加到饲料中的应用可能性，斑节对虾每 7d 投

喂一次该饲料，检测 WSSV 攻毒后 *AMPs* 基因的差异表达，结果表明，crustin1、2 和 3，对虾抗菌肽-3 和对虾抗菌肽-5 等重要的 *AMPs* 基因上调表达。Wang 等（2008）报道了凡纳滨对虾投喂 β-1，3-葡聚糖后重要的 *AMPs* 基因的表达，研究发现，随着 *AMP* 基因 mRNA 的表达，溶菌酶基因和胞质锰超氧化物歧化酶基因出现显著的上调表达，而 *PEN*3 基因则显著下调表达。

二、免疫刺激剂诱导的血细胞免疫活性

血细胞在甲壳动物防御病原方面发挥非常重要的作用，而透明细胞具有非常重要的吞噬功能，通过由 proPO 系统介导的吞噬或者包裹作用清除细菌等外来颗粒物（Söderhäll 等，1992）。Sung 等（1996）研究发现，弧菌在入侵 12h 内可从对虾血淋巴中清除，在 24h 时已检测不到该菌。氧依赖的反应如活性氧产生的超氧阴离子、过氧化氢、氢氧根离子、单态氧和催化次氯酸盐的髓过氧化物酶，以及非氧依赖底物如消化酶，都参与吞噬和消除微生物的过程。Song 和 Hsieh（1994）的研究表明，利用底物硝基蓝四氮唑（NBT）染色检测超氧化物阴离子，正常斑节对虾有 5% 的血细胞被染色，而 β-1，3-葡聚糖饲喂的对虾有 41% 的血细胞被染色；分别用酵母聚糖和佛波醇（PMA）处理对虾后，血细胞染色率分别为 31% 和 9%，基于以上研究结果认为，β-1，3-葡聚糖和酵母聚糖是对虾良好的免疫刺激剂。Sung 等（1996）用 β-1，3-葡聚糖、酵母聚糖和弧菌疫苗浸泡斑节对虾，β-1，3-葡聚糖和酵母聚糖处理的对虾活性氧（ROS）产生量在 3～6h 内显著升高；而弧菌疫苗浸泡 6h 后，ROS 产生量才显著增加。Devaraja 等（1998）研究发现，虾饲料中添加 β-1，3-葡聚糖和弧菌疫苗后，ROS 产生量显著增加，48h 达到峰值，与 β-1，3-葡聚糖和弧菌疫苗单独添加相比，两种成分混合添加到饲料，ROS 活力更强。上述这些结果表明，免疫刺激剂可增强对虾血细胞的抗菌活性。Itami 等（1994）研究发现，日本对虾投喂 β-1，3-葡聚糖，50mg/kg（体重）连续投喂 10d，或 100mg/kg（体重）连续投喂 3d，对虾血细胞吞噬活力增强。日本对虾投喂源自嗜热双歧杆菌（*Bifidobacterium thermophilum*）的肽聚糖后，血细胞的吞噬活性也增强（Itami 等，1998a）。酚氧化酶原系统是甲壳动物防御的重要组成部分，产生并表达重要的抗病原侵染的功能性化合物，如醌类和黑色素（Bachere 等，1995；Smith 等，2003）。甲壳动物使用免疫刺激剂后 PO 活性增强也有报道，Sung 等（1994）体外研究表明，β-1，3-葡聚糖处理对虾后的 5min 到 24h，对虾血细胞 PO 活性显著增强，且于 3h 达到峰值，第 3 天恢复至对照水平（Sung 等，1996）。Devaraja 等（1998）研究了斑节对虾投喂包含弧菌疫苗和 β-1，3-葡聚糖的饲料后，血细胞中 PO 活性的变化。结果发现，口服后 48h，PO 活性达到峰值；弧菌疫苗和 β-1，3-葡聚糖同时投喂对 PO 活性的增强作用更强，表明两种免疫刺激剂之间具有协同作用。这些研究表明，使用微生物细胞壁成分作为免疫刺激剂可增强对虾非特异性防御中的细胞免疫活性。

三、免疫刺激剂诱导对虾的存活率升高

据研究报道，对虾免疫刺激剂处理后，经细菌性或病毒性病原攻毒，存活率有所提高（表 30-2）。Itami 等（1989）发现，日本对虾注射、浸泡或喷淋弧菌疫苗后的第 30 天注射弧菌攻毒，死亡率比未经处理的对照组低。斑节对虾浸泡 β-1，3-葡聚糖后，用创伤弧菌（*V. vulnificus*）攻毒，其存活率较高且保护作用持续至 18d（Sung 等，1994）。日本对虾饲喂

来自裂褶菌（*S. commune*）的β-1，3-葡聚糖后，弧菌攻毒结果显示，实验组对虾出现死亡的时间延迟，存活率显著高于对照组，但最终死亡率在统计学上无显著差异，这可能是因为免疫刺激作用的时间较短暂（Itami 等，1994）。类似的研究表明，对虾连续 7d 投喂嗜热双歧杆菌的肽聚糖，再连续 7d 饲喂不含肽聚糖的饲料，杀虾弧菌或 WSSV 攻毒后，其存活率升高（Itami 等，1998a）。Takahashi 等（1998）使用岩藻多糖处理对虾，WSSV 攻毒后，对虾的存活率提高。β-1，3-葡聚糖和弧菌疫苗联合投喂斑节对虾，WSSV 攻毒后，对虾的存活率提高（Karunasagar 等，1999）。Itami 等（1998b）也报道了类似的研究，投喂肽聚糖的对虾即使携带病毒，仍比对照组存活率高。斑节对虾幼体和仔虾投喂含β-1，3-葡聚糖的饲料 20d，WSSV 攻毒后，实验组对虾比未经处理的对照组免疫力更强且存活率更高（Chang 等，2003）。Chang 等（2012）研究了生姜中含有的抗氧化剂和抗炎特性的活性成分，以及姜酮对凡纳滨对虾幼虾生长、免疫及抗病能力的影响。结果发现，饲料中加入 1mg、2.5mg 和 5mg 的姜酮（每千克饲料）投喂 56d，与对照组相比，虾的体重明显增加，PO、溶菌酶和吞噬活性增加，并且致病性溶藻弧菌攻毒后的存活率升高。甲醇提取的草本免疫刺激剂，如印度铁苋菜（*Acalypha indica*）、狗牙根（*Cynodon dactylon*）、胡黄连（*Picrorrhiza kurrooa*）、南非醉茄（*Withania somnifera*）和生姜（*Zingiber officinalis*）添加到斑节对虾的饲料中，与没有添加草本提取物的对照组比，对虾的血淋巴、生化和其他免疫反应指标均有所提高，并且 WSSV 攻毒后存活率也提高（Yogeeswaran 等，2012）。综上所述，新型免疫刺激剂能够促进甲壳动物防御病原入侵的先天性免疫分子的上调表达，因此具有广阔的应用前景。

表 30-2　已报道的甲壳动物免疫刺激剂及其作用

序号	免疫刺激剂	使用方法	效果	参考文献
1	β-1，3-葡聚糖	口服	WSSV 攻毒后，免疫力提高，存活率显著增加	Chang 等（2003）
2	β-1，3-葡聚糖	浸泡	溶藻弧菌攻毒后 18d，存活率增加	Sung 等（1994）
3	β-1，3-葡聚糖	口服	弧菌攻毒后，吞噬活性以及存活率增加	Itami 等（1994）
4	β-葡聚糖	浸泡	48～120h 后，血淋巴细胞数量和可溶性血细胞蛋白增加	Campa-Cordova 等（2002）
5	β-葡聚糖	口服	酚氧化酶原和活性氧介导的活力增加	Anas 等（2009）
6	β-葡聚糖	口服	总血淋巴数量、酚氧化酶、过氧化物阴离子和超氧化物歧化酶活力增加，直至 27d	Bai 等（2010）
7	β-1，3-葡聚糖，酵母聚糖和弧菌菌苗	浸泡	PO 和氧呼吸暴发活力增加	Sung 等（1996）
8	β-1，3-葡聚糖，裂褶菌	口服	存活率提高，血淋巴细胞活力、细胞黏附以及产生过氧化物阴离子增加	Chang 等（2000）
9	β-1，3-葡聚糖，弧菌菌苗	口服	杀菌力，氧呼吸暴发活力和 PO 活力增强，WSSV 攻毒后存活率增加	Devaraja 等（1998） Karunasagar（1999）
10	中草药免疫刺激剂	口服	WSSV 攻毒后，存活率增加	Yogeeswaran 等（2012）
11	岩藻多糖	口服	WSSV 攻毒后，存活率增加	Takahashi 等（1998）
12	肽聚糖	口服	吞噬活性增加，WSSV 攻毒后存活率增加	Itami 等（1998a，1998b）
13	弧菌菌苗	注射，浸浴	30d 后攻毒，提高存活率	Itami 等（1989）
14	溶藻弧菌菌苗	注射	诱导杀菌肽，延长至第 5 天	Adams（1991）

（续）

序号	免疫刺激剂	使用方法	效果	参考文献
15	酵母多糖	口服	增强酚氧化酶活力，增加血淋巴数量，增强对哈维氏弧菌的杀菌活力	Thanardkit 等（2002）
16	酵母多糖	口服	增加血细胞总数以及颗粒细胞数量	Chotikachinda 等（2008）
17	姜酮	口服	酚氧化酶、溶菌酶和吞噬活力水平增加，溶藻弧菌攻毒后，存活率显著增加	Chang 等（2012）

第四节 致　谢

作者感谢印度政府生物技术部以及印度农业研究委员会对甲壳动物免疫研究提供的资助。

参考文献

Adams, A. 1991. Response ofpenaeid shrimp to exposure to *Vibrio* sp. *Fish Shellfish Immunol* 1: 59-70.

Aguirre-Guzman, G., Sanchez-Martinez, J. G., Campa-Cordova, A., *et al*., 2009. Penaeid shrimp immune system. *Thai J Vet Med* 39: 205-215.

Albores, V. F., Hernández-Lopez, J., Gollas-Galván, T., *et al*., 1998. Activation of shrimp cellular defense functions by microbial products, in *Advances in Shrimp Biotechnology* (ed. T. W. Flegal). Bangkok, National Center for Genetic Engineering and Biotechnology, 161-166.

Amparyup, P., Donpudsa, S. and Tassanakajon, A. 2008. Shrimp single WAP domain (SWD) -containing protein exhibits proteinase inhibitory and antimicrobial activities. *Dev Comp Immunol* 32: 1497-1509.

Amparyup, P., Sutthangkul, J., Charoensapsri, W. and Tassanakajon, A. 2012. Pattern recognition protein binds to lipopolysaccharide and β-1, 3-glucan and activates shrimp prophenoloxidase system. *J Biol Chem* 287: 10060-10069.

Amparyup, P., Charoensapsri, W. and Tassanakajon, A. 2013. Prophenoloxidase system and its role in shrimp immune responses against major pathogens. *Fish Shellfish Immunol* 34: 990-1001.

Anas, A., Lowmann, D. W., Williams, D., *et al*., 2009. Alkali insoluble glucan extracted from *Acremonium diospyri* is a more potent immunostimulants in the Indian White Shrimp, *Fenneropenaeus indicus* than soluble glucan. *Aquac Res* 40: 1320-1327.

Anderson, D. P. 1992. Immunostimulants, adjuvants, and vaccine carriers in fish: application to aquaculture. *Annu Rev Fish Dis* 2: 281-307.

Antony, S. P., Singh, I. S. B., Sudheer, N. S., *et al*., 2011. Molecular characterization of a crustin-like antimicrobial peptide in the giant tiger shrimp, *Penaeus monodon*, and its expression profile in response to various immunostimulants and challenge with WSSV. *Immunobiology* 216: 184-194.

Aspan, A. and Söderhäll, K. 1991. Purification of phenol oxidase from crayfish blood cells and its activation by an endogenous serine proteinase. *Insect Biochem* 21: 363-373.

Aspan, A., Huang, T. S., Cerenius, L. and Söderhäll, K. 1995. cDNA cloning of prophenoloxidase from the freshwater crayfish *Pacifastacus leniusculus* and its activation. *Proc Natl Acad Sci USA* 92: 939-943.

Assavalapsakul, W. and Panyim, S. 2012. Molecular cloning and tissue distribution of the Toll receptor in the black tiger shrimp, *Penaeus monodon*. *Genet Mol Res* 11: 484-493.

Babu, D. T., Antony, S. P., Joseph, S. P., *et al.*, 2013. Marine yeast *Candida aquaetextoris* S527 as a potential immunostimulant in black tiger shrimp *Penaeus monodon*. *J Invertebr Pathol* 112: 243-252.

Bachere, E., Mialhe, E. and Rodriguez, J. 1995. Identification of defence effectors in the hemolymph of crustaceans with particular reference to the shrimp *Penaeus japonicus* (Bate): prospects and applications. *Fish Shellfish Immunol* 5: 597-612.

Bai, N., Zhang, W., Mai, K., *et al.*, 2010. Effects of discontinuous administration of β-glucan and glycyrrhizin on the growth and immunity of white shrimp *Litopenaeus vannamei*. *Aquaculture* 306: 218-224.

Boman, H. 1995. Peptide antibiotics and their role in innate immunity. *Annu Rev Immunol* 13: 61-92.

Bowie, A. and O'Neill, L. A. J. 2000. The interleukin-1 receptor/ Toll-like receptor superfamily: signal generators for pro-inflammatory interleukins and microbial products. *J Leukoc Biol* 67: 508-514.

Bricknell, I. and Dalmo, R. A. 2005. The use of immunostimulants in fish larval aquaculture. *Fish Shellfish Immunol* 19: 457-472.

Campa-Cordova, A. I., Hernandez-Saavedra, N. Y., de Philippis, R. and Ascencio, F. 2002. Generation of superoxide anion and SOD activity in haemocytes and muscle of American White shrimp (*Litopenaeus vannamei*) as a response to β-glucan and sulphated polysaccharide. *Fish Shellfish Immunol* 12: 353-366.

Cerenius, L., Lee, B. L. and Söderhäll, K. 2008. The proPO-system: pros and cons for its role in invertebrate immunity. *Trends Immunol* 29: 263-271.

Chang, C. F., Chen, H. Y., Su, M. S. and Liao, I. C. 2000. Immunomodulation of dietary β-1, 3-glucan in the brooders of the black tiger shrimp *Penaeus monodon*. *Fish Shellfish Immunol* 10: 505-514.

Chang, C. F., Su, M. S., Chen, H. Y. and Liao, I. C. 2003. Dietary beta-1, 3-glucan effectively improves immunity and survival of *Penaeus monodon* challenged with white spot syndrome virus. *Fish Shellfish Immunol* 15: 297-310.

Chang, Y. P., Liu, C. H., Wu, C. C., *et al.*, 2012. Dietary administration of zingerone to enhance growth, non-specific immune response, and resistance to *Vibrio alginolyticus* in Pacific white shrimp (*Litopenaeus vannamei*) juveniles. *Fish Shellfish Immunol* 32: 284-290.

Cheng, W., Liu, C. H., Tsai, C. H. and Chen, J. C. 2005. Molecular cloning and characterisation of a pattern recognition molecule, lipopolysaccharide- and beta-1, 3-glucan binding protein (LGBP) from the white shrimp *Litopenaeus vannamei*. *Fish Shellfish Immunol* 18: 297-310.

Chotigeat, W., Deachamag, P. and Phongdara, A. 2007. Identification of a protein binding to the phagocytosis activating protein (PAP) in immunized black tiger shrimp. *Aquaculture* 271: 112-120.

Chotikachinda, R., Lapjatupon, W., Chaisilapasung, S., *et al.*, 2008. Effect of inactive yeast cell wall on growth performance, survival rate and immune parameters in Pacific white shrimp, (*Litopenaeus vannamei*). *Songklanakarin J Sci Technol* 22: 677-688.

Chou, P. H., Chang, H. S., Chen, I. T., *et al.*, 2009. The putative invertebrate adaptive immune protein *Litopenaeus vannamei* Dscam (LvDscam) is the first reported Dscam to lack a transmembrane domain and cytoplasmic tail. *Dev Comp Immunol* 33: 1258-1267.

Chou, P. H., Chang, H. S., Chen, I. T., *et al.*, 2011. *Penaeus monodon* Dscam (PmDscam) has a highly diverse cytoplasmic tail and is the first membrane-bound shrimp Dscam to be reported. *Fish Shellfish Immunol* 30: 1109-1123.

Chuang, K. H., Ho, S. H. and Song, Y. L. 2007. Cloning and expression analysis of heat shock cognate 70 gene promoter in tiger shrimp (*Penaeus monodon*). *Gene* 405: 10-18.

Cuthbertson, B. J. , Shepard, E. F. , Chapman, R. W. and Gross, P. S. 2002. Diversity of the penaeidin antimicrobial peptides in two shrimp species. *Immunogenetics* 54: 442-445.

Danielli, E. , Ferro, E. A. and Marzoni, R. 1989. Antibacterial activity and induced immunological responses in *Squilla mantis* (Crustacea, stomatopoda) hemolymph. *Eur Aquac Soc Publ* 10: 81-82.

Daquing, A. C. , Sato, T. , Koda, H. , *et al.* , 1999. A novel endogenous inhibitor of phenoloxidase form *Musca domestica* has a cystine motif commonly found in snail and spider toxins. *Biochemistry* 38: 2179-2188.

Da Silva, C. C. A. 2002. Activation of prophenoloxidase and removal of *Bacillus subtilis* from the hemolymph of *Acheta domesticus* L. Orthoptera: Gryllidae. *Neotrop Entomol* 31: 487-491.

Deachamag, P. , Intaraphad, U. , Phongdara, A. and Chotigeat, W. 2006. Expression of a phagocytosis activating protein (PAP) gene in immunized black tiger shrimp. *Aquaculture* 255: 165-172.

Dechklar, M,, Udomkit, A. and Panyim, S. 2008. Characterization of Argonaute cDNA from *Penaeus monodon* and implication of its role in RNA interference. *Biochem Biophys Res Commun* 367: 768-774.

Destoumieux, D. , Bulet, P. , Loew, D. , *et al.* , 1997. Penaeidins, a new family of antimicrobial peptides isolated from the shrimp *Penaeus vannamei* (Decapoda) . *J Biol Chem* 272: 28398-28406.

Devaraja, T. N. , Otta, S. K. , Shubha, G. , *et al.* , 1998. Immunostimulation of shrimp through oral administration of *Vibrio* bacterin and yeast glucan, in *Advances in Shrimp Biotechnology* (ed. T. W. Flegal) . Bangkok, National Center for Genetic Engineering and Biotechnology, 167-170.

Dong, Y. , Taylor, H. E. and Dimopoulos, G. 2006. AgDscam, a hypervariable immunoglobulin domain containing receptor of the *Anopheles gambiae* innate immune system. *PLoS Biol* 4: e229.

Epand, R. and Vogel, H. 1999. Diversity of antimicrobial peptides and their mechanisms of action. *Biochim Biophys Acta* 1462: 11-28.

Evans, E. E. , Weinheimer, P. F. , Painty, B. , *et al.* , 1969. Secondary and tertiary responses of induced bactericidin from West Indian spiny lobster *Panulirus argus*. *J Bacteriol* 98: 943-946.

FAO. 2012. *FAO Yearbook* 2010. *Fishery and Aquaculture Statistics*. Rome, Food and Agriculture Organization of United Nations. Flegel, T. W. 2012. Historic emergence, impact and current status of shrimp pathogens in Asia. *J Invertebr Pathol* 110: 166-173.

Hsieh, S. L. , Chiu, Y. C. and Kuo, C. M. 2006. Molecular cloning and tissue distribution of ferritin in Pacific white shrimp (*Litopenaeus vannamei*) . *Fish Shellfish Immunol* 21: 279-283.

Itami, T. , Takahashi, Y. and Nakamura, Y. 1989. Efficacy of vaccination against vibriosis in cultured Kuruma prawn *Penaeus japonicus*. *J Aquat Animal Health* 1: 238-242.

Itami, T. , Takahashi, Y. , Tsuchihira, E. , *et al.* , 1994. Enhancement of disease resistance of kuruma prawn *Penaeus japonicus* and increase in phagocytic activity of prawn hemocytes after oral administration of β-1, 3-glucan (Schizophyllan), in *Proceedings of the 3rd Asian Fisheries Forum, Singapore* 1992 (eds L. M. Chou, A. D. Munro, T. J. Lam *et al.* ,) . Manila, Asian Fisheries Society, 375-378.

Itami, T. , Kubono, K. , Asano, M. , *et al.* , 1998a. Enhanced disease resistances of Kuruma shrimp *Penaeus japonicus* after oral administration of peptidoglycan derived from *Bifidobacterium thermophilum*. *Aquaculture* 164: 277-288.

Itami, T. , Maeda, M. , Suzuki, N. , *et al.* , 1998b. Possible prevention of white spot syndrome (WSS) in kuruma shrimp, *Penaeus japonicus* in Japan, in *Advances in Shrimp Biotechnology* (ed. T. W. Flegal) . Bangkok, National Center for Genetic Engineering and Biotechnology, 291-295.

Janeway, C. A. and Medzhitov, R. 2002. Innate immune recognition. *Annu Rev Immunol* 20: 197-216.

Jiravanichpaisal, P., Lee, S. Y., Kim, Y. A., *et al.*, 2007. Antibacterial peptides in hemocytes and hematopoietic tissue from freshwater crayfish *Pacifastacus leniusculus*: characterization and expression pattern. *Dev Comp Immunol* 31: 441-455.

Johansson, M. W. 1999. Cell adhesion molecules in invertebrate immunity. *Dev Comp Immunol* 23: 303-315.

Kanost, M. R. 1999. Serine proteinase inhibitors in arthropod immunity. *Dev Comp Immunol* 23: 291-301.

Karunasagar, I. and Karunasagar, I. 1999. Diagnosis, treatment and prevention of microbial diseases of fish and shell fish. *Curr Sci* 76: 387-399.

Karunasagar, I., Pai, R., Malathi, G. R. and Karunasagar, I. 1994. Mass mortality of *Penaeus monodon* due to antibiotic resistant *Vibrio harveyi* infection. *Aquaculture* 128: 203-209.

Kurtz, J. and Franz, K. 2003. Evidence for memory in invertebrate immunity. *Nature* 425: 37-38.

Lai, X., Kong, J., Wang, Q., *et al.*, 2011. Cloning and characterization of a β-1, 3-glucan-binding protein from shrimp *Fenneropenaeus chinensis*. *Mol Biol Rep* 38: 4527-4535.

Lee, S. Y., Wang, R. and Söderhäll, K. 2000. A lipopolysaccharide- and beta-1, 3-glucanbinding protein from hemocytes of the freshwater crayfish *Pacifastacus leniusculus*. Purification, characterization, and cDNA cloning. *J Biol Chem* 275: 1337-1343.

Lehnert, S. A. and Johnson, S. E. 2002. Expression of hemocyanin and digestive enzyme messenger RNAs in the hepatopancreas of the black tiger shrimp *Penaeus monodon*. *J Comp Biochem Physiol B Biochem Mol Biol* 133: 163-171.

Liang, Z., Sottrup-Jensen, L., Aspan, A., *et al.*, 1997. Pacifastin, a novel 155-kDa heterodimeric proteinase inhibitor containing a unique transferrin chain. *Proc Natl Acad Sci USA* 94: 6682-6687.

Lightner, D. V. 2011. Virus diseases of farmed shrimp in the Western Hemisphere (the Americas): a review. *J Invertebr Pathol* 106: 110-130.

Lin, C. Y., Hu, K. Y., Ho, S. H. and Song, Y. L. 2006. Cloning and characterization of a shrimp clip domain serine protease homolog (c-SPH) as a cell adhesion molecule. *Dev Comp Immunol* 30: 1132-1144.

Lin, Y. C., Vaseeharan, B., Ko, C. F., *et al.*, 2007. Molecular cloning and characterization of a proteinase inhibitor, alpha 2-macroglobulin (alpha2-M) from the haemocytes of tiger shrimp *Penaeus monodon*. *Mol Immunol* 44: 1065-1074.

Little, T. J., O'Connor, B., Colegrave, N., *et al.*, 2003. Maternal transfer of strain-specific immunity in an invertebrate. *Curr Opin Biol* 13: 489-492.

Liu, C. H., Tseng, M. C. and Cheng, W. 2007. Identification and cloning of the antioxidant enzyme, glutathione peroxidase, of white shrimp, *Litopenaeus vannamei*, and its expression following *Vibrio alginolyticus* infection. *Fish Shellfish Immunol* 23: 34-45.

Liu, H., Jiravanichpaisal, P., Söderhäll, I., *et al.*, 2006. Antilipopolysaccharide factor interferes with white spot syndrome virus replication *in vitro* and *in vivo* in the crayfish *Pacifastacus leniusculus*. *J Virol* 80: 10365-10371.

Luo, T., Yang, H., Li, F., *et al.*, 2006. Purification, characterization and cDNA cloning of a novel lipopolysaccharide-binding lectin from the shrimp *Penaeus monodon*. *Dev Comp Immunol* 30: 607-617.

Luo, T., Li, F., Lei, K. and Xu, X. 2007. Genomic organization, promoter characterization and expression profiles of an antiviral gene *PmAV* from the shrimp *Penaeus monodon*. *Molecular Immunol* 44: 1516-1523.

Maheswari, R., Mullainadhan, P. and Arumugam, M. 2002. Isolation and characterization of an acetyl group-recognizing agglutinin from the serum of the Indian white shrimp *Fenneropenaeus indicus*. *Arch Biochem Biophys* 402:

65-76.

Maiti, B., Khushiramani, R., Tyagi, A., *et al.*, 2010. Recombinant ferritin protein protects *Penaeus monodon* infected by pathogenic *Vibrio harveyi*. *Dis Aquat Org* 88: 99-105.

Mustaq, S. S. and Kwang, J. 2011. Oral vaccination of baculovirus-expressed VP28 displays enhanced protection against white spot syndrome virus in *Penaeus monodon*. *PloS One* 6: e26428.

Patat, S. A., Carnegie, R. B., Kingsbury, C., *et al.*, 2004. Antimicrobial activity of histones from hemocytes of the Pacific white shrimp. *Eur J Biochem* 271: 4825-4833.

Perazzolo, L. M. and Barracco, M. A. 1997. The prophenoloxidase activating system of the shrimp, *Penaeus paulensis* and associated factors. *Dev Comp Immunol* 21: 385-395.

Phongdara, A., Laoong-u-thai, Y. and Wanna, W. 2007. Cloning of eIF5A from shrimp *Penaeus monodon*, a highly expressed protein involved in the survival of WSSV-infected shrimp. *Aquaculture* 265: 16-26.

Ponprateep, S., Tassanakajon, A. and Rimphanitchayakit, V. 2011. A Kazal type serine proteinase SPIPm2 from the black tiger shrimp *Penaeus monodon* is capable of neutralization and protection of hemocytes from the white spot syndrome virus. *Fish Shellfish Immunol* 31: 1179-1185.

Ponprateep, S., Tharntada, S., Somboonwiwat, K. and Tassanakajon, A. 2012. Gene silencing reveals a crucial role for anti-lipopolysaccharide factors from *Penaeus monodon* in the protection against microbial infections. *Fish Shellfish Immunol* 32: 26-34.

Qiu, L., Jiang, S., Huang, J., *et al.*, 2009. Molecular cloning and mRNA expression of cyclophilin A gene in black tiger shrimp (*Penaeus monodon*). *Fish Shellfish Immunol* 26: 115-121.

Ren, Q., Xu, Z. L., Wang, X. W., *et al.*, 2009. Clip domain serine protease and its homolog respond to *Vibrio* challenge in Chinese white shrimp, *Fenneropenaeus chinensis*. *Fish Shellfish Immunol* 26: 787-798.

Ren, Q., Du, Z. Q., Li, M., *et al.*, 2012. Cloning and expression analysis of an anti-lipopolysaccharide factor-from giant freshwater prawn, *Macrobrachium rosenbergii*. *Mol Biol Rep* 39: 7673-7680.

Rimphanitchayakit, V. and Tassanakajon, A. 2010. Structure and function of invertebrate Kazal-type serine proteinase inhibitors. *Dev Comp Immunol* 34: 377-386.

Rosa, R. D., Bandeira, P. T. and Barracco, M. A. 2007. Molecular cloning of crustins from the hemocytes of Brazilian penaeid shrimps. *FEMS Microbiol Lett* 274: 287-290.

Rusaini and Owens, L. 2010. Insight into the lymphoid organ of penaeid prawns: a review. *Fish Shellfish Immunol* 29: 367-377.

Sakai, M. 1999. Current research status of fish immunostimulants. *Aquaculture* 172: 63-92.

Shanthi, S. and Vaseeharan, B. 2012. cDNA cloning, characterization and expression analysis of a novel antimicrobial peptide gene penaeidin-3 (Fi-Pen3) from the haemocytes of Indian white shrimp *Fenneropenaeus indicus*. *Microbiol Res* 167: 127-134.

Smith, V. J. and Söderhäll, K. 1983. β-1, 3-glucan activation of crustacean hemocytes *in vitro* and *in vivo*. *Biol Bull* 164: 299-314.

Smith, V. J., Brown, J. H. and Hauton, C. 2003. Immunostimulation in crustaceans: does it really protect against infection? *Fish Shellfish Immunol* 15: 71-90.

Snieszko, S. T. 1974. The effect of environmental stress on outbreak on infectious diseases of fishes. *J Fish Biol* 6: 197-208.

Söderhäll, K. and Smith, V. J. 1986. Prophenoloxidase-activating cascade as a recognition and defense system in arthropods, in *Hemocytic and Humoral Immunity in Arthropods* (ed. A. P. Gupta). New York, John Wiley and

Sons, Ltd., 251-286.

Söderhäll, K. and Cerenius, L. 1992. Crustacean immunity. *Annu Rev Fish Dis* 2: 3-23.

Söderhäll, K. and Cerenius, L. 1998. Role of the prophenoloxidase activating system in invertebrate immunity. *Curr Opin Immunol* 10: 23-28.

Söderhäll, I., Kim, Y. A., Jiravanichpaisal, P., *et al.*, 2005. An ancient role for a prokineticin domain in invertebrate hematopoiesis. *J Immunol* 174: 6153-6160.

Söderhäll, I., Wu, C., Novotny, M., *et al.*, 2009. A novel protein acts as a negative regulator of proPO activation and melanization in the freshwater crayfish *Pacifastacus leniusculus*. *J Biol Chem* 284: 6301-6310.

Somnuk, S., Tassanakajon, A. and Rimphanitchayakit, V. 2012. Gene expression and characterization of a serine proteinase inhibitor PmSERPIN8 from the black tiger shrimp *Penaeus monodon*. *Fish Shellfish Immunol* 33: 332-341.

Song, Y. and Hsieh, Y. 1994. Immunostimulation of tiger shrimp (*Penaeus monodon*) hemocytes for generation of microbicidal substances: analysis of reactive oxygen species. *Dev Comp Immunol* 18: 201-209.

Sricharoen, S., Kim, J. J., Tunkijjanukij, S. and Söderhäll, I. 2005. Exocytosis and proteomic analysis of the vesicle content of granular hemocytes from a crayfish. *Dev Comp Immunol* 29: 1017-1031.

Sritunyalucksana, K., Cerenius, L. and Söderhäll, K. 1999. Molecular cloning and characterization of prophenoloxidase in the black tiger shrimp, *Penaeus monodon*. *Dev Comp Immunol* 23: 179-186.

Sritunyalucksana, K., Wongsuebsantati, K., Johansson, M. W. and Söderhäll, K. 2001. Peroxinectin, a cell adhesive protein associated with the proPO system from the black tiger shrimp, *Penaeus monodon*. *Dev Comp Immunol* 25: 353-363.

Sritunyalucksana, K., Wannapapho, W., Lo, C. F. and Flegel, T. W. 2006. PmRab7 is a VP28-binding protein involved in white spot syndrome virus infection in shrimp. *J Virol* 80: 10734-10742.

Sugumaran, M. 1996. Roles of the insect cuticle in host defense reactions, in *New Directions in Invertebrate Immunology* (eds K. Söderhäll, S. Iwanaga and G. R. Vasta). Fair Haven, New Jersey, SOS Publications, 355-374.

Sugumaran, M. and Nellaiappan, K. 2000. Characterization of new phenoloxidase inhibitor from the cuticle of *Manduca sexta*. *Biochem Biophys Res Commun* 268: 379-383.

Sun, C., Du, X. J., Xu, W. T., *et al.*, 2010. Molecular cloning and characterization of three crustins from the Chinese white shrimp, *Fenneropenaeus chinensis*. *Fish Shellfish Immunol* 28: 517-524.

Sung, H. H., Kou, G. H. and Song, Y. L. 1994. Vibriosis resistance induced by glucan treatment in tiger shrimp (*Penaeus monodon*). *Fish Pathol* 29: 11-17.

Sung, H. H., Yang, Y. L. and Song, Y. L. 1996. Enhancement of microbial activity in the tiger shrimp *Penaeus monodon* via immunostimulation. *J Crustacean Biol* 16: 278-284.

Supungul, P., Klinbunga, S., Pichyangkura, R., *et al.*, 2004. Antimicrobial peptides discovered in the black tiger shrimp *Penaeus monodon* using the EST approach. *Dis Aquat Org* 61: 123-135.

Takahashi Y, Uehara K, Watanabe R, *et al.*, 1998. Efficacy of oral administration of fucoidan, a sulphated polysaccharide in controlling white spot syndrome in kuruma shrimp in Japan, in *Advances in Shrimp Biotechnology* (ed. T. W. Flegal). Bangkok, National Center for Genetic Engineering and Biotechnology, 171-173.

Tam, J., Lu, Y. and Yang, J. 2000. Marked increase in membranolytic selectivity of novel cyclic tachyplesins constrained with an antiparallel two-B strand cysteine knot framework. *Biochem Biophys Res Commun* 267: 783-790.

Tan, S. H., Degnan, B. M. and Lehnert, S. A. 2000. The *Penaeus monodon* chitinase 1 gene is differentially expressed in the hepatopancreas during the molt cycle. *Mar Biotechnol* 2: 126-135.

Tassanakajon, A., Amparyup, P., Somboonwiwat, K. and Supungul, P. 2010. Cationic antimicrobial peptides in penaeid shrimp. *Mar Biotechnol* 12: 487-505.

Thanardkit, P., Khunrae, P., Suphantharika, M. and Verduyn, C. 2002. Glucan from brewer's yeast: preparation, analysis and use as a potential immunostimulants in shrimp feed. *World J Microbiol Biotechnol* 18: 527-539.

Tonganunt, M., Phongdara, A., Chotigeat, W. and Fujise, K. 2005. Identification and characterization of syntenin binding protein in the black tiger shrimp *Penaeus monodon*. *J Biotechnol* 120: 135-145.

Tonganunt, M., Saelee, N., Chotigeat, W. and Phongdara, A. 2009. Identification of a receptor for activated protein kinase C1 (Pm-RACK1), a cellular gene product from black tiger shrimp (*Penaeus monodon*) interacts with a protein, VP9 from the white spot syndrome virus. *Fish Shellfish Immunol* 26: 509-514.

Tyagi, A., Khushiramani, R., Karunasagar, I. and Karunasagar, I. 2007. Antivibrio activity of recombinant lysozyme expressed from black tiger shrimp, *Penaeus monodon*. *Aquaculture* 272: 246-253.

Venegas, C. A., Nonaka, L., Mushiake, K., *et al.*, 2000. Quasi-immune response of *Penaeus japonicus* to penaeid rod-shaped DNA virus (PRDV). *Dis Aquat Org* 42: 83-89.

Wang, R., Liang, Z., Hal, M. and Söderhäll, K. 2001a. A transglutaminase involved in the coagulation system of the freshwater crayfish, *Pacifastacus leniusculus*. Tissue localisation and cDNA cloning. *Fish Shellfish Immunol* 11: 623-637.

Wang, R., Lee, S. Y., Cerenius, L. and Söderhäll, K. 2001b. Properties of the prophenoloxidase activating enzyme of the freshwater crayfish, *Pacifastacus leniusculus*. *Eur J Biochem* 268: 895-902.

Wang, X. and Wang, J. 2013a. Pattern recognition receptors acting in innate immune system of shrimp against pathogen infections. *Fish Shellfish Immunol* 34: 981-989.

Wang, X. W. and Wang, J. X. 2013b. Diversity and multiple functions of lectins in shrimp immunity. *Dev Comp Immunol* 39: 27-38.

Wang, Y. C., Chang, P. S. and Chen, H. Y. 2008. Differential time-series expression of immune-related genes of Pacific white shrimp *Litopenaeus vannamei* in response to dietary inclusion of beta-1, 3-glucan. *Fish Shellfish Immunol* 24: 113-121.

Wedde, M., Weise, C., Kopacek, P., *et al.*, 1998. Purification and characterization of an inducible metalloprotease inhibitor from the hemolymph of greater wax moth larvae, *Galleria mellonella*. *Eur J Biochem* 225: 535-543.

Wongprasert, K., Sangsuriya, P., Phongdara, A. and Senapin, S. 2007. Cloning and characterization of a caspase gene from black tiger shrimp (*Penaeus monodon*) infected with white spot syndrome virus (WSSV). *J Biotechnol* 131: 9-19.

Yeh, M. S., Kao, L. R., Huang, C. J. and Tsai, I. H. 2006. Biochemical characterization and cloning of transglutaminases responsible for hemolymph clotting in *Penaeus monodon* and *Marsupenaeus japonicus*. *Biochim Biophys Acta* 1764: 1167-1178.

Yogeeswaran, A., Velmurugan, S., Punitha, S. M., *et al.*, 2012. Protection of *Penaeus monodon* against white spot syndrome virus by inactivated vaccine with herbal immunostimulants. *Fish Shellfish Immunol* 32: 1058-1067.

Zhang, J., Li, F., Wang, Z., *et al.*, 2006. Cloning, expression and identification of ferritin from Chinese shrimp, *Fenneropenaeus chinensis*. *J Biotechnol* 125: 173-184.

Zhao, D., Chen, L., Qin, C., *et al.*, 2009. Molecular cloning and characterization of the lipopolysaccharide and β-1, 3-glucan binding protein in Chinese mitten crab (*Eriocheir sinensis*). *Comp Biochem Physiol B Biochem Mol Biol* 154: 17-24.